Freshwater crayfish
V

Freshwater crayfish

Papers from the Fifth International Symposium on Freshwater crayfish Davis, California, USA, 1981

V

Edited by Charles R. Goldman

AVI PUBLISHING COMPANY, INC.
Westport, Connecticut

Cover Design by Alyne Lavoie-Ruppanner
California or Signal Crayfish: _Pacifastacus leniusculus_ (Dana)

023248

Copyright 1983 by
THE AVI PUBLISHING COMPANY, INC.
Westport, Connecticut

Library of Congress Cataloging in Publication Data

International Symposium on Freshwater Crayfish
    (5th : 1981 : Davis, Calif.)
    Freshwater crayfish--V.

    Bibliography: p.
    1. Crayfish--Congresses. 2. Freshwater invertebrates
--Congresses. I. Goldman, Charles Remington,
1930-   . II. Title. III. Title: Freshwater
crayfish--5. IV. Title: Freshwater crayfish--five.
QL444.M33I57 1983   595.3'841   83-15747
ISBN 0-87055-438-7
Printed in the United States of America

TABLE OF CONTENTS

v

SECTION III: CATION PHYSIOLOGY OF CRAYFISH

SECTION IV: ECOLOGY OF CRAYFISH

SECTION V: PATHOLOGY AND TOXICOLOGY OF CRAYFISH

SECTION VI: PRODUCTION PROBLEMS OF CRAYFISH

SECTION VI continued

SECTION VII: SOVIET CONTRIBUTIONS

SECTION VII continued

SECTION VIII: DESCRIPTIVE PAPERS

LIST OF PARTICIPANTS

Abrahamsson, Maja Sjostrom
Simontorps Akv. Avelslab.
Blentarp S-270 35
Sweden

Adegboye, Duro
Dept. of Biological Sciences
Ahmadu Bello University
Zaria, Kaduna State
Nigeria

*Antsupova, L.V.
The Whole-Union Scientific
Research Institute
Fisheries and Oceanography
V. Krasnaselskaya 17a
Moscow-140, USSR

*Ajaxon, Ragnar
Institute of Physiological Botany
University of Uppsala, Box 540
S-751 21 Uppsala, Sweden

Appelberg, Magnus P.A.
Institute of Limnology
University of Uppsala, Box 557
S-751 22 Uppsala, Sweden

*Arrignon, Jacques C. V.
Consiel Superieur de la Peche
10 Rue Peckt
75025 Paris, France

Avault, James W., Jr.
Fisheries Section
School of Forestry and Wildlife Mgmt.
Louisiana State University
Baton Rouge, LA 70803

Beatty, Kenneth
College of Siskiyous
3631 Summit Drive
Mount Shasta, CA 96067

Berggren, Kurt
1403 Pacific Drive
Davis, CA 95616

Billodeau, Dewey D.
P.O. Box 569
Franklin LA 70538

Bordner, Clark
Bodega Marine Laboratory
Bodega, CA 94923

Brinck, Per
Ekologihuset, Helgonavagen 4
Lund S-223 62, Sweden

*Brodski, S. Ya.
Ukrainian Research
Institute of Fisheries
Kiev, Ukr., USSR

*Burba, A.
Institute of Zoology and Parasitology
Academy of Sciences of the
Lithuanian SSR
Volnius 232600, USSR

*Chien, Yew-Hu
Fisheries Section
School of Forestry and Wildlife Mgmt.
Louisiana State University
Baton Rouge, LA 70803

Christofferson, Jay P.
Biology Department
Cal State University Stanislaus
Turlock, CA 95380

Clark, Wally
Aquaculture Program
University of California
Davis, CA 95616

Coll, M.
Ministerio de Agrucultura y Pesca
ICONA
Gran Via de San Francisco
35 Madrid 5, Spain

Colt, John
Dept. of Civil Engineering
University of California
Davis, CA 95616

Conte, Fred S.
Hutchinson Hall Room 554
University of California
Davis, CA 95616

* Contributor of proceedings article who did not attend conference.

Corriden, Gene
P.O. Box 434
Soquel, Ca 95703

*Cuellar, L.
Instituto Nacional para Conservacion
de la Naturaleza (ICONA)
Gran Via de San Francisco
35 Madrid 5, Spain

*Cukerzis, J. M.
Institute of Zoology and
Parasitology
Academy of Sciences of
Lithuanian SSR
Volnius 232600, USSR

D'Abramo, Louis
Bodega Marine Laboratory
Bodega, CA 94923

Daniel, Gaylord O.
King Lobster of California
4319 Campbell Dr.
Los Angeles, CA 90066

Darrow, Thomas
Biology Department
Lewis & Clark College
0615 S.W. Palatine Hill Rd.
Portland, OR 97219

De la Brettone, Larry, Jr.
202A Kanpp Hall
Louisiana State University
Baton Rouge, LA 70803

*Ekanem, S. B.
Fisheries Section
School of Forestry & Wildlife Mgmt.
Louisiana State University
Bayon Rouge, LA 70803

Eloranta, Pertti V.
University of Jyvaskyla
Department of Biology
SF-40100 Jyvaskyla 10, Finland

*Eriksson, B.
Institute of Freshwater Research
S-170 11 Drottningholm
Sweden

Fanton, Mssr.
c/o Babich Co.
Zone Industrielle de Phare
33700 Nerrignac, Bordeaux, France

Farmer, Mike
East Bay MUD
Oakland, CA

Fernandez, Ramon
Quiñon S.A.
San Esteban de Gormaz
Soria, Spain

Fisher, William
Aquaculture Program
University of California
Davis, CA 95616

Forcella, Anne
Institute of Ecology
University of California
Davis, CA 95616

Fournis, Mssr.
c/o Babich Co.
Zone Industrielle du Phare
33700 Nerrignac, Bordeaux, France

France, Robert L.
Freshwater Institute
Dept. of Fisheries and Oceans
University of Manitoba
Winnipeg, Manitoba R3T 2N6
Canada

Furst, Magnus
Institute of Freshwater Research
S-170 11 Drottningholm, Sweden

Galante, John C.
8 Sea Cliff Ave.
San Francisco, CA 94121

Gaude, Albert Paul
Crawfish Research Center
University of Southwestern Louisiana
Baton Rouge, CA 70810

*Glasgow, L. L.
Fisheries Section
School of Forestry & Wildlife Mgmt.
Louisiana State University
Baton Rouge, LA 70803

Goldman, Charles R.
Division of Environmental Studies
University of California
Davis, CA 95616

Gooch, Donald
Department of Animal Science
University of Southwestern Louisiana
Lafayette, LA 70504

*Graves, Jerry B.
Department of Entomology
Louisiana State University
Baton Rouge, LA 70803

Hamilton, Ragnar
Simontorps Akv. Avelslab
S-270 35 Blentarp, Sweden

Hamilton, Carl
Simontorps Akv. Avelslab
S-270 35 Blentarp, Sweden

Hanson, David
1735 Highland Place, Suite 36
Berkeley, CA 94709

Huner, Jay V.
Department of Biology
Southern University
Baton Rouge, LA 79813

*Järvenpää, T.
Finnish Game and Fisheries
Research Institute
P.O. Box 193
SF-00131 Helsinki 13, Finland

*Johnson, W. B.
Fisheries Section
School of Forestry & Wildlife Mgmt.
Souisiana State University
Baton Rouge, LA 70803

*Jones, Patricia
Department of Natural Resources
Columbus, OH

*Kadniuk, R. P.
The Whole-Union Scientific
Research Institute
Fisheries and Ocenaography
V. Krasnaselskaya 17a
Moscow-140, USSR

Karlsson, A. Stellan
Simontorp
S-270 35 Blentarp, Sweden

Kerridge, David C.
RR2 501 Davis Rd.
Ladysmith, B.C. VOR 2EO, Canada

Klosterman, J. J.
Division of Environmental Studies
University of California
Davis, CA 95616

*Kossakowski, Gustaw
State Fish Farm
Ostrocla, Poland

*Kossakowski, Jozef
Marian Mnich
Inland Fisheries Institute
Olsztyn, Kortowo, Poland

Lahti, Erkki
Department of Zoology
University of Kuopio, Box 138
70101 Kuopio 10, Finland

Laurent, Pierre J.
Station D'Hydrobiologie, INRA
75, Avenue de Corzent
Thonon F 74203, France

Leonhard, Sharon
Freshwater Institute
501 University Crescent
Winnipeg, Manitoba R3T 2N6, Canada

Linqvist, Ossi V.
Department of Zoology
University of Kuopio, Box 138
70101 Kuopio 10, Finland

*Lisovskaya, V.I.
The Whole-Union Scientific
Research Institute
Fisheries and Oceanography
V. Krasnaselskaya 17a
Moscow-140, USSR

*Lopez-Baisson, C.,
Quiñon S.A.
San Esteban de Gormaz
Soria, Spain

Lorena, Andres Salvador Habsburgo
Fuentemilanos 2
Madrid 35, Spain

*Lounatmaa, Kari
Dept. of Electron Microscopy
University of Helsinki
SF-00280 Helsinki 28, Finland

*Lutz, C. G.
Fisheries Section
School of Forestry & Wildlife Mgmt.
Louisiana State University
Baton Rouge, LA 70803

*Mackaviciene, G.
Institute of Zoology & Parasitology
Academy of Science of the
Lithuanian SSR
Vilnius 232600, USSR

McClain, W. Ray
Rte. 6, Box 189
Orange, TX 77630

McCormick, Thomas B.
366 Callan Ave.
San Leandro, CA 94577

McGriff, Darlene
California Dept. of Fish & Game
1701 Nimbus Rd., Suite C
Ranch Cordova, CA 95670

McMahon, Brian R.
Department of Biology
University of Calgary
Calgary, Alberta T2N 1N4, Canada

*Mickeniene, L.
Institute of Zoology & Parasitology
Academy of Science of the
Lithuanian SSR
Vilnius 232600, USSR

*Miltner, Michael R.
Fisheries Section
School of Forestry & Wildlife Mgmt.
Louisiana State University
Baton Rouge, LA 70803

Momot, Walter T.
Lakehead University, RR 6
Thunder Bay, Ontario P7B 5E1
Canada

Morales, Helen
Institute of Ecology
University of California
Davis, CA 95616

Morera, Joachim M.
King Lobster of California
4319 Campbell Drive
Los Angeles, CA 90066

*Morgan, D. O.
Department of Biology
University of Calgary
Calgary, Alberta T2N 1N4, Canada

*Morris, H.
Feed and Fertilizer Lab
Louisiana State University
Baton Rouge, LA 70803

*Morrissy, N. M.
Western Australial Marine Research Lab
State Dept. of Fisheries & Wildlife
North Beach 6020,
Western Australia

*Nikinmaa, M.
Department of Zoology
University of Hilsinki
Arkadiankatu 7
SF-00100 Helsinki 10, Finland

Nolfi, James
Assoc. in Rural Development, Inc.
P. O. Box 897
Burlington, VT 05402

Norman, Karen
Bodega Marine Laboratory
Bodega, CA 94923

Nuzum, Robert
2130 Adeline Street
Oakland, CA 94623

*Nylund, Viljo
Finnish Game & Fisheries Research
Institute, P.O. Box 193
SF-00131 Helsinki 13, Finland

Odelstrom, Tommy
Institute of Limnology
University of Uppsala, Box 557
S-751 22 Uppsala, Sweden

O'Keefe, Ciaran
Department of Zoology
University of Dublin
Trinity College
Dublin 2, Ireland

Patterson, Pat
6201 S.E. 145th Avenue
Portland, OR 97236

Payne, James F.
Memphis State University
Memphis, TN 38152

Persson, Mikael
Institute of Physiological Botany
University of Uppsala, Box 540
S-751 21 Uppsala, Sweden

*Petkevich, T. A.
The Whole-Union Scientific
Research Institute
Fisheries and Oceanography
V. Krasnaselskaya 17a
Moscow-140, USSR

*Pfister, V.
Fisheries Section
School of Forestry & Wildlife Mgmt.
Louisiana State University
Baton Rouge, La 70803

*Price, James O.
Memphis State University
Memphis, TN 38152

Provenzano, Anthony J., Jr.
Department of Oceanography
Old Dominion University
Norfolk, VA 23508

*Pursiainen, M.
Evo Inland Fisheries and
Aquaculture Research Station
SF-16970 Evo, Finland

*Ramos, L.
Quiñon S.A.
San Esteban de Gormoz
Soria, Spain

*Reynolds, Julian D.
Department of Zoology
University of Dublin
Trinity College
Dublin 2, Ireland

Richards, Kenneth J.
Riversdale Farm
Stour Provost
Gillingham, Dorset, U.K.

Roe, Pamela
Department of Biological Science
California State College Stanislaus
Turlock, CA 95380

*Roché, Bernard
Service Regional d'Amenagement
des Eaux de Corse
20200 Bastia, France

Romaire, Robert
Fisheries Section
School of Forestry & Wildlife Mgmt.
Louisiana State University
Baton Rouge, LA 70803

Rundquist, Jane
Division of Environmental Studies
University of California
Davis, CA 95616

Schultz, David
1735 Highland Place, Suite 36
Berkeley, CA 94709

Shimizu, Steven
Division of Environmental Studies
University of California
Davis, CA 95616

*Söderhäll, Kenneth
Institute of Physiological Botany
University of Uppsala, Box 540
S-751 21, Uppsala, Sweden

*Soivio, A.
Department of Zoology
University of Helsinki
Arkadiankatu 7
SF-00100 Helsinki 10, Finland

Sommer, Ted
Institute of Ecology
University of California
Davis, CA 95616

Spitzy, Reinhard
Hinterthal 20
A-5761 Maria Alm
Salzburg, Austria

*Stepaniuk, I. A.
The Whole-Union Scientific
Research Institute
Fisheries and Oceanography
V. Krasnaselskaya 17a
Moscow-140, USSR

*Supronovich, A. V.
The Whole-Union Scientific
Research Institute
Moscow-140, USSR

xv

Tamayo, David
5338 College Avenue #C
Oakland, CA 94681

Taylor, Dene
Tauranga Rd., R. D. 2
Waihi, New Zealand

Thomas, J. J.
Department of Biological Science
University of London
Goldsmiths' College
London SE14 6NW, U. K.

*Vey, Alain
Station de Recherches de
Pathologie Compareé
30380 Saint-Christol-les-Ales,
France

Wang, Mary Y.
9326 Carmel Rd.
Atascadero, CA 93422

Washizu, Makoto
Division of Environmental Studies
University of California
Davis, CA 95616

Watson, Terri
Santa Cruz, CA

Westman, Kai
Finnish Game and Fisheries
Research Institute
P. O. Box 193
SF-00131 Helsinki 13, Finland

Wheatly, Michele G.
Department of Biology
University of Calgary
Calgary, Alberta T2N 1N4, Canada

*Witzig, J. F.
Department of Zoology
North Carolina State University
Raleigh, NC 27067

Wright, John S.
P.O. Box 4278
Santa Barbara. CA 93103

Wright, William O.
6252 Parkhurst
Goleta, CA 93017

PREFACE

The International Association of Astacology (I.A.A.) was organized by Sture Abrahamsson, and its First Congress was held in Austria in 1972 with the assistance of a number of interested individuals, including Charles R. Goldman, who organized the Fifth Congress. There were 50 scientists and interested laymen at the First Congress. The purpose of the I.A.A. is to encourage the scientific study of crayfishes for the benefit of mankind, to provide for the dissemination of research findings related to crayfish, and to develop an international forum for free discussions of problems relevant to crayfish.

The Second Congress, organized by James Avault, was held at the University of Louisiana, U.S.A., in 1974, with participants from 15 countries; the Third Congress, organized by Ossi Lindqvist, was held in Kuopio, Finland, with participants from 13 countries attending; and the Fourth Congress, organized by Pierre Laurent, was held in Thonon-les-Bains, France, in 1978, with representatives from 16 different countries.

The Fifth Congress attracted participants from Austria, Canada, Finland, France, Ireland, New Zealand, Nigeria, Spain, Sweden, and the U.K. as well as the U.S. to the University of California, Davis. A number of additional papers sent by scientists, unable to attend, are included in this Congress's proceedings. The Congress brought together an excellent mixture of basic scientists, aquaculturists, students, and those interested in marketing. The papers that were presented were reviewed and have been edited for publication in this volume.

The meetings, beyond the valuable exchange of ideas among the participants, created the proceedings which remain the most important product of the five congresses. They continue to represent the only international collection of contemporary crayfish research in existence. The volumes of the proceedings are of steadily increasing scientific value in the growing body of published research on these important animals. Supplies of the previous editions have already been exhausted and are now to be found only in libraries or in the possession of I.A.A. members. In additions to the scientific interest in crayfish, their economic value as a food delicacy for harvest and sale in Europe, south and western United States, Africa, and Australia is well documented.

The Fifth Congress, held from August 16 to 21, 1981, followed a format similar to the previous congresses, and papers have been grouped by subject matter accordingly. Members arrived on Sunday afternoon and evening, and were welcomed to the Davis campus of the University of California by a buffet dinner at the Dining Commons. The Congress opened Monday morning with short welcoming addresses by professors Charles Hess, Dean of the School of Agriculture, and Francisco Ayala, Associate Dean of the Division of Environmental Studies and Director of the Institute of Ecology.

Professor Per Brinck of the University of Lund was invited by the Organizer to inaugurate the Sture Abrahamsson Memorial Lecture, which appears in this volume of the proceedings. The lecture summarized the important crayfish research accomplished by Dr. Abrahamsson in Sweden with Per Brinck and in the United States with the Organizer. Particular emphasis was given to the careful research leading to the introduction of Pacifastacus leniusculus into Swedish waters, and the research and management program directed toward reestablishing the commercial crayfishery in Europe with this plague-resistant species. The Memorial lecture is to be included in future Congresses.

A special luncheon was held at the Memorial Union, and in the evening recreational swimming was followed by a traditional California steak barbecue. Tours for accompanying persons were organized for Folsom, California on Tuesday, with the I.A.A. banquet at the University Faculty Club that evening. An enormous ice sculpture of a crayfish was created by Mr. Ray Rose for this event. Thirty-five kilograms of Pacifastacus leniusculus from the Sacramento River were collected by Steve Shimizu and cooked under the expert direction of Evelyne de Amezaga Goldman for the first course of the banquet. California wine, Aquavit, short speeches, live music and dancing highlighted the evening.

Following two days of meetings the Congress adjourned for the mid-Congress field trip to the famous Napa Valley where four of the finest wineries kindly hosted our tour. A picnic lunch, organized by Jane Rundquist, which featured wine, smoked salmon, and an abundance of California grapes was served in the Park at Saint Helena. Participants returned to Davis for an informal social hour before retiring. A short tour of the Sacramento River crayfishery was conducted by Steve Shimizu for interested participants.

The program continued on Thursday, and concluded with the election of officers for the Society and acceptance of the Swedish delegation's invitation to hold the Sixth Congress in Lund, Sweden, in 1984. A post-Congress tour took participants to Lake Tahoe where they had a special luncheon at the lakeside home of Joan Lundquist, followed by a boat tour of the lake led by Robert C. Richards, and a crayfish and cocktail party at Star Harbor hosted by the Congress Organizer. That evening the post-Congress tour stayed in Reno, Nevada, where they attended the famous M.G.M. Grand Casino show "Hello Hollywood Hello." The tour included a visit to Nevada's historic silver mining town of Virginia City before returning to Davis.

This Fifth Congress would not have been possible without the devoted office and field trip assistance of Mr. George Malyj and Mrs. Meryllene Smith. Mrs. Beth Clark and Mrs. Shirley Bell from Conference Services of the University provided extensive logistic support for the Congress. Dr. Wally Clark, Director of the Davis Aquaculture Program, hosted one of the evening social hours.

Evelyne de Amezaga played the major role for food and beverage selection for the Congress as well as the mid- and post-Congress tours. Jane H. Rundquist and Ted Sommer were particularly helpful in the day-to-day operation of the Congress.

Help with the task of assembling and editing these proceedings for publication deserves special acknowledgment. The editor was assisted by Ted Sommer in the initial stages, and by Assistant Editor, Anne C.

Forcella in the difficult final stages. Carol Barnes, also of the Division of Environmental Studies, did an outstanding job of formating and producing the photoready copy for publication. William V. Sleuter, Assistant Manager of University Repro Graphics, provided valuable advice on the illustrations.

Finally, Alyne L. Ruppanner drew the cover illustration of the California Crayfish <u>Pacifastacus</u> <u>leniusculus</u> on natural stream substrata. Look closely and you may see the gold they still guard in the foothill streams of California.

Charles R. Goldman, Editor
Professor of Environmental Studies
University of California Davis 95616

STURE ABRAHAMSSON MEMORIAL LECTURE

AN ECOLOGIST'S APPROACH TO DEALING WITH THE LOSS OF
ASTACUS ASTACUS

Per Brinck
Ecology Building
University of Lund
Lund, Sweden

As a tribute to the late Dr. Sture Abrahamsson of Lund University Professor Goldman has instigated the Sture Abrahamsson lecture - to be held for the first time at this symposium. Hopefully, it will be a regular event at future symposia.

Professor Goldman has invited me to give the first Sture Abrahamsson lecture and the main theme will be the work of Sture, its general background and the development of systems for culture of crayfish and prawns. The story is an interesting case study of how a country which lost an essential part of its productive stock of indigenous crayfish, succeeded in successfully replacing it with an exotic species.

Sture's premature death in January 1973 put an end to his promising achievements in crayfish research and development. He had a great interest in the International Association of Astacology. The organization was created in 1972 as a result of the first symposium on crayfish which he initiated. During its first decade its activities have been kept closely in line with the program established by Sture and a few others.

Early fieldwork on development of Crustacea

In most West European countries postgraduate students who intend to proceed to a PhD have their interest in pure science. Through the years I have had some 80 such students and there have been only two who have shown an interest in applied zoology already when they first came to me. One of them was Sture Abrahamsson.

After he had graduated in 1959 he approached me and told me that he wanted to start on development of Crustacea and would like to do his introductory scientific task abroad. At that time Dr. Helge Backlund was investigating food and feeding of young brown trout. In his thesis on the Wrack Fauna of Sweden and Finland he had demonstrated that there was a high production of terrestrial Amphipoda in the wrack beds. These amphipods made excellent food for the trout. The problem was whether they could be easily collected in the field or cultured in sufficient amounts to be used as food for young trout. For that purpose we needed more information about their life history. We had good contacts with research laboratories in the Netherlands, and after studies in Utrecht Sture went to the field station at Den Helder. There he carried out an experimental investigation of the relations between _Talitrus saltator_ and three species of _Orchestia_ and their environments. A manuscript on the results was written in 1960. But feeding trout with talitrids was possible only on a small scale and probably only a non-profit basis and Sture changed his interest.

In 1960 he had been working for some months in a project under the Swedish Board of Fisheries and its Institute of Freshwater Research at Drottningholm. He worked at the Mörsil hydroelectric power station. He found it attractive and afterwards, when we were discussing the program for his doctoral thesis, he expressed a wish to remain in contact with the Fisheries Board and its research laboratories.

Crayfish and food habits in Sweden

At that time, the late 1950s and early 1960s, there was one all-pervading problem in the commercial handling of Crustacea in Sweden. That was how to procure enough crayfish to meet the high demand.

How could it be that there was such an interest in crayfish in Sweden? Documents report that in the old days in the 16th century, it was an esteemed dish at the court. In spite of lack of detailed records, there is no doubt that crayfish were an easily available food in part of Sweden. In some early cookery-books there is an abundance of recipes for preparation of crayfish, which indicates that it was a well known food item in the 18th century.

Crayfish plague and its consequences

Around the turn of the century crayfish had become economically important in Sweden. The estimated catch was about 200 tons per year, about half of which was exported. In 1908 a change became evident. The export dropped. The reason was that in 1907 the crayfish plague had appeared in Sweden, transmitted with crayfish brought from Finland to Stockholm. Nobody at that time was aquainted with the disease. The lethal crayfish were thrown in Lake Mälaren from where the plague spread rapidly. At the end, it had destroyed about half the Swedish stock of the native species Astacus astacus Linnaeus. As the indigenous stock could no longer provide enough crayfish to meet the demand, Sweden gradually increased its imports.

In the 1950s the situation was complicated. While production decreased and imports continued to increase, attractive substitutes for Astacus astacus were few and delivery slow. The pressure on the authorities to act was strong. But was there satisfactory information about the different species of crayfish to act?

Early studies on crayfish and Aphanomyces

The dispersal of the crayfish disease was followed anxiously by landowners as well as fishery authorities. There are numerous reports on the changing status of various crayfish populations, and there was an increasing pressure on the authorities to intensify research on the plague. In 1930, Professor O. Nybelin made a substantial contribution by culturing Aphanomyces astaci from dying crayfish. He demonstrated that the fungus was the pathogen attacking and ravaging the Swedish crayfish, while previously various bacteria had been suggested to be the cause of death.

In the 1950s, the investigations of the native crayfish and possible substitutes to be introduced were entrusted to Professor G. Svärdson of the Freshwater Research Institute at Drottningholm who by close contacts with experienced fishermen and by his own field work considerably broadened our knowledge on crayfish in Swedish waters and prepared for coming actions.

The patogen and its dispersal

It is true that the taxomonic status of the pathogen had been settled and details of its biology were known. But still, there were large gaps in our knowledge about its reproduction, dynamics and transmittance - so important for the understanding of the dispersal of the disease. Again, its physiology was essentially unknown. It had appeared for the first time in Europe in Italy in 1860, but only recently - after its endemicity with the North American crayfish had been definitely settled - did we realize that its appearance in Europe was a result of an early introduction. To this early stock of the parasite in Europe fresh material has been repeatedly added by introducting individuals of North American crayfish of different species and provenance. The general idea was held that the North American species were resistant to the disease. Numbers of scientists - some of them represented in this audience - were at that time puzzled by the fact that the European species of crayfish were dying out after American species had been introduced in native crayfish habitats. In the 1950s and 1960s when many introductions took place, the extinction was thought to be related to the high competitive ability of the American crayfish.

Actions for restoration of the native crayfish stock

In Sweden different opinions were voiced. While conservationists wanted to retain the indigenous species - best of all with a gene for resistance to the plague added to the stock - others suggested that a North American species should be introduced to Sweden. A dense population of any crayfish in a lake gave the owner an ever increasing revenue, in fact higher than the income from all the fish species of the lake together.

The authority which had the immediate responsibility was the governmental Board of Fisheries. The Board found the basis for a final decision weak, so thought it wise to pursue both lines of action: to support a detailed investigation of the plague fungus and to consider the introduction of a North American species. In Uppsala, in 1960 Torgny Unestam started an investigation of Aphanomyces and its relations to the host animals.

First introduction of North American crayfish to Sweden

In 1960, Professor Svärdson imported individuals of Orconectes virilis (Hagen) and Pacifastacus leniusculus (Dana) to Sweden. They were released in two separate lakes near the Baltic coast in the central part of the country.

Dr. Svärdson, who took the initiative to introduce these two species in Sweden, was a champion of introductions of foreign species to perturbed or destroyed aquatic ecosystems. In 1958 he had visited the U.S. in order to find crayfish species which could take the place of Astacus astacus in plague-stricken areas. After careful considera- tion of available information the governmental authorities decided in

1959 that O. virilis and P. leniusculus be introduced provisionally. The introduction of Orconectes failed for unknown reasons, while Pacifastacus established a local population.

In both projects, it was soon found that fundamental knowledge about the crayfish was lacking. Applying the results of laboratory studies of the fungus to field conditions in crayfish populations was as impossible as answering questions about specific relations, reproduction and growth of the introduced species in relation to the native Astacus.

Such was the situation when Sture Abrahamsson decided to start his investigations on crayfish.

Projects without funds - times of insecurity and economic restraints

At that time, there was a very positive attitude to research on crayfish and Aphanomyces. There was only one snag - which is usual when research is initiated in an undeveloped field. There was no money available for those who were to do the work. While the Swedish research councils generously supported pure science, there were still considerable difficulties in obtaining money for applied science. Crayfish investigations, ultimately aimed at restoring the Swedish stock of commercial crayfish, seemed outside the terms of reference of the Natural Science Research Council. It is true that there was a Council for Technical Research but it was not interested in the crayfish projects as such because they were too far from technological development. The Board of Fisheries had no funds for support of research projects in freshwater fisheries.

Despite this situation, a first application was presented to the Board of Fisheries in March 1961, giving a plan and listing expenses for field work. The Board gave a grant out of their funds for restoring water systems exploited by hydroelectric schemes. For 12 years the Board granted such money for basic expenses. The salaries of the project leaders and their assistants had, however, to be obtained by other means. In 1961, Lund University gave a grant for two years towards the cost of the degree of licentiate of philosophy which Sture passed in 1965. At Uppsala University Torgny Unestam got the same support for two years.

In 1962, two bills had been introduced in the Riksdag, the Swedish Parliament, in support of research and development on crayfish in order to increase production. They failed, but Dr. Bengt Lundholm, secretary to the Research Committee for Natural Resources, referring to the bills, suggested that a research team of a botanist and a zoologist plus technical assistance be established. The plans were presented to the Natural Science Council in 1963 but did not meet with enthusiasm and an application was turned down at the prepatory meetings. I succeeded, however, in persuading the full Council meeting in May 1963 that the project should be adopted in principle and run from 1964 but that money should be granted for Torgny Unestam already for 1963. After a further year of support from the Board of Fisheries the zoological part was taken over by the Research Council, where the project stayed for three years. Then the temporary Commission for Natural Resources added the crayfish project to its program and paid salaries for one year (1967). It handed over its recommendation to the newly established Nature Conservancy Board, later transformed to the Swedish National Nature Conservancy Office, which supported the project for three years, until July 1971. In spite of the fact that it was generally agreed that the project was important and attractive and that studying populations and carrying out laboratory and field experiments over a series of years was a long-term program, necessary funds were only given for one year. At times the applications were passed from hand to hand between the secretariats of the various organizations and grants were received, only after remonstrance.

Although it may be argued that the University departments in Lund and Uppsala had poor contacts with the Research Councils and other authorities, it was certainly not so. I think the principle reason for the difficulties and the stubborn resistance was the weak position of biology, particularly development in biology, in the 1950s and 1960s. In the scientific community, physics, astronomy and chemistry were very strong, and the crayfish projects were only a small part of the battle which was fought until the attitudes changed in the late 1960s and 1970s. Still funds for research and development in freshwater fishery remain very inadequate.

These facts add nothing to research on crayfish. But they do tell much about the insecure circumstances in which the project teams were working, and they show the depressing atmosphere in which the data on crayfish ecology had to be obtained.

Life history of Astacus astacus

Let us turn now to the research done by Sture Abrahamsson. Already in the first application to the Board of Fisheries there was a detailed description of important areas of work. It was expressly stated that proposals to introduce foreign crayfish species should not be adopted until we knew the ecology of the indigenous species better. Therefore, field investigations of selected populations ought to be carried out, relating to population dynamics, activity, and food and feeding. Growth and reproduction in relation to climate and water conditions were also important items. Sture had selected a series of sites from the southernmost province of Skåne to the mountains in the central Swedish province of Jämtland, to which were later added localities in the far north.

It was soon found that studies on growth, density and migration could not be performed without a reliable method for marking large numbers of individuals. A method was invented which was fast and efficient, using cauterisation by a soldering gun. It made marking of up to 6,392 individuals of the same population possible and the marking persisted for two subsequent moults. This method was first published in 1963 and made possible comprehensive quantitative investigations of crayfish.

One of the localities examined in detail was the Rögle ponds at Lund. The results were published in 1966. In a pond of 25,000 m^2 the catchable population was calculated to be 12,400 individuals at the end of August and in early September. An estimation of the total population gave about 50,000 individuals, i.e. 2 crayfish m^{-2}. Of these 24% were juveniles, 42% males and 34% females. About 75% of the population consisted of individuals less than three years old. Owing to the density, growth was very slow. There was a large surplus of males and this was particularly marked along the shores which were the most attractive feeding places. The littoral vegetation was heavily grazed, particularly in August when crayfish were grazing by night on land up to 0.5 m from the water's edge.

As is well known by this audience, crayfish ecology was not then a favoured topic among biologists. Of the approximately 450 nominal species and subspecies of freshwater crayfish of the world, about 3/4 are North American, while very few species live in the Old World. Most data refer to Nearctic species, but still detailed life history studies exist for very few species and there seems to be such a variety of patterns that it is difficult to draw conclusions by analogy. Therefore, much of Sture's early work was devoted to inventing menthods for ecological work on crayfish in the laboratory and the field. Working up life histories is time-consuming. On the other hand, it gave him a unique knowledge of the material which was instrumental for the invention of his method for culturing crayfish.

Preparing for transplantation of Astacus astacus to the north

It is evident from a report published in 1964, that Sture at that time had the intention of finding populations of Astacus astacus suited for transplantation to the numerous streams, rivers and lakes in northern Sweden where there had been no crayfish. One particularly interesting habitat were those river systems which had been exploited by hydroelectric schemes. Impoundment had created numerous lakes and others had been changed. New ecosystems were to be established or old ones adjusted to new environments. Would crayfish be productive there?

Investigation of some populations in the north which had been deliberately implanted showed that the growth rate was fairly high and the number of large individuals considerable. Being perhaps too optimistic, he concluded that "after relevant temperature and other environmental data had been collected" a regional plan for transplantation of crayfish to the north would be made and the introduction executed in such a way that the spreading of the crayfish plague would be impeded. Because of the physiographic peculiarities of North Sweden, with river systems running from west to east, the northern region is separated from the southern plague-affected area. This is also one reason why indigenous crayfish have not spread to northern Sweden.

Dr. Svärdson's great knowledge of biological productivity and introductions of fish in northern waters made him question the feasibility of such a project. Within the framework of the grants from the Board of Fisheries Sture therefore continued his investigations in the north. Established crayfish populations at eight sites were examined closely and samples of individuals transferred to ten other geographically separated waters. Fecundity, hatching, survival, growth of juveniles and adults were compared for the different populations in relation to latitude, altitude, and temperature. Also frequency of moulting was studied in relation to temperature and food supply which were regulated experimentally. The results - published in 1972 - were conclusive.

Sture found that although stocks of Astacus astacus can live and grow in waters up to 68° N in Sweden, breeding is confined to water bodies where the temperature during the three summer months averages 15° or more. Therefore, the northernmost breeding populations of Astacus are located in the coastal region, close to the Baltic. The harsh climate slows down growth and sexual maturity and the long pre-reproductive period incurs greater mortality.

During the course of these investigations Sture soon realized that although populations of Astacus astacus in the north could give some compensation for the loss by plague in the southern part of the county, the productivity of the new areas would never equal that of the southern populations. Therefore, restoration of the crayfish production in the south had to be accomplished and he was convinced that introduction of a species which could withstand the plague was the best solution.

The studies of Astacus astacus were continuously being broadened. Besides studies in population ecology, investigations were carried out on the autecology of the species at different densities and at different environmental conditions: densities, temperature and food were changed experimentally. At the same time he added ethological observations of aggression, behaviour at mating and mechanisms controlling the mortality of the juveniles. Numerous data important for culturing crayfish were collected.

Life history of Pacifastacus leniusculus in Sweden

After the groundwork had been done, the time had come to include Pacifastacus in the investigation. In the plans for 1964 and 1965 studies were proposed of its activity, growth and longevity and of its relations to Astacus astacus, in terms of competition. Finally, it was suggested that methods for culturing juveniles in ponds should be examined closely. It was concluded that if the investigations showed that the species could be introduced without negative consequences, it should replace Astacus astacus in water systems affected by Aphanomyces.

In fact the first implantation of Pacifastacus in southernmost Sweden had been carried out in October 1963 in the isolated pond where Sture had for three years studied Astacus astacus. After a second introduction in 1964, the development of the population was studied for five years. In the very same environment, studies of P. leniusculus gave comparative information about density, growth and reproduction of the species which definitely showed that Pacifastacus was superior in a number of ways. The most important qualities are that it sexually matures one year earlier, the egg production is about 90% higher, the number of egg-bearing females is higher, the population density increases much faster, the growth is much faster, the annual number of moults is higher, and finally, it is heavier with better flesh production than Astacus. While adult Astacus males grew by about 5 mm per moult and the females by 2 mm, the corresponding values for Pacifastacus were approximately 10 mm for males and 8 mm for females. After these data had been made public, there was little doubt that Pacifastacus leniusculus was the species to replace the plague-stricken species in Europe.

Studies of Pacifastacus leniusculus in California in cooperation with Charles R. Goldman

Although these studies in the development of populations of Pacifastacus in Sweden were important, a good knowledge of the specific reactions to local conditions was also necessary for future introductions. On the other hand, studies of the species in its natural habitats in North America would give much information about population dynamics, behaviour and relations to the environment of populations in their native ecosystem. This kind of information could not be obtained in Sweden until after populations had been established for several years. Therefore, Sture planned to stay one year (1967) in the United States. The California Fish and Game Department was very helpful. Field work was planned to be carried out in three areas, viz. Lake Tahoe, Natoma Lake in the American River system and Berryessa Lake of Putah Creek in the Inner Coast Range. The crayfish which had been imported to Sweden in 1959 came from Natoma Lake. Berryessa Lake was, however, found to be unsuitable and was replaced by Lake Hennessey in the Coast Range from which crayfish were introduced to Sweden in 1967. This introduction failed, however, so in 1969 and 1970 the founder populations were taken from Lake Tahoe where the climatic conditions were more similar to the conditions in South Sweden. Natoma Lake and Hennessey Lake are both manmade reservoirs with fluctuating water level. They were of particular interest, because of the previous discussion of stocking impounded lakes in North Sweden with crayfish.

Sture located his headquarters at the Institute of Ecology at the University of California, in Davis. Soon after his arrival in January 1967 he came in contact with the Director, Professor Charles Goldman who became interested in his work and supported it generously. In fact, Sture found working with Charles Goldman so stimulating and fruitful that at the end of the year he took up a discussion with me by letter whether he should remain in California and continue his studies of crayfish there. The main part of his work was performed at Lake Tahoe under very favourable conditions: the crayfish population was enormous - estimated to be over 56 million adults - the resources for limnological research substantial, and the collaboration with Professor Goldman very successful, as can be seen from the publications and report it generated. In 1970, Abrahamsson and Goldman jointly published a paper on the distribution, density and production of Pacifastacus leniusculus in Lake Tahoe. And in 1971, Sture finished a paper on "Trappability, locomotion and diel pattern of Astacus astacus and Pacifastacus leniusculus". It was provisionally accepted for publication in Ecology, but the editors' lengthy handling of the matter was responsible for the fact that the final editing was still not complete when Sture died, in January, 1973. The manuscript is an interesting and valuable contribution to astacology, and has been included in the proceedings of this symposium after some revision.

Pacifastacus leniusculus accepted for transplantation to lakes and streams in Sweden

Sture returned to Sweden in May 1968. He wrote a series of reports and gave verbal presentations about the results which aroused great interest. The Research Committee for Ecology of the Natural Science Research Council arranged a symposium on Astacus astacus, Pacifastacus leniusculus and the plague which was held in Lund 27

Figure 1. Sture Abrahamsson fishing <u>Pacifastacus</u> <u>leniusculus</u> on the effluent of Lake Tahoe. Stellan Karlsson photograph.

August 1968. The Swedish authorities after consideration of the situation decided that Pacifastacus leniusculus should be accepted for transplantation to Sweden. In 1969, Sture was again in the U.S. with Professor Goldman for crayfish research as well as for supervision of trapping and transport of 60,000 crayfish from Lake Tahoe which were distributed to about 69 lakes in Sweden after having been properly quarantined. Even though several of these introductions were failures, the numbers of individuals and sites were still much too small for a rapid increase of the production of crayfish in Sweden. As a result, there were proposals that Pacifastacus should be introduced on a larger scale. On the other hand, opponents emphasized the obvious risks for introduction of parasites and fish diseases. At this time something happened which changed the situation completely. A method for culturing crayfish was developed.

Cooperation on crayfish with Ruben Rausing

Sture Abrahamsson started his work on crayfish at the old Institute of Zoology in Lund, where he joined the team of freshwater ecologists. Besides the facilities at the department building, they needed a field laboratory. In 1963, part of their activities was located at an old water mill, which belonged to the Swedish Salmon and Trout Association, the Håstad Mill on the Kävlinge river. Another old water mill was, however, found to be more suitable. It was placed at our disposal after rebuilding and in 1965 the team moved part of the experimental field work to the Stampen mill at the Simontorp Estate which belonged to the Rausing family. Dr. Ruben Rausing took a lively interest in the crayfish studies and contributed to the studies in various ways. In his doctoral dissertation (1971), Sture acknowledged that Dr. Rausing's "Original ideas, conveyed to me in our many stimulating discussions, may in the future be of great importance for the restoration of European crayfish waters."

One such idea, first discussed in September 1966, aimed at an investigation of the crayfish as consumers of submerged vegetation and other organic debris in the littoral zone. This is a very efficient way to convert such matter to animal protein, as was shown by Charles Goldman in 1974. At the same time, the crayfish control the macrophyte vegetation which is a problem in many parts of Europe. These questions, including the general influence of a crayfish population on the fauna and flora of the habitat, were worked on at Simontorp from 1966 and elaborated in a report early in 1969.

The other proposal was the production of young of Pacifastacus for introduction to plague-infested waters.

Methods for crayfish breeding

The contacts between Dr. Ruben Rausing, Sture, and Professor Goldman resulted in cooperation. As a consultant, Sture worked with Dr. Rausing in order to find a method for culturing crayfish in a closed system. Although there were already methods for culture of crayfish in ponds - Sture had been interested in these things since 1964 and frequently commented on them - there remained, however, many problems to be solved in artificial breeding. Some of these increased with the crowding resulting from mass-production, e.g. cannibalism, competition and various forms of stress, especially in connection with moulting. Others were independent of the density, e.g. food and feeding, water and temperature regimes and substrate which may diminish loss of eggs and young. Transport and delivery of the juveniles intended for introduction presented a particular problem. This is easy in the laboratory and on a small scale. But finding a system which is economically feasible on a large scale is more difficult. A special container was developed which accommodated 100 juveniles and could be dropped over suitable crayfish habitats. It sinks to the bottom and functions as an initial protection shelter for the young.

Establishment of the Simontorp hatchery

Studies on rearing crayfish were much advanced when on 16 October 1969 the Board of Fisheries gave the Simontorp Estate a permit for building a plant for aquaculture of crayfish. Arrangements were made through the Fisheries Research Board and the California Department of Fish and Game to import 50,000 crayfish from Lake Tahoe to serve as a breeding base after the hatchery technique had been developed. On 24 March, 1970 the Simontorp hatchery announced that juveniles of Pacifastacus had been produced and were available for distribution. The first series of juveniles hatched in Sweden in the culture system were transferred to natural waters in July 1970. In 1970 and 1971, 200,000 juveniles altogether were produced and distributed. This was the end of the importation of adult crayfish to Sweden for immediate introduction in lakes and rivers. Dr. Ruben Rausing in recognition of his contribution was elected an honorary member of IAA in 1976.

Regulations for implantation of Pacifastacus leniusculus in Swedish waters

The successfully breeding of Pacifastacus and the production of juveniles for introduction were received with great satisfaction. This did not mean, however, that implantation of Pacifastacus was uncontrolled. A permit was necessary, and the regulation remained strict, so that Pacifastacus should not interfere with Astacus in regions or water systems where there was a productive stock of the latter species. Therefore, introductions were and still formally are restricted to water systems where plague has been reported, and to closed ponds.

This cautious policy is a compromise between two conflicting opinions, as is evident from the minutes of comments on the wording of the new statutes on controlling crayfish plague, in 1974. While the Freshwater Research Laboratory of the Board of Fisheries wanted to have Pacifastacus as widely distributed as possible, Dr. Unestam recommended very heavy restrictions in order to block the dispersal of the plague as much as possible. While the background of the policy is evident, the implication remains unclear. In the official notes the Board stated that the "Board of Fisheries and the National Nature Conservancy Office agree that the native stock of Astacus astacus must be retained as long as possible for conservancy as well as economic reasons." But zoologists and animal ecologists agree with the Freshwater Laboratory that Pacifastacus will disperse widely irrespective of restrictions, and retaining present conditions with irregular outbreaks of plague in Astacus is not economical. That was precisely the reason why Pacifastacus was first introduced. In practice, it is impossible to control the distribution of a species of crayfish which is sold alive for food without any restrictions, and freely transported.

The investigations by Dr. Unestam and his team of the plague fungus contributed greatly to our knowledge of its physiology and ecology. In one way the results were negative: in <u>Astacus astacus</u> there is a complete lack of individuals resistant to the fungus and no methods were found to produce such a resistance or else to prevent or delay the establishment of the parasite in the crayfish. The reason for the heavy restrictions recommended by Dr. Unestam was, I think, the fact that the only way to stop outbreak of the disease in <u>Astacus</u> is to block contact between crayfish and fungus, and the only way to do this is to eradicate the fungus. For that purpose T. Unestam, G.-G. Nestell and Sture Abrahamsson introduced in 1973 "an electrical barrier for preventing migration of freshwater crayfish in running water" as "a method to stop the spread of the crayfish plague." Electric current kills or paralyzes crayfish. In a double electrical barrier a buffer zone without crayfish is formed. Combined with intentional infection of crayfish downstream of the barriers, the method may be a tool to exterminate the crayfish plague in exposed water systems - provided that all the aggregates of the local population of crayfish can be infected. As is well known, this is problematic.

Prospects for future development of crayfish production

The geographic range and the commercial production of crayfish have for quite a long time been concentrated in the southeastern parts of Sweden where most introductions of <u>Pacifastacus</u> have been made. One reason has been the heavy predation by eels in the southwestern part to which may today be added the acidification of surface waters caused by industrial pollution. Therefore, <u>Pacifastacus</u> will gradually take over in the southeast. But there are other, potential production areas. In a report of 3 February 1971 Sture dealt with such possibilities, focusing on implantation of cold-adapted <u>Astacus</u> from North Finland populations in northern Sweden. He refrained from proposing introduction of <u>Pacifastacus</u> in that region, although one of his interests during visits to the U.S. in 1971 was to find cold-adapted populations of <u>Pacifastacus</u> whence individuals could be brought to selected sites in north Sweden for studies. It is a pity that the possibility of exploiting this vast potential has not been explored any further.

At times, a few more such potential production areas have been discussed. For the southwestern region, populations adapted to the lower part of the pH gradient would be welcome. And finally, there are populations of <u>Pacifastacus</u> in the western U.S. adapted to the coastal brackish water. One of the largest brackish water areas in the world, the Baltic Sea, extends all along the eastern coast of Sweden. Stocking the Baltic Sea with crayfish would be a venture of possibly enormous annual yields, but would also raise problems beyond imagination. This is a tempting field for experimental work.

Thesis on crayfish ecology presented at Lund University

Besides scientists and customers there were others interested in the activities and achievements of Simontorp hatchery. In September 1970, The Swedish Ministry of Foreign Affaires sent a letter to the Board of Fisheries asking whether Sture Abrahamsson had taken crayfish in Lake Tahoe for scientific or commercial purposes. The original enquiry came from the Department of Fish and Game in California. The problems brought up were whether the large collections of crayfish made by Sture and his associates had any influence on the lake's population of crayfish or the ecological balance of the lake. Because of Dr. Goldman's strong conservation posture in the Tahoe basin, political forces attempted to discredit the crayfish transports to Sweden on the grounds that the stock would be depleted. Since the trapping was less than 0.01% of the adult population, this allegation was absurd.

In the U.S. the case was soon closed, but in Sweden the consequences were more far-reaching. Rough letters came from the Board of Fisheries and the National Nature Conservancy Office. The latter had in May 1970 announced that they would not support the crayfish project after 30 June 1971. They found the investigations "scientifically valuable and practically important" but outside their ordinary field of action and referred the case to the Board of Fisheries. There were certainly several reasons for the final negative decision. At that time there was strong evidence that the work done by Sture had been and still was of great importance for restoring the European crayfish populations. In spite of these facts grants for the personnel were refused.

After discussion with Dr. Ruben Rausing an application was sent to the National Swedish Board for Technical Development proposing a program of aquaculture in closed freshwater and marine systems for development of industrial production of Crustacea and Mollusca. It was a stimulating and interesting program. But there was no interest. Again, the grants for personnel were refused.

This seemed to be the end of ecological studies of crayfish which had been carried out at Lund University since 1961. In this situation Dr. Ruben Rausing came to the rescue and connected Sture more closely to the Simontorp hatchery.

The first international symposium on crayfish

In the meantime there was a boom in Pacifastacus interest in Europe and Sture made numerous contacts with people who wanted to solve their local problems. He discussed with me whether we should arrange a symposium on freshwater crayfish in Lund. I was doubtful, because I did not think the interest was that great. In 1971, however, he convinced me and suggested that it should be located somewhere in Central Europe. Early in 1972, he told me that he had found a very able man who was ready to serve as co-ordinator. I received information about that person and indeed was impressed. He had the most unusual curriculum vitae I had ever seen in connection with scientific activity. The man was Herr Reinhard Spitzy, in 1976 elected an honorary member of the Association. The symposium was held in Hinterthal in Austria from 12 to 15 September 1972 and was enjoyed by many crayfish enthusiasts from Europe and the U.S. It was a great success. The proceedings were published early in 1973, after the manuscripts had been edited by Sture who worked on the volume until the last days of his life.

A true champion in astacology...

Sture Abrahamsson's list of publications contains 25 items. When he died he also left a large manuscript, intended to form the basis of a number of descriptive and analytical reports and papers, relating to laboratory and field work on crayfish in Sweden and California. As a scientist he had an unusual gift: he worked his ideas straight through to practical development. He was tenacious and positive about his research. Difficulties were many, but he had a unique ability to overcome them. With the death of Dr. Abrahamsson astacology lost much of its momentum.

... and the realization of his plans and intensions

In the introductory remarks to this lecture I mentioned that Sture Abrahamsson was exceptional among a multitude of students, because of his keen interest in applied zoology and his wish to contribute to its development. After 12 years of successful work he died in January 1973 - a short time after the establishment of the Simontorp hatchery, where his method of culturing crayfish was put into practice. It was left to the personnel of the hatchery and its laboratory to carry out and expand his plans for the replacing of the plague-stricken species of crayfish indigenous in Europe. What has been done?

Results of the Swedish restoration program

Since the start of the program in 1969 about 300 waters have been stocked with Pacifastacus from the Simontorp hatchery. Generally, it takes a long time until deliberately introduced species have developed dense populations, but Pacifastacus is an exception. Several waters already have populations as dense as those of Astacus during the pre-plague period. As a whole, the restoration has been a success, despite the fact that there are great differences in the stocking program, because the freshwaters in Sweden are privately owned and managed.

The physiography of the water system of Sweden implies that there is a concentration of good crayfish habitats in the southeast of the country. This is well known from catches of Astacus astacus, before the crayfish plague spread. A selection of 20 such crayfish habitats, now stocked with Pacifastacus, was reported for 1980 to show an average yield of 5.4 crayfish per effort at test-fishing, i.e. the number of crayfish per trap per night. The change of the catches through time from the introduction of Pacifastacus is shown in Fig. 2, which also gives information about the general development of the catches in all Pacifastacus waters which have been tested by controlled trapping. With this rapid development in mind it is reasonable to assume that Pacifastacus leniusculus will be the dominating crayfish species in Sweden in the near future.

Already one year after the hatchery had started its distribution of young crayfish, requests for material came in from several countries and at present at least 17 countries in Europe have populations of Pacifastacus, developed from juveniles imported from Simontorp.

As a sideline, pond production of crayfish for consumption has been started in Sweden and elsewhere.

The techniques used by the laboratory can easily be adapted to other crayfish species than Pacifastacus.

Simontorp also has an interest in aquaculture of warmwater species, but in this case it is a freshwater prawn, Macrobrachium rosenbergii, for which a technique for culture through its complete

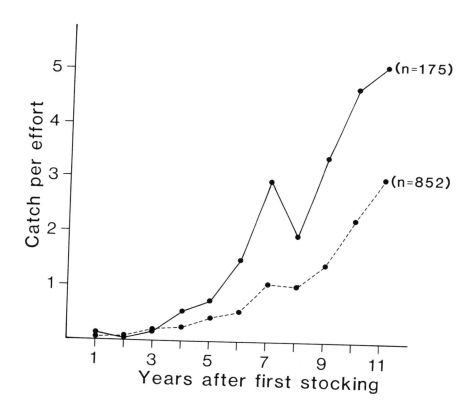

Figure 2. The change of the catches of _Pacifastacus leniusculus_ through time in (top curve) 20 good _Astacus_ habitats in southeastern Sweden and (bottom curve) all waters where _Pacifastacus_ has been introduced and followed by controlled trapping.

life cycle in a closed system has been developed. Similarly, this technique can be adapted to other species of prawns.

These acheivements represent a major contribution to the aquaculture of Crustacea. They show that the goals that were defined by Dr. Abrahamsson were both far-sighted and realistic.

LIST OF DR. STURE ABRAHAMSSON'S PUBLICATIONS

1963. Flodkräftundersökningen - Norrlandsaspekter. Svensk Fiskeri Tidskrift 1963: 172-176.

1964. Pågående och planerad forskning rörande ekologien hos Astacus astacus (Linné) och Pacifastacus leniusculus (Dana). (English summary.) Svensk naturvetenskap 17:306-316. Stockholm.

1964. Kräftfisket i Mälaren och Hjälmaren. Fiske 1964:64-67. Örebro.

1964. Forskning rörande flodkräftans och signalkräftans ekologi. Svenska Lax- och Laxöringsföreningen u.p.a. Arsbok 1963:13-16. Malmö.

1965. Flodkräftan och signalkräftan i skånska vattendrag. Skånes Natur 52:131-139. Malmö.

1965. Den lägre djurvärlden i Höje å vid Lund. In Höje å - Zoologisk utredning. PP. 50-56. Lund.

1966. A method of marking crayfish Astacus astacus Linné in population studies. Oikos 16(1965):228-231. Copenhagen.

1966. Dynamics of an isolated population of the crayfish Astacus astacus Linné. Oikos 17:96-107. Copenhagen.

1968. Flodkräftan - historik och allmän översikt. In Ecol. Bull. 3:15-18. Lund 1968.

1968. Erfarenheter av signalkräftan (Pacifastacus leniusculus) i Nordamerika mot bakgrunden av en eventuell inplantering i Sverige. In Ecol. Bull. 3:40-49.

1969. Historik, allmän översikt och reformförslag beträffande fångst och handel med flodkräfta. Fauna och flora 1969:98-104. Märsta.

1969. Signalkräftan - erfarenheter från USA och aspekter på dess inplantering i Sverige. 1969:109-116. Märsta.

1970. & C.R. Goldman. 1970. Distribution, density, and production of the crayfish Pacifastacus leniusculus Dana in Lake Tahoe, California-Nevada. Oikos 21:83-91. Copenhagen.

1970. Restaurering av Sveriges Kräftbestånd. Informationsskrift utgiven av Simontorps Akvatiska Avelslaboratorium. 5 jpp. Värnamo.

1971. Ett kräftpestvatten och dess restaurering. Svenskt Fiske 1-2:44-47 and 52. Stockholm

1971. Signalkräftan acklimatiserad i kräftpesthärjat vatten. Fauna och flora 71:2-10. Märsta.

1971. Density, growth and reproduction of the crayfish Astacus astacus and Pacifastacus leniusculus in an isolated pond. Oikos 22:373-380. Copenhagen.

1971. Population ecology and relation to environmental factors of Astacus astacus Linné and Pacifastacus leniusculus Dana. Dissertation. II pp. Lund.

1971. Restaurering av kräftbestånd med signalkräfta av Simontorpsstammen. 23 sid. Ill. Värnamo.

1972. Fecundity and growth of some populations of Astacus astacus Linné in Sweden. Inst. of Freshw. Res. Bulletin 52:23-27. Drottningholm.

1972. & T. Unestam & C.G. Nestell. An electrical barrier for preventing migration of freshwater crayfish in running wter. A method to stop the spread of the crayfish plague. Inst. of Freshw. Res. Bulletin 52:199-203. Drottningholm.

1972. Ergebnisse der Erneuerung der schwedischen Krebsbestände mit der amerikanischen Krebsart Pacifastacus leniusculus. Osterreich. Fischerei 25:21-24.

1973. The Freshwater crayfish Astacus astacus in Sweden and the introduction of the American crayfish Pacifastacus leniusculus. In Freshwater crayfish 1:27-40. Lund.

1973. Methods for restoration of crayfish waters in Europe. The development of an industry for production of young of Pacifastacus leniusculus. In Freshwater Crayfish 1:203-210. Lund.

1982. Trappability, Locomotion and diel pattern of activity of the crayfish Astacus astacus Linné and Pacifastacus leniusculus Dana. In Freshwater Crayfish 5.

Editor of: Freshwater Crayfish 1. Papers from the First International Symposium of Freshwater Crayfish, Austria 1972. 252 pp. Lund 1973.

I
GENERAL PHYSIOLOGY AND
BIOLOGY OF CRAYFISH

ON THE SEXUAL DIMORPHISM AND CONDITION INDEX IN THE CRAYFISH ASTACUS ASTACUS L. IN FINLAND

Ossi V. Lindqvist and Erkki Lahti
Department of Zoology, University of Kuopio
P.O. Box 138, 70101
Kuopio 10, Finland

ABSTRACT

The weight/length relationships were compared in crayfish caught during the course of the year. There appears a strong dimorphism in the bodily dimensions in mature crayfish. There are large differences in the condition index in crayfish in different lakes and in different seasons. The crayfish increase their relative weights rapidly in early summer, but in August there is a dip in their condition due to the molting event. The water temperatures in summer seem to greatly affect the condition of crayfish.

INTRODUCTION

Sexual dimorphism is a common phenomenon in a number of higher crustaceans, including crayfishes (cf. Huxley, 1878). We would expect the evolutionary reasons for such a phenomenon would be in differences in the sexual strategies of the sexes and/or differential exploitation of available resources. Thus, the features involved in sexual dimorphism are important also for the developmnt of proper cropping strategy by man (cf. Rhodes and Holdich, 1979), be it for crayfish in the wild or in aquaculture.

For fish, a special measure called condition index has been developed for measurement of the state of the animal in terms of weight/length relationship (Tesch, 1971; Weatherley, 1972; Royce, 1973). This index should be reasonably good indicator of, say, the environmental effects on the overall state of the animal. With the crustaceans the index has been rarely used (cf. Haefner, 1973), though it certainly has some application with crustaceans also. Yet its apparent limitations have to be born in mind.

In this study we first depict the degree of sexual dimorphism in Astacus astacus with regard to the weight/length relationship and various bodily dimensions, and then measure the condition index in different water bodies during the course of the summer. Some comparative material is already available from studies by Järvekulg (1958), Abrahamsson (1971) and Lindqvist and Louekari (1975).

METHODS

The crayfish used in measurement of sexual dimorphism were caught by trapping, in the years 1973-1976, from the following lakes and rivers around Kuopio (ca. 62°C): Tallusjärvi, Liesjärvi, Liesjoki, Palosenjoki, Rytkynjärvi, Matkusjärvi, Valkealampi, Huttujärvi, Viitajärvi, and Heinikkajärvi. Trapping was conducted throughout the year, although a major part of the yield is from a period between June and October. (In this area the lakes are ice-covered from November until mid-May, on the average.) The total catch was 2090 mature crayfish, consisting of 961 females and 1129 males.

For measurement of the weight/length relationship as well as for calculation of the condition index we used crayfish that were collected by trapping from Lake Tallusjärvi in the summers of 1973 and 1974: the catch consisted of 533 females and 715 males. In 1974 we obtained 1410 crayfish (602 females and 808 males) from Lake Rytkynjärvi for comparative measurements.

The linear regressions between weight and length were calculated by

$$\log W = \log a + b \log L$$

Where W = weight, and L = length.

The condition index was calculated using

$$K = 25 \times \frac{W}{L^b}$$

Where W = weight, and L = length. The value for the exponent (b) was variable (> 3) as the growth curve in crayfish is allometric in form. All statistical calculations were performed at the Computer Center of the University of Kuopio.

Body length of crayfish was measured from the tip of the rostrum to the tip of the telson, excluding setae. The chelae length was measured from the carpal joint to tip of propodus. Animals with missing appendages were excluded.

RESULTS

The A. astacus male is heavier than the female in all size classes (> 70 mm) (Fig. 1). The heavier male weight is mostly the result of heavier chelae (Fig. 2). The width of the abdomen is larger in the female (Fig. 3), and also in the relative length of the abdomen there is a slight difference in favor of the females (Fig. 4). The longer relative length of the carapace in males is statistically significant. In both chelae width and carapace width there was no apparent sexual dimorphism.

In the summer of 1973 the water temperatures were relatively high (for Lake Tallusjärvi, see Fig. 6), and 1973 was generally regarded as a "good" crayfish year. Molting in the lakes and rivers around Kuopio was completed by early August. In 1974 the temperatures were generally lower, especially early in the summer, which also delayed molting until the first half of August.

The conditions indices calculated for the crayfish obtained from Lake Tallusjärvi (Fig. 5) show marked differences between the years 1973 and 1974. In both years the index goes up in May and June, but the rise is steeper in 1973. (The water temperatures rose faster in early summer in 1973, which also showed in better trapping result in May and June of 1973. The ice break-up occurs around May 11 in this lake.) We do not have data from July 1973, but the index was probably much higher in July 1973 than in July 1974. The differences in the

4

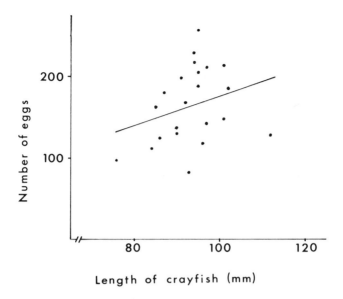

Figure 1. The dependence of the number of pleopodal eggs in winter and early summer on the female body length. The regression equation is $Y = 1.62X + 11.1$ (N = 22).

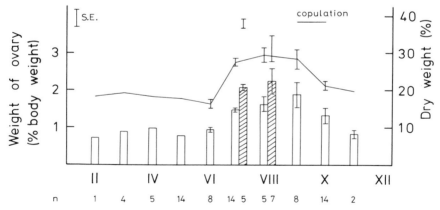

Figure 2. The proportion of the wet weight of the two ovaries out of each female's body weight, and the dry weight of the ovaries, during the course of the year in 1975. The Roman numerals refer to months: only every other month is indicated. The crayfish are from the following lakes and rivers: II and III, Palosenjoki; IV, Liesjoki; V, Matkusjärvi (1976); VI, Liesjärvi; VII, Liesjärvi (open column) and Viitajärvi (hatched column); VIII, Liesjärvi (open column) and Valkeinen (hatched column); IX, Liesjärvi; X, Heinikkajärvi; XI, Liesjoki.

5

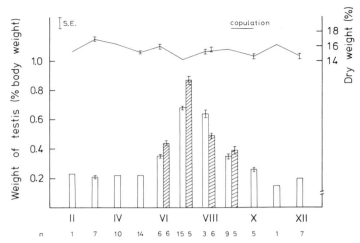

Figure 3. The proportion of the wet weight of the two testes out of each male's body wieght, and the dry weight of the testes, during the course of the year in 1975. The Roman numerals refer to months: only every other month is indicated. The crayfish are from the following lakes and rivers: II, III, IV, Liesjoki; V, Matkusjärvi (1976); VI, Liesjärvi (open column) and Liesjoki (hatched column); VII, Liesjärvi (open column) and Viitajärvi (hatched column); VIII, Liesjärvi (open column) and Valkeinen (hatched column); IX, Liesjärvi (open column) and Huttujärvi (hatched column); X, Pirttijärvi; XI, Liesjärvi, XII, Valkeinen.

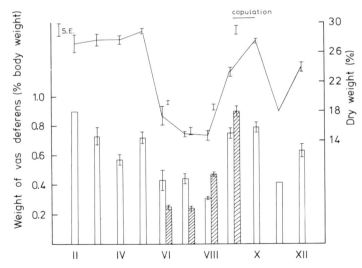

Figure 4. The proportion of the wet weight of the two vas deferens out of each male's body weight, and dry weight of the vas deferens, during the course of the year in 1975. For other explanations, see Fig. 3.

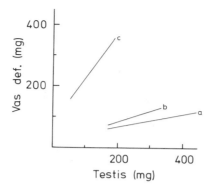

Figure 5. The relationship in wet weight between testes and vas deferens in male crayfish caught at Lake Liesjärvi in 1975. a=July (N=15); b=August (N=3); C=September (N=9).

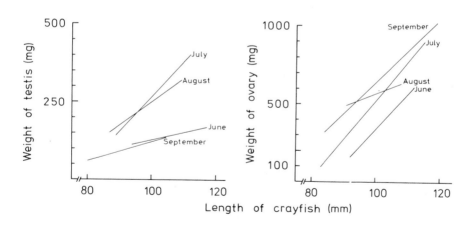

Figure 6. The correlations between body length and gonad weight in different summer months in crayfish caught from Lake Liesjärvi in 1975. Males. June: r=0.57 (N=8), July: r=0.84 (N-16), August: r=0.93 (N=3), September: r=0.73 (N=9). Females. June: r=0.94 (N=6), July: r=0.90 (N=15), August: 4=0.17 (N=6), September: r=0.41 (N=8).

condition indices are less pronounced in males than in females.

The results of statistical comparisons of linear regressions of the weight/length relationship in the summer months are shown in Table 1. In 1973, the relative weight per unit length in males remained unchanged from June to July, but the relative weight decreased from July to August, increased from August to September, and again remained unchanged from September to October. In the females in 1973, the relative weight per unit length decreased from June until August, but increased again from August to September. After September there was little if any change.

In the "cold" summer of 1974 there were few statistically significant changes in the weight/length regressions (Table 1), i.e the peaks and troughs were more level throughout the summer. Table 2 depicts a comparison of the condition indices of crayfish in two nearby lakes in two months in 1974. There are differences, especially with the females. These differences might be somehow related to differences in the local conditions: Lake Tallusjärvi is shallow, Lake Rytkynjärvi relatively deep.

DISCUSSION

Sexual dimorphism, with males relatively heavier than the females in Astacus astacus is well established (Järvekülg, 1958; Abrahamsson, 1971; Lindqvist and Louekari, 1975). The same probably applies to Pacifastacus leniusculus (cf. Mason, 1974), though Flint (1975) did not observe such a difference between males and females in Lake Tahoe. Generally, the growth rate of the males is faster (Abrahamsson 1971, 1972), and the females have also to invest in their eggs as well as to guard them. The females also feed less during the incubation period (Abrahamsson, 1972; cf. Bennet, 1974), though it is doubtful whether winter-time feeding plays any significant role in Finnish populations of A. astacus. Flint and Goldman (1977) emphasized the differences in growth rates in different habitats. In addition to temperature, the availability of protein for food may be a key factor in explaining such differences (cf. Castell and Budson, 1974).

In comparison to Estonian populations (Järveklüg, 1958), the Finnish A. astacus seem to be somewhat heavier for the same size. The Rögle Pond population in southern Sweden does not differ from Finnish crayfish (Abrahamsson, 1971), though an explanation might be the high crayfish density present in the pond. Heavier weight in northern populations of A. astacus, if it is real, might be an indirect result of the slower growth rate if in such a situation the weight of the cuticle increases faster than that of the internal organs (cf. Stein and Murphy, 1976). There exist no data on this.

In the male A. astacus the allometric growth of the chelae is stronger than in the female. The size and form of the chelae seems to be different in different areas of the distribution in A. astacus (Abrahamsson, 1972), and, for instance, in Orconectes virilis (Weagle and Ozburn, 1970).

The width of the female abdomen bears some relation to her egg-carrying activities (Abrahamsson, 1971): In Pacifastacus

Table 1. Statistical comparisons of linear regressions of the weight/length relationship between different months at Lake Tallusjärvi in 1973 and 1974. The arrows indicate either increasing (ascending) or decreasing (descending) relative weight per unit length in the crayfish.

Males	1973		1974	
June vs. July	N.S.	f = 93	N.S.	f = 107
July vs. August ↓	P< 0.001	114	P< 0.01	191
Aug. vs. Sept. ↑	P< 0.001	128	N.S.	251
Sept. vs. Oct.	N.S.	94	N.S.	122
Females				
June vs. July ↓	P< 0.01	59	N.S.	46
July vs. August ↓	P< 0.01	84	N.S.	186
Aug. vs. Sept. ↑	P< 0.001	100	N.S.	285
Sept. vs. Oct.	N.S.	70	N.S.	124

Table 2. The condition indices of crayfish in two nearby lakes during two months in 1974.

Males	July	October
Lake Tallusjärvi	0.33±0.03 (N = 41)	0.40±0.03 (n = 22)
Lake Rytkynjärvi	0.39±0.04 (N = 115)	0.33±0.02 (n = 35)
Females		
Lake Tallusjärvi	0.27±0.02 (N = 16)	0.96±0.05 (N = 10)
Lake Rytkynjärvi	0.74±0.06 (N = 69)	0.74±0.04 (N = 45)

leniusculus the abdomen is relatively wider than A. astacus, but so does Pacifastacus females carry relatively more eggs attached to her pleopods.

Both the weight/length regressions and the condition indices calculated for A. astacus indicate the following pattern. The crayfish build their bodily reserves early in the summer, and molting in July or August (depending on the year) comsumes some of those reserves. After molting the crayfish tend to regain their reserves in late summer. The early summer temperatures may be crucial for the molting and growth in summer, while the reserves obtained after molting may contribute to the survival over the winter. The changes observed in the gonad weights are so small a fraction of the total body weight (Lahti and Lindqvist, 1981) that they do not show up in the condition index. Furthermore, an increase in the weight of ovaries is offset by a diminishing hepatopancreas at the same time (Lindqvist and Louekari, 1975).

The use of the condition index is useful in comparative studies, but it also has some limitations. (However, it is more sensitive as a measure than the weight/length regressions use, though they measure very much the same thing.) The first limitation is that it tells nothing about the relative weights of different internal organs, of course (Lindqvist and Louekari, 1975). Thus, it is a rather coarse measure of the condition, but it is easily and rapidly obtained. Second, the condition index only measures the weight, but it tells nothing about the composition of the body and different organs. For instance, the crayfish experience changes in their water content in relation to molting (cf. Stein and Murphy, 1976; Lahti and Lindqvist, 1981). So molting brings a certain bias to the interpretation of the condition index. Thus the condition index is only one measure of the physiological condition of the animal, and measurement of physiological and biochemical parameters can add greatly to the understanding of the life cycle of the crayfish (cf. Stewart and Li, 1969; Castell and Budson, 1974).

ACKNOWLEDGMENT

Financial assistance for this study was obtained from the Research Foundation for Natural Resources (Suomen Luonnonvarain Tutkimussäätiö).

LITERATURE CITED

Abrahamsson, S. 1971. Density, growth and reporduction in populations of Astacus astacus and Pacifastacus leniusculus in an isolated pond. Oikos 22:373-380.

Abrahamsson, S. 1972. Fecundity and growth of some populations of Astacus astacus Linné in Sweden. Inst. Freshwater Res. Drottningholm, Rep. 52:23-37.

Bennet, D.B. 1974. Growth of the edible crab (Cancer pagurus L.) off southwest England. J. mar. Biol. Ass. U.K. 54:803-923.

Castell, J.D. and S.D. Budson. 1974. Lobster nutrition: The effect on Homarus americanus of dietary protein levels. J. Fish. Res. Bd

Can. 31:1363-1370.

Flint, R.W. 1975. Growth in a population of the crayfish Pacifastacus leniusculus from a subalpine lacustrine environment. J. Fish. Res. Bd Can. 32:1433-2440.

Flint, R.W. and C.R. Goldman. 1977. Crayfish growth in Lake Tahoe: effects of habitat variation. J. Fish. Res. Bd Can. 34:155-159.

Haefner, P.A. Jr. 1973. Length-weight relationship of the sand shrimp. Crangon septemspinosa. J. Chesapeake Sci. 14:141-143.

Huxley, T.H. 1878. On the classification and the distribution of the crayfishes. Proc. Zool. Soc. London, 1878, pp. 752-788.

Järvekülg, A. 1958. Joevähk Eestis. Biologia ja töönduslik tähtsus. 188 p. Tartu.

Lahti, E. and O.V. Lindqvist. 1981. On the reproductive cycle of the crayfish Astacus astacus L. (Crustacea, Astacidae) in the trapping season in Finland. Freshwater Crayfish 5:00-00.

Lindqvist, O.V. and K. Louekari. 1975. Muscle and hepatopancreas weight in Astacus astacus L. (Crustacea, Astacidae) in the trapping season in Finland. Ann. Zool. Fenn. 12:237-243.

Mason, J.C. 1974. Crayfish production in a small woodland stream. Freshwater Crayfish 2:449-479

Royce, W.E. 1973. Introduction to the Fishery Sciences. 351 p. New York and London.

Stein, R.A. 1976. Sexual dimorphism in crayfish chelae: functional significance linked to reproductive activities. Can. J. Zool. 54:220-227.

Stein, R.A. and M.L. Murphy. 1976. Changes in proximate composition of the crayfish Orconectes propinquus with size, sex, and life stage. J. Fish. Res. Bd Can. 33:2450-2458.

Stewart, J.E. and M.F. Li. 1969. A study of lobster (Homarus americanus) ecology using serum protein concentrations as an index. Can. J. Zool. 47:21-24.

Tesch, F.W. 1971. Age and Growth. In: Ricker, W.E. (Ed.): Methods for Assessment of Fish Production in Fresh Waters. - Oxford and Edinburg.

Watson, J. 1970. Maturity, mating, and egg laying in the spider crab, Chionoecetes opilio. J. Fish. Res. Bd Can. 27:1607-1616.

Weagle, K.V. and G.W. Ozburn. 1970. Sexual dimorphism in the chela of Orconectes virilis. Can. J. Zool. 48:1041-1042.

Weatherley, A.H. 1972. Growth and Ecology of Fish Population. 293 p., London and New York.

DIMORPHISM IN THE BRITISH CRAYFISH
AUSTROPOTAMOBIUS PALLIPES (LEREBOULLET)

W. J. Thomas
Department of Biological Sciences, University of London,
Goldsmiths' College, London S.E. 14 6 NW

ABSTRACT

The size and form of the first two pairs of abdominal appendages in A. pallipes are obvious features distinguishing male from female. During development from hatchling stages to adults many more such distinguishing features appear. These include differences in setal armature, integumentary openings, pleopod propotions and gross body form. The life cycle of A. pallipes appears strictly rhythmic with periods of low and high dimorphism clearly linked with the reproductive behaviour.

INTRODUCTION

Differences in the gross body structure between males and females of the European Astacidae are well documented in the general works of Lereboullet (1858); Huxley (1860); Bott (1950) and Karaman (1963). The first detailed study of dimorphism in A. pallipes is that of Rhodes and Holdich (1979). Focussing their attention on chelar and abdominal form they present a quantitative picture of the earlier empirical records, confirming that in A.pallipes the males possess larger clelae within a given size group after attaining sexual maturity, also they make clear the differences in abdominal size that exist between sexually mature males and females. The aim of this study is to show that dimorphism in A. pallipes involves additional differences of anatomy, some of which develop gradually, others appearing in phase with sexual maturity.

MATERIALS AND METHODS

Specimens of A. pallipes were collected from the River Darent at points between numbers 539657 and 533656 Kent, England of the National grid reference. A total of three hundred specimens were examined. Observations on the body and appendages were made at optical and electron microscope levels. Material used for examination under the scanning electron microscope was first freeze-dried and stored in a dessicator until required for further treatment. The latter consisted of fixing the specimen to a standard stereoscan stub with colloidal silver in iso-butyl methyl ketone (Achesons "dag" 915) directly on the stub or as an addition to sellotape to provide better mechanical fixation and conductivity. The specimens were then coated with carbon. All the observations were made using a Cambridge Instrument Company Mark II Stereoscan scanning electron microscope.

RESULTS AND DISCUSSION

General

Four weeks after hatching anatomical features distinguishing the sexes began to appear. After one year of growth sexual differentiation

on the basis of external anatomy is complete. This dimorphism includes
differences in gross body form, certain of the appendages and the setal
armature. Just prior to and during the first reproductive season
(Autumn) additional features appear further distinguishing the sexes.
These include glair glands, integumentary pores with accompanying behavioural changes all of which reflect a period of high sexual dimorphism
in A. pallipes. This phase of the life cycle commences in the latter
half of September and extends through to the begining of November by
which time all sexually mature females which bear spermatophores bear
eggs.

Specific

1. **The abdomen and its appendages.** During the hatchlings' 10-12 day
period of attachment to the mother the outline and proportions of the
abdomen is identical in all the individuals. In the three month old
crayfish differences in the appearance of the abdomen have appeared
between males and females. The abdomen in females becoming broader in
outline and more domed in transverse section (Fig. 1a). This is due to
the increased tergal and epimeral breadth combined with the greater
depth of the pleural plates on abdominal segments 2, 3, 4, 5, and 6.
Segment 1 remains small and lacks pleural plates in both sexes.

With the onset of maturity these differences in abdominal
dimension become more marked giving a "wasp-waisted" effect to the
females at its' junction with the cephalotothorax whilst in the male
the abdomen tapers gently from segment 1 to the telson (Fig. 1a and b).
This differentiation of the abdomen between the sexes quickly becomes
apparent even though the pattern of growth seems to be essentially the
same for both sexes until sexual maturity is reached. Experiencing
their second autumn the crayfish of both sexes are mature and display a
peak of dimorphism which thereafter is repeated annually. After
sexualy maturity a slower growth rate is apparent in the females.

In the River Darent the mature females moult in July prior to
copulation and spawning in the autumn. As the ovigerous females carry
their eggs from October to June they have no opportunity to moult again
until the July of the following year.

The pleural edges bear a fringe of pappose setae directed
obliquely inwards towards the sternal plates. Females possess longer
setae and the fringes themselves are denser. This difference of
setation is manifest early in the life of the animals. Three months
old individuals can be sexed on this basis with a more emphatic
difference appearing as sexual maturity approaches. All but the first
and last sternal ridges in the females bear a bundle of setae in the
mid-line (Fig. 1c). This bundle consists of between 5 and 15 setae all
set very closely together and radiating out fanwise from their sockets.
Juvenile females possess the same pattern of distribution only the
number of setae are less. The sternal ridges in males of all ages are
glabrous.

By mid-September sexually mature females are made conspicuous by
the presence of creamy-white patches on the pleura, epimera and sternum
of the abdomen, and on the bases of the pleopods and uropods, Thomas
(1977). The integument overlying these areas is perforated by groups
of pores arranged in roughly circular and oval plates (Fig. 1d and e).

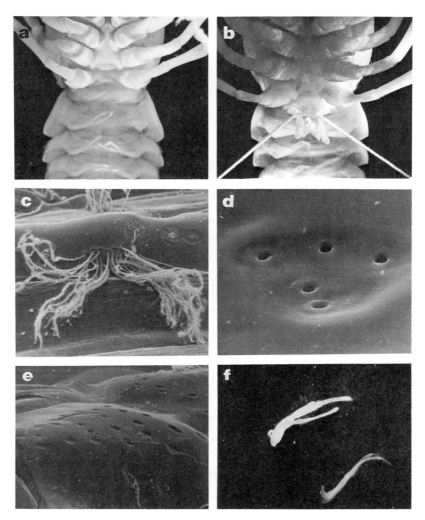

Figure 1: (a)Ventral view of the posterior thoracic and anterior abdominal region of a female <u>A. pallipes</u>. Note the "wasp-waist" effect at the junction of abdomen and thorax (actual size). (b) Ventral view of the posterior thoracic and anterior abdominal region of a male <u>A.p allipes</u>. Note the gradual reduction in breadth of the abdominal segments (actual size). (c) A group of oosetae on the sternal plate of segment 4 of a sexually mature female. Note the plumose distal ends of the setae to which the eggs are attached. X 700. (d) A "pore-plate" lying above the glair gland region of the pleuron in a sexually active female. X 2.2K. (e) The pleuron surface of a sexually active female showing the pore-plates which appear after the "sexual moult." X30. (f) A pleopod from abdominal segment 3 of a mature female (upper figure). Note the setal fringe and relatively broad face of the appendage. The lower figure is that of a pleopod taken from the same location, in a male. X1.5.

Their distribution is constant, being numerous on the anterior faces of the pleura and epimera facing the openings of the oviducts. Glair glands and pores first appear in the females of A. pallipes during the second September of their lives, the pores appearing after the final July moult. So these pores and glands can be looked upon as belatedly appearing secondary characters, and the July moult as the "sexual maturity moult." After egg laying the creamy white colouration of the glair glands persists until the spring, disappearing in late April-early May, to reappear in late September.

Even the first independent hatchling stages can be distinguished as either males or females. On what are future male specimens the second abdominal segment bears pleopods with thickened endopods, whilst individuals bearing four pairs of identically formed pleopods can be adjudged as females. A pair of papillae on the first abdominal segment of two month old males also differentiates them from the females. An increase in size and change of form of these appendages occurs with successive moults until the end of October. Subsequent moults result in the fully formed gonopods occuring in the males within the 14 month old age group. At this stage the crayfish vary in size from between 14-22mm carapace length.

Only amongst the pleopods is there a difference of setal armature between the sexes. It is possible to distinguish male from female pleopods not only on the basis of size and proportions - being broader faced and longer in the females (Fig. 1f) - but also on the density and variety of setae present. All the pleopods of females bear specialized egg-bearing setae, the appearance of those oosetae early in the life of the females and their continued presence makes them an obvious secondary sexual characteristic. These oosetae are semiplumose, although Kukenthals (1927) stresses that in decapods the eggs are attached to non-plumose setae, and Chantran (1870) describing egg laying in A. fluviatilis, describes the eggs as being attached to non-plumose setae. The egg bearing setae of A. pallipes thus seem to be different from those of other decapods so far examined. Once they appear they are a permanent feature of the female pleopods, there being no periodic transformation of the setae as in other decapods. During the breeding season the size difference between the pleopods of males and females is accentuated by the development of the glair glands and the longer, denser fringes of setae, emphasizing this period of high demorphism so characteristic of A. pallipes, it also fits the idea of a "sexual moult" at their first appearance.

2.   The Cephalothorax and its appendages. No specifically "male setae" are found in A. pallipes and only one specialization of position occurs. A closely set group of setae present on the apex of an integumentary outgrowth situated on the inner ventral face of the coxopods of the fourth pair of walking legs (Fig. 2a). On both the basipods and coxopods of the thoracic appendages the setae are longer and in greater profusion than in the females (Fig. 2b). The three posterior sterna of the thorax in females are perfectly smooth and depressed dorsalwards to form a cup-like receptacle for the spermathecae of the males conveniently near the oviducal openings (Fig. 2c). In the males the corresponding region is covered with setae. After copulation almost all the spermathecae are deposited in the thoracic receptacle of the female, but sometimes isolated threads of spetmathecal material are found elsewhere on the ventral surface of the abdomen.

15

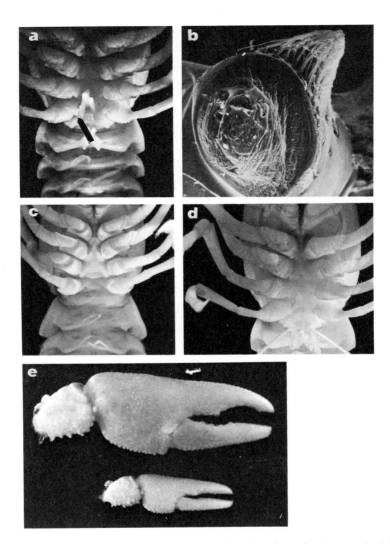

Figure 2: (a) Ventral view of a male at the junction of thorax and abdomen. Note the outgrowth from the inner face of the coxopods of the 4th walking legs (arrow) also the deeply cleft nature of the ventral thoracic wall (actual size). (b) The male genital opening with its adjacent fringe of long plumose setae. X200. (c) Ventral surface of a sexually active female at the junction of thorax and abdomen. Note the triangular shaped depression between walking legs 3, 4 and 5; also the female genital openings on the coxopods of the third pair of walking legs (actual size). (d) Ventral view of the male at the thoracia-abdominal junction. Note the closeness of the coxopods of the walking legs as compared with the female (actual size). (e) The 'outer' surface view of chelae taken from a large male (upper) and large female (lower). Note the biting edges of the male chela as compard with the smooth edged condition of the female (actual size).

Measurements of the distances between the coxopods of the third and fourth pairs of walking legs of both males and females clearly indicate a greater width between the opposing walking legs of the females as compared with males, on average twice the width (Fig. 2c and d). A group of setae set between the bases of the fourth pair of walking legs is characteristic of males only, a clear dimorphic feature developing early and before sexual maturity.

The chelae are one of the most obvious dimorphic features of A. pallipes Stein (1976). At an early age (14 months) the chelae of the males are consistently larger and at sexual maturity this difference is emphatic (Fig. 2e) Rhodes and Holdrich (1979). As growth and sexual maturity occur the chelae change roles, becoming more linked with social behaviour than as aids to feeding Stein (1976). Even so the chelae of large males and females do differ, the "biting edges" of the chelae in males bearing well developed integumentary outgrowths whilst those of the females remain relatively smooth. The chelae in both males and females lose setae with increasing age illustrating the phenomenon of setal replacement, noted in other decapods.

ACKNOWLEDGMENTS

I would like to acknowledge the Research Committee of the University of London, Goldsmiths' College for their financial assistance enabling the purchase of equipment and the meeting of travel expenses.

LITERATURE CITED

Bott, R. 1950 Die flusskrebse Europas (Decapoda) Astacidae, p 1-36 Senckenbergiana Biologica, 483.

Chantran, S. 1870. Observations sur l'histoire naturelle des Écrevisses p 201-202 Compte Rendu Hebdomadaire des Seances des Sciences 74D.

Huxley, T.H. 1880 The Crayfish. An introduction to the study of Zoology p 1-14 Kegan Paul and Co., Ltd. 371 p.

Karaman, M.S. 1963 Studies der Astacidae (Crustacea Decapoda), p 111-132 Hydrobiologica 22.

Kukenthal, W. 1927 Decapoda latreille 1802: Zehnfusser p 840-1038 Handbuch der Zoologie, Bd3. Halftel. Berlin & Leipzig.

Rhodes, C.P. and D.M. 1979 On size and sexual dimorphism in Austropotamobius pallipes (Lereboullet). A step in assessing the commercial exploitation potential of the native British freshwater crayfish. p 345-358 Aquaculture, 17.

Stein, R.A. 1976 Sexual dimorphism in crayfish chelae, functional significance linked to reproductive activities. Pp. 220-227 Canadian Journal of Zoology 54.

ON THE REPRODUCTIVE CYCLE OF THE CRAYFISH
ASTACUS ASTACUS L. IN FINLAND

Erkki Lahti and Ossi V. Lindqvist
Department of Zoology, University of Kuopio
P.O. Box 138, 70101 Kuopio 10, Finland

ABSTRACT

Astacus astacus were caught from different lakes and rivers in the district of Kuopio, Finland, during the course of the year. The changes in the weight and water content of the gonads were recorded. The peak weight of testes occurs in late June and the vas deferens reach their peak in October, i.e. before or at the copulation. The ovaries attain their highest weight in July through August. The eggs in these populations produced per female are more numerous and their diameter is smaller than in A. astacus populations in both Estonia and southern Sweden.

INTRODUCTION

To develop means of more effective exploitation of the crayfish Astacus astacus in the wild in Finland as well as to assess its potential for aquaculture, we need more information on its reproductive biology.

Some information on the egg production is available (Lindqvist and Louekari, 1975): in nearby countries the studies by Järvekülg (1958) and Abrahamsson (1971, 1972) are valuable. Comparative studies of other crayfish species, mostly from North America, refer to Pacifastacus leniusculus (Abrahamsson, 1971; Mason, 1978), P. trowbridgii (Mason, 1969), Orconectes virilis (Momot, 1967), O. propinquus (Capelli and Magnuson, 1974), and O. obscurus, O. sanborni and O. propinquus (Fielder, 1972).

In this study we have measured the seasonal changes in gonad weights, which are a measure of the gonadal activity, as well as the egg production in Astacus astacus in lakes and rivers in the Kuopio district in Finland (about 62°N).

MATERIALS AND METHODS

The crayfish used in measurement of egg production were caught by traps from Lake Liesjärvi and River Palosenjoki and River Liesjoki in 1975, and from River Liesjoki and Lake Matkusjärvi and Lake Valkealampi in 1976. These lakes and rivers are all situated within 5 km from Kuopio. This material consisted of 69 females and with 132 males, all with the total body length over 75 mm (i.e. mature individuals only were used). The eggs present in female pleopods were counted, and the average diameter of eggs on each female was measured from a sample of 10 eggs.

Gravimetric measurements of gonads were performed on crayfish caught in 1975 from six lakes (Heinikkajärvi, Huttujärvi, Pirttijärvi,

18

Valkealampi, Viitajärvi and Liesjärvi) and from two rivers (Liesjoki and Palosenjoki). The trapping period extended from February through December. Some additional material was obtained from Lake Matkusjärvi in May, 1976. The total number of mature crayfish was 206, consisting of 94 females and 112 males. Immediately after trapping the crayfish were put on ice, and transported to the laboratory for measurement. The weight of gonads was recorded to the nearest 0.1 mg with dry wieght and 1.0 mg with wet weight.

RESULTS

Among the crayfish caught by traps in winter and early summer there appears a definite bias in the sex ratio in favor of the males. This is probably not real for the population but rather reflects a difference in the susceptibility of the sexes towards traps. The females comprise only one third of the total catch, and out of these, less than 50% were egg-bearing (Table 1). There was no significant correlation between female size and egg diameter (Fig. 1), though truly large females were not present in this material. The egg diameter will increase at least from February onward (Table 1).

The major increase in the size of the ovaries takes place in July (Fig. 2), and the maximum is reached in September before the time of copulation. In November the ovarian weight is down to a level where it remains until the next summer. Interlake differences in the peak ovarian weight are 25%. Nearly all of the weight increase in ovaries is due to an increase in the dry weight (Fig. 2).

The maturation and increase in weight of the testes begins in June (Fig. 3), and the peak is reached in July and August. In October, after copulation, the weight is down to a level where it remains until the next reproductive season. The interlake differences in peak weight were relatively smaller than those in the weights of ovaries. The dry weight of the testes remains relatively constant throughout the year (Fig. 3) indicating that, in contrast to the ovaries, the weight changes are mostly due to changes in the water content.

The weight of vas deferens was at its lowest in June through August (Fig. 4), which indicates that the functional maturation of these organs takes place only after the testes have reached their peak size. The weight increases in September and October are accompanied by a transfer of sperm cells into vas deferens. The changes in weight of vas deferens are mostly due to changes in the dry weight (Fig. 4). Interlake differences were large, indicating the possible effects of variable environments on gonad maturation and copulation.

The relationship between testicular weight and the vas deferens weight (Fig. 5) is such that the testes are relatively heavier in June through August, i.e. at the time of spermiogenesis (Lahti, in preparation).

The correlation of body size with the gonad weight is highly positive in all summer months in females, while in males such a correlation exists only in July and August, i.e. during the peak of spermiogenesis (Fig. 6).

19

Table 1. The proportion of the female sex in the total catch, the proportion of egg-bearing females among all females, and the average egg diameter in <u>A. astacus</u>. The data are a compilation from both 1975 and 1976.

	Females per total catch		Egg-bearing females among all females		Average egg diameter	
	%	N	%	N	mm	♀♀
February	30	10	100	3	2.30	1
March	37	30	45	11	2.69	5
April	30	37	36	11	2.60	4
May	37	89	33	33	---	---
June	31	35	56	11	2.85	6
$\bar{x}$	34		42			
Total		201		69		16

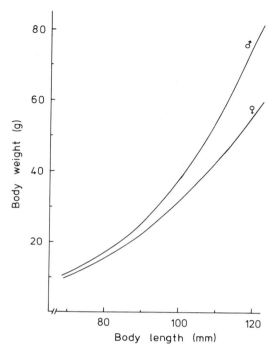

Figure 1. A computerized fit of the relation of body weight to the total length in _Astacus_ _astacus_. N = 1073 for males, N = 908 for females.

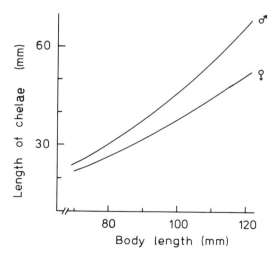

Figure 2. A computerized fit of the relation of chelae length to body length in _Astacus_ _astacus_. N = 1070 for males, N = 905 for females.

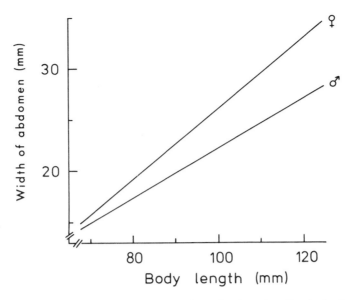

Figure 3.   The width of the abdomen in relation to the body length in
Astacus astacus. Males: Y = 0.243X - 2.038   (N = 128)
Females: Y = 0.343X - 8.274   (N = 156)

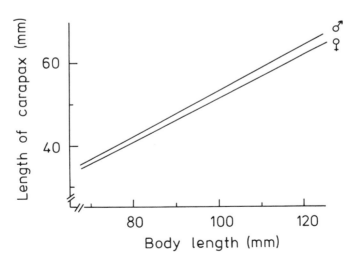

Figure 4.   The length of the carapace in relation to the body length in
Astacus astacus. Males: Y = 0.550X - 1.886   (N = 1143)
Females: Y = 0.521X - 0.768   (N = 947)

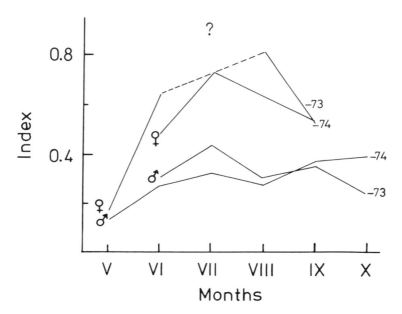

Figure 5. The monthly averages of the condition index in <u>Astacus</u> <u>astacus</u> at Lake Tallusjärvi in 1973 and 1974. The question mark indicates that the data for July 1973 are lacking.

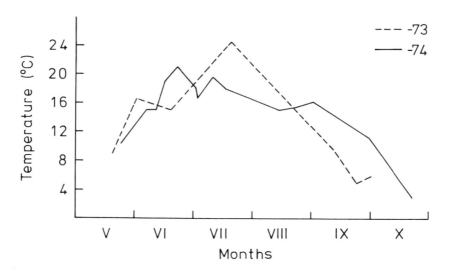

Figure 6. The surface temperatures in Lake Tallusjärvi in 1973 and 1974.

The dormant, winter-time ovaries are light yellow in color, but in July the ovaries change to a greenish color, and later in August they turn dark brown. The testes change little in their color during the course of the year. The vas deferens are still empty in July and August and they have a transparent light color: later they change into milt white in appearance (with the transfer of sperm into them).

DISCUSSION

One of the reasons for the bias in the sex ratio in the winter- and spring-time trappings is a difference in the moving activity between the sexes, although the spatial distribution of the sexes in different parts of the lake may also be variable. Field observations indicate that egg-bearing females are especially less active than both other females without eggs and males. For these reasons the share of egg-bearing females (42%) among the female population may be unrealistically low. In an earlier study the proportion of egg-bearing females among all females in August was 91.8% (Lindqvist and Louekari, 1975). Some of the females may simply lose their pleopodal eggs, though some resorption may also take place in the ovaries (Mason, 1969). In Abrahamsson's (1971) study the corresponding figure in November was 81.5%, but only 65.0% in May.

In winter and early spring we found an average of 166 pleopodal eggs per female, though in the fall the ovarial egg count has been found to be 248 on the average (Lindqvist and Louekari, 1975). Similarly, in Estonian populations of A. astacus the number of eggs in the ovaries was 182 (Järveküelg, 1958). In southern Sweden, Abrahamsson (1971) counted an average of only 100 pleopodal eggs in A. astacus females. Järvelküelg (1958) estimated that only about 2/3 of the eggs present in the ovaries ever hatch. This estimate applies very well with the results obtained in the present study: the number of pleopodal eggs (166) is 2/3 (67.3%) of the ovarial count of 248 (Lindqvist and Louekari, 1975). For Orconectes virilis Momot (1967) found the corresponding figure as 58%.

The positive correlation between female body size and both ovarial and pleopodal egg count has been established in numerous studies for the crayfish (cf. Järveküelg, 1958; Abrahamsson, 1971, 1972; Lindqvist and Louekari, 1975) as well as for crabs (cf. Haynes et al. 1976). The differences in egg counts between individual females seem to be large (cf. Abrahamsson, 1971).

Compared to both Estonian and southern Swedish populations, the eggs of A. astacus in the Kuopio district in Finland are both smaller in diameter and more numerous in number per female. This may be predictable when we consider the Finnish populations of A. astacus living in more unpredictable enviroments than the more southern populations (cf. Stearns, 1976). In Finland A. astacus reaches the northern limit of its distribution. According to Abrahamsson (1971), the average egg diameter in A. astacus in southern Sweden was 2.5-3.5 mm, and the diameter did not correlate with the female size. In Pacifastacus leniusculus the eggs were smaller but their number was nearly twice as large as A. astacus. According to Mason (1978), large females of P. leniusculus produce larger eggs, but individual differences were also large, which is probably due to differences in

feeding conditions during the period of ovarian development. Also in
P. leniusculus the egg diameter tends to increase towards hatching
(Mason, 1978).

We can infer from the changes in the weight of the ovaries as well
as from histological studies (Lahti, in preparation) that the ova
mature by mid-September, on the average, and the egg laying takes place
in October; this has been confirmed also with field observations. The
maturation process of the sperm in the testes begins earlier, in June
and July. After this the sperm cells are transferred to the vas
deferens, which shows up as an increase in the weight of vas deferens
in September and October. Copulation occurs in September through
October. Later in the winter some new sperm moves into vas deferens,
which shows up in the weight as well as the color of this organ.

The dry weight of the testes remains almost constant throughout
all seasons: increase in water content seems to be accompanied by
elevated metabolic level during the spermiogenesis. There appear large
species-specific differences in the seasonal development of the testes
in different crayfish species (cf. Wielgus, (1973) for Orconectes
limosus and Astacus leptodactylus).

By comparing the minimum and maximum weights of the crayfish
gonads during the course of the year we can obtain some estimate of the
level of investment in reproduction by the crayfish. In the female A.
astacus the peak weight is 2 times the minimum weight, while in the
male the testes' weights varies 4.5-fold (Figs. 3 and 4). However,
mere weight changes may be misleading, since we know little about the
metabolic levels in, and of the composition of, the gonads during the
course of gametogenesis. In any case, the gonads comprise a rather
small percentage of the total body weight, and the changes in gonad
weights bear little if any significance for the seasonal changes in
condition index (Lindqvist and Lahti, 1981).

In the allocation of resources between reproduction and growth
there are two main strategies available for the crayfish. It may
invest heavily in immediate reproductive effort, or it may invest more
in growth as a larger size in the future means higher reproductive
potential. Thus in populations under heavy trapping pressure we may
expect that selection favors higher immediate reproductive effort as
well as a smaller size at first reproduction. Exact data on such
possible effects are still lacking.

In this study we paid little attention to the possible differences
between crayfish populations in different lakes and rivers, though
there are several indications that such differences may be quite
pronounced. Also different years experience differences in the molting
schedule, which also affects the reproductive cycle. (The main
determinant is the water temperature in summer.) Such year-to-year
differences can be assessed by anyone who has tried to trap crayfish in
Finnish waters in some consecutive years.

ACKNOWLEDGMENT

Financially this study was aided by the Research Foundation for
Natural Resources (Suomen Luonnonvarain Tutkimussäätiö).

25

LITERATURE CITED

Abrahamsson, S. 1971. Density, growth and reproduction in populations of Astacus astacus and Pacifastacus leniusculus in an isolated pond. Oikos 22:373-380.

Abrahamsson, S. 1972. Fecundity and growth of some populations of Astacus astacus Linné in Sweden. Inst. Freshwater Res. Dropttningholm, Rep. 52:23-37.

Capelli, G.M. and J.J. Magnuson. 1974. Reproduction, molting and distribution of Orconectes propinquus (Girard) in relation to temperature in a northern mesotrophic lake. Freshwater Crayfish 2:415-427.

Fielder, D.D. 1972. Some aspects of the life histories of three closely related crayfish species, Orconectes obscurus, O. sanborni, and O. propinquus. The Ohio Journal of Science 72:129-145.

Haynes, E., J.F. Karinen, J. Watson and D.J. Hopson. 1976. Relation of number of eggs and length to carapace width in the brachyuran crabs Chionoecetes bairdi and C. opilio from the southeastern Bering Sea and C. opilio from the Gulf of St. Lawrence. J. Fish. Res. Bd Can. 33:2592-2595.

Järvekülg, A. 1958. Joevähk Eestis. Biologia ja Töönduslik tähtsus. 185 pp. Tartu.

Lindqvist, O.V. and K. Louekari. 1975. Muscle and Hepatopancreas weight in Astacus astacus L. (Crustacea, Astacidae) in the trapping season in Findland. Ann. Zool. Fenn. 12:237-243.

Lindqvist, O.V. and E. Lahti. 1981. On the sexual dimorphism and condition index in the crayfish Astacus astacus L. Freshwater Crayfish 5:00-00.

Mason, J.C. 1969. Egg-laying in the western North American crayfish, Pacifastacus trowbridgii (Stimpson) (Decapods, Astacidae). Crustaceana 19:37-44.

Mason, J.C. 1978. Significance of egg size in the freshwater crayfish, Pacifastacus leniusculus (Dana). Freshwater Crayfish 4:83-92.

Momot, W.T. 1967. Population dynamics and productivity of the crayfish, Orconectes virilis, in a marl lake. Am. Midl. Nat. 78:55-81.

Stearns, S.C. 1976. Life history tactics. A review of the ideas. Quart. Rev. Biol. 51:3-47.

Wielgus, E. 1973. Structure of the testes in indigenous crayfishes. Folia Morph. (Warsz.) 32:115-124.

METHODOLOGICAL CONSIDERATIONS FOR QUANTITATIVE DETERMINATION OF CONSUMPTION FOR THE CRAYFISH PACIFASTACUS LENIUSCULUS

Jane C. Rundquist and Charles R. Goldman
Division of Environmental Studies. University of California.
Davis, California 95616, U.S.A.

ABSTRACT

The importance of quantifying crayfish consumption is discussed within the framework of bioenergetics. A laboratory method to measure crayfish consumption using brineshrimp as a food source is presented and analyzed. This method was used for diurnal experiments with juvenile crayfish where it was found that at 20°C they eat at a steady rate through 24 hours with only a slight preference for eating in the dark. Whether the brineshrimp were alive or dead did not influence consumption rates. Other examples of application of the method were included. They illustrated the need to consider potential cyclic eating patterns, the nutritional state, and the high innate variablity of crayfish when designing experiments involving consumption.

INTRODUCTION

It has been said "We are what we eat!" A similar claim could be made for a particular organism, species, commmunity, or ecosystem. Eating and the relationships involving eating define and influence the biotic world at physiologoical, ecological, and evolutionary levels. If we know when, where, and how an organism obtains its food, what the food is, how much is eaten, how its body breaks down and utilizes the food, and what the waste products are, we not only know a great deal about that individual but also how it functions as a living unit in its environment. Similarily, who is eating what, when, where, and at what rates will largely define how a community operates and many of the selection pressures on species within that community.

It is not surprising, then, that major portions of the biological literature are devoted to detailed elucidations of eating related topics. This is particularily true for recent years where much research has been focused by trophic dynamic and bioenergetic concepts. From physiologist to animal breeder, from geneticist to botanist, from nutritionist to ecologist, it is clear that understanding the details of energy and materials exchange whether at a molecular, cellular, individual, or trophic level is a key to understanding and perhaps controlling living systems.

If we are to understand crayfish, we must understand their bioenergetics. This is especially important for the ecologist or aquaculturist interested in crayfish growth and production. Unfortunately, the development of detailed energy and nutrional budgets like those that have contributed so profoundly to the efficient production of domestic animals has not been accomplished for crayfish. In fact comprehensive energetics work has been carried out for only a few aquatic organisms with most of these being either fish (e.g. Winberg, 1956; Brett, 1964, 1965, 1967, 1971a, 1971b; Brett and Glass,

1973; Brett et al 1969; Fischer, 1970; Davies, 1964; Paloheimo and Dickie, 1965, 1966) or zooplankton (e.g. Marchant, 1978; Corner, 1961; Corner and Newall, 1967; Corner et al. 1965, 1967; Cowey and Corner, 1963). The paucity of comprehensive energetics information on aquatic organisms compared to land animals is due not only to the generally shorter termed and more recent research efforts in this direction, but also to the inherent difficulties of quantifying bioenergetic factors in the aquatic medium. The classic energy budget equation used by Winberg (1956) and others is I= G + R + E where I= ingestion, G= growth, R= respiration, and E= egestion. The measurement of each of these factors is complicated by the addition of water, but the measurement of ingestion (consumption) is especially problematic because of the difficulty of fully retreiving uneaten rations from the water. Nonetheless the importance of understanding and quantifying feeding relationships has spurred investigators working with aquatic animals to devise many ingenious methods of measuring consumption, the specifics dependent on the peculiar characteristics of the animal used and the information sought.

Literature that focuses specifically on crayfish consumption is extremely limited. What is available generally discusses crayfish feeding habits (e.g. Flint, 1975), impact of crayfish feeding (e.g. Flint and Goldman, 1975; Magnuson et al., 1975), or what has been experimentally fed to crayfish (e.g. Huner et al., 1974) without quantifying intake. Moshiri and Goldman (1969) did measure consumption for the crayfish _Pacifastacus_ _leniusculus_ in order to estimate assimilation efficiency. They fed preweighed dried portions of raw chicken and raw vascular plants to starved crayfish, allowed them to feed an average of 65 minutes, and then filtered the holding water onto preweighed milipore filters. These were dried and then weighed on a Cahn electrobalance allowing the calculation of consumption by difference. Algae counts from water samples taken before and after a feeding period have also been used to estimate algal consumption. This technique was applied by Budd et al. (1979) to provide evidence of filter feeding by the crayfish _Orconectes_ _propinquus_. Clearly there is a need for more detailed work regarding crayfish feeding and consumption.

This paper offers a beginning. In it a relativley quick and easy method to determine consumption by juvenile or adult crayfish is described. Its strengths and limitations are discussed. In addition data from several experiments using this method are presented. The results of these experiments suggest factors for consideration when designing research concerned with crayfish feeding.

METHOD FOR ESTIMATING CRAYFISH CONSUMPTION IN THE LABORATORY

Crayfish, like most benthic invertebrates, feed slowly and except through starvation are not easily persuaded to eat food immediately upon presentation. In addition they eat either by picking up small discrete food particles or by tearing large pieces apart before grasping and lifting them to the mandibles. These two behavior traits greatly limit the use of prepared diets for studies of crayfish growth and feeding behavior where consumption must be estimated. Meyers (1971; et al. 1970) pointed out that most available crude diets have been designed for actively feeding fish and are not very water stable. Even diets especially formulated for crustaceans using alginate as a binder to provide water stability will leach nutrients. Huner et al. (1974) reported that their crustacean diet promoted long-term maintenance and growth of crayfish, but that crayfish would not accept it after it had been in the water for more than two hours.

In addition to the problem of changing palatiblity over 24 hours, prepared diets introduce a host of difficulties for calculating how much was consumed because uneaten portions must be recovered by filtration. First of all, a portion of the diet will leach into solution preventing full recovery. While the fraction lost can be estimated experimentally and a correction factor applied, there will an unknown error in its application. The rate of leaching is dependent in part upon ration surface area which will vary depending on the physical handling and tearing of the food by the crayfish. Also the nutritional value of the portion consumed may vary depending on when during the 24 hours it was eaten. The difficulty of separating fecal material, which often appears in the form of suspended particles, from leftover food is another major problem with the filtration method. A third area of concern for experiments requiring high replication is the time factor. For each consumption determination four weights must be recorded: that of the food container, the food container plus food, the filter, and the filter plus remaining food. Also the filters must be dried before final weighings and the filtration itself takes time and requires disturbing the crayfish. Overall this method is unwieldy and expensive for a large number of animals over a long period of time.

Estimation of adult crayfish consumption using filtration can pose additional problems. A 5μ filter will clog if a major portion of the food is not consumed requiring either additional filters or the use of aliquots and their potential errors. At the other end of the crayfish weight scale is the need for high method sensitivity. For most growth experiments juvenile crayfish will be used requiring a high resolution to detect the small amounts they consume. A 20 mg juvenile given a generous diet of 5% by weight per day, requires a level of detection below 1 mg. While a standard laboratory scale has a resolution of .1 mg, tests with 2.4 cm filters weighing from 17 – 23 mg depending on the type used, showed variations of up to .5 mg on actual reweighings on subsequent days. A Cahn electrobalance has a resolution of .008 mg in the weight range of the filter, but its use can more than double the time involved for each weighing. Additionally, fluctuations in filter weights due largely to variation in air humidity remains a problem even with this sensitive instrument and can overwhelm the slight change in filtrate weight respresented by intake for a juvenile crayfish.

It was concluded that the best way to circumvent some of these problems was to offer the crayfish precounted rather than preweighed portions of food. At the very least this would reduce the time factor for each determination and eliminate any error due to fecal contamination. After testing a variety of discrete particulate foods ranging from Daphnia to tapioca pellets, live San Francisco Bay brineshrimp, Artemia salina, appeared the most likely candidate for quantitative dietary use with crayfish. Artemia salina average approximately .5 mg dry weight/individual and could be estimated by eye to 1/4 of a brineshrimp or .1 mg dry weight making the limits of detection within the desired range. In addition Artemia would provide a complex and nutrious food source important for growth studies. Because

29

Artemia have been used extensively in crustacean and other research, they have been the subject of considerable study including detailed composition analysis (Gallager and Brown, 1975). The availablity of such information potentially enhances the utility of data obtained from their use in feeding and growth experiments. On a practical level live Artemia could be obtained at a reasonable cost and with regularity at the local aquarium store. Culture techniques for Artemia are also available which could allow greater uniformity of size and composition than achieved for this study. Most importantly, even second stage Pacifastacus captured and ate live brineshrimp with apparent relish.

Having decided to use live brineshrimp as a food source, the method developed to estimate crayfish consumption was straightforward. Fresh live brineshrimp were purchased from the local aquarium store twice weekly on the day of delivery and stored in aerated seawater at 4-5 C until needed. Using an eyedropper medium sized non-egged brineshrimp were picked out by eye in order to increase the uniformity of the food particle being offered the crayfish. Based on preliminary feeding tests, an excess of the brineshrimp were counted and offered to the crayfish on the desired time schedule, normally once a day. At the end of the time period what remained was counted and the number eaten calculated by difference. The addition and removal of the brineshrimp was easily facilitated with the eyedropper and caused minimal disturbance of the animal.

A series of analyses on the brineshrimp were conducted to determine average dry weight, percent carbon, percent nitrogen, and C:N ratios per individual so counts could be converted to these measures. The variability of these measures for brineshrimp subjected to experimental conditions were also evaluated. The details of these tests are reported elsewhere (in manuscript) but the results are summarized here.

Table 1 summarizes the analyses for average dry weight, percent carbon, percent nitrogen, and C:N ratios per individual brineshrimp. 95 % confidence intervals for the estimated total dry weight of increasing numbers of brineshrimp were calculated with the statistic $n\mu + 2ns$ where n= number of brineshrimp, μ = mean dry weight per brineshrimp, and s= standard deviation. This relationship shows that for a given brineshrimp population mean and variance the relative accuracy of total dry weight (or some other measure) is a function of the number of brineshrimp in the sample. This means it is possible to control to some degree the accuracy of a consumption estimate using brineshrimp by controlling the number eaten. For example, to estimate 95% of the time within 10% accuracy of the total dry weight consumed, at least 20 brineshrimp must be eaten.

A one way analysis of variance on dry weight, percent carbon, percent nitrogen, and C:N for brineshrimp sampled through a six month period showed that these measures can vary significantly from one batch to another. This means that careful attention must be paid to the frequency of sampling brineshrimp for determining mean values to be used as conversion factors. Analysis of brineshrimp held for 5 days under the described holding conditions showed no trend for changes in dry weight, percent carbon, percent nitrogen, or C:N. For experimental purposes brineshimp from day 1 can be assumed to be equivalent to those used on day 5.

30

Table 1. Data Summary For Dry Weight, Carbon, And Nitrogen Analysis
Of Brineshrimp (BS)

	ANALYSIS	N	MEAN	STD.DEV.	RANGE
Mg Dry Wt/BS	individuals	62	.518	.111	.382-.812
	groups	13	.513	.054	.461-.595
% Nitrogen	individuals	64	10.0	1.1	6.8-13.7
	groups	13	9.4	.7	7.8-10.4
% Carbon	individuals	64	44.7	4.1	34.8-50.2
	groups	13	43.8	1.6	40.8-46.8
C:N	individuals	65	4.50	.44	3.91-6.67
	groups	13	4.69	.36	4.27-5.57

The individual data was obtained from five different batches of
brineshrimp with the analyses run on single brineshrimp. The group data
was compiled from thirteen batches of brineshrimp where anlyses were run
in triplicate on ground subsamples of groups of 25-100 brineshrimp.

31

Further tests were conducted to determine the stability of brineshrimp as a food source under potential experimental conditions. Analyses showed that brineshrimp held 24 hours in a salinity or temperature gradient showed a dry weight loss compared to samples taken at time 0. In addition the amount of the dry weight loss increased with increasing temperature (range: 5-24 C) and decreased with increasing salinity (range: 0-50% seawater). Percent carbon, percent nitrogen, and C:N varied only slightly through 24 hours and were unaffected by temperature or salinity changes. This was interpreted to mean that the weight loss that occurred was generally proportional in terms of the major nutritional indicators (C and N). Dry weight, then, is a reasonably steady indicator of nutritional value of the brineshrimp and any correction factor for weight loss will also correct for changes in nutritional value. Table 2 shows correction factors calculated for changes in dry weight through 24 hours at different temperatures and salinities. A final set of tests evaluated brineshrimp mortality through 24 hours at different temperatures and salinities. These showed increasing mortality after 12 hours with increasing temperature and decreasing salinity.

Based on the results of these analyses as well as actual feeding trials with crayfish, it was concluded that counting live brineshrimp is an effective method for estimating crayfish consumption for some kinds of experiments. These would include experiments to evaluate growth and growth efficiencies under a wide range of conditions and work on crayfish feeding patterns and behavior. While brineshrimp do not provide as versatile a food source as a prepared diet where exact rations can be formulated, they can be analyzed for calories, protein, carbohydrate, ash free dry weight, or whatever parameter is of interest for energy and nutrient budget development. As with prepared diets brineshrimp nutrient loss through time remained a problem, especially in freshwater at higher temperatures, but straightforward correction factors can mitigate this difficulty. Also some experiments, for example studying the effect of crayfish density on feeding patterns, only need relative counts eaten by one group compared to another and the problems associated with conversion to some other paramater become moot.

A major advantage of the outlined method is its speed and ease of application. With some practice brineshrimp counting goes quite rapidly. Approximately 1000 medium sized non-egged brineshrimp can be separated from the full population in an hour and 2500 previously separated brineshrimp counted in an hour. This makes it possible to design and practically handle experiments requiring detailed information on consumption and/or a large number of individuals. This is especially true when using juvenile crayfish who normally consume 10 or less brineshrimp per day. The method becomes more tedious for adults where large numbers of brineshrimp can be eaten in a single day. The maximum number of brineshrimp observed eaten in a single day was 600+ (all that were offered) by a female weighing 33 grams who had been starved for 11 days. A rate of 100 to 200 per day would be more normal for this size crayfish. The rule of thumb used in our experiments was to offer adults at least 10% and juveniles at least one brineshrimp more than their previously recorded high eaten. After a molt, the number offered was automatically increased due to a finding that crayfish appetite often increased at this point. This approach seemed to balance the conflicting needs to offer food in excess while minimizing extra counting.

Table 2. Correction Factors For Brineshrimp Dry Weight For Application To Consumption Experiments

TEMPERATURE C	REGRESSION LINE Y=PREDICTED DRY WT X=HOURS	R^2	P(TAIL) FOR F TEST ON SLOPE	CORRECTION FACTOR (C) APPLIED AFTER		
				6 HRS	12 HRS	24 HRS
12	Y=5.27 - .0505X	.7459	.00003	.972	.943	.884
16	Y=5.07 - .0463X	.6025	.00067	.934	.907	.860
20	Y=5.16 - .0659X	.8583	.00000	.941	.903	.828
24	Y=4.93 - .0617X	.7066	.00009	.898	.864	.794
% SEAWATER (16°C)						
12.5	Y=5.12 - .0439X	.6697	.00020	.947	.921	.871
25.0	Y=5.04 - .0206X	.2769	.04389	.943	.930	.907
50.0	Y=5.09 - .0161X	.1071	.23370	.956	.947	.928

The regression lines were calculated from brineshrimp dry weight measurements collected every 6 hours for 24 hours under the specified conditions. The correction factors (C) were calculated as C= W_a/W_i where W_a = brineshrimp dry weight at the midpoint of the named time period (i.e. average weight during that period) predicted by the regression line and W_i = mean measured initial brineshrimp dry weight. The correction factor (C) would be applied to estimates of consumption using brineshrimp as follows: I=NW$_i$C where I=total dry weight of brineshrimp ingested during given time period, and N= number of brineshrimp consumed.

Counting brineshrimp to estimate crayfish consumption does have limits in its application. In cases where counts must be converted to some other measure, e.g. mg N, the estimate of amount consumed is calculated by multiplying the number of brineshrimp eaten by the mean quantity of N per brineshrimp, that is, it is estimated indirectly rather than directly measured. While this approach will on the average give a good estimate of N consumed, especially at high numbers of brineshrimp, it might not be accurate enough in each individual case for some kinds of determinations, for example estimating assimilation. The previously discussed direct method of Moshiri and Goldman (1969) would certainly be more reliable for this kind of experiment. In addition to limits of the method itself at least two important areas of inquiry involving consumption are not readily approached by using brineshrimp as the sole diet. Research directed towards crayfish nutrition and field studies must rely on other methods to estimate consumption.

DIURNAL CRAYFISH FEEDING BEHAVIOR

Experiments were conducted to determine crayfish eating patterns through 24 hours and whether these patterns were influenced by light or the condition of the food offered. These experiments were intended to elucidate crayfish feeding behavior and provide data regarding consumption estimates using brineshrimp. As shown in the prevous section, under experimental conditions brineshrimp showed both differential mortality and loss of mass over a 24 hour period. A straightforward correction factor for this weight loss (Table 2) could be applied only if it were assumed that crayfish ate brineshrimp at a steady rate and showed no behavioral preference for live or dead food.

Two groups of 15 juvenile crayfish matched for weight were held for 24 days in individual containers at 20 C. A light regime provided by incandescent lights of 12 hours light and 12 hours dark was applied. Group 1 was fed fresh live brineshrimp twice a day (every 12 hours) for three days and four times daily (every 6 hours) on the fourth day. At each feeding the previous brineshrimp were removed, the number consumed calculated by difference, and fresh ones added. Nearly all uneaten brineshrimp were alive. Group 2 were fed live brineshrimp once daily and consumption checked on the same schedule as for Group 1. After 24 hours the old brineshrimp were removed, many of them already dead, and fresh ones added.

Five crayfish from Group 2 died and none of their feeding data was included in the data analysis. Counts of brineshrimp eaten during the 12 hour period of light and dark were converted to percent eaten during light and dark for each of the 24 days. Similarily, for the six days that counts had being taken every six hours, counts were converted to percent eaten for each of four periods. The total number of brineshrimp eaten each day was converted to number eaten per mg crayfish to standardize any differences in consumption due to size difference. Numbers of brineshrimp eaten were also converted to dry weight eaten using using the correction factors for brineshrimp weight loss through time. Statistical analyses of this data was consistent with the analysis based on counts only, and only the latter is reported here. All crayfish were weighed at the beginning and end of the experiment and three days after each molt with the most recent weight used for each conversion. Weight gain expressed as per cent wet weight increase from the beginning to the end of the experiment was calculated for each individual.

A summary of eating patterns of crayfish in the light and dark are shown in Table 3. Statistical analysis using two way analysis of variance for days of feeding and light regime showed that the percent of food eaten by crayfish in Group 1 during 12 hours of light was significantly different than during 12 hours of dark (alpha = .0007). No differences were detected between light and dark for Group 2. Not surprisingly the percent of food eaten in the light (or dark) was also significantly different between the two groups (alpha =.02). While the preference for eating in the dark is highly significant for Group 1, a mean of 54.7% of total food consumed in the dark does not indicate a strong behavioral difference between light and dark. This is confirmed by Table 4 which analyzes the percentage of days a preference was shown for eating in the light or dark.

In Table 4A the data is compiled into three groups: a preference for eating in the light, a preference for eating in the dark, or no preference. For these calculations it was assumed that a crayfish who ate 50 $\pm$ 5% of its food in the light or dark did not show a preference for either. Under this assumption crayfish in Group 1 showed an actual preference for eating in the dark only 50% of the time. A more detailed analysis is given in Table 4B where the data is compiled into five equal groups with the middle range (50 $\pm$ 10% of the total food eaten) assumed to show no preference for light or dark. This distribution clearly shows no strong preference for eating in the dark by either group as the mid 20% of the distribution (no preference) accounts for 45% of the days of Group 1 and 34% for Group 2. It does show, however, a slight skewering towards the dark compared to the light for Group 1. This is consistent with the other analyses.

A closer look at juvenile crayfish eating patterns was provided by the six days where consumption was measured every six hours (Table 3). A one way analysis of variance on this data between periods within each group showed no differences for Group 1 but significant difference for Group 2. Students t-tests showed differences between periods 2 and 4 (alpha=.04) for that group. Students t-tests matched for periods between Groups 1 and 2 showed no differences. For both groups the most consumption occured during the last six hours of dark (Period 4) and the least during Period 1 for Group 1 and Period 2 for Group 2. Despite the statistical difference found between Periods 2 (mean % consumed = 21.9) and 4 (mean % consumed = 29.6), the consumption data given in Table 3 is generally consistent with the hypothesis that under laboratory conditions juvenile crayfish feed at a steady rate througout 24 hours. There are no large deviations from the expected 25% consumption level for any period. The slight increase in feeding found in Period 4 is consistent with the previously noted slight preference for feeding in the dark.

In addition to comparing patterns of consumption, the actual numbers of brineshrimp eaten per crayfish and per mg crayfish were totaled for each individual and differences between groups were tested using the Student's t-test. Percent growth was also calculated for each group and compared using this statistic. No differences were found in either case between Group 1 and Group 2 (Total brineshrimp eaten/crayfish: t=.2920, p=.7714, df=24; Total brineshrimp eaten/mg

Table 3. Feeding Patterns of Juvenile Crayfish During 24 Hours

| | | | PERCENT EATEN | | | |
| | | # DAYS | GROUP 1 | | GROUP 2 | |
PERIODS	LIGHT REGIME	OF DATA	MEAN	STD.DEV.	MEAN	STD.DEV.
1 & 2	LIGHT	24	45.3	4.2	51.0	7.38
3 & 4	DARK	24	54.7	4.2	49.1	7.38
1	LIGHT	6	23.6	8.3	26.6	8.5
2	LIGHT	6	24.5	7.3	20.0	8.6
3	DARK	6	24.5	8.3	24.1	7.5
4	DARK	6	27.4	8.3	29.3	7.1

Crayfish in Group 1 were fed live brineshrimp at the beginnings of periods 1 and 3 for 3 days and at the beginning of all periods on the 4th day. Crayfish in Group 2 were fed at the beginning of period 1 only. The standard deviation is based on variation between individuals (Group 1, n=15; Group 2, n=11).

Table 4. Juvenile Crayfish Eating Preferences in Light and Dark

| | | PERCENT OF TOTAL FEEDING DAYS | |
PERCENT OF TOTAL FOOD EATEN IN LIGHT	PREFERENCE CATEGORY	GROUP 1 (360 DATA DAYS)	GROUP 2 (239 DATA DAYS)
A.			
55.1–100	Prefer light	27	23
45.1–55	No preference	23	20
0.0–45	Prefer dark	50	42
B.			
80.1–100	Strongly prefer light	15	6
60.1–80	Prefer light	10	25
40.1–60	No preference	45	34
20.1–40	Prefer dark	30	28
0.0–20	Strongly prefer dark	10	7

Group 1 and 2 are the same as described in Table 3 above. Note that the middle range assumed to show no feeding preference for the light or the dark is defined as 50 % ± 5% for A and 50% ± 10% for B.

crayfish: t=.2355, p=.8147, df=24; Percent growth: t=.2355, p=.8146, df=24).

Based on these results, it was concluded that all else being equal (Group 1) juvenile crayfish exhibit a slight behavioral preference for eating in the dark. Average consumption was least during the first six hours of light and most during the last six hours of dark. The preference, however, is not strong and appears easily overcome by other factors. When live food was offered only at the beginning of the light period (Group 2), no preference for eating in the light or dark was found apparently indicating a stimulatory response to the fresh food. This is consistent with the finding that the first six hours of light had the second highest feeding level for Group 2 (after the last six hours of dark). The lowest level was found for the second six hours of light.

While the trend of these data suggest that the feeding pattern of juvenile crayfish may be influenced by the state of the brineshrimp offered, no direct statistical evidence to support this conclusion was found. For the purpose of applying a correction factor for brineshrimp weight loss when calculating crayfish consumption, it can be assumed that the crayfish eat brineshrimp at a steady rate through 24 hours whether they are alive or dead. That both groups received essentially the same nutrional input whether eating all live or some live and some dead brineshrimp is confirmed by the similar total consumption and growth data for the two groups. It should be pointed out that these experiments were conducted at 20°C which creates a fairly high metabolic demand for Pacifastacus. Whether the same eating patterns and responses would be found at lower metabolic rates is unknown.

CRAYFISH CONSUMPTION EXPERIMENTS; OTHER CONSIDERATIONS

We have conducted numerous other experiments using brineshrimp to estimate crayfish consumption. Portions of the results of three of these are offered here as examples of application of the described method and because they illustrate factors which should be considered when designing crayfish experiments involving consumption.

Figure 1 follows eating patterns of adult crayfish held individually at 16°C in either freshwater or 50% seawater for 32 days. A light regime of 12 hours light and 12 hours dark was applied. While the graphs show no apparent influence of the increased salinity on crayfish consumption, of more interest here are the apparent 3-6 day eating cycle found for these and numerous other individuals studied. Clearly, such patterns if commonly found must be taken into account when estimating the length of time a crayfish feeding experiment should be run.

In Figure 2 the eating patterns of two individual crayfish that have been starved 11 days is shown. The gorging behavior the first two days food is offered is very evident. Less obvious is that the average number of brineshrimp consumed per unit weight over the next 12 days remained considerably higher than that observed for crayfish that had not been starved (mean =7.08±2.19 for starved, mean = 2.70±1.51 for non-starved). This result is not too surprising but it does serve as a reminder that the nutritional state of the crayfish entering an

Figure 1. Eating Patterns of Adult Crayfish at Four Salinities

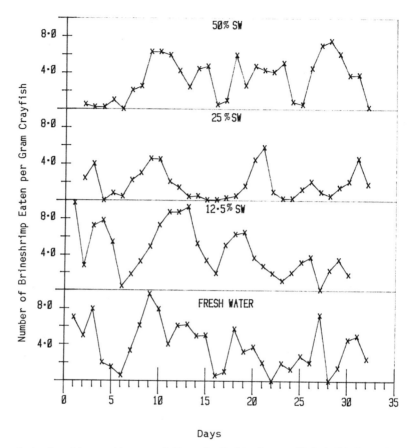

Days

Examples of eating patterns of four individual crayfish held in water of differing salinities.

Figure 2. Eating Pattern of Adult Crayfish Following 11 Days of
Starvation

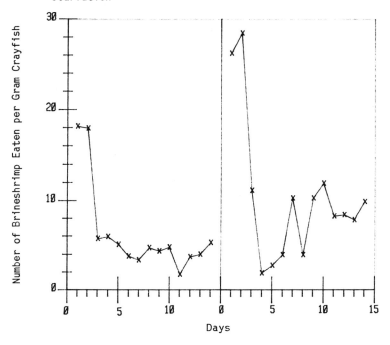

Figure 3. Eating Pattern of Adult Crayfish in Fresh and 50% Seawater

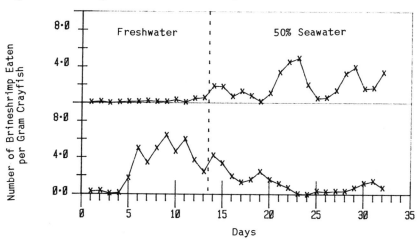

experiment may signifcantly influence the results. Care must be taken
to adequately acclimate the crayfish not only for temperature and other
environmental factors but also for nutritional stability.

The final example, shown in Figure 3 illustrates two individual
crayfish responding in exactly opposite ways to an experimental
variable, in this case moving from fresh to 50% sea water. Levels of
consumption of the first crayfish increased rather dramatically when
moved to 50% sea water, while those of the second crayfish tapered off
to almost nothing. Such mixed responses are typical of results found
for many types of experiments involving crayfish. They seem to have an
innate high individual variablity which expresses itself in many ways
including levels of consumption. While preliminary experiments can
provide valuable information for the design of most studies, they are
especially important for crayfish studies where high replication may be
required to obtain statistically meaningful results. When possible,
preliminary work to provide some estimate of crayfish variability for
the factor under study should be carried out. This will allow the
estimation of a reasonable sample size so that a real response to that
factor can be detected despite possible high variablity.

LITERATURE CITED

Budd, T.W., J.C. Lewis, and M.L. Tracey, 1979. Filtration feeding in
Orconectes propinquus and Cambarus robustus (Decapoda,Cambaridae).
Crustaceana, Suppl. 5: 131-134.

Brett, J.R. 1971. Satiation time,appetite, and maximum food intake of
sockeye salmon (Oncorhynchus nerka) J. Fish. Res. Bd. Can.
28: 409-415.

Brett, J.R. 1971. Energetic responses of salmon to temperature. A
study of some thermal relations in the physiology and freshwater
ecology of of sockeye salmon (Oncorhynchus nerka). Am.
Zoologist, 11: 99-113.

Brett, J.R. 1967. Swimmming Performance of sockeye salmon
(Oncorhynchus nerka) in relation to fatigue time and temperature.
J. Fish Res. Bd. Can. 24(3): 1731-1741

Brett, J.R. 1964. The respiratory metabolism and swimming performance
of young sockeye salmon. J. Fish. Res. Bd. Can, 21(5):
1183-1226

Brett J.R. and N.R. Glass. 1973. Metabolic rates and critical
swimming speeds of sockeye salmon (Oncorhynchus nerka) in relation
to size and temperature. J. Fish. Res. Bd. Can. 30:
379-387.

Brett, J.R., J.E. Shelbourn, and C.T. Shoop. 1969. Growth rate and
body composition of fingerling sockey salmon, Oncorhynchus nerka,
in relation to temperature and ration size. J. Fish. Res. Bd.
Can. 26: 2363-2394.

Corner, E.D.S. 1961. On the nutrition and metabolism of zooplankton.
I. Preliminary observations on the feeding of the marine copepod,

Calanus helgolandicus (Claus). J. Mar. Biol. Ass. U.K. 41: 5-16.

Corner, E.D.S. and B.S. Newell. 1967. On the nutrition and metabolism of zooplankton. IV. The forms of nitrogen excreted by _Calanus_. J. Mar. Biol. Ass. U.K. 47: 113-20.

Corner, E.D.S, C.B. Cowey, and S.M. Marshall. 1967. On the nutrition and metabolism of zooplankton. V. Feeding efficiencies of _Calanus Finmarchicus_. J.Mar.Biol. Ass.U.K. 47:259-270.

Corner, E.D.S., C.B. Cowey, and S.M. Marshal. 1965. On the nutrition and metabolism of zooplankton. III. Nitrogen excretion by _Calanus_. J. Mar. Biol. Ass. U.K. 45: 429-442.

Cowey, C.B. and E.D.S. Corner. 1963. On the nutrition and metabolism of zooplankton. II. The relationship between the marine copepod _Calanus_ helgolandicus and particulate material in Plymouth sea water in terms of amino acid composition. J. Mar. Biol. Ass. U.K. 43: 495-511.

Davies, P.M.C. 1964. The energy relations of _Carassius_ auratus _L_.-- I. Food input and energy extraction efficiency at two experimental temperatures. Comp. Biochem. Physiol. 12: 67-79.

Flint, R.W. 1975. The natural history, ecology, and production of the crayfish _Pacifastacus leniusculus_ in a subalpine Lacustrine environment. Thesis. University of Calififornia, Davis.

Flint, R. W. and C.R. Goldman.(1975) The effects of a benthic grazer on the primary productivity of the littoral zone of Lake Tahoe. Limnol. Oceanogr. 20(6): 935-944.

Fischer, Z. 1970. The elements of energy balance in grass carp (_Ctenopharyngodon idella Val_.). Part I. Pol. Arch. Hydrobiol. 17(30): 421-434.

Gallager, M. and W.D. Browne. 1975. Composition of San Francisco Bay brine shrimp (_Artemia salina_). J. Agriculture and Food Chem. 23(4) 630-631.

Huner, J.V., S.P. Meyers,and J.W. Avault, Jr. 1974. Response and growth of freshwater crawfish to an extruded, water-stable diet. Freshwater Crayfish. J.W. Avault,ed. Louisiana State Univ. Baton Rouge, Louisiana.

Kitchell, J.F. and J.T. Windell. 1968. Rate of gastric digestion in pumpkinseed sunfish, _Lepomis gibbosus_. Amer. Fish. Soc. Trans. 97: 489-492.

Magnuson, J.J., G.M. Capelli, J.G. Lorman, and R.A. Stein. 1975. Consideration of crayfish for macrophyte control. Water Quality Management through Biological Control, Symposium Volume. Dept. of Environ. Engineering Sciences. Univ. of Florida and U.S. E.P.A.

Marchant, R. 1978. The energy balance of the Australian brine shrimp,

Parartemia _zietziana_ (Crustacea:Anostraca). Freshwat. Bio. 8:481-489.

Meyers, S.P., J. Avault, D. Butler, and J.S. Rhee. 1970. Development of rations for economically important aquatic and marine invertebrates. Louisiana State Univ. Coastal studies Bulletin 5: 157-172.

Meyers, S.P. 1971. Crustacean ration formulation research. Feedstu. 43:27.

Moshiri, G.A. and C.R. Goldman. 1969. Estimation of assimilation efficiency in the crayfish _Pacifastacus_ _leniusculus_ (Dana) (Crustacea:Decapoda). Arch. Hydrobiol. 66(3): 298-306.

Paloheimo, J.E. and L.M. Dickie. 1966. Food and growth of fishes. III. Relations among food, body size, and growth efficiency. J. Fish. Res. Bd. Can. 23(8): 1209-1248.

Paloheimo, J.E. and L.M. Dickie. 1965. Food and growth of fishes. II. Effects of food and temperature on the relation between metabolism and body weight. J. Fish. Res. Bd. Can. 23(6): 869-908.

Persson, L. 1979. The effects of temperature and different food organisms on the rate of gastric evacuation in perch (_Perca_ _fluviatilis_). Freshwat. Bio. 9: 99-104.

Tyler, A.V. 1970. Rates of gastric emptying in young cod. J. Fish. Res. Bd. Can. 27: 1177-1189.

Winberg G.G. 1956. Rate of metabolism and food requirements of fishes. Nauch. Trudy Belorussk gos Univ. Minsk. 251 pp. (Fish. Res. Bd. Can. Trans. Serv., No. 199).

RESPIRATION AND IONOREGULATION IN THE EURYHALINE CRAYFISH PACIFASTACUS LENIUSCULUS ON EXPOSURE TO HIGH SALINITY: AN OVERVIEW

Michèle G. Wheatly and B. R. McMahon
Department of Biology, University of Calgary
2500 University Drive NW, Calgary, Alberta
Canada T2N 1N4

ABSTRACT

Respiratory gas exchange, ventilation, heart rate and acid-base balance were studied in a freshwater euryhaline crayfish and correlated with osmo- and ionoregulation during 48h exposure in 25, 50 and 75% seawater. Despite increases in haemolymph ionic levels some degree of regulation was evident. Respiratory and cardiac frequency decreased in 25 and 50% seawater but were elevated in 75%. Concomitantly O_2 uptake decreased as the medium approached isosmicity with the blood. Blood gas analysis demonstrated reductions in O_2 tension and content but the a-vO_2 content difference was maintained except in 50% SW. An acidosis was measured in 75% seawater which was partially offset by hyperventilatory alkalosis and may have been coupled with the decrease in blood ammonia. Evidence that haemocyanin acts as a source of free amino acids for elevation of osmolality will be presented. The opposing effects of pH and ionic concentrations on the respiratory pigment were assessed: in vivo analysis indicated an increase in O_2 affinity but the existence of another dialysable cofactor was implicated in vitro. These changes resulted in reduced CO_2 buffering capacity.

INTRODUCTION

Variation in ambient salinity is known to exert profound effects on osmo- and ionoregulation, respiratory gas exchange and acid-base balance in decapod crustaceans. Previous studies, mainly concentrating on the effects of external dilution in marine species have examined the processes of ionoregulation (Shaw, 1961; Siebers et al., 1972), gas exchange (Taylor, 1977; Taylor et al., 1977), acid-base status (Truchot, 1981), oxygen transport (Truchot, 1973; Weiland and Mangum, 1975) and intermediary metabolism (Huggins and Munday, 1968; Schoffeniels and Gilles, 1970 for reviews). The purpose of the present investigation was to integrate each of these functions in order to define the overall physiological response to hypersaline exposure.

Consideration of the respiratory as well as ionic responses to hypersaline exposure is pertinent since the gill is a complex exchanger whose ionoregulatory mechanisms also affect nitrogenous regulation (i.e., Na^+/NH_4^+) and via acid-base balance (Na^+/H^+, Cl^-/HCO_3^- or OH^-) gas exchange. Because of these interactions the physiological problems of migration into hypersaline media are multifarious and complex. As external ion concentrations rise, the resulting changes in O_2 uptake may be partially ascribed to changes in the amount of work needed for ionoregulation as well as reflecting gross morphological changes at the exchange epithelium in response to increased osmoregulatory demand (e.g., changes in thickness, functional area, etc.). These changes in O_2 uptake and demand may be effected by ventilatory and/or circulatory adjustments which will simultaneously affect CO_2 excretion, and thereby pH. Alternatively acid-base disturbances may be related to modifications in ion exchange by alteration of the strong ion difference (S.I.D.), another major determinant of pH.

The species selected for study was the euryhaline crayfish *Pacifastacus leniusculus* (Dana) which has been reported in brackish water along the west coast of North America (Miller, 1960) and was shown in a study by Kerley and Pritchard (1967) to be capable of regulating body ions over a wide range of salinities.

MATERIALS AND METHODS

A. Animals and maintenance:

Experiments were performed on adult *Pacifastacus* of either sex and mean mass 78.7± 17.2 g, obtained from commercial sources (Monterey Bay Hydroculture, California USA) and held prior to experimentation in tanks (120 x 62 x 11 cm deep) provided with a continuous flow (740 ml min^{-1}) of dechlorinated tap water and fed twice weekly on chopped smelt.

B. Experimental protocol:

Crayfish were housed individually in chambers (21 x 14 x 7 cm deep) supplied with recirculated aerated water at a rate of 40 mls min^{-1}. Physiological measurements were made in freshwater (control) and after 48h equilibration in experimental salinities equivalent to 25, 50 and 75% seawater made up from commercial sea salt (Instant Ocean).

C. Analytical procedures:

(i) Haemolymph osmolality and inorganic ion concentrations:

Haemolymph was sampled as described below. Osmolality was determined using a vapour pressure osmometer (Wescor model 5100B). Analysis of inorganic ions was performed on frozen samples, Na^+, K^+, Mg^{2+}, Ca^{2+} and Cu^{2+} concentrations using atomic absorption spectrophotometry (Jarrell-Ash 850) and Cl^- on a digital chloridometer by coulometric titration with silver ions (Searle Buchler 4-2500). Similar measurements were made on the experimental media.

(ii) Respiratory and cardiovascular variables:

Rate of oxygen consumption (hereafter designated $\dot{M}O_2$) was measured by continuous flow respirometry i.e., by monitoring both incurrent and excurrent O_2 tension and the rate of flow through of water. Branchial water flow rate ($\dot{V}_w$) was estimated utilizing the Fick principle and mean expired and inspired O_2 tension together with $\dot{M}O_2$. The former was obtained via an open ended mask positioned anterior to the exhalent openings. Respiratory frequencey (f_{sc}) was recorded as fluctuations in the hydrostatic pressure associated with ventilation by insertion of indwelling branchial catheters connected to differential pressure transducers (Hewlett-Packard 267 BC) amplified and displayed on an ink-writing oscillograph (Narco IVP). Heart rate (f_H) was measured as the impedance change across a pair of insulated stainless steel wires inserted through the carapace above the heart.

(iii) Haemolymph gas and acid-base analysis:

Pre- (venous, v) and postbranchial (arterial, a) haemolymph were sampled respectively from the ventral or pericardial sinus and analysed for oxygen tension (PO_2) and content (CO_2), pH, and total carbon dioxide content (ΣCO_2). PO_2 of haemolymph or water was measured using an oxygen electrode and acid-base analyser (Radiometer E5047, PHM 71/72) while CO_2 was determined automatically (Lexington Instruments). pH was measured with a liquid junction capillary electrode (Radiometer G 299a) and ΣCO_2 by the micromethod of Cameron (1971). CO_2 tension (PCO_2) was calculated from pH and ΣCO_2 via the Henderson-Hasselbalch equation using recently derived values for the constants pK_1 and αCO_2 corrected for salinity acclimation (see Wheatly and McMahon, 1982a).

(iv) Haemolymph ammonium, protein and free amino acids:

Ammonium ion (NH_4^+) concentration was measured spectrophotometrically by a phenol hypochlorite method (Solórzano, 1959) and excretion rate was assayed by the same technique in a closed system. Circulating protein was determined using the principle of protein-dye binding using Coomassie Brilliant Blue G-250 (Bradford, 1976). α amino nitrogen was measured using the ninhydrin micromethod of Clark (1964) and concentration of free amino acids (FAA) estimated after correction for NH_4^+ concentration. Values are expressed throughout as mean $\pm$ 1 SEM and significance at the 5% level as calculated by Student's t test is denoted.

RESULTS

No mortality was observed during exposure for 48h to 25, 50 or 75% seawater.

Osmolality and inorganic ion concentrations

Variations in haemolymph osmolality and inorganic ion concentrations with respect to levels in the external milieu are shown in Fig. 1, which clearly demonstrates the considerable regulatory potential of this crayfish. No significant change in either osmolality or concentration of any ion occurred in 25% SW, and whilst significant increases in all ions (except K^+) were measured in 50% and also in osmolality in 75% SW, the observed increases were modest in comparison with ambient changes. Both hypo- and hyperosmotic regulation are involved with the transition point occurring at approximately 44% SW (osmolality of 427 mOsm kg $^{-1}$). The medium became isoionic with respect to [Na^+], [Cl^-], [K^+], [Mg^{2+}] at external concentrations of 254.6, 204.0, 4.6 and 2.5 mM l^{-1} respectively. Calcium regulation departed from the general pattern, being maintained 3-4 mM above ambient throughout.

In the results which follow the responses in 25 and 50% SW often differ from those in 75% indicating the varying effects of hypo- and hyperionic conditions.

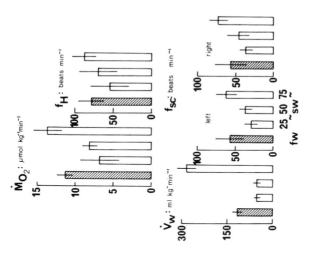

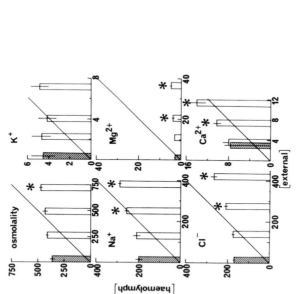

Figure 1. Variation of osmolality and inorganic ion concentrations in postbranchial haemolymph (ordinate - did not differ significantly from levels in venous blood) with external concentration (abscissa) for *Pacifastacus* in freshwater (crosshatched) or after 48h acclimation in 25, 50 and 75% SW. Except for osmolality (mOsm kg⁻¹), the units of all axes are mM l⁻¹. Values expressed as mean ± 1 SEM and asterisks denote significance from FW values. Diagonal lines represent points of equal osmotic/ionic strength.

Figure 2. The effect of exposure to varying degrees of hypersalinity on O_2 uptake ($\dot{M}O_2$), ventilation volume ($\dot{V}_W$) and cardiac and respiratory frequencies (f_H and f_{SC} respectively).

Respiratory and cardiovascular variables

f_H was reduced in 25 and 50% SW from 79 $\pm$ 16 to 55 $\pm$ 27 and 70 $\pm$ 25 beats min^{-1} respectively (Fig. 2) but elevated in 75% SW to 88 $\pm$ 16 beats min^{-1}. The ventilatory response was essentially similar, both scaphognathites exhibiting a reduction in f_{SC} in 25 and 50% SW from 57 $\pm$ 19 to 34 $\pm$ 9 and 42 $\pm$ 11 beats min^{-1} (combined mean branchial response) and an elevation in 75% SW to 68 $\pm$ 13 beats min^{-1}. Although variation around mean values was large, these changes were consistent in magnitude and direction for individual animals and therefore are significant on paired comparison analysis. The measured changes in respiratory frequency caused a significant reduction in $\dot{V}_W$ in 25 and 50% SW from 117 $\pm$ 16 to 53 $\pm$ 8 and 52 $\pm$ 10 ml kg^{-1} min^{-1} respectively. In 75% SW there was a marked increase in branchial water flow rate (2.5 fold). The outcome of these changes in cardiac and ventilatory pumping was a significant reduction in $\dot{M}O_2$ from 11.3 $\pm$ 1.1 to 6.9 $\pm$ 2.6 and 7.1 $\pm$ 1.0 μMol kg^{-1} min^{-1} in 25 and 50% SW, and an elevation to 13.6 $\pm$ 1.9 μMol kg^{-1} min^{-1} in 75% SW.

Acid-base status

Postbranchial haemolymph values of pH and ΣCO_2 at each acclimation salinity are listed in Table 1. Prebranchial values are omitted but differed slightly as would be predicted due to CO_2 excretion at the gill. No change in pH_a was measured after acclimation to 25 and 50% SW (suggesting complete compensation of any acid-base disturbance). However a significant acidosis resulted from exposure to 75% SW. PCO_2 decreased progressively with increasing acclimation salinity.

Plotting these data on a diagram relating changes in HCO_3^-, pH and PCO_2 (modified from Davenport, 1974) allows estimation of the origin (i.e., respiratory or metabolic) of the acidosis observed in 75% SW. Fig. 3 combines two such diagrams obtained in FW and 75% SW since both the PCO_2 isopleths and the haemolymph buffer values (i.e. slope of the relationship $\Delta HCO_3^-/\Delta pH$ pq = FW, rs = 75% SW) are altered by the ionic changes associated with hypersaline acclimation. In Fig. 3 a respiratory disturbance would be characterised by translation along a buffer line while any metabolic contribution to an acid-base disturbance is seen by translation along a PCO_2 isopleth.

In the present study both metabolic and respiratory disturbances are displayed. An acidosis of metabolic origin is seen when points of equal PCO_2 are compared on buffer lines from FW and 75% SW (A-X). Perhaps in partial compensation for this acidosis a respiratory alkalosis (translation along buffer line from X to final pH at B) can be observed. Similar but smaller shifts apparently allow full compensation for metabolic acidosis following acclimation to 25 and 50% salinities.

O$_2$ transport

The characteristics of O_2 binding by the respiratory pigment haemocyanin (HCy) were investigated at a range of pH values and acclimation salinities by measuring CO_2 of haemolymph equilibrated _in vitro_ with humidified gas mixtures of varying PO_2. Fig. 4 demonstrates that Bohr and salt effects act in opposition, increasing [H$^+$] effecting a reduction in affinity (elevation in half saturation pressure, P_{50}) whilst elevation in haemolymph inorganic ion levels causes an increase.

47

Table 1. Effect of hypersaline exposure on postbranchial haemolymph acid-base status in <u>Pacifastacus</u>. Values expressed as mean ± 1 SEM with number of observations in parentheses. Asterisks denote significance from freshwater values. PCO_2 was calculated from mean values.

	FW	25% SW	50% SW	75% SW
pH_a	7.954 ±0.024 (9)	8.000 ±0.033 (8)	7.964 ±0.037 (7)	7.830* ±0.015 (6)
ΣaCO_2 mEq 1^{-1}	8.77 ±1.21 (11)	5.92 ±1.31 (5)	4.92* ±0.95 (10)	2.93* ±1.47 (9)
PCO_2 torr	2.79	1.34	1.03	0.66

Table 2. Post- and prebranchial haemolymph O_2 tensions and contents and content difference in <u>Pacifastacus</u> during hypersaline exposure.

	FW (n=9)	25% SW (n=8)	50% SW (n=7)	75% SW (n=6)
PaO_2 (torr)	34.4 ±5.5	15.3* ±3.1	12.1* ±1.9	14.3* ±2.1
PvO_2 (torr)	17.2 ±1.3	9.0* ±1.7	9.9* ±1.0	10.0* ±1.3
CaO_2 (vols %)	1.33 ±0.14	0.73* ±0.08	0.58* ±0.04	0.66* ±0.06
CvO_2 (vols %)	0.86 ±0.13	0.45* ±0.05	0.39* ±0.06	0.32 ±0.04
$CaO_2 - CvO_2$ (vols %)	0.53 ±0.12	0.32 ±0.03	0.18* ±0.03	0.34 ±0.08

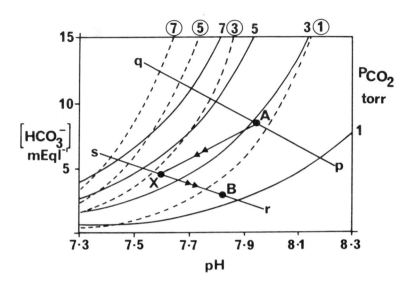

Figure 3. Postbranchial [HCO_3^-], pH and PCO_2 for <u>Pacifastacus</u> in FW (A) and after 48h acclimation in 75% SW (B). The curved isopleths are lines of equal CO_2 tension and the diagonal lines represent CO_2 buffer values which are observed to alter with salinity acclimation, solid and broken (PCO_2 levels encircled) isopleths and lines pq and rs corresponding in each case to FW and 75% SW respectively. For explanation consult text.

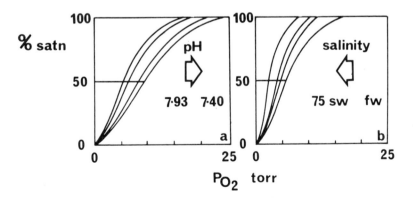

Figure 4. Oxygen dissociation curves constructed <u>in vitro</u>: a) for FW acclimated <u>Pacifastacus</u> at pH 7.93, 7.78, 7.56 and 7.40. b) on haemolymph from animals acclimated in FW, and 25, 50 and 75% SW at constant pH (ca 7.95). The horizontal line at 50% saturation provides an indication of the changing affinity for O_2.

In vivo significant reductions in pre- and postbranchial tension and contents were measured (Table 2). However except in 50% SW, the a-v content difference was maintained by depletion of the reserve held in venous blood. The increase observed in 75% compared with 50% SW at similarly low tension values may indicate an increase in the affinity of the blood pigment for O_2.

[NH_4^+], protein and [FAA]

There was a progressive reduction in both circulating [NH_4^+] and excretion rate during hypersaline exposure (Table 3) concomitant with which was a 2.5 fold increase in FAA. To investigate the possible source of these FAA the concentration of haemolymph protein was assayed (Table 4). Knowing that, in crustaceans, haemocyanin constitutes 96% of circulating protein (van Holde and van Bruggen, 1971) other indicators of [HCy] such as circulating [Cu^{2+}] and $HCyO_2^{max}$ (which is the maximum amount of O_2 bound by HCy) were also determined.

Significant reductions in all three variables and hence [HCy] were observed with increasing acclimation salinity (Table 4).

DISCUSSION

The present investigation has demonstrated that the euryhaline N. American crayfish Pacifastacus leniusculus can tolerate at least 48h of hypersaline exposure in as much as 75% seawater. The European species Potamobius astacus can withstand 50% SW for up to 6 weeks but cannot tolerate higher salinities (Hermann, 1931). Similarly Kentall and Schwarz (1964) recorded an LD_{50} of 72-96h at this salinity for the North American species Orconectes virilis and Cambarus bartonii. In contrast the present data together with an earlier report of LD_{50} of 8 days in 90% SW (Kerley and Pritchard, 1967) confirm that this species of crayfish is highly tolerant of salinity stress and possesses a dual ability for hyper/hypo ionic and osmotic regulation. For this reason the physiological modifications which occur vary according to the degree of hypersalinity imposed.

In dilute media the crayfish solves the inherent problems of diffusive salt loss and osmotic water loading by reduction in boundary permeability, possession of ion transporting systems with a high affinity for specific ions and by production of hyposmotic urine (Kirschner, 1979). Conversely during hypersaline exposure, hyporegulation involves active NaCl extrusion and water uptake, the mechanisms for which in invertebrates are controversial but may involve branchial Cl^- secreting cells (Copeland, 1967) and reversal of the Na^+ ion exchanges which operate in fresh water (Henry and Mangum, 1980).

The present investigation confirmed the generally held conclusion that $\dot{M}O_2$ and ventilatory and circulatory work are least in media isosmotic with the blood (Potts and Parry, 1964) since they partially reflect the metabolic work involved in the processes of osmoregulation. However other modifications in intermediary metabolism (Gilles, 1973) and locomotor activity (A.C. Tayulor, 1977) may also contribute to overall changes in gas exchange. A number of investigators have reported elevated $\dot{M}O_2$ on external dilution in regulating marine species (A.C. Taylor, 1977; E.W. Taylor et al., 1977). Such changes are not

Table 3. NH_4^+ and FAA concentrations in postbranchial haemolymph and NH_4^+ excretion rates during exposure to hypersalinity in _Pacifastacus_.

	FW	25% SW	50% SW	75% SW
$[NH_4^+]a$ mM l^{-1}	1.19 ±0.09 (8)	0.85* ±0.10 (11)	0.76* ±0.09 (11)	0.88* ±0.03 (14)
$[FAA]a$ mM l^{-1}	1.29 ±0.16 (8)	2.31* ±0.34 (11)	3.10* ±0.60 (11)	3.07* ±0.16 (14)
NH_4^+ excretion µM $kg^{-1} h^{-1}$	200.4 ±27.4 (8)	57.3* ±24.6 (8)	78.4* ±17.2 (8)	49.6* ±24.7 (8)

Table 4. Variation in circulating protein, Cu^{2+} and haemocyanin bound O_2 during hypersaline exposure in _Pacifastacus_.

	FW	25% SW	50% SW	75% SW
[Protein] mg ml^{-1}	63.6 ±6.2 (15)	37.3* ±5.4 (15)	44.6* ±2.2 (16)	46.8* ±2.7 (15)
$[Cu^{2+}]$ µg ml^{-1}	77.5 ±3.9 (15)	45.0* ±7.4 (5)	49.0* ±2.9 (16)	54.9* ±2.3 (15)
$C_{HCyO_2}^{max}$ vol %	1.35 ±0.16 (4)	0.80* ±0.01 (4)	0.53* ±0.04 (4)	0.46* ±0.02 (4)

evident in osmoconforming decapods (Remane and Schlieper, 1958).

Changes in environmental salinity also affect haemolymph acid-base balance. In the present study on acidosis was observed in 75% SW which mirrors the increase in pH recorded on dilution in <u>Callinectes sapidus</u> (Weiland and Mangum, 1975) and in <u>Carcinus maenas</u> (Truchot, 1973). PCO_2 levels calculated in all of these studies indicate that the acid-base disequilibrium is relatively independent of ventilation and results largely from modifications in ion exchange and thereby strong ion difference. To this end a reduction in bicarbonate concentration was observed. The resulting acidosis may be exacerbated by the loss of buffering capacity for H^+ from NH_4^+ and circulating protein, the prominent non-bicarbonate buffers.

It is well known that alteration of tissular levels of FAA participate in intracellular isosmotic regulation during salinity acclimation (Gilles, 1979). In <u>Pacifastacus</u> the coupling of elevated FAA with reduced $[NH_4^+]$ appears indicative of reductive amination. The buffering capacity generally afforded by protonation of NH_3 to NH_4^+ will be reduced (Mangum et al., 1976). Additionally a decrease in NH_4^+, since it can be substituted as a counterion for Na^+ exchange (Towle, 1974) may contribute to the reduction in S.I.D.

The enlargement of the FAA pool extracellularly in the present investigation together with measured reductions in protein levels suggest that degradation of haemolymph proteins may contribute to the FAAs transported to the intracellular compartment in addition to <u>de novo</u> synthesis. A similar phenomenon was observed during hypersaline exposure in the mactrid clam <u>Rangia cuneata</u> which is an osmoconformer except that the FAA were retained in the ECF in this case (Henry et al., 1980).

The observed reduction in $[Cu^{2+}]$, the ligand to which molecular O_2 is bound, and yet a maintenance of the Cu^{2+}: protein ratio at around 0.11 - 0.12% which is characteristic or pure HCy would suggest that HCy constituted a large proportion of the protein broken down in preference or addition to other blood proteins. This implies that secondary to its respiratory function it can provide a source of osmolytic effective molecules during hyperosmotic shock. Conversely, HCy synthesis was recently reported by Boone and Schoffeniels (1979) during hypo-osmotic stress in <u>Carcinus</u>.

The inevitable corollary of HCy breakdown is its effect on O_2 transport as was evidenced by a reduction in $C^{max}_{HCyO_2}$. The increases in affintiy of O_2 binding <u>in vivo</u> (75% SW c.f. 50% SW) and with progressive increases in acclimation salinity <u>in vitro</u> are partly due to elevation in haemolymph ion concentrations, notably of divalent cations (Mangum and Lykkeboe, 1979; Truchot, 1975). In addition recent work in our laboratory points to the existence of as yet unidentified dialysable cofactor(s) other than H^+ and divalent cations which can modulate O_2 binding (Wheatly and McMahon, 1982b). The acidosis observed <u>in vivo</u> in 75% SW would partially offset the salt effect. The increase in O_2 affinity allows greater delivery of O_2 for a small tension difference.

Close examination of the O_2 combining characteristics indicates that the reduction in O_2 binding is not problematical <u>per se</u>. In no case does $C^{max}_{HCyO_2}$ fall below the settled (FW) a-v CO_2 difference and thus a small O_2 reserve remains for utilization under conditions of increased O_2 demand. As a result the pigment becomes increasingly important in O_2 transport during hypersaline exposure as P_aO_2 descends from the shoulder to the functional range of the dissociation curve.

ACKNOWLEDGMENTS

The authors gratefully acknowledge financial support provided by a Postdoctoral Research Fellowship awarded by University of Calgary to MGW and NSERC grant A5762 to BRM.

LITERATURE CITED

Boone, W.R. and E. Schoffeniels. 1979. Hemocyanin synthesis during hypo-osmotic stress in the shore crab Carcinus maenas (L). Comp. Biochem. Physiol. 63B, p. 207-214.

Bradford, M.M. 1976. A rapid and sensitive method for the quantitation of microgram quantities of protein utilizing the principle of protein-dye binding. Anal. Biochem. 72, p. 248-254.

Cameron, J.N. 1971. Rapid method for determination of total carbon dioxide in small blood samples. J. appl. Physiol. 31, p. 632-634.

Clark, M.E. 1964. Biochemical studies on the coelomic fluid of Nephtys hombergi (Polychaeta:Nephtyidae) with observations on changes during different physiological states. Biol. Bull. 127, p. 63-84.

Copeland, D.E. 1967. A study of salt secreting cells in the brine shrimp (Artemia salina). Protoplasma 63, p. 363-384.

Davenport, H.W. The ABC of Acid-Base Chemistry. 6th edn. Chicago: The University of Chicago Press (1974).

Gilles, R. 1973. Oxygen consumption as related to the amino acid metabolism during osmoregulation in the blue crab Callinectes sapidus. Neth. J. Sea. Res. 7, p. 280-289.

Gilles, R. Intracellular organic osmotic effectors. In R. Gilles (ed.). Mechanisms of Osmoregulation in Animals, John Wiley and Sons LTD, Chichester, New York, Brisbane, Toronto (1979) 111 p.

Henry, R.P. and C.P. Mangum. 1980. Salt and water balance in the oligohaline clam Rangia cuneata. I. Anisosmotic extracellular regulation, J. exp. Zool. 211, p. 1-10.

Henry, R.P., C.P. Mangum and K.L. Webb. 1980. Salt and water balance in the oligohaline clam Rangia cuneata. II. Accumulation of intracellular free amino acids during high salinity adaptation. J. exp. Zool. 211, p. 11-24.

Hermann, F. 1931. Über den Wasserhaushalt des Flusskrebses (Potamobius astacus Leach). Z. vergl. Physiol. 14, p. 479-524.

Huggins, A.K. and K.A. Munday. Crustacean metabolism. In O. Lowenstein (ed) Advances in Comparative Physiology and Biochemistry 3. Academic Press. New York and London (1968) 271 pp.

Kentall, A. and F.J. Schwarz. 1964. Salinity tolerances of two Maryland crayfishes. Ohio J. Sci. 64, p. 403-409.

Kerley, D.E. and A.W. Pritchard. 1967. Osmotic regulation in the crayfish Pacifastacus leniusculus, stepwise acclimated to dilutions of seawater. Comp. Biochem. Physiol. 20, p. 101-113.

Kirschner, L.B. Control mechanisms in crustaceans and fishes. In R. Gilles (ed). Mechanisms of Osmoregulation in Animals. John Wiley and Sons LTD, Chichester, New York, Brisbane and Toronto (1979) 157 pp.

Mangum, C.P. and G. Lykkeboe. 1979. The influence of inorganic ions and pH on oxygenation properties of the blood in the gastropod mollusc Busycon canaliculatum. J. exp. Zool. 207, p. 417-430.

Mangum, C.P., S. Silverthorn, J.L. Harris, D.W. Towle and A.R. Krall. 1976. The relationship between blood pH, ammonia excretion and adaptation to low salinity in the blue crab Callinectes sapidus. J. exp. Zool. 195, p. 129-136.

Miller, G.C. The taxonomy and certain biological aspects of the crayfish of Oregon and Washington, Master's thesis, Oregon State University, Corvallis, Oregon (1960).

Potts, W.T.W. and G. Parry. Osmotic and Ionic Regulation in Animals. Pergamon Press. Oxford, London, New York and Paris (1964). 330 pp.

Remane, A. and C. Schlieper. Die Biologie des Brackwassers. In Die Binnengewässer von Prof. Dr. August Thienemann. Band XXII. E. Schweizerbartsche Verlagsbuchhandlung, Stuttgart (1958).

Schoffeniels, E. and R. Gilles. Osmoregulation in aquatic arthropods. In M. Florkin and B.T. Scheer (eds) Chemical Zoology: Arthropoda. Vol. 5A. Academic Press, New York, (1970) 255 pp.

Shaw, J. 1961. Studies on ionic regulation in Carcinus maenas (L). J. exp. Biol. 38, p. 135-52.

Siebers, D., C. Lucu, K.R. Sperling and K. Eberlein. 1972. Kinetics of osmoregulation in the crab Carcinus maenas. Mar. Biol. 17, p. 291-303.

Solórzano, L. 1959. Determination of ammonia in natural waters by the phenolhypochlorite method. Limnol. Oceanogr. 14, p. 799-801.

Taylor, A.C. 1977. The respiratory responses of Carcinus maenas (L) to changes in environmental salinity. J. exp. mar. Biol. Ecol. 29, p. 197-210.

Taylor, E.W., P.J. Butler and A. Al-Wassia. 1977. The effect of a decrease in salinity on respiration, osmoregulation and activity in the shore crab Carcinus maenas (L) at different acclimation temperatures. J. comp. Physiol. 119, p. 155-170.

Towle, D.W. 1974. Equivalence of gill Na^+ and K^+ ATPase from blue crabs acclimated to high and low salinity. Amer. Zool. 14, p. 1259.

Truchot, J.-P. 1973. Fixation et transport de l'oxygène par le sang de Carcinus maenas: variations en rapport avec diverses conditions de température et de salinité. Neth. J. Sea. Res. 7, 482-495.

Truchot, J.-P. 1975. Factors controlling the in vitro and in vivo oxygen affinity of the haemocyanin in the crab Carcinus maenas (L.). Respir. Physiol. 24, p. 173-189.

Truchot, J.-P. 1981. The effect of water salinity and acid-base state on the blood acid-base balance in the euryhaline crab Carcinus maenas (L.). Comp. Biochem. Physiol. 68A, p. 555-561.

van Holde, K.E. and E.F.J. van Bruggen. The haemocyanins. In S.N. Timosheff and G.D. Fasman (eds). Subunits in Biological Systems, A, Marcel Dekker, NY (1971).

Weiland, A.L. and C.P. Mangum. 1975. The influence of environmental salinity on haemocyanin function in the blue crab, Callinectes sapidus. J. exp. Zool. 193, p. 265-274.

Wheatly, M.G. and B.R. McMahon. 1982a. Responses to hypersaline exposure in the euryhaline crayfish, Pacifastacus leniusculus. I. The interaction between ionic and acid-base regulation. J. Exp. Biol. 99:425-556.

Wheatly, M.G. and B.R. McMahon. 1982b. Responses to hypersaline exposure in the euryhaline crayfish Pacifastacus leniusculus. II. Modulation of haemocyanin oxygen binding in vitro and in vivo. J. Exp. Biol. 99:447-467.

II
ACIDIFICATION
PROBLEMS OF CRAYFISH

RESPONSE OF ACID STRESS UPON THE OXYGEN UPTAKE IN EGGS OF THE CRAYFISH ASTACUS ASTACUS L.

Magnus P.A. Appelberg
Institute of Limnology
University of Uppsala
Box 557, S-751 22
Uppsala, SWEDEN

ABSTRACT

A small respirometer for measuring the oxygen uptake in single eggs of the crayfish Astacus astacus L. has been developed. In a 0.38 ml chamber the oxygen uptake has been recorded for 10 and 20 minutes with an oxygen electrode as a sensor. Measurements have been made during different stages of development of the eggs and the effect of acid stress upon the oxygen uptake has been investigated. A mean oxygen uptake at 15.0 °C of 0.031 μg O_2 mg $dw^{-1}h^{-1}$ in undeveloped eggs rose to 0.733 μg O_2 mg $dw^{-1}h^{-1}$ in eggs just about to hatch. Respiration rate decreased when lowering the pH from 7.0 to 3.0 with adaption times of 24 h or less. When using acclimation times of 48 h at each pH level, significant rises in the respiration rate have been recorded at pH between 4.0 and 6.0. The inhibition of the oxygen uptake due to acid stress is shown to be a reversible process under some circumstances.

INTRODUCTION

The acidification of Swedish waters has under recent years become a serious threat to the native crayfish species Astacus astacus L. In the life cycle the reproduction seems to be the stage most susceptible to low pH. In waters with pH less than 6.0 and with an alkalinity less than 0.1 mekv 1^{-1} the reproduction may often fail (Appelberg unpublished).

For several aquatic animals there has been shown an effect of acid stress upon the ion exchange (e.g. Shaw 1960, Kirchner 1970, Leivestad et al. 1976, Leivestad et al. 1980, Malley 1980, Vangenechten and Vanderborght 1980). Aluminium-poisoning may have the same effect and it is not always possible to separate this from the acid stress (Muniz and Leivestad 1980).

Many studies have been performed on the O_2 uptake in crayfish under various treatments but few concerning the O_2 uptake in relation to pH (Cukerzis 1968, Moshiri et al. 1971, Sutcliffe and Carrick 1975, Wolwekamp and Waterman 1960 for review). Cukerzis (1968) obtained a change in O_2 uptake rate at different pH levels in adult Astacus astacus and A. leptodactylus. However, reproduction and the youngest stages in the life cycle often are the more sensitive to environmental stress. Therefore, it must be of interest to study the earliest stage in the life cycle of the crayfish when looking at effects of acidification.

The present study is an attempt at obtaining a picture of the O_2 uptake in eggs of the crayfish A. astacus exposed to low pH during different times of exposure. Since very few data on the O_2 uptake in eggs of crayfish are available, the O_2 uptake has also been followed from undeveloped eggs until hatching.

MATERIALS AND METHODS

Females of Astacus astacus from Lake Rottnen and Lake Vallsjon were held in the laboratory for spawning. After egg-laying, egg-bearing females were placed in plastic pools supplied with running tap water at 8-12 °C. Six egg-bearing females were held in experimental water at pH 6.0 and 8-14 °C from spawning until the eggs were hatched. The pH was adjusted daily with H_2SO_4.

At spring 1981 egg-bearing females were collected at Lake Holmsjon. Eggs were stripped at the lake and the O_2 uptake in the eggs was measured one day after stripping.

Unless otherwise mentioned, the experimental water consisted of 9/10 distilled water and 1/10 tap-water. This mixture resulted in a pH of 6.8, alkalinity of 0.5 mekv 1^{-1} and conductivity was 7.0 mS m^{-1}. In the water from Lake Holmsjon pH was 6.7, alkalinity 0.10 mekv 1^{-1} and conductivity 3.9 mS m^{-1}. The concentration of aluminium in the experimental water and lake water was about 200 µg 1^{-1} and 100-150 µg 1^{-1} respectively.

The eggs were stripped from females just before the start of each experiment. When eggs were going to be used several times they were, in long run tests kept in 200 ml glass bottles with air supply at 15.0 °C, and in short time tests kept in 4 ml test tubes at 15.0 °C.

The respirometer used in this study is constructed for measuring the O_2 uptake in single eggs of crayfishes. It consists of two circular chambers made out of plexi glass (Figure 1). The outer chamber is coupled to a water bath for temperature control and is separated from the inner one which acts as the respiration chamber. As sensor an oxygen electrode filled with KCl and with a platina wire, 1.0 mm in diameter, is used. The electrode is coupled to a penrecorder via an amplifier. Constructor of the electrode is Dr. O. Tottmar at the Institute of Zoophysiology, University of Uppsala. To supply a steady current past the membrane of the electrode, a magnetic follower is rotated with the help of a magnetic stirrer.

The volume of the test chamber is 383 µl. When correction for the volume of the magnetic follower and egg is made, the media volume is 360 ± 0.047 µl.

All measurements were made at 15.0 ± 0.02 °C and with constant illumination. The O_2 consumption of the electrode was 0.151 ± 0.047 µg h^{-1} in water saturated with air at 15.0 °C. Before each time of measurment of the eggs, the respirometer was run with medium only, the corrections for the O_2 consumption of the electrode and that originating from microbiological activity in the medium could be done. All measurments lasted for 10 to 20 minutes in a medium which at the start was saturated with air. The O_2 saturation in the medium never fell below 60% during the time of measurments. After the experiments, each egg was dried for 24 h in 60 °C and weighed.

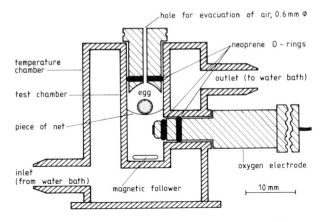

Figure 1.   Vertical section through the respirometer.

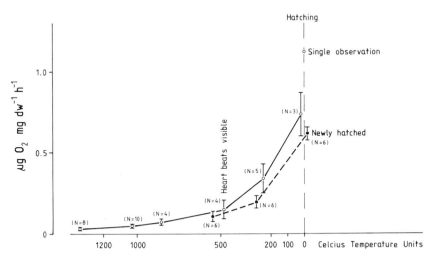

Figure 2. O_2 uptake in developing eggs of <u>Astacus astacus</u> at $15^{o}C$.
o Laboratory-reared eggs at pH 6.0; ● eggs naturally developed in Lake
Holmsjön. Mean $\pm$ 1 SD.

RESULTS

Oxygen uptake in developing eggs.

Figure 2 shows the O_2 uptake in eggs of A. astacus in relation to dry weight from undeveloped eggs to hatching. The O_2 uptake increased from $0.031 \pm 0.007 \mu g\ O_2$ mg $dw^{-1}h^{-1}$ (N=8) to $0.733 \pm 0.134\ \mu g\ O_2$ mg $dw^{-1}h^{-1}$ (n=3) in eggs reared at pH 6.0 in the laboratory. During hatching the O_2 uptake reached a maximum. One juvenile that hatched in the respirometer increased its O_2 uptake to $1.123\ \mu g\ O_2$ mg $dw^{-1}h^{-1}$ during hatching.

The O_2 uptake in eggs that were stripped from females living under natural conditions in Lake Holmsjon (pH between 6.5 and 7.0) were measured at three different times for comparison. Their uptake rates were slightly lower but less scattered than in the eggs reared in the laboratory.

During the last 30 days of development the O_2 uptake rose more than fourfold in the eggs. The rapid increase began after the time heartbeats were visible.

Effect of acid stress upon the oxygen uptake.

In order to study the importance of time of exposure to acid water on the uptake of O_2, four series of experiments were carried out. In each series the pH in the medium was lowereed from pH 7.0 to 3.0 with one unit at a time (Figure 3a-d, Table 1). With an exposure time of 30 minutes at each pH level there was no significant difference in the O_2 uptake between the pH levels. When using an exposure time of 100 minutes the O_2 uptake was significantly higher at pH 7.0, 6.0 and 4.0 than at pH 3.0 ($P<0.05$). One egg showed a slight increase at pH 5.0 which made the standard deviation higher at that pH level.

With 24 h of exposure time at each pH level the O_2 uptake at pH 7.0 and 6.0 was significantly higher than the uptake rate at pH 4.0 and 3.0 ($P<0.05$). The uptake rate at pH 5.0 was significantly higher than at pH 3.0 ($P<0.01$). There was a slight elevation in the uptake rate at pH 6.0, compared to that at pH 7.0, but the difference was not significant at the 95% level.

When using 48 h of exposure time there was a significant increase ($P<0.05$) in the O_2 uptake at pH 5.0 compared to the uptake at pH 7.0, 4.0 and 3.0. Also at pH 6.0 there was an increase (not significant).

Recovery of the oxygen uptake inhibited by acid stress.

When eggs which had been exposed for 24 hours at each pH level were transferred from pH 3.0 back to pH 7.0, the O_2 uptake after 72 hours was significantly lower than at pH 3.0 ($P<0.01$) (Figure 4). After 120 h in pH 7.0 there had been a recovery, and the uptake rate was nearly as high as at pH 7.0 in the start of the experiment series.

In order to compare laboratory-reared eggs to naturally reared, eggs were stripped from females collected in Lake Holmsjon one day before the start of an experiment series. After 24 h in lake water of pH 7.0 the eggs were transferred to lake water acidified with H_2SO_4 to

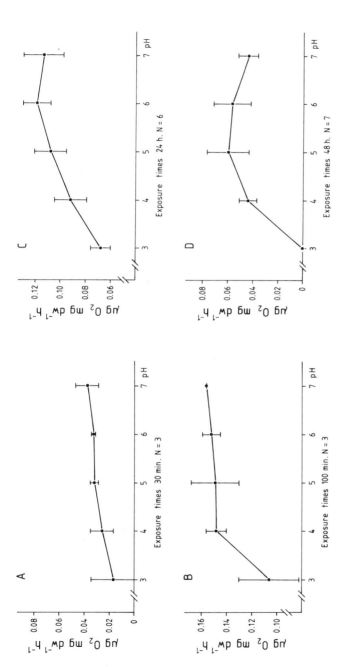

Figure 3. O$_2$ uptake in eggs of Astacus astacus reared in the laboratory at 15°C. Mean ± 1 SD.

Table 1 - O_2 uptake in eggs of <u>Astacus astacus</u> during different times of exposure to different pH. In µg O_2 mg $dw^{-1}h^{-1}$ at 15.0 °C. Mean ± 1 SD.

Acclimation time	N	pH 7.0	pH 6.0	pH 5.0	pH 4.0	pH 3.0
30 min	3	0.038±0.009	0.033±0.001	0.32±0.003	0.026±0.009	0.017±0.018
100 min	3	0.156±0.000	0.152±0.007	0.149±0.019	0.148±0.008	0.106±0.024
24 h	6	0.114±0.016	0.119±0.011	0.108±0.013	0.092±0.012	0.068±0.008
48 h	7	0.044±0.008	0.057±0.015	0.060±0.017	0.044±0.007	dead

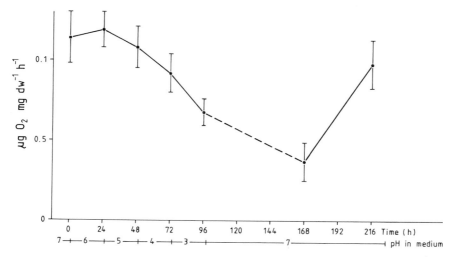

Figure 4. O_2 uptake in eggs of <u>Astacus astacus</u> reared in the laboratory at 15° C. Mean ± 1 SD; N = 6.

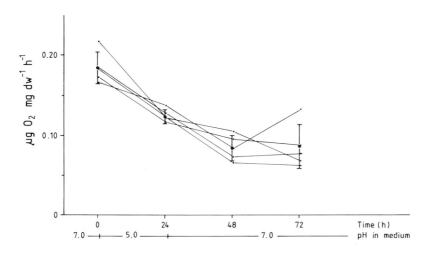

Figure 5. O_2 uptake in eggs of <u>Astacus astacus</u> naturally developed in Lake Holmsjön at 15° C.● Mean ± 1 SD; • single eggs.

pH 5.0 and then after another 24 h transferred back to pH 7.0 (Figure 5). The O_2 uptake had decreased significantly (P<0.01) after 24 h in pH 5.0 and were still decreasing after 24 h in pH 7.0, 48 h after the eggs had been transferred to pH 7.0 from 5.0, the O_2 uptake began to rise in some of the eggs and stabilize in others.

During these series of tests there was a slight difference in stage of development of the eggs. It was assumed that differences in the response to acid stress due to stage of development would be of less importance within the used range of developemnt. In each series a single egg could be followed individually throughout the series (Figure 5). It was found that the response in O_2 uptake to acid stress differed to a higher extent between eggs from different females than between eggs from the same female.

DISCUSSION

The only previous data on the O_2 consumption in eggs of crayfishes that have been available originate from Kossakowski (1975). He reports that the O_2 uptake in eggs of Orconectes limosus were 150 mm^3 g $ww^{-1}h^{-1}$, although the stage of development of the eggs or the temperature was not mentioned. Nevertheless, this value lies in the same range as those found in the present study. When a comparison is made between the O_2 uptake in eggs found here, and the O_2 uptake in juveniles of A. astacus obtained by Cukerzis (1968), it may be suggested that the metabolism in relation to weight is at its maximum in the moment of hatching.

There are few data published on the sublethal effects of acid stress on the crayfish. Studies on the lethal limits of pH for instance have been performed by Jay and Holdich (1977) (Austropotamobius pallipes), Newcombe (1975) (Parastacoides tasmanicus) and Morgan and McMahon (1982). Data from Svardson (1974), Furst (1977) and Appelberg (unpublished) indicate that the species A. astacus seldom is found in waters with pH below 5.5-6.0.

High H^+ concentration in the water interferes with the ion exchange in several aquatic animals. Kirchner (1970) describes the uptake of Na^+ and Cl^- as an active process which is balanced by H^+ and HCO_3^-. Studies on fish show that the concentrations of Na^+ and Cl^- in the blood plasma decrease when pH falls (e.g. Leivestad et al. 1976, Leivestad et al. 1980). Vangenechten and Vanderborght (1980) report a higher level of both the influx and the efflux of Cl^- and a reduced influx of Na^+ in Corixa punctata at low pH. Shaw (1960) showed that the uptake of Na^+ in crayfish Austropotamobius pallipes is reduced when pH falls below 6.0. Also the uptake of Ca^{2+} in crayfish may be disturbed by low pH (e.g. Malley 1980, Appelberg in press).

In the present study neutral water was acidified with H_2SO_4. When acidifying neutral water, the content of free CO_2 will momentarily rise, because of a changed equilibrium of the carbonate buffering system. Since CO_2 was not measured in the different test waters, it may be an artifact in some of the experiments. In the experiments with adaptation times of 24 and 48 hours, the water was supplied with air, and it is therefore probable that the CO_2 in the water soon became in equilibrium with the air and did not affect the respiration rate of the eggs. In the 30 and 100 minutes experiments, when there was only air supply to the water just before each recording, the CO_2 content may have been of importance. On the other hand, there is no significant deviation in oxygen uptake rate between different pH levels in the 30 minutes experiment. Still, the lower uptake rate at pH 3.0 in the 100 minutes experiment may be due to a high content of free CO_2 in the experimental water.

A pH of 3.0 seems to be critical to eggs of Astacus astacus. In the 24 h experiment of O_2 uptake was reduced and in the 48 h experiments the eggs died at pH 3.0. It is probable that this effect is caused by a severe acidosis in the eggs. Morgan and McMahon (1982) obtained a haemolymph acidiosis in Procambambarus clarkii exposed to pH 3.8, and for brook trout, exposed to pH 3.0, Packer (1979) recorded a reduced O_2 consumption and a reduced blook oxygen capacity together with blood acidosis.

The increased in O_2 uptake at pH 5.0 in the 48 h experiment may be explained as a stimulation of the tissue respiration due to a mild acid stress as has been discussed for fish (Packer 1979). A possibility may be that the energy cost to maintain the ion balance during a mild acid stress, is increased. But since the metabolic cost for osmoregulation is calculated to be less than 10% of the total metabolic energy (Robertson 1960, Sutcliffe and Carrick 1975) this explanation may not be the only one.

Obviously, a pH of 3.0 may be lethal to eggs of A. astacus when exposed for 48 h or more, but sublethal with an exposure time of 24 h or less. The return of the O_2 uptake rate to a nearly normal level in eggs that had been exposed to pH 3.0 for 24 h and had been transferred to pH 7.0, indicates that the change in O_2 uptake rate due to acid stress may be a reversible process.

The significantly reduced O_2 uptake at pH 5.0 that was obtained when using naturally developed eggs in acidified lake water may be explained as a difference in resistibility to acid stress between different populations of A. astacus. Eggs from Lake Holmsjon may have been more sensitive to acid stress than the eggs from Lake Rottnen and Lake Vallsjon which were laboratory-reared.

High concentration of aluminium in combination with low pH may have an additive effect on the acid stress to adult fish (Muniz and Leivestad 1980). Baker and Schofield (1980), on the other hand, have shown that the presence of aluminium may mitigate toxic effects of low pH to fish eggs. It seems therefore most probably that the changes in O_2 uptake obtained in the present study may be due to low pH and not an effect of aluminium-poisoning.

Slow growth of brown trout, Salmo trutta, in acid water has been explained as an effect of high metabolic cost for the disturbed ion exchange (e.g. Rosseland 1980). Since Buck and Siewert (1980) have obtained a positive correlation between pH and weight gain in crayfish, it may be possible that acid stress could have a similar effect on the growth of crayfish to some extent.

In conclusion, acid stress alone or in combination with toxic effects of aluminium does have an effect upon the O_2 comsumption in eggs of A. astacus. A slight decrease in pH may give an increased O_2 uptake rate under some circumstances, but a more severe acid stress leads to a reduced uptake rate. A reduction in the uptake rate, such as that found at pH 3.0 after 24 h, may be a reversible process.

ACKNOWLEDGMENTS

I wish to thank Dr. O. Tottmar for the help with the constructing the oxygen electrode and the respirometer. Also professor B. Pejler, Dr. M. Furst and Dr. P. Nyberg are thanked for critical reviewing.

LITERATURE CITED

Appelberg, M.A.P. in press. The effect of low pH on Astacus astacus L. during moult. In The second Scandinavian Symposium on Freshwater Crayfish. Finnish Fisheries Research.

Baker, J.P. and C.L. Schofield. 1980. Aluminium toxicity to fish as related to acid precipitation and Adirondack surface water quality, p. 292-293. In D. Drablos and A. Tollan (Eds.) Proc., Int. conf. ecol. impact acid precip., Norway 1980, SNSF-project. 383 p.

Buck, J. and H.F. Siewert. 1980. Effects of low pH levels on body weight of crayfish, p. 232-233. In B. Moulton (Ed.) Proc. Indiana Academy of Science 89. 445 p. (Abstract only).

Cukerzis, J. 1968. Interspecific relations between Astacus astacus L. and Astacus leptodactylus ESCH. Ekologia Polska, Ser A, 31. p. 1-6.

Fürst, M. 1978. Försurningens inverkan på flodkräftan Astacus astacus L. p. 90-94. In M. Fürst (Ed.) Nordiskt kräftsymposium 1977. Information från Sötvattenslaboratoriet, Drottningholm 14, 1978. 95 p. (Summary in English).

Jay, D. and D.MN. Holdich. 1977. The pH tolerance of the crayfish Austropotamobius pallipes (Lereboullet), p 363-370. In O.V. Lindqvist (Ed.) Freshwater Crayfish III, Kuopio, Finland 1976, 504 p.

Kirschner, L.B. 1970. The study of NaCl transport in aquatic animals. American Zoologist 10, p. 365-376.

Kossakowski, J. 1975. Crayfish Orconectes limosus in Poland, p. 31-47. In J.W. Avault Jr. (Ed.) Freshwater Crayfish II, Baton Rouge, Louisiana, U.S.A. 1974, 676 p.

Leivestad, H., G. Hendrey, I.P. Muniz and E. Snekvik. 1976. Effects of acid precipitation on freshwater organism, p. 86-111. In F.H. Braekke (Ed.) Impact of Acid precipitation on Forest and Freshwater Ecosystems in Norway. SNSF FR 6/11 111 p.

Leivestad, H., I.P. Muniz and B.O. Rosseland. 1980. Acid stress in trout from a dilute mountain stream, p. 318-319. In D. Drablos and A. Tollan (Eds.) Proc., Int. conf. impact acid precip., Norway 1980, SNSF-project. 383 p.

Malley, D.F. 1980. Decreased survival and calcium uptake by the crayfish Orconectes virilis in low pH. Canadian Journal of Fisheries and Aquatic Sciences. Vol 37, 3. 1980. p. 364-372.

Morgan, D.O. and B.R. McMahon. 1982. Acid tolerance and effects of sublethal acid exposure on iono-regulation and acid-base status in two crayfish Procambarus clarki and Orconectes rusticus. J. exp. Biol. 97, p. 241-252.

Moshiri, G.A., C.R. Goldman, D.R. Mull, G.L. Godshalk, and J.A. Coil. 1971. Respiratory metabolism in Pacifastacus leniusculus (Dana) as related to its ecology. Hydrobiologia 37, 2. p. 183-195.

Muniz, I.P. and H. Leivestad. 1980. Toxic effects of aluminium on the brown trout, Salmo trutta L. p. 320-321. In D. Drablos and A. Tollan (Eds.) Proc., Int. conf. impact acid precip., Norway 1980, SNSF-project. 383 p.

Newcombe, K.J. 1975. The Ph tolerance of the crayfish Parastacoides tasmanicus (Erichson) Decapoda, Parastacidae). Crustaceana 29, p. 231-234.

Packer, R.D. 1979. Acid-base balance and gas exchange in brook trout (Salvelinus fontinalis) exposed to acidic environments. J. exp. Biol. 79, 127-134.

Robertson, J.D. 1960. Osmotic and ionic regulation. p. 317-339. In T.H. Waterman (Ed.) The Physiology of Crustacea Vol 1 Academic Press, New York. 670 p.

Rosseland, B.O. 1980. Physiological responses to acid stress in fish. 2. Effects of acid water on metabolism and gill ventilation in brown trout Salmo trutta L. and brook trout Salvelinus fontinalis Mitchill. p. 348-349. In D. Drablos and A. Tollan (Eds.) Proc., Int. conf. ecol. impact acid precip., Norway 1980, SNSF-project. 383 p.

Shaw, J. 1960. The absorption of sodium ions by the crayfish Astacus pallipes Lereboullet. III The effect of other cations in the external solution. Journal of Experimental Biology 37, p. 548-556.

Sutcliffe, D.W. and T.R. Carrick. 1975. Respiration in relation to ion uptake in the crayfish Austropotamobius pallipes. Journal of Experimental Biology 63, p. 689-699.

Svärdson, G. 1974. Översikt över laboratoriets verksamhet med plan för år 1974. Information från Sötvattensloboratoriet, Drottningholm 1, 1974, 26 p.

Vangenechten, J.H.D. and O.L.J. Vanderborght. 1980. Effect of acid pH on sodium and chloride balance in an inhabitant of acid freshwaters: the waterbug Corixa punctata (Illig.) (Insecta, Hemipters. p. 342-343. In D. Drablos and A. Tollan (Eds.) Proc., Int. conf. ecol. impact acid precip., Norway 1980, SNSF-project, 383 p.

Wolwekamp, H.P. and T.H. Waterman. 1960. Respiration , p. 35-100. In T.H. Waterman (Ed.) The Physiology of Crustacea Vol 1, Academic Press, New York. 670 p.

ACID TOXICITY AND PHYSIOLOGICAL RESPONSES TO SUB-LETHAL ACID EXPOSURE IN CRAYFISH

B. R. McMahon and D. O. Morgan
Department of Biology, University of Calgary
2500 University Drive, Calgary, Alberta, Canada T2N 1N4

ABSTRACT

Although crayfish commonly inhabit water systems affected by acid precipitation relatively little is known of their tolerance or of their physiological responses. At least in acute (96 h) exposures, crayfish are substantially more acid (H_2SO_4) tolerant than most fish species (LS_{50} for <u>Procambarus clarki</u> and <u>Orconectes rusticus</u> were pH 2.8 and 2.5 respectively). Physiological responses to sublethal exposure (4 days at ambient pH = 3.8) included the development of severe hemolymph acidosis. The acidosis was largely metabolic in origin and was associated with marked depression of hemolymph bicarbonate levels. This degree of acid exposure produced little disturbance of hemolymph Na^+, CL^-, K^+, or MG^{++} ion levels. No increase in hemolymph $SO_4^=$ resulted. Hemolymph Ca^{++} levels rose significantly suggesting that some invading H^+ ions were being buffered by dissolution of exoskeletal or other carbonate stores. With the exception of the rise in Ca^{++} the physiological responses observed are qualitatively similar to those of trout tested at similar external Ca^{++} levels but the crayfish were exposed to substantially higher acid loads. The reasons for the greater tolerance of crayfish are not clear at this time.

INTRODUCTION

Release into the atmosphere of sulphur and nitrogen dioxides resulting from industrial and urban development causes airborne formation of sulphuric and nitric acids. As a result acidic precipitation (acid rain) has been noted over wide areas of N. America and Europe (Likens and Bormann 1974, Jeffries et al. 1979, Hesslein 1979). The resulting acidification of natural waters has caused mortality in freshwater fish (Leivestad and Muniz 1976, Beamish and Harvey 1972) as well as in several invertebrate populations (Almer et al. 1974, Sprules 1975).

Recently a number of studies investigated the physiological effects of, particularly sublethal, acid exposure on several fish species. Specific effects predictably include disturbance of blood acid-base status (Packer 1979, Neville 1979a, McDonald et al. 1980, Hobe et al. 1980) but also include disturbance of acid-base regulation (Packer and Dunson 1970, 1972b, Leivestad and Muniz 1976, McDonald et al. 1980, Hobe et al. 1980) as well as cardiovascular performance (Milligan and Wood, 1980, 1982) and oxygen depletion and transport (Packer 1979). Physiological responses recorded in the literature are somewhat varied but Neville (1979b, c) has shown that some confusion may have resulted from the unrecorded presence of hypercapnia, McDonald et al. (1980) demonstrate marked differences resulting from variation in water calcium levels while Graham and Wood (1981) demonstrate differences in response to different acid types.

Although invertebrates obviously constitute an important part of aquatic ecosystems affected by acid precipitation very little is known of either tolerance levels or basic physiological responses to acid exposure. Several crayfish inhabit waters of acid stressed areas, and their basic respiratory acid-base and osmo-iono regulatory processes are reasonably well described, two crayfish, a northern species Orconectes rusticus and a southern species Procambarus clarki were selected for study of acid tolerance and physiological responses to short-term sublethal levels of acid stress.

MATERIALS AND METHODS

Adult Procambarus clarki and Orconectes rusticus of either sex were obtained from commercial suppliers in Louisiana and Southern Ontario respectively; animals were held for at least 2 weeks prior to experimental use in flowing Calgary tap water at 15°C. Analysis of the water using techniques outlined below yielded the following ionic composition: Na^+, 0.08; K^+, 0.02; Ca^{++}, 1.1; Mg^{++}, 0.6; Cl^-, 0.5; $SO_4^=$, 0.03; HCO_3^-, 2.14 (all values in mM). Animals were fed regularly with chopped smelt prior to but not during the experiments. All crayfish used were judged to be in an intermolt stage as determined by skeletal rigidity (Vranx and Durliat 1978). For the physiological studies adult crayfish (15-30 g Orconectes, 35+ g Procambarus) were prepared for hemolymph sampling by drilling a pair of small holes in the exoskeleton, above the pericardial cavity adjacent to the heart. Holes were sealed with rubber dam glued to the carapace with cyanoacrylate cement, and the animals left to recover at least 4 d in normal tap water prior to commencement of the experimental regime.

Toxicological Testing

Groups of 9-10 Procambarus and Orconectes (15-30 g) were exposed to several concentrations of sulphuric acid (Reagent grade) for 4 day periods to allow estimation of LC_{50}, the pH or H^+ ion concentration at which 50% mortality occurred. Sulphuric acid (H_2SO_4) was used as it is the more common environmental pollutant. Beamish (1972) for White Sucker, Packer and Dunson (1972) for Brook Trout and Graham and Wood (1981) for Rainbow Trout have all shown that H_2SO_4 is actually slightly less toxic than HCl, although the latter authors point out that the difference is actually somewhat variable depending on the levels of ambient $[Ca^{++}]$ and possibly other substances.

For toxicity testing animals were placed in glass tanks containing 12 litres of recirculated, well aerated and filtered tap water (15 ± 0.5°C). The pH levels in all tanks (except control) were adjusted to the desired pH level 24 h prior to the addition of animals to allow removal of CO_2 (aeration) resulting from the titration of water bicarbonate. Orconectes were exposed to pH 2.0, 2.3, 2.6, 3.0 (10.0, 4.6, 3.6, 3.1 mM H^+) and Procambarus to pH 2.0, 2.5, 3.0 and 3.5 (8.9, 4.8, 2.2, 1.6 mM H^+) following preliminary testing to determine a lethal level. pH values were monitored several times daily and were maintained within 0.05 pH unit. No significant decrease in oxygen tension (PIO_2) occurred but ammonia concentration had risen slightly in these static tanks after 4 d exposure.

Mortality was recorded at 24, 36, 48 and 96 h. Animals were judged dead if all movement had ceased and reflex movements of eyes and tail could not be elicited following repeated stimulation. In both species no mortality occurred in similar groups of control animals at natural pH (7.5-8.0). LC_{50} was estimated for each species using the Reed-Muench method as described by Woolf (1968).

Analytical Procedures

Methods of sampling and analysis are exactly similar to those described by Morgan and McMahon (1982). Briefly water pH was measured on samples at experimental temperature using either Beckman 4001 or Fisher Accumet pH meters with combination electrodes calibrated with Fisher buffers at pH 4.0 and 6.8; additional calibration with 1.00×10^{-2} M HCl was employed during measurements of the acid water samples. Hemolymph pH was measured using a micro pH electrode (Radiometer, G299A liquid junction) and $[CO_2]$ using the micro-method of Cameron (1971). Cation concentrations were measured by atomic absorption spectrophotometry (Jarrell-Ash 850) following appropriate dilutions and addition of reagents for suppression of interference. Chloride was determined coulometrically (Buchler 4-2500 chloridometer) and sulphate by a modified version of the turbidometric technique of Berglund and Sörbo (1960). Osmolality of hemolymph was determined on 5 μl samples using a vapour pressure osmometer (Wescor 5100 B).

Using measured values of pH and $[CO_2]$ the carbon dioxide tension (PCO_2) was calculated using the method described by Wilkes et al. (1980) for Orconectes rusticus.

Statistical Analysis

Differences in toxicity and between means of physiological samples were tested by Student's t-test of the means. Throughout the analysis $P < 0.05$ is judged to indicate a significant difference.

RESULTS

Toxicity Tests

This preliminary study allowed calculation of the LC_{50} values at periods from 24-96 h, at 15°C and comparison between species (Fig. 1, Table 1). LC_{50} values at 24-36 h were similar but on longer exposure Procambarus appeared slightly more susceptible such that at 96 h LC_{50} = 1.6 mM H_2SO_4 (pH = 2.8) and 3.1 mM H_2SO_4 (pH 2.5) were calculated for Procambarus and Orconectes respectively. At no time, however, was the difference between the two species significant. In 4 days exposure LC_{50} was still declining and thus these levels are not definitive (i.e. incipient LC_{50}) values as defined by Sprague (1969). The purpose of these tests was not to provide a detailed analysis of acid toxicity but rather to provide an estimate of 96 h sublethal levels of acid exposure for use in subsequent physiological studies. pH values 1 unit above (H^+ concentrations 10x below) LC_{50}, i.e. pH 3.8 ($[H^+]$ = 0.16 mM) and pH 3.5 ($[H^+]$ = 0.32 mM) were chosen for Procambarus and Orconectes respectively.

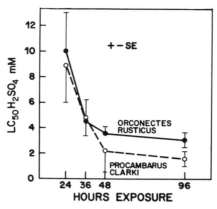

Figure 1. LC$_{50}$ for _Procambarus clarki_ and _Orconectes rusticus_ exposed to sulphuric acid for 24, 36, 48, and 96 h.

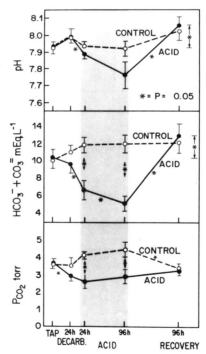

Figure 2. Changes in acid-base status resulting from a 4 day exposure to sublethal concentration (= pH 3.8) in _Procambarus clarki_. Carbon dioxide tension (P_{aCO_2}) calculated from pH_a and CO_2 content by method of Wilkes et al. (1980).

Table 1. LC_{50} values for crayfish in sulphuric acid expressed in terms of pH and concentration of H^+ in mM.

LC_{50} at	Orconectes rusticus		Procambarus clarki	
	pH $\overline{X}$ ± SE	$[H^+]$ M x 10^{-3}	pH $\overline{X}$ ± SE	$[H^+]$ M x 10^{-3}
24 hours	2.0 ± 0.13	10.0 ± 3	2.05 ± 0.14	8.9 ± 2.9
36 hours	2.34 ± 0.09	4.6 ± 1.1	2.32 ± 0.12	4.8 ± 1.4
48 hours	2.45 ± 0.06	3.6 ± 0.5	2.66 ± 0.14	2.2 ± 0.7
96 hours	2.51 ± 0.08	3.1 ± 0.06	2.79 ± 0.12	1.6 ± 0.05

Table 2. Comparison of acid-base and ionic changes resulting from exposure to sulphuric acid in relatively hard water for 2 crayfish and 2 fish species.

	Orconectes rusticus	Procambarus clarki		White Sucker[2]	Rainbow Trout[3]
		1 day accl.	14 day accl.[1]		
External $[H^+]$	0.31	0.16	0.16	0.05	0.06
mM H_2SO_4 pH	3.5	3.8	3.8	4.3	4.2
$\Delta H^+{}_b$ m.Eq.	13.4	6.6	19.3	4.3	10.9
ΔHCO_3^- mM	-8.9	-5.3	-14	-2.2	-6.3
ΔPCO_2 torr	+1.1	-0.5	-1.2	±0.81	-0.7
ΔNa^+ mM	-34	-18	-28	-10.3	-3.7
ΔCl^- mM	-18	-22	-7	-8.6	-25.3
ΔK^+ mM	/	-0.4	+0.5	+0.4	+1.21
ΔCa^{++} mM	+9	0.5	6.1	No Δ	No Δ

Data from present study except:
[1]Morgan and McMahon 1982.
[2]Hobe et al. 1980.
[3]McDonald and Wood 1981

Physiological Effects of Acid Exposure

Physiological responses to acid exposure were qualitatively similar in both species and for the sake of brevity the present paper concentrates on the results for Procambarus clarki while major points of difference are listed in Table 2.

In order to avoid complicating effects of a transient external hypercapnia resulting from the titration of water bicarbonate, and to provide an equivalent ionic environment for both control and acid treated crayfish, both sets of animals were transferred to neutral (pH 7.5-8.0) but previously decarbonated water for 24 h at the start of the experiment. A slight alkalosis (Figs. 2 and 3) accompanies this transfer probably resulting from changes in ambient Na^+ and HCO_3^- levels (see Morgan and McMahon 1982).

Transfer of Procambarus to acidified, previously decarbonated water resulted in a significant decrease in postbranchial hemolymph pH (pHa) at 24 h exposure (Fig. 2a). pH continued to fall but more slowly and no further significant decrease occurred over the remainder of the exposure period. In this species, postbranchial hemolymph carbon dioxide tension ($PaCO_2$) also decreased to levels significantly below controls during acid exposure (Fig. 2c). Significant decreases in $PaCO_2$ also resulted on acid exposure in Procambarus acclimated to decarbonated water (Morgan and McMahon 1982), but $PaCO_2$ is slightly elevated in acid stressed Orconectes. The differences may be associated with the somewhat greater acidosis observed in the latter species. In both species the acidosis is associated with dramatic loss of hemolymph CO_2 reserves, largely bicarbonate (Figs. 2, 3) such that the animals' hemolymph bicarbonate stores are depleted 40% in the first 24 h and more than 50% over the 4 day acid exposure.

Replotting the acid-base data in the form of a diagram linking the changes in pH and $[CO_2]$ via the Henderson-Hasselbalch equation (Fig. 3) allows further analysis of the acidosis. Essentially the diagram plots the changes in bound CO_2 ($HCO_3^- + CO_2^=$) and pH against a format of lines of equivalent PCO_2 (CO_2 isopleths) demonstrating the interrelationships involved. Simplifying, the observed decrease in pH can be of two causes, an increase in CO_2, i.e. carbonic acid, termed a "respiratory" acidosis or an increase in H^+ produced by other metabolic means, e.g. with lactate during anaerobiosis, this being termed a "metabolic" acidosis. As displayed in Fig. 3 a purely respiratory acidosis would be seen as a movement up and along a line representing the relationship $\Delta HCO_3/\Delta pH$ (representing non-bicarbonate buffering of H^+ ions) while a purely metabolic acidosis would occur at constant PCO_2, i.e., along the isopleth of the original CO_2 tension. In displaying the data for Procambarus (Fig. 3) it is apparent that the alkalosis resulting on entry into decarbonated water is purely respiratory (i.e. loss of CO_2) while the acidosis resulting from consequent exposure to acid is entirely metabolic, involving no increase in PCO_2. The H^+, however, need not be generated internally but could simply diffuse into the hemolymph from the environment.

Following return to "neutral" waters, animals show relatively rapid recovery from the effects of acid (Figs. 2, 3). In Orconectes rusticus pre-acid exposure levels of both pH and $[CO_2]$ are reached and indeed exceeded within 24 h after return to neutral water. This

76

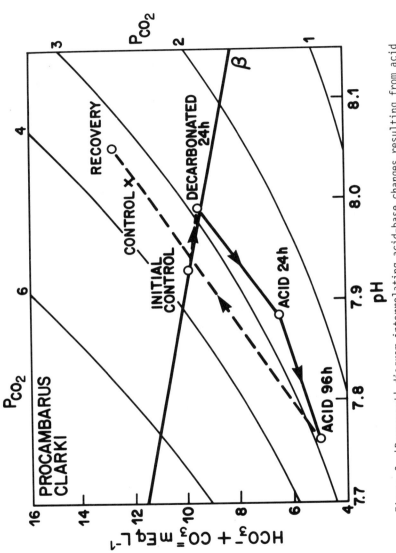

Figure 3. 'Davenport' diagram interrelating acid-base changes resulting from acid exposure. Isopleths calculated using method outlined in Wilkes et al. (1980). β = the slope of the relationship $\Delta HCO_3^-/\Delta pH$ (non-bicarbonate buffer line).

"overshoot" in acid-base regulation is apparently quite persistent, as both pH and $[CO_2]$ remain elevated even 96 h later (Fig. 2). Recovery from the acidosis also seems to be an entirely metabolic event involving no significant change in PCO_2.

Ionic and Osmotic Effects of Acid Exposure

Considerable variability occurred in measured ion concentrations recorded either from control or acid treated groups. Additionally, ion levels decreased persistently throughout the experimental period in all animals tested. Consequently, to better display these trends and to show differences resulting from acid exposure, the data are presented in histogram form in Fig. 4. Two comparisons are included for the acid stressed group. Comparison is made between mean data at each measurement point and both the initial and current control values. The former comparison allows assessment of progressive changes occurring during the experimental regime while the latter allows interpolation of ongoing changes in control ion status.

Hemolymph osmolality decreased progressively in both groups throughout the treatment period possibly as a result of hemolymph lost during repetitive sampling. The decrease, however, was greater in the acid stressed animals in which osmolality was significantly depressed at both 24 and 96 h acid exposure. As would be expected from the data in Figs. 2 and 3, $[H^+]$ increased progressively in acid exposed animals and had decreased below control levels (alkalosis) 96 h after return to neutral water (Fig. 4).

Changes in the other measured ions tend to follow one or other of the patterns above. Hemolymph $[Cl^-]$ closely follows hemolymph osmolality decreasing progressively in both acid and control groups. The rate of decrease is, however, greater in acid exposed animals which are significantly depressed below initial control values at 24 h acid exposure and show progressive decrease even in the recovery period (Fig. 4). Hemolymph $[HCO_3^-]$ also decrease markedly and significantly during acid exposure. HCO_3^- levels thus seem to be adjusted in compensation for the change in H^+ (compare in Fig. 4) rather than associated with the general trend towards decreased osmolality. Hemolymph $[Na^+]$ also decreased slightly (not significantly) throughout acid exposure but, unlike $[Cl^-]$ increased in recovery. No significant change occurs in hemolymph $[K^+]$ of acid stressed animals but again an overall decrease in $[K^+]$ occurred in the acid treated group.

Despite the massive increase (over 100 fold) in external $[SO_4^=]$ resulting from acid exposure, hemolymph $[SO_4^=]$ rose only slightly (0.1 mM) during acid exposure, perhaps demonstrating a basic impermeability of crayfish gills to this ion. $[SO_4^=]$ in hemolymph of control animals also showed a slight but progressive decreasing trend throughout the experimental regime. No significant overall change in $[Mg^{++}]$ concentration occurred in either control or acid treated group but hemolymph $[Ca^{++}]$ rose significantly above control levels throughout acid exposure decreasing significantly in recovery. In the present experimental series the increase in Ca^{++} was small (Fig. 4), but in this species acclimated to decarbonated water (Morgan and McMahon 1982) and in <u>Orconectes</u> <u>rusticus</u> (Fig. 5 and Table 2) marked elevation of hemolymph $[Ca^{++}]$ was observed. With this exception the effects of acid exposure on hemolymph ion concentrations are similar in both species.

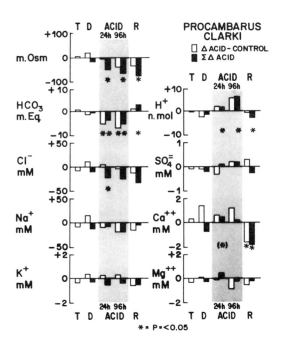

Figure 4. Changes in hemolymph osmolality and concentration of several ions resulting from 24, 36, 48, and 96 h exposure to sulphuric acid (= pH 3.8) in Procambarus clarki. Closed columns compare mean data from acid exposed animals at each sampling period with the initial control value. Open columns compare acid treated with control levels recorded at that time.

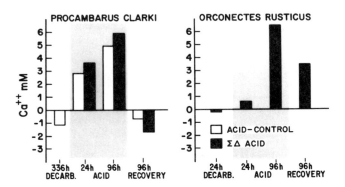

Figure 5. Changes in hemolymph calcium levels recorded in Procambarus clarki and Orconectes rusticus exposed 4 days to pH 3.8 and pH 3.5 respectively. Details as per legend in Fig. 4. Data for Procambarus from Morgan and McMahon (1982).

DISCUSSION

Physiological changes occurring in response to acid exposure seem qualitatively similar, not only between the two crayfish species but also between crayfish and the several fish species listed in Table 2. Quantitative differences are seen in both comparisons but these are hard to interpret physiologically since the levels of acid stress are not identical and major changes in response may result from small differences in external Ca^{++} (McDonald et al. 1980, Graham and Wood 1981). Comparison of the present results with those of Morgan and McMahon (1982) additionally reveal significant differences associated with the degree of acclimation to experimental conditions. Quantitative comparison of the data in Table 2 thus should be made with caution.

Comparison of the degree of acidosis induced is particularly difficult. Some literature reports compare levels of acidosis as ΔpH but this comparison can be misleading due to the logarithmic nature of the pH concept. For instance, ΔpH of 0.30 units is equivalent to a $\Delta[H^+]$ of 6.3 n.moles between pH 8.2 and 7.9, but to 50 n.moles $\Delta[H^+]$ for the interval pH 7.6-7.3. Comparison of $\Delta[H^+]$ accurately compares the state of hemolymph acidosis at a particular sampling interval but gives little information as to the rate of acquisition or production of H^+, since H^+ buffered or excreted by the animal are not included. Better comparison can be made by calculation of ΔH^+_b, the amount of H^+ added to serum from non-respiratory acids (McDonald et al 1980), which considers H^+ buffered within blood or hemolymph systems. Although this calculation does not allow consideration of H^+ buffered by other systems or excreted, it is, in the absence of direct flux measurements, perhaps the best comparative measure and is included in Table 2. ΔH^+_b for Procambarus in the present study thus lies within the range presented for fish, but overall, crayfish seem to exhibit the higher acidosis. This may result from the considerably greater acid-load imposed (Table 2).

In Procambarus and to a lesser extent in Orconectes, the acidosis is largely metabolic (Fig. 3, little or no increase in PCO_2). This could originate endogenously, i.e. metabolic production of H^+, or exogenously, by influx of H^+ down the tremendous H^+ gradient which develops during acid exposure. Reduction of O_2 transport has been implicated in acid stressed fish (Packer 1979) but may result only from severe (i.e. lethal) acid exposure, since McDonald et al. (1980) were unable to record increase in lactate from trout exposed to sublethal acid levels. Hemolymph lactate levels were not measured in the present study but the animals remained quiescent throughout their sublethal acid exposure, and thus elevated lactate levels probably played little role in the development of acidosis in either species. Increased H^+ influx, on the other hand, seems more likely to be the major route, since increased influx of H^+ has been shown to accompany acid exposure both in Rainbow Trout (McDonald and Wood 1981) and in crayfish (Johnson and Wood, pers. comm.).

Several mechanisms may be involved in buffering the H^+ influx. Consideration of the data of Figure 3 indicates that within the hemolymph considerable buffering occurs, mostly by titration of hemolymph bicarbonate with a smaller contribution from hemolymph protein, largely hemocyanin. Several additional mechanisms may be involved. DeFur et al. (1980) demonstrated that exoskeletal (or other) carbonate stores may be mobilized to allow buffering of the acidosis generated by air exposure in the crab Cancer productus. Some evidence implicating a similar mechanism in crayfish is evident in the significant elevation of hemolymph Ca^{++} which accompanies acid exposure (Figs. 4 and 5). Buffering by skeletal dissolution has also been postulated for fish (Beamish 1974, Heisler 1980, McDonald and Wood 1981).

Several other mechanisms involve transport of H^+ out of the extracellular fluid either into the cells (Heisler 1980) or back into the environment via exchange mechanisms operating across the branchial or renal epithelium. Neither mechanism has been studied in crustacean species but both are involved in fish. Transport of H^+ into the intracellular compartment (presumably involving cation exchange) becomes the principal buffering mechanism in trout by 4 days acid exposure (McDonald and Wood 1981) buffering an estimated 77% of the H^+ influx. Renal excretion of H^+ has recently been shown to be important in hydrogen ion excretion in fish (Wood and Caldwell 1978, Kobayashi and Wood 1980). However, although marked increase in renal excretion occurs in acid exposed trout, excretion by this route can probably only remove 30-50% of H^+ influx resulting from exposure to pH 4.2. Significant excretion across the gills seems unlikely under acid conditions. Excretion routes involving Na^+/H^+ and Na^+/NH_4^+ exchange are well known for both fish and crustaceans, but Na^+ uptake appears to be inhibited at pH 4.0 (Packer and Dunson 1980, McWilliams and Potts 1978, McWilliams 1980). Branchial ammonia excretion increases only slightly during acid exposure in trout (McDonald and Wood 1981) and it is likely that below pH 5.0 in the ambient water the great majority of NH_3 is excreted by nonionic diffusion.

As with the acid-base effects discussed above sublethal exposure also affects iono and osmoregulatory balance similarly in crustaceans and fish species tested (Table 2). Net loss, principally of Cl^- and Na^+ ions, occurs resulting in loss of total osmolality (Fig. 4). The loss of these ions probably results both from inhibition of their normal uptake routes in addition to an increase in passive efflux caused by the high acid conditions (McWilliams and Potts 1978, Heisler 1980, McDonald et al. 1980, McWilliams 1980). The ion flux studies of McDonald and Wood (1981) demonstrate that ion loss is not restricted to plasma but that intracellular ions are also significantly depleted. As pointed out by McDonald et al. (1980) external Ca^{++} levels play a dramatic role in regulating the extent of the ionoregulatory disturbance associated with acid exposure in fish. Ionic disturbance is considerably greater in low calcium media and acid-base stress is minimal, while at the higher external Ca^{++} levels typical of hard water the acid-base disturbance predominates. The affect of the Ca^{++} ion is not fully understood but presumably involves its effects on the passive permeability of the gill system, since Na^+ uptake mechanisms seem largely independent of $[Ca^{++}]$ in freshwater fish (McWilliams 1980).

LC$_{50}$ values reported in the present study plus similar values from the literature (Baker 1979, Newcombe 1975) indicate that crayfish are considerably more resistant to acid exposure (Table 2) than fish species studied at least in the short term. The LC$_{50}$ values of 1.6 and 3.1 mM H_2SO_4 (pH 2.8 and 2.5) for Procambarus and Orconectes respectively are substantially higher (lower in pH) than those commonly reported for fish, e.g. Beamish (1972) reports a 100 h LC$_{50}$ of pH 3.9

([H$^+$] = 0.13 mM), Daye and Garside (1975) report LC$_{50}$ of pH 3.5 ([H$^+$] = 0.32 mM) at 167 h for Brook Trout, while McDonald et al. (1980) and Graham and Wood (1981) report LC$_{50}$'s of pH 4.0-4.2 (0.10-0.06 mM) and pH 4.1-4.5 (0.08-0.03 mM) at 96 h and 167 h respectively for Rainbow Trout. Amongst fish, Cardinal Tetras seem to be relatively acid resistant since Dunson et al. (1977) report that they live indefinitely at pH 3.5. The above comparison actually points out that the inadvisability of using pH values for LC$_{50}$ determinations. Due to the logarithmic nature of the pH concept these results can be particularly misleading in toxicological studies. For example expressed as pH the LC$_{50}$ values for Orconectes and Procambarus are apparently similar but actually differ in H$^+$ content by a factor of 2 (Table 1).

The reason for the apparently greater tolerance of crayfish to acid exposure is not understood. Table 2 shows that both species of crayfish exhibit similar physiological responses, both of hemolymph acid base and ionic status as fish exposed to much lower acid concentrations. Possibly some of the greater tolerance of the crayfish may be attributed to the presence of the chitinous and calcified exoskeleton. This structure extends as a thin chitinous layer over the gills and may impede both the influx of H$^+$ ions and the passive efflux of Na$^+$ and Cl$^-$ across the gills, in addition to acting as a reservoir of carbonate ions for use as a buffer for the remaining H$^+$ influx. R. France (this volume), however, reports that crayfish from Lake 223 (pH = 5.2) of the Experimental Lakes area of Northwestern Ontario show significantly reduced exoskeletal rigidity suggesting that the greater acid tolerance may actually only apply in the relatively short term. The apparently higher tolerance of intermolt animals may also be offset by a greater sensitivity of postmolt animals (Malley 1980), while Baker (1979) for hatchling O. rusticus (LC$_{50}$ = pH 4.0) and R. France (this volume) for juvenile O. virilis both show the greater sensitivity of juvenile stages to acid exposure. Additionally, the latter author shows reduction of female reproductive stress at pH 5.6 in a natural population of O. virilis. Abrahamsson (1972) noted that mortality occurred in the wild for Astacus astacus when animals were introduced in lakes at pH 5.6 ([H$^+$] = 0.003 mM). These levels are far below the LC$_{50}$ values measured in this and other acute studies. Clearly more long term studies are needed before a reasonable estimate of acid tolerance may be made for crayfish populations.

ACKNOWLEDGMENTS

This work was supported by an NSERC summer fellowship to D. O. Morgan and by NSERC Grant No. A5762 to B. R. McMahon. The authors wish to express their thanks to H. Hobe, P. R. H. Wilkes and A. Pinder for advice and assistance, and to Dr. M. G. Wheatly for a critical reading of the manuscript.

LITERATURE CITED

Abrahamsson, S. 1972. Fecundity and growth of some populations of Astacus astacus LINNE in Sweden, pp. 23-37. In: Report no. 52 from the Institute of Freshwater Research, Drottingholm, Sweden.

Almer, B., W. Dickson, G. Ekstrom, E. Hornstrom and U. Miller. 1974. Effects of acidification on Swedish lakes. Ambio 3:30-36.

Baker, M. J. 1979. The effect of acidic precipitation in the crayfish. B.Sc. Hons. Thesis, Trent University. Peterborough, Canada.

Beamish, R. J. 1972. Lethal pH for the white sucker Catastomus commersoni (Lacepede). Trans. Am. Fish. Soc. 101:355-358.

Beamish, R. J. 1974. Growth and survival of white suckers (Catastomus commersoni) in an acidified lake. J. Fish. Res. Bd. Can. 31:49-54.

Beamish, R. J. and Harvey, H. H. 1972. Acidification of the La Cloche Mountain Lakes, Ontario, and resulting fish mortalities. J. Fish. Res. Bd. Can. 29:1131-1143.

Berglund, F. and Sorbo, B. 1960. Turbidimetric analysis of inorganic sulphate in serum plasma and urine. Scand. J. Clin. and Lab. Invest. 12:147-153.

Cameron, J. N. 1971. Rapid method for determination of total carbon dioxide in small blood samples. J. Appl. Physiol. 31:632-634.

Daye, P. G. and Garside, E. T. 1975. Lethal levels of pH for brook trout, Salvelinus fontinalis (Mitchill). Can. J. Zool. 53:639-641.

deFur, P. L., P. R. H. Wilkes and B. R. McMahon. 1980. Non-equilibrium acid-base status in C. productus: role of exoskeletal carbonate buffers. Resp. Physiol. 42:247-261.

Dunson, W. A., F. Swartz and M. Silvestri. 1977. Exceptional tolerance of low pH of some tropical blackwater fish. J. Exp. Zool 201: 157-162.

France, R. 1982. Paper in this volume, page 98.

Graham, M. S. and C. M. Wood. 1981. Toxicity of environmental acid to the rainbow trout. Interactions of water hardness, acid type and exercise. Can. J. Zool. (submitted).

Heisler, N. 1980. Regulation of the acid-base status in fishes, pp. 123-162. In: M. Ali (ed.) Proceedings of NATO-ASl Symposium on Physiology of Fishes.

Hesslein, R. H. 1979. Lake acidification potential in the Alberta Oil Sands Environmental Research Program Study Area. Project Report HY 2.2. Alberta Oil Sands Environmental Research Program.

Hobe, H., P. R. H. Wilkes, R. L. Walker, C. M. Wood and B. R. McMahon. 1981. Effects of low pH on the physiology of the benthic cyprinid Catostomus commersoni. Am. Zool. 20:591A.

Jeffries, D. S., C. M. Cox and P. J. Dillon. 1979. Depression of pH in lakes and streams in central Ontario during snowmelt. J. Fish. Res. Bd. Can. 36:640-646.

Kobayashi, K. A. and C. M. Wood. 1980. The response of the kidney of the freshwater rainbow trout to true metabolic acidosis. J. Exp. Biol. 84:227-244.

Leivestad, H. and I. P. Muniz. 1976. Fish kill at low pH in a Norwegian river. Nature 259:391-392.

Likens, G. E. and F. H. Bormann. 1974. Acid rain: A serious regional environmental problem. Science 184:1176-1179.

Malley, D. F. 1980. Decreased survival and calcium uptake by the crayfish Orconectes virilis in low pH. Can. J. Fish. Aquat. Sci. 37:364-372.

McDonald, D. G., H. Hobe and C. M. Wood. 1980. The influence of calcium on the physiological responses of the rainbow trout, Salmo gairdneri, to low environmental pH. J. Exp. Biol. 88:109-131.

McDonald, D. G. and Wood, C. M. 1981. Branchial and renal acid and ion fluxes in the rainbow trout, Salmo gairdneri, at low environmental pH. J. Exp. Biol. 93:101-118.

McWilliams, P.G. 1980. Effects of pH on sodium uptake in Norwegian brown trout (Salmo trutta) from an acid river. J. Exp. Biol. 88:259-267.

McWilliams, P. G. and W. T. W. Potts. 1978. The effects of pH and calcium concentrations on gill potentials in the brown trout, Salmo trutta. J. Comp. Physio. 126:277-286.

Milligan, C. L. and C. M. Wood. 1980. The influence of low environmental pH on cardiovascular function in the rainbow trout Salmo gairdneri. The Physiologist 23(4):913A.

Milligan, C.L. and C.M. Wood. 1982. Disturbances in haematolosy, fluid volume distribution and circulatory function associated with low environmental pH in the rainbow trout Salmo gairdneri.

Morgan, D. O. and B. R. McMahon. 1982. Acid tolerance and effects of sublethal acid exposure on ionoregulation and acid-base status in two crayfish Procambarus clarki and Orconectes rusticus. J. Exp. Biol. 97:241-252.

Neville, C. M. 1979a. Sublethal effects of environmental acidification on rainbow trout (Salmo gairdneri). J. Fish Res. Bd. Can. 36:84-87.

Neville, C. M. 1979b. Influence of mild hypercapnia on the effects of environmental acidification on rainbow trout (Salmo gairdneri). J. Exp. Biol. 83:345-349.

Neville, C. M. 1979c. Ventilatory response of rainbow trout (Salmo gairdneri) to increased H$^+$ ion concentration in blood and water. Comp. Biochem. Physiol. 63a:373-376.

Newcombe, K. J. 1975. The pH tolerance of the crayfish _Parastacoides_ _tasmanicus_ (Erichson) (Decapoda, Parastacidae). Crustaceana 29:231-234.

Packer, R. K. 1979. Acid-base balance and gas exchange in brook trout (_Salvelinus fontinalis_) exposed to acidic environments. J. Exp. Biol. 79:127-134.

Packer, R. K. and W. A. Dunson. 1970. Effects of low environmental pH on blood pH and sodium balance of brook trout. J. Exp. Zool. 174:65-72.

Packer, R. K. and W. A. Duson. 1972. Anoxia and sodium loss associated with the death of brook trout at low pH. Comp. Biochem. Physiol. 41A:17-26.

Sprules, W. G. 1975. Midsummer crustacean zooplankton communities in acid-stressed lakes. J. Fish. Res. Bd. Can. 32:389-395.

Vranx, R. and M. Durliat. 1978. Comparison of the gradient of setal development of uropods and of scaphognathites in _Astacus_ _leptodactylus_. Biol. Bull. 155:627-639.

Wilkes, P. R. H., P. L. deFur and B. R. McMahon. 1980. A new operational approach to P_{CO_2} determination in crustacean haemolymph. Resp. Physiol. 42:17-28.

Wood, C. M. and F. H. Caldwel. 1978. Renal regulation of acid-base status in a freshwater fish. J. Exp. Zool. 205:301-307.

Woolf, C. M. 1968. Principles of Biometry. D. Van Nostrand Co., Toronto.

EFFECTS OF HYPOXIA ON THE HAEMOLYMPH OF THE FRESHWATER CRAYFISH, ASTACUS ASTACUS L., IN NEUTRAL AND ACID WATER DURING THE INTERMOULT PERIOD

T. Järvenpää[1], M. Nikinmaa[2],
K. Westman[1] and A. Soivio[2]

ABSTRACT

During the intermoult period crayfish were able to adapt to an oxygen saturation level of 30% (ca. 3.2 mg 1^{-1}) at 11-12 °C in neutral water (pH 7.2). The adaptation was mainly respiratory; the haemolymph O_2 affinity increased markedly in hypoxia. This increase was probably due to the simultaneous increase in the haemolymph Ca^{2+} and Mg^{2+} concentrations. Acidification of the water alone (pH 4), without a reduction in the O_2 saturation level, did not affect the haemolymph constituents appreciably. In acid water (pH 4) depleted of oxygen (30% of sat.), however, respiratory adaptation did not seem to take place. On the contrary, difficulties in pH- and osmoregulation were apparent. The latter observations are of considerable importance in Finland, since construction of engineering works on the good crayfish rivers often results in simultaneous drops in both the oxygen content and the pH value.

INTRODUCTION

In Finland, radical changes in water quality as a consequence of various types of construction activities, especially the dredging, clearing and embanking of rivers, cutting of river banks, and the construction of reservoirs, together with the regulating and polluting of water, caused particulary during the 1960s and 1970s, marked economic losses in the country's sole endemic crayfish species Astacus astacus, (e.g. Westman 1973, Niemi 1976). Construction operations in rivers along with crayfish plague, continue to pose the greatest threat to Finnish crayfish stocks.

Clearing and dredging of rivers in a number of other countries, e.g. Sweden (Vallin 1964), Poland (Kossakowski 1973) and the United States (Hobbs & Hall 1974), similarly have been shown frequently to be the cause of a reduction in, or of the extinction of, crayfish populations. Polluting of the water is held to be an important factor leading to crayfish fall-off in a large number of European countries (Kossakowski 1973, Laurent 1973, Erencin & Köksal 1977).

Construction operations, in particular dredging and clearing, frequently cause long-term turbidity in the water. With this increase

[1]Finnish Game and Fisheries Research Institute, Fisheries Division, P.O. Box 193, SF-00131 Helsinki 13, Finland

[2]Department of Zoology, University of Helsinki, Arkadiankatu 7, SF-00100 Helsinki 10, Finland

in solid matter and iron content. a decrease in oxygen content and, especially in sulphide-rich soils, produce acidification as well.

The adverse effects, observable as crayfish mortality, emigration, reduction in growth rate, and impediment to reproduction, do not generally result from a change in one particular environmental factor, rather several factors are usually responsible. This makes the reason for the damage very difficult to determine, thus impeding the planning of management measures for the crayfish stocks. Furthermore, information relating to the combined effect of different environmental factors on crayfish stocks is very sparse.

Some information is available on the oxygen requirements of Astacus and its kin (Lindroth 1950, Lahti & Lindqvist 1979) and also on its ability to withstand an increase in acidity (Abrahamsson 1972, Cukerzis 1970, Appelberg 1979, Malley 1980). On the other hand very few details are available on the ability of crayfish to acclimatize to a simultaneous decrease in oxygen content and pH. Cukerzis (1973) has demonstrated that both a very low and a very high pH increases the oxygen consumption of crayfish (<u>Astacus</u> <u>astacus</u> and <u>A. Leptodactylus</u>).

Information respecting the effect of simultaneous acidification and hypoxia is urgently needed, since construction activities and other environmental changes often lead to a simultaneous decrease in oxygen and pH. Thus a research program was started at the Finnish Game and Fisheries Research Institute to study the effects of water quality changes on the physiology of crayfish. The aim of this study was to investigate, under laboratory conditions, the effects of hypoxia and acidification on the haemolymph of the freshwater crayfish <u>Astacus</u> <u>astacus</u> are presented. The main emphasis in this study, conducted during the intermoult period, was on the respiratory function of the haemolymph. Since in hypoxia the major physiological problem lies in how to ensure an adequate uptake and transport of oxygen to the tissues, acidification of the environment may further aggravate this problem.

MATERIALS AND METHODS

Test animals

The experiments were carried out at the University of Helsinki, Department of Zoology, Division of Physiology in October-November 1980. The crayfish, <u>Astacus astacus</u> (32 ♂♂, 19,8 ± 9.4 g carapace length 43.5 ± 0.3 mm and 32 ♀♀, 22.5 ± 0.8 g carapace length 45.7 ± 0.6 mm) were in the intermoult stage. They were brought from Evo Inland Fisheries Research Station three weeks before the experiment to be acclimatized to Helsinki dechlorinated tap water (O_2 sat. 90%, pH 7.2 - 7.5, T 12 °C).

Test group

After the acclimatization period the animals were divided into 4 experimental groups (8 males and 8 females in each) which were placed in 60 liter aquaria. The first group was allowed to stay in normoxic water, the second was subjected to hypoxia (O_2 sat. 30%, ca. 3.2 mg l^{-1}), the third to acid water (pH 4) and the fourth both to hypoxia and acid water (O_2 sat. 30%, pH 4). In all cases the exposure lasted for a week after which the sampling of haemolymph took place. The oxygen saturation of water was lowered by bubbling nitrogen into the aquarium and the pH by adding sulphuric acid to the water. The pH was allowed to stabilize overnight before the animals were transferred into the aquarium. Both the oxygen tension and the pH of the water were monitored throughout the study and, when necessary, the pH adjusted by further additions of H_2SO_4. The whole water body in each aquarium was changed on the fourth day of the test. The experimental temperature varied from 11 - 12 °C. The photoperiod was natural. The animals were fed rotting alder leaves throughout the experiments and pieces of ceramic drain tubings served as shelters for them on the bottom of aquaria.

Haemolymph sampling

Haemolymph samples (0,9 ml) were taken in air with 1 ml disposable Tuberculine syringes through a 23 gauge hypodermic needle from prebranchial sinuses at the base of the last pair of walking legs. For the sampling the animals were out of water for c. 1 minute. Immediately after sampling the portion of haemolymph for protein determination was piptetted into Biuret reagent, the portion for ion (except for Cu^{2+}) determinations was deproteinized in 2 volumes of 10% TCA, the portion for Cu^{2+} determinations in one volume of 1 N HCl, one volume of 10% TCA was added 25 minutes later and the sample centrifuged at c. 12,000 g for 3 minutes. The portion for pH, oxygen affinity, and oxygen capacity determinations was centrifuged with Eppendorf Microfuge at c. 12,000 g for 1 minute to prevent the clotting of the sample. The oxygen affinity was determined by the mixing method (Edwards and Martin 1966), and the oxygen capacity with modified Tucker's method (cf. Bridges et al. 1979) using 0.08 M KCN for the liberation of O_2. The Mg^{2+} and Ca^{2+} concentrations were determined by atom absorption spectrophotometer (Perkin-Elmer, USA). The Cu^{2+} concentration was determined spectrophotometrically with the Boeringer Mannheim GmbH Cu-test. Total proteins were determined spectrophotometrically by the Biuret method.

RESULTS

The values of the physiological parameters for the four groups, normoxic control, hypoxic, acid, and hypoxic acid group, are given in Figures 1-5. From the figures it is clear that, compared to the normoxic control values, hypoxia, acidification, and acidification in hypoxic water cause several changes in the haemolymph of crayfish. The changes can be grouped into three categories: The effects of hypoxia and acidification on 1) the oxygen carrying capacity; 2) the haemplymph oxygen affinity (Fig. 4) and 3) the magnesium, calcium and lactate concentrations of the haemolymph (Fig. 5, text).

1.   Oxygen carrying capacity of the haemolymph

The oxygen capacity of haemolymph was significantly higher in the females than in the males in normoxia (Fig. 1). In hypoxia the oxygen capacity decreased in the females but not in the males. Hypoxia occurring together with acidification caused a further decrease in the O_2 capacity in the females, but not in the males, although a slight decreasing trend was apparent.

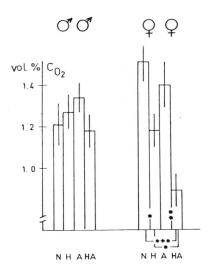

Figure 1. The haemolymph oxygen capacities (CO_2) of male and female
Astacus astacus in normoxic (N), hypoxic (H), acid (A) and hypoxic acid
(HA) water. The bar diagrams give the mean$\pm$ SEM, N = 8. The asterisks
indicate the statistical significance of the differences between the
groups (*** $P<0,001$, ** $P<0,01$, * $P<0,05$), the student's t-test
being used for comparisons.

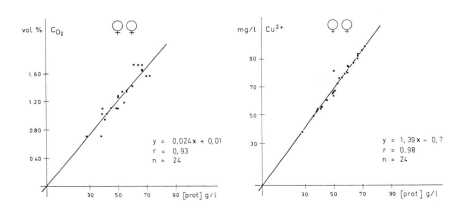

Figure 2. The correlations between the haemolymph oxygen capacities
(y-axis), Cu^{2+} (y-axis) and protein concentrations (x-axis) in female
crayfish. There was a very significant positive correlation between
the parameters (P 0.001).

89

In this study the haemocyanin concentration as such was not determined, but since we reached extremely good correlations in the present experiments between the oxygen capacities, protein and copper concentrations (Fig. 2), any of these three parameters gives a good idea about the changes in the haemocyanin concentration. Clearly, the changes in the haemolymph protein and copper concentrations (Fig. 3) were similar to those in oxygen capacity.

2. Haemolymph oxygen affinity

The oxygen affinity of haemolymph is inversely proportional to the P_{50} value. In hypoxia the oxygen affinity increased significantly in both the males and the females (Fig. 4). But in hypoxia in acid water the opposite took place in the females; their oxygen affinity decreased significantly. In males a similar change did not take place. It is notable that the acidification of water alone did not cause great changes in the haemolymph oxygen affinity of either sex.

3. Haemolymph magnesium, calcium and lactate concentrations

The changes in the Mg^{2+} and Ca^{2+} concentrations were similar in both sexes (Fig. 5). In hypoxia both the Mg^{2+}, and especially the Ca^{2+}, concentration increased. The changes were proportional to the changes in the oxygen affinity. The following equation describes the relation between the oxygen affinity and the Mg^{2+}, Ca^{2+} and lactate concentrations:

$$1/P_{50} = 0.042 \ (Ca^2 + Mg^{2+} + lactate) - 0.637$$
$$r = 0.774, \ n = 45$$

Again, hypoxia in acid water caused the opposite, a significant decrease in the Ca^{2+} and Mg^{2+} concentrations in females but not in males.

The changes in lactate concentrations were similar in both the males and the females. In hypoxia the lactate concentration increased significantly (P 0.01), from 0.074 ± 0.004 g/l ($\bar{x}\pm$ SEM, n=16) to 0.230 ± 0.057 g/l. In acid water the lactate concentration decreased slightly to 0.066 ± 0.004 g/l, and in hypoxic acid water to 0.057 ± 0.001 g/l, the latter change being significant (P 0.01) as compared to normoxia. The statistical testing of the lactate concentrations was carried out with Mann-Whitney's U-test as the lactate concentrations of the hypoxic group did not follow normal distribution.

Discussion

1. Normaxic situation; considerable differences between sexes

Andrews (1967) found only very limited sexual variation in the haemolymph of Orconectes limosus. In Astacus astacus differences in concentrations of haemolymph constituents are considerable (Järvenpää et al 1979). For this reason sexes were examined separately in this study. In normoxia the females had much higher oxygen capacities than the males, this giving the former much greater scope for activity, as there is a direct relationship between the haemocyanin concentration and the scope for activity, at least in the lobster, Homarus gammarus

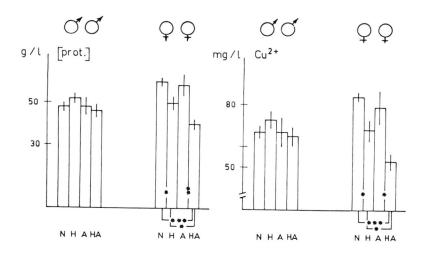

Figure 3. The protein and Cu^{2+} concentrations in the haemolymph of male and female crayfish in the different experimental groups. Legend as in Figure 1.

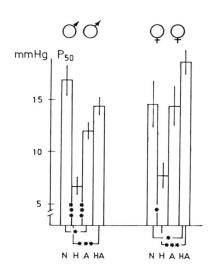

Figure 4. The P_{50} values (P_{50}) of male and female crayfish in the different experimental groups. Legend as in Figure 1.

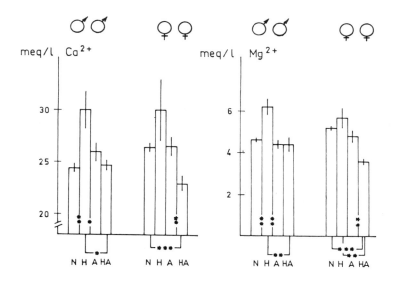

Figure 5. The Ca^{2+} and Mg^{2+} concentrations in the haemolymph of male and female crayfish in the different experimental groups. Legend as in Figure 1.

Table 1. The statistical differences between the means in some haemolymph parameters between sexes of the crayfish <u>Astacus astacus</u> (xxx $P<0.001$, xx $P<0.01$, x $P<0.05$, o $P<0.1$, - <u>no significant</u> difference). Student's t-test was used for comparisons, n = 8 in both sexes.

	CO_2	[prot]	Cu^2	P_{50}	Ca^2	Mg^{2+}
Normoxia	x	xxx	xxx	--	xx	xx
Hypoxia	--	--	--	--	--	--
Acid water	--	--	--	--	--	--
Hypoxia in acid water	x	--	x	x	o	--

(Spoek, 1974). Also, the haemolymph magnesium and calcium concentrations were higher in the females than in the males.

2. Effects of environmental changes on the sex differences in the
    respiratory pigments of haemolymph.

From Table 1 it is apparent that any of the tested environmental
changes make the sex differences smaller, as no significant differences
were seen between the sexes in either 30% O_2 saturation or pH 4.
Further, when hypoxia and acidification occurred simultaneously, the
females had signifiantly lower oxygen capacities and oxygen affinities,
thus having a smaller scope for activity. Therefore, it seems that the
effects of environmental changes are much more serious for females than
for males, and a greater effect on the survival of the females would be
expected.

3. Hypoxia acclimation

Astacus astacus is able to maintain its optimal O_2 intake down to
5 mg O_2/l concentration, and survive down to 2 mg/l concentration at
$+15^0$ C (Lindroth, 1950). Therefore the oxygen saturation of this study
(30%, 3.2 mg O_2/l) is within the so called zone of capacity adaptation,
in which meaningful physiological adaptations can be expected.

As stated in the introduction, the adaptation to a hypoxic
environment must contain a solution to the problem of how to maintain
an adequate oxygen uptake and transport in the face of diminished O_2
supply. One solution to this problem is to decrease the oxygen
consumption. However, this solution, taking place in several
crustaceans (McMahon & Wilkens, 1975; Taylor, 1976; Bridges & Brang,
1980), seems to be a resistance adaptation as it involves a switch to
anaerobic metabolism, and the resultant oxygen debt has to be paid back
(McMahon & Wilkens, 1975; Bridges & Brand, 1980). Another solution, an
increase in the respiratory water flow, occurs e.g. in Orconectes
virilis (McMahon et al., 1974), Homarus vulgaris (McMahon et al., 1978)
and Austropotamobius pallipes (Wheatly & Taylor, 1981). However, this
mechanism is in terms of energy quite costly. Indeed, in long term
experiments the ventilation volume returns towards the prehypoxic level
(McMahon et al., 1974, 1978), showing that a less costly adaptation
starts to function.

It is likely that the most economical solution from the energy
point of view, the increase in the oxygen affinity of haemolymph, is
the most important long term adaptation to hypoxia. In this study the
P_{50} value decreased from c. 16 mmHg in normoxia to 6-7 mmHg in hypoxia,
so the oxygen affinity doubled in a week's hypoxia. Also Wheatly &
Taylor (1981) have shown that the oxygen affinity increases in hypoxia
in Austropotamobius pallipes. The increase in oxygen affinity is of
major importance to the hypoxic animals, as in hypoxia (PO_2 30-60 mmHg)
80-90% of the oxygen delivered to the tissues is carried bound in the
haemocyanin (McMahon & Wilkens, 1975; Taylor, 1976).

In vitro studies with decapod haemocyanins have shown that
increases in the Mg^{2+}, Ca^{2+} (Truchot, 1975) and lactate (Truchot, 1980)
concentrations and pH (Angersbach & Decker, 1978) increase its oxygen
affinity. In hypoxia the pH of the haemolymph goes up (McMahon et al.,
1978; Nikinmaa et al., 1981) for two reasons: the initial hyperventila-

tion causes respiratory alkalosis (Wheatly & Taylor, 1981) and later a non-ventilatory increase in blood buffer base concentration increases the pH (Truchot, 1975a). The base is most likely CO_3^-, and it is possibly liberated from exoskeletal components, as simultaneously with the HCO_3^- increase Ca^{2+} concentration also increased (DeFur et al., 1980). $CaCO_3$ is the primary exoskeletal inorganic compound (Huner et al., 1978).

The increases in the lactate magnesium and calcium concentration also increase the haemocyanin oxygen affinity _in vivo_, as shown by the very significant correlation between these concentrations and $1/P_{50}$. Calcium and magnesium ions may be exoskeletal in origin, or calcium may be liberated from gastroliths or the hepatopancreas (Huner et al., 1978). The third possibility is that calcium is taken up from water. At least Carcinus maenas is very permeable to the calcium ion (Greenaway, 1976); the time constant for Ca^{2+} influx is only a half of that of Na^+ influx (Shaw, 1961).

The role of lactate in increasing the oxygen affinity of haemolymph may be twofold: it increases the oxygen affinity as such (Truchot, 1980), but lactic acid may also cause the dissolution of the exoskeleton and thus cause the liberation of Ca^{2+}, Mg^{2+} and HCO_3^- in the haemolymph. This function is fulfilled by succinic acid in the bivalve mollusc Mercenaria mercenaria (Crenshaw & Neff, 1969).

4. Lack of adaptation in hypoxic acid water

The problems the water breathing animals encounter in hypoxic waters are presented above. When in addition the water is acidified, there are several additional problems. Na^+ tends to be washed out from the animal (Nikinmaa et al., 1981) and the haemolymph pH tends to decrease (Dejours & Armand, 1980; Nikinmaa et al., 1981). The decrease of pH decreases the oxygen affinity of the haemocyanin. Further, the decrease in water pH inhibits the influx of Ca^{2+} (Malley, 1980); at the pH of this study (pH 4) the uptake of Ca^{2+} ceased altogether in Orconectes virilis. In Astacus astacus (Appelberg, 1979) and in Procambarus clarkii (Huner et al., 1978) the postmolt calcification is retarded by low pH. Also this effect on calcium metabolism is likely to have an effect on the respiratory properties of haemolymph.

Although acidification alone in this phase of the annual cycle had practically no effect on the haemolymph parameters determined in this study, the effects of acidification in hypoxic water were marked. Indeed, in hypoxic acid water the adaptations to hypoxia did not take place. On the contrary, the oxygen affinity of the haemolymph decreased significantly in the females due to the simultaneous decreases in the pH (Nikinmaa et al., 1982) and the calcium concentration. Simulataneous with the pH drop a marked washout of monovalent ions took place (Nikinmaa et al., 1982). Further, the oxygen capacity also decreased, especially in the females. Both these effects decrease the oxygen binding and transport to the tissues. None of these changes took place in acid water alone. This is thus an excellent example of how an animal can compensate for or adapt to one environmental stressor, but when two such stressors occur in concert a regulatory/compensatory breakdown is apparent. The apparent lack of adaptation in hypoxic acid water may lead to a decreased survival in longer term exposures, when still another environmental stressor exerts its influence. Also, the effect of all the changes in water quality examined in this study may be far more dramatic when occurring during a more sensitive period in the lifecycle of the crayfish.

ACKNOWLEDGMENT

This study was supported by a grant from the Academy of Finland, Research Council for the Natural Sciences.

LITERATURE CITED

Abrahamsson, S. 1970. Fecundity and growth of some populations of Astacus astacus Linné in Sweden. Rep. Inst. Freshw. Res. Drottningholm 52:23-37.

Andrews, P. 1967. Über den Blutchemismus des Flusskrebses Orconectes limosus und seine Veränderungen im Laufe des Jahres. Z. vergl. Physiol. 57:7-43.

Angersbach, D., H. Decker. 1978. Oxygen transport in crayfish blood: effect of thermal acclimation, and short-term fluctuations related to ventilation and cardiac performance. J. Comp. Physiol. 123, 105-112.

Appelberg, M. The effect of low pH on Astacus astacus. L. during moult. The second Scand. Symp. Freshwater Crayfish., Lammi, Finland 1979, manuscript.

Bridges, C.R., A.R. Brand. 1980. The effect of hypoxia on oxygen consumption and blood lactate levels of some marine crustacea. Comp. Biochem. Physiol. 65A, 399-409.

Bridges, C.R., J.E.P.W. Bicudo and G. Lykkeboe. 1979. Oxygen content measurement in blood containing haemocyanin. comp. biochem. Physiol. 62A:457-462.

Crenshaw, M.A., J.M. Neff. 1969. Decalcification at the mantle-shell interface in molluscs. Am. Zool. 9, 881-885.

Cukerzis, J. 1970. Placiaznplio vezio biologia (Astacus astacus L.). (Summary: The biology crayfish (Astacus astacus L.) 204 pp. Vilnus.

Cukerzis, J. 1973. Biologische Grundlagen der Methode der kunstlichen Aufzucht der Brut des Astacus astacus L. In: Abrahamsson, S. (ed.), Freshwater Crayfish 1:187-201. Lund.

DeFur, P.L., P.R.H. Wilkes, and B.R. McMahon. 1980. Non-equilibrium acid-base status in Carcinus productus: role of exoskeletal carbonate buffers. Respir. Physiol. 42, 247-261.

Dejours, P., and J. Armand. 1980. Haemolymph acid-base balance of the crayfish Astacus leptodactylus as a function of the oxygenation and the acid-base balance of the ambient water. Respir. Physiol. 41, 1-11.

Edwards, M.J., R.J. Martin. 1966. Mixing technique for the oxygen-hemoglobin equilibrium and Bohr effect. J. App. Physiol. 21, 1898-1902.

Erencin, Z. and G. Köksal. 1977. On the crayfish, Astacus leptodactylus, in Anatolia. In: Lindqvist O.V. (ed.), Freshwater Crayfish 3:187-192. Kuopio.

Greenaway, P. 1976. The regulation of haemolymph calcium concentration of the crab Carcinus maenas (L). J. Exp. Biol., 64, 149-157.

Hobbs, H.H. and E.T. Hall. 1974. Crayfishes (Decapoda: Astacidae). In: Hart, C.W. and S.L.H. Fuller (eds.), Pollution ecology of freshwater invertebrates: 195-214. New York.

Huner, J.V., J.G. Kowalczuk, and J.W. Avault. 1978. Postmolt calcification in subadult red swamp crayfish, Procambarus clarkii (Girard) (Decapoda, Camparidae). Crustaceana 34, 275-280.

Järvenpää, T., K. Westman, and A. Soivio. 1979. Sampling and analysing of the haemolymph of the freshwater crayfish Astacus astacus (L.). The Second Scand. Symp. Freshwater Crayfish., Lammi, Finland 1979, manuscript.

Kossakowski, J. 1973. The freshwater crayfish in Poland. In Abrahamsson, S. (ed.), Freshwater Crayfish 1:17-26. Lund.

Lahti, E. and O.V. Lindqvist. 1979. On the survival of the crayfish (Astacus astacus L.) in lakes with different oxygen regimes. The Second Scand. Symp. Freshwater Crayfish., Lammi, Finland 1979, manuscript.

Lindroth, A. 1950. Reactions of Crayfish on low oxygen pressure. Inst. Freshw. Res. Drottningholm Rep. 31, 110-112.

Malley, D.F. 1980. Decreased survival and calcium uptake by the crayfish Orconectes virilis in low pH. Can. J. Fish. Aquat. Sci., 37, 364-372.

McMahon, B.R, J.L. Wilkens. 1975. Respiratory and circulatory responses to hypoxia in the lobster Homarus americanus. J. Exp. Biol. 62, 637-655.

McMahon, B.R., W.W. Burgren, J.L. Wiklens. 1974. Respiratory responses to long-term hypoxic stress in the crayfish Orconectes virilis. J. Exp. Biol. 60, 195-206.

McMahon, B.R., P.J. Bulter, E.W. Taylor. 1978. Acid base changes during recovery from disturbance and during long-term hypoxic exposure in the lobster Homarus vulgaris. J. Exp. Zool. 205, 361-370.

Nikinmaa, M., T. Järvenpää, K. Westman, A. Soivio. 1981. Acid base balance of Astacus astacus haemolymph in hypoxic and acid waters. Submitted for publication.

96

Shaw, J. 1961. Studies on ionic regulation in Carcinus maenas (L.) I. Sodium Balance. J. Exp. Biol. 38, 135-152.

Spoele, G.L. 1974. The relationship between blood haemocyanin level, oxygen uptake, and the heart-beat and scaphognathite-beat frequences in the lobster Homarus gammarus. Neth. 7. Sea Res. 8, 1-26.

Taylor, A.C. 1976. The respiratory responses of Carcinus maenas to declining oxygen tension. J. Exp. Biol. 65, 309-322.

Truchot, J.P. 1975a. Factors controlling the in vitro and in vivo oxygen affinity of the haemocyanin in the crab Carcinus maenas (L.). Respir. Physiol. 24, 173-189.

Truchot, J.P. 1975b. Changements de le´etat acide-base du sang en fonction de l´oxygenation de l´eau chez le crabe, Carcinus maenas (L.). J. Physiol. 70, 583-592.

Truchot, J.P. 1980. Lactate increases the oxygen affinity of crab hemocyanin. J. Exp. Zool. 214, 205-208.

Vallin, S. 1964. Kräftan, Potamobius astacus (Linné). In: Andersson, K.A. (ed.), Fiskar och fiske i Norden: 505-512. Stockholm

Wheatly, M.G., and E.W. Taylor. 1981. The effect of progressive hypoxia on heart rate, ventilation, respiratory gas exchange and acid-base status in the crayfish Austropotamobius pallipes. J. Exp. Biol. 92, 125-142.

Westman, K. 1973. The population of the crayfish Astacus astacus L. in Finland and the introduction of the american crayfish Pacifastacus leniusculsu Dana. In: Abrahamsson, S. (ed.), Freshwater Crayfish 1:41-55. Lund.

RESPONSE OF THE CRAYFISH <u>ORCONECTES VIRILIS</u> TO EXPERIMENTAL ACIDIFICATION OF A LAKE WITH SPECIAL REFERENCE TO THE IMPORTANCE OF CALCIUM.[1]

Robert L. France

Freshwater Institute/University of Manitoba
Department of Fisheries and Oceans
501 University Crescent
Winnipeg, Manitoba Canada R3T 2N6

ABSTRACT

Life history characteristics of the crayfish <u>Orconectes virilis</u> were examined during 1979-81 in four small Canadian Shield basins in the Experimental Lakes Area, northwestern Ontario. One of these lakes, Lake 223, has been undergoing experimental acidification since 1976 to simulate the effects of acid precipitation. Carapace rigidity and Ca^{++} content were significantly lower in L223 crayfish, possibly as a result of postmolt inhibition of Ca^{++} uptake. Growth in L223 has not been affected by acidification to pH 5.35. Incomplete hardening of the cuticular glair-cement compound forming the egg capsule membrane and stalk has resulted in a failure of secure pleopod egg attachment causing the L223 population to suffer severe recruitment failure. Lake 223 crayfish have responded to acidification-related remobilization of heavy metals with an increased bioaccumulation of Mn and Hg. <u>Thelohania</u> sp. infection in the L223 population was 1.7% during 1979 compared to 0.3% for the control lakes. In 1980 this prevalence had risen to 6.5% in 1980 this prevalence had risen to 6.5% in L223 with no increases in the control populations. It is suggested that loss of crayfish populations to gradual lake acidification will be brought about by reproductive impairment and possibly increased susceptibility to parasitic infection before those acid levels are reached that result in direct toxic damage to crayfish stocks. Present findings are summarized and incorporated with a literature discussion forming a thesis on the critical importance of the often insidious interaction between H^+ and environmental Ca^{++} in determining the future of crayfish populations exposed to cultural acidification.

INTRODUCTION

Anthropogenic emissions of SO_2 from fossil fuel burning and metal smelting have led to increases in the acidity of preciptiation in Scandinavia (Oden 1976), northeastern United States (Cogbill and Likens 1974) and eastern Canada (Dillon et al. 1978). Consequently, this has led to the acidification of many watercourses, often causing irreversible damage to their resident biota (Hendrey et al. 1976; Leivestad et al. 1976). Numerous studies have been performed concerned with the ecological effects of acidification upon fish populations

---

[1]for more detailed presentation of methodology, results, and discussion refer to France, R.L. in prep a. Life history of the crayfish <u>Orconectes</u> <u>virilis</u> in the Experimental Lakes Area, with special reference to the experimental acidification of Lake 223. (To be submitted to Can. J. Fish. Aquat. Sci.).

(reviewed in Spry et al. 1981), but little work has been undertaken to investigate the field response of invertebrate populations to acid pollution.

The continued acidification of softwater Scandinavian and North American lakes and rivers poses a serious threat to the long-term survival of crayfish populations. Already, there are numerous citing of Astacus astacus populations that have suffered from increasing water acidity over recent years (Abrahamsson 1972; Hultberg 1976; Almer et al. 1978; Furst 1977a; Appelberg 1980). Despite a considerable amount of money and effort devoted to restocking crayfish, there have been reports of unsuccessful attempts due to lowered pH conditions (Furst 1977b). In one case (Abrahamsson 1972) the cessation of liming by fishery biologists to an inflowing watercourse during a single month period resulted in a 96% mortality of the lake crayfish population. Svardson's (1974) survey of crayfish distribution related to the pH of Swedish waters (Fig. 1) suggests that crayfish may serve as an important indicator of the early stages of lake acidification. Furst (1977a) experimentally found a limit of pH 5.6 for the successful repeoduction of A. astacus. Malley (1980) and Appelberg (1980) determined that depressed pH can exert sublethal effects upon crayfish molting and calcification. It is not know if the results of these studies can be extrapolated to natural populations.

Orconectes virilis is the most widely distributed and abundant crayfish species in Canada (Crocker and Barr 1968). Its range extends from north- central Alberta (Aiken 1968) throughout the geologically acid-sensitive Precambrian Shield including the Haliburton-Muskoka region of south-central Ontario (Berrill 1977), perhaps the most heavily acidified area of the globe (Dr. D.W. Schindler, Freshwater Institute, pers. comm.), and to the northeastern United States where it is often the dominant crayfish species (Aiken 1965). O. virilis is of commercial importance and supports an extensive fishery for bait and biological specimens (Crocker an Barr 1968). The commercial harvest of O. virilis has been undertaken in northern Wisconsin for decades (Threinen 1958) and recently the economic feasibility of opening a new fishery in Vermont has been investigated (Nolfi and Miltner 1978). Further, the importance of this crayfish species in maintaining an efficient energy flow, especially within those low productivity glacial lakes most vulnerable to acid precipitation (Momot et al. 1978), suggests that its disappearance could generate serious disturbances within lake ecosystem dynamics.

The purpose and scope of the present study were to investigate the life history and ecophysiological response of population of O. virilis to the experimental acidification of its lake habitat. The result of this work in conjunction with laboratory studies of physiological mechanisms (Malley 1980; unpublished data), life stage tolerances (France in prep b), behavorial modifications and acclimatization abilities (France in prep c), and consecutive whole-lake population estimates (Davies in prep) is hoped to accurately describe the reaction of this crayfish species to acidification. Present findings are incorporated with a literature discussion forming a dialectic thesis on the critical importance of enviromental calcium in determining the future of crayfish populations exposed to cultural acidification.

STUDY AREA

The Experimental Lakes Area (ELA) is a headwater region containing small glacially-formed basins in the Canadian Precambrian Shield at 93°30'-94°00'W and 49°30'-49°45'N, approximately 300 km east of Winnipeg, Manitoba. Detailed physical, chemical and biological charateristics of the ELA region can be found in the three "ELA volumes": J. Fish. Res. Bd. Canada 28:121-304, 1971; J. Fish. Res. Bd. Canada 30:1409-1552, 1973; and Can. J. Fish. Aquat. Sci. 37:311-559, 1980.

Lake 223 (L223) has been undergoing experimental acidification with sulfuric acid since 1976 to simulate the effects of acid preciptiation. As a result, epilimnion pH has decreased at a rate of approximately 0.25 pH units per annum; from 6.7-7.0 in 1976, to 6.0-6.2 in 1977, 5.7-5.9 in 1978, 5.5-5.7 in 1979, 5.2-5.4 in 1980, and to 4.9-5.1 in 1981. Methods of acid addition and detailed chemical and biological results are presented in Schindler et al. (1980), Schindler (1980), and Schindler and Turner (1982).

RESULTS AND DISCUSSION

Of all the inorganic ions contributing to physiological maintenance in crustacean biology, Ca^{++} is of leading importance (Robertson 1941). Because of this, the distribution of crayfish is to some degree controlled by the level of Ca^{++} in potentially habitable waters (Robertson 1941; Couegnas in Macan 1961; Rhoades 1962; Greenway 1974; Capelli 1975). The lakes at ELA average 1.6 mg 1^{-1} Ca^{++}, among the lowest levels for this ion recorded anywhere in the world (Armstrong and Schindler 1971). The species O. virilis at ELA is existing close to the lower environmental limits for Ca^{++} for crayfish and relatively minor affects on Ca^{++} metabolism may therefore be critical.

CARAPACE RIGIDITY

Mechanisms of Ca^{++} metabolism throughout the crayfish molt cycle have been described by McWhinnie (1962) and Greenway (1974). Because the exoskeleton has to be recalcified and hardened as quickly as possible following every molting event, the Ca^{++} requirments of aquatic crustacea are high. Postmolt crayfish depend largely upon the uptake of Ca^{++} from the aquatic environment to calcify the new exoskeleton as the amount stored in the gastroliths is of insufficient quantity (Adegboye et al. 1974). Fishery management studies have shown that the proportion of soft exoskeletons is much higher in softwater experimental pools, due to differences in the magnesium-calcium salt content of the exoskeletons (Bretonne et al. 1969). It has long been known in crayfish aquaculture practice that crayfish in acid waters tend to have thinner shells (LeCaze 1970). Borgstrom and Hendrey (1976) hypothesied that the effect of low pH on aquatic crustaceans may be to interfere with the Ca^{++} uptake mechanisms following molting. Malley (1980), using ELA O. virilis, found that Ca^{++} uptake was in fact inhibited below pH 5.75 with a slower progression of molt cycle stages and of exoskeleton calcification at pH 5.0. It was thought, Malley (1980), that the effect of low pH on blocking O. virilis postmolt Ca^{++} uptake may operate through (1) the scarcity of HCO_3 to accompany Ca^{++} for electrical neutrality, (2) interference with the exchange of internal H^+ for external Ca^{++} due to elevated concentration of H^+ in the ambient

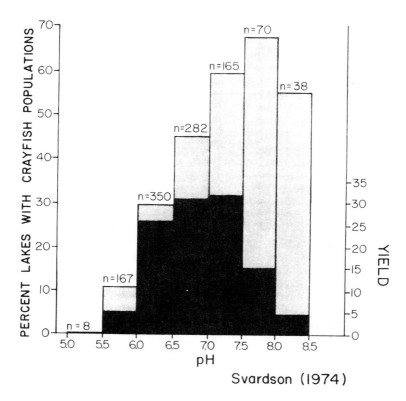

Svardson (1974)

Figure 1.   Distribution of <u>Astacus astacus</u> in 1,080 Swedish lakes.
Yield (dark shading) represents number of crayfish per hectare from
survey of 113 lakes.   Data from Svardson, G. 1974.   Oversikt av
Laboratoriets verksamhet med plan for ar 1974.   Information fran
Sotvattenslaboratoriet (1). 27 p.

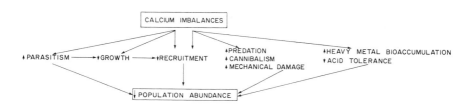

Figure 2.   Conceptual interrelationships between the life history
variables influencing crayfish population abundance in lakes subjected
to acidification.

environment, or (3) the direct effect of H^+ on the active transport system such as by altering molecular configuration. McMahon and Morgan (this conference) suggested that a dissolution of calcium carbonate from the exoskeleton to provide the hemolymph with more bicarbonate buffer could also result in soft-shelled crayfish at low pH levels. Appelberg (1980) found that the calcifying of A. astacus carapaces was delayed at low pH levels resulting in a significant decrease in carapace dryweight and Ca^{++} content in those crayfish molting in laboratory acid water.

Subjective estimates of carapace rigidity imply that the L223 population is undergoing problems in postmolt exoskeleton calcification and/or carapace breakdown-resorption. Values of carapace rigidity in turn were strongly correlated with carapace dryweight and Ca^{++} content. Carapace Ca^{++} content was significantly lower in L223 crayfish (mean % dryweight+ 95% C.I. of 13.90+0.89) than in animals from three control lakes (19.82+0.55; 20.34+1.04; 22.18+0.84).

Delay or absolute limitation in successful completion of calcification, thereby producing prolonged periods of soft exoskeletons, would make populations more vulnerable to mechanical damage and cannibalism, both of which can be a significant factor limiting crayfish distribution and abundance in softwater environments (Abrahamsson and Goldman 1970; Bretonne et al., 1969). Although small crayfish may be predated on by fish at any molt cycle stage, larger individuals are taken only after ecdysis during the period when exoskeletons are still soft (Scott and Duncan 1967), suggesting that it is likely an indirect effect of acidification on crayfish populations will be an increased predation pressure. Incomplete hardening of the form I male copulatory appendages may eventually interfere with mating by leaving the males impotent. Finally, Ca^{++} uptake imbalances may delay, interfere, or even interrupt the molting cycle itself such that individual growth will be retarded and onset of sexual maturity stayed.

GROWTH

Crayfish population abundance in oligotrophic ELA lakes is regulated by the density-independent environmental control of growth, operating through reproduction (France in prep d). The effects of acidification upon growth can therefore be expected to be of profound importance. A reduction in growth rate of the crayfish in the control L239 population by only 10% would result in a 17% decrease in popultion size; further growth decreases to 30% of the initial value produces a population of only half the original size. Laboratory studies have shown that at low pH the progression of crustacean molt cycle stages and length of the intermolt period may be changed (Borgstrom and Hendrey 1976; Malley 1980; Appelberg 1980).

Despite severe molt-related problems with exoskeleton calcification and hardening, and species specific alterations in abundance and diversity of food resources, the timing of molting events and growth of O. virilis in L223 has not been affected by acidification to pH 5.35. During 1979 the growth rate of the L223 population is what one would expect it to have been on the basis of its relationship to lake phytoplankton productivity. Continued acidification from the mean epilimnion pH of 5.60 in 1979 to a value of 5.35 in 1980 did not change the ordering of growth rates among the lakes. Further, the better climatic growing conditions during 1980 actually served to increase growth in L223 corresponding to similar increases observed in the other study lakes. Acidification related decreases in crayfish abundance will therefore be brought about by other processes long before any alteration in growth dynamics will pose a significant threat.

REPRODUCTION

Depressed environmental pH, mediated through Ca^{++} disturbances, can potentially affect crayfish reproduction in two major ways. Beamish et al. (1975) proposed that a long-term failure of normal Ca^{++} metabolism may be the primary cause for the observed extinction of fish populations in several Ontario lakes. The failure of mature female fish inhabiting an acid lake to spawn and release their ova was coincident with unelevated levels of maternal serum Ca^{++} which should normally occur. Stephens (1952) has shown that the eggs of O. virilis may be easily resorbed under experimental conditions with a resulting mobilization of substances that are released into the blood. Atresia has been demonstrated to occur in natural populations of crayfish subjected to stressful conditions (Morrissy 1975; Vey 1977). One might expect that an organism such as crayfish would be more susceptible to egg resorption and Ca^{++} retention in acid water than fish, due to the increased prevalence of this element in general crustacean biology. Despite this, there is litte indication of more pronounced ovarian egg resorption in L223 crayfish at a lake mean pH of 5.35 compared to crayfish in control lakes. Atresia therefore seems to be an unimportant factor influencing crayfish recruitment in acid environments relative to other more serious reproductive disturbances.

Oviposition in crayfish has been closely studied (Andrews 1904, 1906; Mason 1970). The eggs are laid into an abdominal pouch containing a mucus secretion from the cement glands referred to as glair. Following a period of special coordinated activity by the female and the subsequent hardening of this surrounding glair compound, the eggs become firmly cemented to the pleopods. All studies report little egg loss during this period.

Failure to secure attachment of pleopod eggs in L223 females (measured as the mean percentage +95% C.I. of the total brood easily removed of 54.1+5.9, significantly higher than values of 27.8+6.4 and 30.5+7.0% for control crayfish) brought about by incomplete hardening of the glair-cement, has resulted in a recruitment failure in the L223 population. The percentage of females with only a partial egg compliment in L223 was 19.0% in 1979, 42.2% in 1980, and 47.7% in 1981 in contrast to levels of approximately 5% over the three years in the control lakes. The percentage of L223 females with dead eggs ranged from 6.9% to 16.4% during the study period compared to a complete absence of this condition in the control populations. On an egg production basis, the percent reproductive failure in the three control populations averaged 3.20+1.76 (95% C.I.) while levels of 18.67, 36.24, and 29.37% were obtained during 1979-81 in L223.

Cano in Herrick (1896) was the first to investigate the gradual hardening of decapod egg cement: "fixation could not be explained without the interaction of the water .... it would seem that the water might explain the chemical change which the cement undergoes, a change analogous to that which is observed in the exoskeleton after the molt."

Yonge (1937) found that the outer egg membrane and binding cement have properties indistinguishable from those of the cuticle. Yonge (1946) further stated: "this élasticity (of the outer membrane) is therefore probably a property of the cuticular constituent of the integument also. In the case of the eggs this has biological significance. The outer membrane attaches the eggs to the pleopods in the decapod crustacea, but it also protects them and has to withstand the pressure of adjacent eggs and the effect of the constant beating of the pleopods needed to produce the respiratory current around the developing mass of eggs. In this connection its elasticity as well as its firm consistency will be of real value. Moreover, during development the egg increases in size, becoming oval with the long axis 50% greater than the original diameter .... Certainly the outer membrane could not remain around the developing egg securing it to the pleopods and protecting it were it not capable of stretching greatly." The demonstrated ease in which L223 eggs can be removed is a result of an absence of this crucial property of elasticity caused by incomplete cuticular hardening. Yonge (1936) investigated the hardening mechanism of decapod cuticle and found it to be sensitive to acids. The cuticle reaches its lowest permeability at pH 5.2, its iso-electric point. Malley (1980) has shown that the uptake of C^{++} by postmolt _O. virilis_ was inhibited below pH 5.75 resulting in incomplete exoskeleton hardening. It is probable that a somewhat similar process involved with cuticular phenolic hardening and calcification has prevented a secure egg attachment in L223 crayfish. _A. astacus_ ovigerous females from acidified Swedish waters have been found to show a similar condition (Magnus Appelberg, pers. comm). The present work then supports the tentative hypothesis put forward by Furst (1977a) that it will be the sensitivity of reproduction to pH, particularly during the egg-laying and brooding stages, that will determine the long-term survival of crayfish populations to lake acidification.

HEAVY METAL UPTAKE

As a result of the whole-lake acidification of L223 several increases have been observed in the volume-weighted average concentrations of heavy metals due to sediment remobilization (Schindler et al. 1980; Schindler and Turner, in press); the most notable effect being a steady increase in the Mn concentration from $12.5 \ \mu g \cdot l^{-1}$ in 1976 to $127.0 \ \mu g \cdot l^{-1}$ in 1980. Crayfish have responded with an increased bioaccumulation of Mn (L223 value of $240 \ \mu g \cdot g^{-1}$ dry wt compared to a mean for the control lakes of $61 \ \mu g \cdot g^{-1}$ wt) and Hg (L223 value of $0.52 \ \mu g \cdot g^{-1}$ dry wt compared to control mean of $0.26 \ \mu g \cdot g^{-1}$ dry wt).

Calcium can play a modifying role in the uptake of heavy metals. Franzin and McFarlane (1980) and McFarlane and Franzin (1980) found that discrepancies in bioaccumulation for fish and macrophytes could be explained on the basis of the lake Ca^{++} concentration. Laboratory studies have shown the Ca^{++} provides an ameliorative protection against the uptake and toxicity of heavy metals in fish (Kinkade and Erdman 1975; Zitko and Carson 1976) and invertebrates (Bryan 1967; Wright 1980). It is though that processes may involve (1) a competition for active cellular binding sites between the heavy metals and Ca^{++}, (2) a complexing of Ca ions to metals thereby inhibiting their uptake, or (3) an interference in the "accidental" metal uptake via a Ca^{++} regulatory mechanism. The range of Ca^{++} concentrations in the ELA study lakes

$(1.74-2.39 \text{ mg} \cdot 1^{-1})$ is too narrow to explain differences in metal uptake, indicating that elevated tissue levels of Mn and Hg in L223 were caused by acidification, not differences in water quality.

PARASITISM

Thelohania sp. parasitism in the L223 population was 1.7% during 1979 and increased to 6.5% (range 4.3-13.3% for individual collections) during 1980 with a correspondingly constant prevalence of <1% in the three control lakes.

The higher prevalence of Thelohania sp. parasitism in the L223 population leads itself to a number of interesting hypotheses, the most simple of which is, if as was previously believed (cf. France et al. in prep.), the disease is spread via cannibalism, then it is possible that the general reduction in exoskeleton rigidity observed in L223 has allowed for an abnormally high degree of cannibalism to take place. Bretonne et al. (1969) found a higher mortality rate in softwater ponds than in those with higher water hardness, suggestive of an elevated rate of cannibalism among the softer shelled crayfish.

Little is known about how different environmental changes may enhance this particular infection in the host, i.e. are crayfish defense reactions lowered under acid stress. Crayfish are susceptible to a wide variety of parasite species (Johnson 1975; Unestam 1973) and display a multi-level system of defense reactions of which the outer cuticular layer is the first point of resistance. Ambroski et al. (1974) have stated that much of the apparent resistance to bacterial infection reflects the inability of these organisms to penetrate the inert cuticle. Bacterial problems therefore develop only in localized areas of the softer parts of the cuticle. Continuous mechnaical erosion resulting in exposure of the calcified endocuticle greatly increases the susceptibility of crayfish to bacterial infection (Ambroski et al. 1974). Hyphal growth of the fungus Aphamnomyces astaci, known as the "crayfish plague," is also restricted to only the soft, non-calcified parts of the cuticle, the other areas being impermeable perhaps due to inhibitory compound(s) present on the cuticle surface (Unestam 1974; Unestam et al. 1977). The fungus penetrates the exoskeleton by utilizing a combination of mechanical pressure and cuticle dissolving extracellular enzymes. Crayfish resistance to the plague may be substantially lowered through mechanical damage to the carapace of loss of appendages (Persson and Soderhall; Vey et al. this conference) both of which have occurred more frequently in the L223 population. The markedly thinner exoskeletons of crayfish inhabiting acidified waters could therefore act synergistically by increasing the probability of successful bacterial or fungal entry. The exposure of Scandinavian crayfish populations already suffering from crayfish plague and a variety of other parasites of which only one in Thelohania contejeani, to acid pollution could have dramatic and far reaching consequences on the etiology of such diseases.

SUMMARY

Crustaceans are among the most sensitive organisms to lake acidification an a full understanding of complex relationships between their biology and depressed pH is therefore of paramount importance. Caution must be applied, though, by future researchers to avoid simply considering the effects of acid precipitation upon crayfish solely in terms of hydrogen ion concentration in isolation from the influence of other external modifying chemical factors. The indirect sublethal interaction of H^+ with a single element - Ca^{++}, often operating in an insidious fashion, may strongly influence the eventual success or extinction of crayfish, and perhaps other similar crustaceans, subjected to acid stress (Fig. 2). This study suggests that the loss of crayfish populations to gradual lake acidification will be brought about by reproductive impairment and possibly increased susceptibility to parasitic infection before those acid levels are reached that result in direct toxic damange to crayfish stocks.

ACKNOWLEDGMENTS

P. Olesiuk provided assistance during several crayfish Scuba dives. Calcium determinations and heavy metal tissue analyses were performed by the Freshwater Institute chemistry and toxicology units respectively. Microporidian parasite was identified by L. Graham, University of Manitoba. D.W. Schindler, I.J. Davies and D. Malley provided helpful suggestions on manuscript preparation.

LITERATURE CITED

Abrahamsson, S. 1972. Fecundity and growth of some populations of Astacus astacus Linne in Sweden. Rep. Inst. Freshwater Res. Drottningholm 52:23-37.

Abrahamsson, S.A.A., and C.R. Goldman. 1970. Distribution, density and production of the crayfish Pacifastacus leniusculus Dana in Lake Tahoe, California-Nevada. Oikos 21:83-91.

Adegboye, D., I.R. Hagadorn, and P.F. Hirsch. 1974. Variations in hemolymph calcium associated with the moulting cycle in the crayfish. In Proc. 2nd Int. Symp. Freshwater Crayfish. p. 227-247.

Aiken, D.E. 1965. Distribution and ecology of three species of crayfish from New Hampshire. Amer. Midl. Natur. 73:240-244.

Aiken, D.E. 1968. Further extension of the known range of the crayfish Orconectes uirilis (Hagen). Natl. Museum Can. Bull. No. 223. Contrib. Zool. 4:43-48.

Almer, B., W. Dickson, C. Ekstrom, and E. Horstrom. 1978. Sulfur pollution and the aquatic ecosystem, Ch. 7. In J. Nriagu (ed.) Sulfur in the Environment, part 2., John Wiley and Sons, N.Y. 464 p.

Amborski, R.L., G. Lopiccolo, G.F. Amborski, and J. Huner. 1974. A disease affecting the shell and soft tissues of Louisiana crayfish Procambarus clarkii. In Proc. 2nd Int. Symp. Freshwater Crayfish. p. 299-316.

Andrews, E.A. 1904. Breeding habits of crayfish. Amer. Natur. 38:165-206.

Andrews, E.A. 1906. Egg-laying of crayfish. Amer. Natur. 40:348-356.

Appelberg, M. 1980. The effect of low pH on *Astacus astacus* L. during moult. Scandinavian crayfish symposium 1979.

Armstrong, F.A.J., and D.W. Schindler. 1971. Preliminary chemical characterization of waters in the Experimental Lakes Area, northwestern Ontario. J. Fish. Res. Bd. Canada. 28:171-187.

Beamish, R.J., W.L. Lockhart, J.C. Vanloon, and H.H. Harvey. 1975. Long-term acidification of a lake and the resulting effects on Fishes. Ambio. 4:98-102.

Berrill, M. 1978. Distribution and ecology of crayfish in the Kawartha Lakes region of southern Ontario. Can. J. Zool. 56:166-177.

Borgstrom, R., and G.R. Hendrey. 1976. pH tolerance of the first larval stages of *Lepidurus arcticus* (Pallas) and adult *Gammarus lacustrus* G.O. Sars. Internal. Rep. Norw. Inst. Water Res. Oslo, Norway. 37 p.

Bretonne, Jr. de la, L., J.W. Avault, Jr., and R.O. Smitherman. 1969. Effects of soil and water hardness on survival and growth of the red swamp crawfish *Procambarus clarkii* in plastic pools. Proc. 23rd. Conf. SEast. Ass. Game Fish Comm. 23:629-633.

Bryan, G.W. 1967. Zinc regulation in the Freshwater crayfish (including some comparative copper analysis) J. Exp. Biol. 46:281-296.

Capelli, G.M. 1975. Distribution, life history, and ecology of crayfish in northern Wisconsin, with emphasis on *Orconectes propinquus* (Girard). Ph.D. Thesis, Univ. of Wisconsin. 220 p.

Cogbill, G.V., and G.E. Likens. 1974. Acid precipitation in the northeastern United States. Wat. Resour. Res. 10:1133-1137.

Crocker, D.W., and D.W. Barr. 1968. Handbook of crayfishes of Ontario. Univ. of Toronto Press.

Davies, I.J. In prep. Effects of an experimental whole-lake acidification on a population of the crayfish *Orconectes virilis* (Decapoda).

Dillon, P.J., D.G. Jeffries, W. Synder, R. Reid, N.D. Yan, D. Evans, J. Moss, and W.A. Scheider. 19768. Acidic precipitation in south-central Ontario: recent observations. J. Fish. Res. Bd. Canada. 35:809-815

France, R.L. In prep a. Life history of the crayfish *Orconectes virilis* in the Experimental Lakes Area, with special reference to the experimental acidification of Lake 223.

107

France, R.L. In prep b. Experimental response of the crayfish *Orconectes virilis* to low pH. I. Life Cycle. (To be submitted to Aquat. Toxicol.)

France, R.L. In prep c. Experimental response of the crayfish *Orconectes virilis* to low pH. II. Acclimatization and avoidance. (To be submitted to Aquat. Toxicol.)

France, R.L. In prep d. Growth of the crayfish *Orconectis virilis* in small oligotrophic Canadian Shield Lakes in the Experimental Lakes Area, northwestern Ontario. (To be submitted to Can. J. Fish. Aquat. Sci.).

France, R.L., L. Graham, and T. Wiens. In prep. Microsporidian *Thelohania* sp. parasitism of the crayfish *Orconectes virilis* in Canada. (To be submitted to Con. J. Fish. Aquat. Sci.)

Franzin, W.G., and G.A. McFarlan. 1980. An analysis of the aquatic macrophyte, *Myriophyllum exalbescens*, as an indicator of metal contamination of aquatic exosystems near a base metal smelter. Bull. Environ. Contam. Toxicol. 24:597-605.

Furst, M. 1977a. Forsurninings inverkan pa Flodkraften *Astacus astacus*. In Nordiskt kraftsymposium 1977. ed. M. Furst. Information Fran Sotvattenslaboratoriet, Drottningholm 14:90-94.

Furst, M. 1977b. Introduction of *Pacifastacus leniusculus* (Dana) into Sweden: methods, results and management. In Proc. 3rd Int. Symp. Freshwater Crayfish. p. 229-247.

Greenway, P. 1974. Calcium balance at the postmolt stage of the Frestwater Crayfish *Austropotamobius pallipes* (Lereboullet). J. Exp. Biol. 61:35-45.

Hendrey, G.R., K. Baalstrud, T.S. Tragen, M. Loake, and G. Raddum. 1976. Acid precipitation; some hydrological changes. Ambio. 5:224-227.

Herrick, F.H. 1896. The american lobster: a study of its habits and development. Bull. U.S. Fish. Comm. 15:1-252.

Hultberg, H. 1976. Thermally stratified acid water in late winter -- a key factor inducing self-accelerating processes which increase acidification. Water, Air and Soil Pollut. 7:279-294.

Johnson, S.K. 1975. Crawfish and freshwater shrimp diseases. Texas A. and M. University, Texas Agricultural extension service. 18 p.

Kinkade, M.L., and H.E. Erdman. 1975. The influence of hardness components (Ca^{2+} and Mg^{2+}) in water on the uptake and concentration of cadmium in a simulated freshwater ecosystem. Environ. Res. 10:308-313.

LeCaze, C. 1970. Crawfish farming. Bull. Louisiana Wildlife and Fisheries Comm. No. 7.

Leivestad, H., G. Hendrey, I.P. Muniz, and E. Snekvik. 1976. Effects of acid precipitation on Freshwater organisms. p. 87-111. In F. Braekke (ed.) Impact of acid precipitation on forest and freshwater ecosystems in Norway. Sur Nedbors Virkning Pa Skag og Fisk Res. Rep. 6/76.

Macan, T.T. 1961. Factors that limit the range of freshwater animals. Biol. Rev. 36:151-198.

Malley, D.F. 1980. Decreased survival and calcium uptake by the crayfish Orconectes virilis in low pH. Can. J. Fish. Aquat. Sci. 37:364-372.

Mason, J.C. 1970. Egg-laying in the western North American crayfish, Pacifastacus trowbridgii (Stimpson) (Decapoda, Astacidae). Crustaceana 19:37-44.

McFarlane, G.A., and W.G. Franzin. 1980. An examination of Cd, Cu, and Hg concentrations in livers of northern pike, Esox lucius and white suckers, Castostomus commersoni, from five lakes near a base metal smelter at Flin Flan, Manitoba. Can. J. Fish. Aquat. Sci. 37:1573-1578.

McMahon, B.R. and D.O. Morgan. (this conference). Physiological responses to sublethal acid exposure in crayfish.

McWhinnie, M.A. 1962. Gastrolith growth and cacium shifts in the freshwater crayfish, Orconectes virilis. Comp. Biochem. Physiol. 7:1-14.

Momot, W.T., H. Gowing, and P.D. Jones. 1978. The dynamics of crayfish and their role in ecosystems. Amer. Midl. Natur. 99:10-35.

Morrissy, N.M. 1975. Spawning variation and its relationship to growth rate and density in the marron Cherax tenuimanus (Smith). Fish Res. Bull. West. Aust. 16:1-32.

Nolfi, J.R., and M. Miltner. 1978. Preliminary studic on a potential crayfish fishery in Vermont. Proc. 4th Int. Symp. Freshwater Crayfish p. 312-322.

Oden, S. 1976. The acidity problem -- an outline of concepts. Water, Air and Soil Pollut. 6:137-166.

Persson, M. and K. Soderhall. (This conference). Pacifastacus leniusculus and its resistance to the crayfish plague parasite, Aphanomyces astaci.

Rhoades, R. 1962. Further studies on Ohio crayfishes. Cases of sympatry of stream species in southern Ohio. Ohio J. Sci. 62:27-33.

Robertson, J.D. 1941. The function and metabolism of calcium in the invertebrata. Biol. Rev. 16:106-133.

Schindler, D.W. 1980. Experimental acidification of a whole-lake: a test of the oligotrphication hypothesis. Proc. Int. conf. ecol. impact acid precip., Norway 1980, SNSF project.

Schindler, D.W., R. Wageman, R.B. Cook, T. Ruszczynski, and J. Prokopowich. 1980. Experimental acidification of Lake 223, Experimental Lakes Area: background data and the first three years of acidifaction. Can. J. Fish. Aquat. Sci. 37:342-354.

Schindler, D.W. and M.A. Turner. In press. Phyiscal, chemical and biological responses of lakes to experimental acidification.

Scott, D., and K.W. Duncan. 1967. The function of freshwater crayfish gastroliths and their occurrence in perch, trout, and shag stomachs. N.Z.J. mar. Freshwat. Res. 2:99-104.

Spry, I.JK., C.M. Wood, and P.V. Hodson. 1981. A literature review: the effects of environmental acidification on fishes with reference to heavy metals. Can. J. Fish. Aquat. Sci. Tech. Rep. in press.

Stephens, G.J. 1952. Mechanisms regulating the reproductive cycle in the crayfish Cambarus. 1. The female cycle. Physiol. Zool. 25:70-83.

Svardson, G. 1974. Oversikt av Laboratoriets verksamhett med plan ar 1974. Information from Sotvattenslaboratoriet (1). 27 p.

Threinen, C.W. 1958. A summary of observations on the commercial harvest of crayfish in northwestern Wisconsin. Wis. Conserv. Dp. Fish. Manage. Div. Misc. Rep. No. 2. 14 p.

Unestam, T. 1972. Significance of diseases on freshwater crayfish. In Proc. 1st Int. Symp. Freshwater Crayfish. p. 135-150.

Unestam, T. 1974. Defence reactions in crayfish towards microbial parasites, a review. In Proc. 2nd Int. Symp. Freshwater Crayfish. p. 327-336.

Unestam, T., Soderhall, K., Nyhlen, L., Svensson, E., and R. Ajaxon. 1977. Specialization in crayfish defence and fungal aggressiveness upon crayfish plague infection. In Proc. 3rd Int. Symp. Freshwater Crayfish. p. 321-331.

Vey, A. 1977. Studies on the pathology of crayfish under rearing conditions. In Proc. 3rd Int. Symp. Freshwater Crayfish. p. 311-319.

Vey, A., K. Soderhall and R. Ajaxon. (This conference). Susceptibility of Orconectes limosus to the crayfish plague Aphanomyces astaci.

Wright, D.A. 1980. Cadmium and calcium interactions in the freshwater amphipod Gammarus pulex. Freshwater Biology 10:123-133.

Yonge, C.M. 1936. On the nature and permeability of chitin II -- the permeability of the uncalcified chitin lining the foregut of <u>Homarus</u>. Proc. Roy. Soc. Lond. Ser. B. 611:15-41.

Yonge, C.M. 1937. The nature and significance of the membranes surrounding the developing eggs of <u>Hamarus vulgaris</u> and other Decapods. Proc. Zool. Soc. Lond. Ser. <u>A. 107:499-517.</u>

Yonge, C.M. 1946. Permeability and properties of the membranes surrounding the developing egg of <u>Homarus vulgaris</u>. J. Mar.Biol. Ass. V.K. 26:432-438.

Zitko, V., and W.G. Carson. 1976. A mechnaism of the effects of water hardness on the lethality of heavy metals to Fish. Chemosphere 5:299-303.

III
CATION PHYSIOLOGY OF CRAYFISH

TABLE SIZE AND PHYSIOLOGICAL CONDITION
OF THE CRAYFISH IN RELATION TO CALCIUM ION ACCUMULATION

Duro Adegboye
Department of Biological Sciences
Ahmadu Bello University
Zaria, Kaduna State, Nigeria

ABSTRACT

Calcium ion concentration in the tissues of the crayfish is a major factor which influences the length, body weight as well as the physiological robustness of the crayfish. Since the state of health of the crayfish is an important criterion in crayfish research and industry, the results are discussed in relation to the possible methods of improving crayfish yields through calcium ion supplementation. It has been estimated that a 3 mg % increase in calcium concentration in each of the tissues would result in about 20 percent increase in crayfish biomass.

INTRODUCTION

It is an established fact that the addition of calcium salts to either the ponds of or the diets of the crayfish greatly enhanced the growth rate and ultimately the table size of the crayfish (Westman, 1972; de la Bretonne et al., 1971; Avault, et al., 1970; Avault, 1972). The positive effect of supplementary calcium salts on the growth rate of the crayfish could be due to a number of physiological reasons. The availability of extra calcium ions in the ponds might reduce the unavoidable competition for calcium ions by the crayfish specimens maintained in the ponds. The excessive loss of calcium ions from the hemolymph into the environment reported by Greenaway (1972, 1974) could be corrected for by the rapid uptake of calcium from ponds to which calcium salts have been added (Adegboye et al., 1978a). The additional calcium salt might also serve as a buffer in raising the pH and total hardness of the water to a level which significantly enhances the growth and development of the crayfish (see Avault, 1972). Apart from the ultimate increase in the biomass of the crayfish following the addition of calcium salts to the ponds and diets, certain calcium-dependent processes within the crayfish tissues might receive a boost from the supplementary calcium ions. The energy normally expended by the crayfish in the uptake of calcium ions from freshwaters is drastically reduced when more calcium ions are added to the pond or the diets of the crayfish. The energy thus saved could be utilized for growth reproduction and development of the crayfish.

For practical purposes, the nutrients supplied to an organism may be conveniently divided into a utilizable portion as well as the unutilizable portion. The useful portion is that which is absorbed, transported, stored in the tissues and later metabolized in the tissues. The unutilizable portion can be further subdivided into two parts, namely, a part that is absorbed but which is immediately excreted from the tissues, and the unabsorbed portion which forms part of the feces. Although supplementary nutrients enhance growth and development, it is only the utilizable portion of the nutrients that is actually involved in crayfish metabolism. Therefore, while the quantity of calcium salts added to ponds and diets may be important in the overall growth increases, it is only the calcium ions that are accumulated in the tissues that are actually involved in the calcium-dependent growth and development in the crayfish. Since it is impracticable to cut up the crayfish and determine the concentration of calcium in the tissues and later determine the increase in size following ecdysis in the same animal, the inferences are based on the premoult size of the crayfish. Incidentally, it has been shown by various workers that the increase in size following moult bears a direct relationship to the premoult size. According to Farmer (1973), the growth per moult in decapods is of the progressive geometric type. While the percentage increase in size following a given moult (ecdysis) falls off with the age or preecdysial size of the decapod, the absolute (measurable) increase in size at each moult increases with the size (age) of the decapod. In other words, the size of the postecdysial decapod is a reflection of the size attained by the preecdysial decapod. Thus, any factor that influences (or is related to) the preecdysial size of the crayfish will of necessity affect the postecdysial size of the animal. In this report, therefore, any significant relationship observed between the tissue calcium concentration and body size of the preecdysial crayfish is interpreted to mean the relationship between the tissue calcium concentration and the postecdysial size of the crayfish:

Calcium Concentration is proportional to Preecdysial Size

Preecdysial Size is proportional to Postecdysial Size

Calcium Concentration is proportional to Postecdysial Size

MATERIALS AND METHODS

The specimens of crayfish used in this study were obtained from Carolina Biological Supply Company of Burlington, North Carolina, U.S.A. The stages in the mouting cycle of the crayfish were determined with the methods of Stevenson (1972). The male form I crayfish used in the analysis were identified with the criteria laid down by Hobbs (1973). Methods for determining the concentration of calcium ions in the tissues have been described elsewhere (Adegboye et al., 1975a, 1978b). Hemolymph calcium was determined by a microfluorometric method, while calcium in the ash of the hepatopancreas, exoskeleton, and gastrolith was determined by automated colorimetric spectrophotometry.

The Crayfish Constant was estimated by dividing the wet body weight by the cube of the carapace length (W/L^3). As explained in a previous work, the stages in the moulting cycle were assigned numerical units to facilitate statistical analysis: The period immediately following ecdysis (=A) was assigned "1"; the late postmoult stage (=B) was given "2"; Early intermoult stage (=C_1-C_2), "3"; Late intermoult stage "4"; Early premoult stage (=D_1-D_2), "5"; and the Late premoult stage (=D_3-D_4) was assigned the number "6" (Adegboye et al., 1975b). The Physiological Condition Factor of the crayfish was estimated by multiplying the Crayfish Constant by the numerical equivalence of the stage in the moulting cycle.

Measurements of the physical parameters as well as chemical analyses were carried out in the Departments of Zoology and

Pharmacology of the University of North Carolina, Chapel Hill, U.S.A. Statistical analyses of the data were carried out at the Computer Centre, Ahmadu Bello University, Zaria, Kaduna State, Nigeria. Multiple regression analyses were performed on the data using SPSS Version 6.50 of April, 1976 from the Northwestern University, U.S.A. Coefficient of Determination (=R^2) was used in the case of multiple regression analysis to test the level of significance of the various relationships examined in this study.

Body size, the moulting cycle, the availability of calcium in the environment, as well as the relationships between the concentrations of calcium ions in the tissues were considered in the construction of a model of calcium homeostasis in the crayfish.

RESULTS

Table 1 shows the relationship between the carapace length and the concentration of calcium in the tissues of the crayfish. During the late intermoult stage, the carapace length was direcly related to the concentration of calcium in the exoskeleton. The carapace length had no significant relationship to the level of calcium in either the hemolymph or hepatopancreas at this stage.

At the early premoult stage, the carapace length was directly related to the concentration of calcium in the hemolymph, old exoskeleton and the hepatopancreas of the crayfish. However, during the premoult stage, the carapace length was not related to the concentration of calcium in anyone of the hemolymph, old exoskeleton, hepatopancreas or gastrolith. The relationship between the carapace length and the concentration of calcium in the newly formed exoskeleton was significant.

During the late intermoult stage (C_4), the body weight, carapace length and the crayfish constant were all significantly correlated with the concentration of calcium ions in the exoskeleton. These observations are summarized in Table 2.

There was a highly significant relationship between the size of the crayfish and a function of calcium concentrations in the hemolymph, exoskeleton and hepatopancreas during the early premoult stage (Table 3a). Similarly, the body weight or the crayfish constant or the physiological condition factor was also significantly related to a function of tissue calcium concentrations.

As shown in Table 3b, the relationship between the physical parameters of the crayfish and tissue calcium concentration was such that the tissue calcium concentration could be estimated from a function of carapace length, body weight, crayfish constant and physiological condition factor during the early premoult stage.

The carapace length of the precdysial crayfish in the late premoult stage was highly correlated with a function of calcium concentrations in the tissues of the crayfish (Table 4). While the concentration of calcium in the hemolymph, the gastrolith and the new exoskeleton appeared to contribute to the size of the crayfish, it seems that the higher the concentration of calcium in the

117

Table 1. Relationships between the Carapace Length and the concentration of calcium ions in the tissues of the crayfish.

| | Dependent Variable = Carapace Length cm CORRELATION COEFFICIENTS | | |
	Late Intermoult C_4	Early Premoult D_1-D_2	Early Premoult D_3-D_4
Hemolymph (Ca)	-0.18	0.91	0.15
Old Exoskeleton (Ca)	0.84	0.95	0.35
Hepatopancreas (Ca)	-0.14	0.92	-0.52
Gastrolith (Ca)	---	---	0.05
New Exoskeleton (Ca)	---	---	0.66
	(n=10)	(n=10)	(n=7)

Table 2. Relationships between the carpace length and the concentrations of calcium in the tissues of the crayfish during the late intermoult stage (C_4).

n = 10

Dependent Variables

BW = Body Weight
CL = Carapace Length
K = Crayfish Constant
F = Physiological Robustness

| | CORRELATION | | COEFFICIENTS (R) | |
	BW	CL	K	F
Hemolymph Ca mg/100 ml	-0.30	-0.18	-0.30	-0.36
Exoskeleton Ca mg % ash	0.73	0.84	0.73	0.41
Hepatopancreas Ca mg % ash	0.03	-0.14	0.03	0.07
	BW	CL	K	F

Table 3. Relationships between the Physiological Robustness and Tissue Calcium Concentrations in the Crayfish (early premoult stage D_1/D_2).

A = Hemolymph Ca mg/100 ml
B = Exoskeleton Ca mg % ash
C = Hepatopancreas Ca mg % ash
CL = Carapace Length cm
BW = Body Weight gm
K = Crayfish Constant gm/cm^3
F = Physiological Condition (gm/cm^3) stage 3a.

3a. Estimation of Physical Parameters

R^2

Carapace
Length = 2.69 + 0.79 E-01 "A" + 0.10 E-02 "B" + 0.56 E-01 "C" 0.97

Body
Weight = 6.76 + 0.38 E-01 "A" + 0.62 E-01 "B" + 1.07 "C" 0.96

Crayfish
Constant = 0.38 - 0.30 E-02 "A" + 0.54 E-04 "B" + 0.64 E-02 "C" 0.86

Physiological
Condition = 1.92 - 0.15 E-01 "A" + 0.19 E-03 "B" + 0.32 E-01 "C" 0.86

3b. Estimation of Tissue Calcium

R^2

Hemolymph
Ca = 103.60 - 2.83 "CL" + 1.78 "BW" + 1368.79 "K" - 329.01 "F" 0.96

Exoskeleton
Ca = 237.45 + 95.91 "CL" - 5.00 "BW" + 7726.49 "K" - 1544.34 "F" 0.96

Hepatopancreas
Ca = 4.49 - 3.49 "CL" + 0.88 "BW" + 43.38 "K" - 9.81 "F" 0.93

Table 4. Relationship between carapace length and the concentrations of calcium in the tissues of the crayfish during the late premoult stage (D_3/D_4).

Dependent Variable = Carapace Length cm

Mean Response = 2.74 cm

Standard Deviation = 0.51

Variables: A = Hem Ca = Hemolymph Ca mg/100 ml

B = Hep Ca = Hepatopancreas Ca mg % ash

C = Gas Ca = Gastrolith Ca mg % ash

D = New Ca = New exoskeleton Ca mg % ash

E = Exo Ca = Old Exoskeleton Ca mg % ash

Carapace Length =

0.18 E-10 "A" - 0.14 "B" + 0.11 E-10 "C" + 0.32 "D" - 0.22 E-02 "E"

Multiple R = 0.993

R^2 = 0.987

n = 7

P < 0.001

Table 5. The effect of increasing the concentration of calcium in every tissue (by 1 mg %) on body size.

Variables	Mean Value
Carapace Length cm	3.76
A = Hemolymph Ca mg/100 ml	45.77
B = Exoskeleton Ca mg % ash	51.42
C = Hepatopancreas Ca mg % ash	3.20

Carapace Length = 2.688 + 0.00789 "A" + 0.0103 "B" + 0.0565 "C"

R^2 = 0.97

Increase in Tissue (Ca) By	Estimated Carapace Length	% Change In Carapace Length
0 mg %	3.76 cm	0 %
1 mg %	3.83 cm	1.86 %
2 mg %	3.91 cm	3.99 %
3 mg %	3.98 cm	5.85 %

Thus for every 1 mg % increase in tissue calcium, there is a 2% increase in carpace length.

hepatopancreas and in the old exoskeleton, the smaller was the size of the crayfish.

DISCUSSION

The size attained by an organism as well as the physiological condition of that animal are products of several factors. Such factors must of necessity include the availability of nutrients for the construction of growing parts. Organic and inorganic substances along with special cofactors constitute the materials required for the growth metabolism. Unlike other groups of animals, the arthropods (the group to which the crayfish belongs) are clad with an external exoskeleton which must be shed at ecdysis before growth can take place (Russell-Hunter, 1969). Following ecdysis and during the period of hardening of the new cuticle, the arthropod increases in size with the uptake of water and air into the newly created internal spaces. The water and air are subsequently replaced by new tissues. According to Russell-Hunter (1969), while the actual increase in size occurs at ecdysis (due to uptake of water and air), the actual tissue growth takes place during the intermoult period when no increase in size is detectable.

Size increase is a multifaceted phenomenon in living organisms and the available nutrients feature prominently in determining the growth rate. While the quantity of ingested food is important, the utilizable portion of the ingested food is even more important as far as growth and increase in size are concerned. The proportion of food that is utilizable is determined by processes such as digestion, absorption, transportation, storage and metabolism of the food item. Growth, or increase in size, is a life-long problem in the life of decapod crustaceans. The whole of the life processes of the arthropods, in general, appear to be under the direct influence of the growth phenomenon. The moult-related accumulation of metabolic reserves is a very important feature of the physiology of decapod crustaceans (Passano, 1960).

In the crayfish, as in all other arthropods, linear growth is preceded by moulting. Any factor or set of factors that inhibits moulting will consequently inhibit linerar gowth. Starvation and physiological demands on metabolic reserves are factors which reduce the frequency of moulting, with the resultant limitation on the rate of linear growth. This study has clearly demonstrated that the levels of metabolic reserves in the tissues of the crayfish is reflected in the size of the animal prior to ecdysis. Farmer (1973) has also shown that the pre-ecdysial size influences to a large extent the size attained by the decapod following moulting. Furthermore, several carcinologists have shown that in the period of "starvation" immediately before and after ecdysis, the decapod depends on the food reserves in the tissues (Passano, 1960; Russell-Hunter, 1969). Therefore, the concentration of metabolic reserves in the premoult crayfish does not only determine the size of the crayfish before moulting, but also the size of the crayfish following ecdysis.

Since the postmoult size can be predicted from the premoult size, and the premoult size is dependent on the concentration of metabolites in the tissues, it is reasonable to believe that the concentration of metabolites in the premoult crayfish has a great influence on the size attained by the crayfish following moulting. Thus, our findings, which

121

are based entirely on statistical analyses, explain why the supplementary diets fed to decapods greatly increase, not only the biomass and the frequency of moulting, but also the resultant increase in linear dimensions (Smitherman et al., 1967; Loyacano, 1967; de la Bretonne et al., 1969, 1971; Avault et al., 1970). Thus, while crayfish can grow in the wild without supplemental feeding (Avault, 1972), the addition of utilizable items of food to the ponds would greatly increase the concentration of premoult metabolic reserves (Smitherman et al., 1967) which would in turn enhance the frequency of moulting rates and consequently absolute increase in size following ecdysis. Perhaps, as in Rhodnius, high concentration of metabolites might serve as a trigger for the moulting process in the crayfish as well (see Wigglesworth, 1965).

The importance of supplemental calcium in the culturing of crayfish can be illustrated with the example in Table 6. An addition of 1 mg calcium % to the concentration of calcium in each of the hemolymph, exoskeleton and hepatopancreas would lead to a 2 percent increase in the carapace length prior to moult. Since the postmoult size is larger than, and proportional to premoult size, it is expected that the increase in tissue calcium concentration by 1 mg % would lead to significant increases in body size following moult.

To the consumer of crayfish, the physiological robustness which is an index of the state of health of the crayfish, is an important factor to be considered in selecting the right specimen for the table. As we have clearly shown, even this "state of health" of the crayfish is dependent on the concentration of calcium ions in the tissues, at least during the early premoult stages. The physiological condition of the crayfish is also a very important factor in selecting specimens of crayfish for research. Therefore, to obtain a physiologically fit specimen in the laboratory, the specimens should be maintained in tanks with adequate supply of supplemental calcium.

On the basis of the works of de la Bretonne et al. (1969), Westman (1972) and our statistical analyses of length-calcium relationships in the crayfish, we would like to recommend that calcium salts and sources of calcium ions such as bone, crayfish waste, snail shells and egg shells be added to the diet and ponds of crayfish to improve yields.

ACKNOWLEDGMENTS

The author is grateful to the African-American Institute, New York and the Ahmadu Bello University, Zaria, Nigeria for the financial support required for this work. I acknowledge with gratitude the valuable pieces of advice I received from Prof. Irvine R. Hagadorn and Professor Philip F. Hirsch throughout the period of my studies in the U.S.A. Discussions held with Prof. James Avault, Jr., and Prof. Jay Huner were most useful. The manuscript was typed by Mrs. M.O. Obilana.

Table 6. The effect of increasing the concentration of tissue calcium (by 1 mg %) on body weight of the crayfish.

Body Weight gm	15.109
Hemolymph Ca mg/100 ml	45.77
Exoskeleton Ca mg % ash	51.42
Hepatopancreas Ca mg % ash	3.20

Body Weight = 6.757 + 0.0385 "A" + 0.0617 "B" + 1.0671 "C"
R^2 = 0.9586 $P < 0.001$

Increase In Tissue	Estimated Body Weight gm	% Change In Body Weight
0 mg %	15.11	0 %
1 mg %	15.21	0.67 %
2 mg %	17.44	15.42 %
3 mg %	18.61	23.15 %

Thus, a 1 mg % increase in tissue calcium concentration would result in 0.67% increase in tissue weight while a 3 mg % increase in tissue calcium would result in 23% increase in crayfish production.

LITERATURE CITED

Adegboye, J.D., I.R Hagadorn and P.F. Hirsch. 1975a. Microfluoro-metric determination of calcium in the hemolymph and other tissues of the crayfish. 2nd Int'l Crayfish Symposium; Baton Rouge, La: U.S.A. Freshwater Crayfish (2):211-225.

Adegboye, J.D., I.R. Hagadorn and P.F. Hirsch. 1975b. Variations in hemolymph calcium associated with the moulting cycle in the cray-fish. 2nd Int'l Crayfish Symposium; Baton Rouge, LA; U.S.A., Freshwater Crayfish (2):227-247.

Adegboye, J.D., I.R. Hagadorn and P.F. Hirsch. 1978a. Metabolism of calcium in the crayfish. 4th Int'l Crayfish Symposium, Thonon, France. Freshwater Crayfish 4:25-34.

Adegboye, J.D., I.R. Hagadorn and P.F. Hirsch. 1978b. Factors modify-ing calcium concentration in the hemolymph of the crayfish Procambarus actus acutus Girard. 4th Int'l Crayfish Symposium. Thonon, France. Freshwater Crayfish, 4:1-14.

Avault, J.W. JR., L. de la Bretonne, Jr. and E. Jaspers. 1970. Cul-ture of crawfish, Louisiana's crustacean king. The American Fish Farmer 1(10):8-14 and 27.

Avault, J.W. JR. 1972. Crayfish farming in the United States. Freshwater Crayfish 2:240-250.

de la Bretonne, L., Jr., J.W. Avault, Jr. and R.O. Smitherman. 1969. Effects of soil and water hardness on survival and growth of red swamp crawfish, Procambarus clarkii, in plastic ponds. Proceed-ings of the 23rd Annual Conference of Southeastern Association of Game and Fish Commissioners 23:626-633.

de la Bretonne, L., JR. and J.W. Avault, Jr. 1971. Liming increases crawfish production. Louisiana Agriculture 15(1):10.

Farmer, A.S. 1973. Age and growth in Nephrops norvegicus (Decapoda: Nephropidae). Marine Biology 23:315-325.

Greenaway, P. 1972. Calcium regulation in the freshwater crayfish Austropotamobius pallipes (Lereboullet). 1. Calcium balance in the intermoult animal. J. Exp. Biol. 57:471-487.

Greenaway, P. 1974a. Calcium balance at the pre-moult stage of the freshwater crayfish Austropotamobius pallipes (Lereboullet) J. Exp. Biol. 61:27-34.

Hobbs, H.H. Jr. 1973. Crayfishes (Astacidae) of North and Middle America Biota of Freshwater Ecosystems: Identification Manual No. 9. Environmental Protection Agency. Washington, D.C.

Loyacano, H. 1967. Some effects of salinity on two populations of red swamp crawfish, Procambarus clarkii (Girard). Proceedings 21st Annual Conference Southeastern Association of Game and Fish Commissioners 21:423-435.

Passano, L.M. 1960. Moulting and its control. In: The Physiology of Crustacea: Volume I. London: Academic Press.

Russell-Hunter, W.D. 1969. _A Biology of Higher Invertebrates_. The Macmillan Company. London.

Smitherman, R.O., J.W. Avault, JR., L. de la Bretonne, Jr.. and H.A. Loyacano. 1967. Effects of supplemental feed and fertilizer on production of red swamp crawfish, _Procambarus clarkii_, in pools and ponds. Proceedings 21st Annual Conference Southeastern Association of Game and Fish Commissioners 21:452-458.

Stevenson, J.R. 1972. Changing activities of the crustacean epidermis during the moulting cycle. American Zoologist 12:373-380.

Westman, K. 1972. Cultivation of the American crayfish _Pacifastacus leniusculus_. Freshwater Crayfish 1:212-220.

Wigglesworth, V.B. 1965. _The Principles of Insect Physiology_. Methuen and Co. Ltd., London.

INDICATIONS OF AGE-DEPENDENT METABOLISM OF CALCIUM
IN THE CRAYFISH (CRUSTACEA, DECAPODA)

Duro Adegboye
Department of Biological Sciences
Ahmadu Bello University
Zaria, Nigeria

ABSTRACT

A review of previous works on calcium metabolism in the crayfish Procambarus acutus acutus (Adegboye, 1975), indicates that the regulation of calcium ion concentration in the tissues is under the influence of age-related processes of the crayfish. In certain instances, older crayfish store a greater proportion of calcium in certain tissues than younger crayfish. However, the influence of aging on hemolymph calcium varies with the stage in the moulting cycle. The observed age-dependent metabolism of calcium in the crayfish corroborates age-dependent processes of decapod crustaceans: These processes include, the frequency of moulting, increase in size per moult, onset of sexual maturity, fecundity, and rate of embryonic development.

INTRODUCTION

Several studies have established that there are certain aspects of calcium metabolism which are related to age in mammalian species. In their study of thyrocalcitonin and aging in rats, Orimo and Hirsch (1973) concluded that there was a reduced rate of bone calcium metabolism with advancing age. It is now an established fact that younger decapod crustaceans moult more frequently than the older specimens of the same species (Passano, 1960). Apart from the occasional reference to the age-dependent rate of moulting, there has been no deliberate effort to study the influence of age on moult-related processes in decapod crustaceans.

Age, which is estimated from body size, is one of the major factors which is correlated with certain physiological processes of decapod crustaceans. Such age-dependent processes include the metabolism of moulting, the frequency and duration of moulting (Passano, 1960; Kamiguchi, 1971; Speck et al., 1972; Ennis, 1972; Travis, 1954; McWhinnie, 1962; Brown and Powell, 1972) the onset of sexual maturity (Newman and Pollock, 1971; Silbergbauer, 1971; Haefner, 1972; Harris et al., 1972; Brown and Powell, 1972); weight of ovaries and absolute fecundity (Stypinska, 1978); female size and egg count (Payne, 1971); the rate of embryonic development and the rate of larval development (Price and Chew, 1972), and percent increase in size per moult (Travis, 1954).

Previous work on Procambarus acutus acutus have shown that during the late premoult stage when calcium is resorbed from the exoskeleton, older crayfish concentrated more calcium in the hemolymph than younger specimens (Adegboye et al., 1975b). Furthermore, during the late intermoult stage the concentration of calcium in the exoskeleton is higher in older crayfish than in younger ones (Adegboye et al., 1975b, 1978a). In view of the close association between age and the moulting process on the one hand (Passano, 1960), and highly significant relationship between the moulting cycle and the metabolism of calcium on the other (Travis, 1955, 1957; Greenaway, 1974a, b; McWhinnie, 1962; Adegboye et al., 1975b, 1978a); it is reasonable to suggest that the moment by moment regulation of calcium in the crayfish might also be affected by the age of the crayfish. This paper reports on some of our preliminary observations which lend support to the idea that the regulation of calcium is an age-dependent process in the crayfish. The observations were culled from our previous works on calcium metabolism in the crayfish <u>Procambarus</u> <u>acutus</u> <u>acutus</u> (Adegboye, 1975).

OBSERVATIONS

1. Age-dependent response of the crayfish to cooling:

Figure 1 shows the effect of transferring 6 specimens of crayfish from a room temperature of 24°C to a cold room set at 4°C, there is a general increase in the concentration of calcium in the hemolymph of the crayfish. The maximum increase in hemolymph calcium concentration is 67.60% while the minimum increase is 2.80%. For crayfish weighing between 11.74 mg to 18.78 g (inclusive), the percent change in hemolymph calcium increases with the age (size) of the crayfish. However, above 18.78 g, the percent change in hemolymph calcium falls slightly with age.

2. Age-dependent response of the crayfish to Crustecdysone Injection:

A dose of 0.67 µg/gram body weight of ecdysterone injected into 7 crayfish resulted in an increase in hemolymph calcium 120 hours after injection. The minimum increase in hemolymph calcium was 14.48 percent change per gram-body-weight while the maximum change was 30.75 percent change in hemolymph calcium per-gram-body-weight. For small specimens of crayfish weighing between 6g and 9g (inclusive) the percent change in hemolymph calcium per-gram body-weight increased with the age of the crayfish. On the other hand, above 9g body weight there was a gradual decline in the percent change in hemolymph calcium per-gram body-weight with increasing age (Figure 2). The oldest crayfish had the least change in hemolymph calcium in response to crustecdysone injection.

3. Age-dependent relationship between tissue calcium:

When the concentration of calcium in the hemolymph was regressed against the concentration of calcium in the exoskeleton of the crayfish in the late intermoult stage, the correlation coefficient was not significant at the 5 percent level.

Similarly, there was no direct relationship between hemolymph calcium and hepatopancreas calcium ($P>0.05$). There was also no relationship between the concentrations of calcium in the hepato-pancreas and the exoskeleton ($P>0.05$). However, when the concentration of calcium in the hemolymph was regressed against a function of age and the concentration of calcium in the exoskeleton the correlation coefficient ($r = 0.88$) was highly significant at the 0.1 percent level. There was also a significant correlation between hepatopancreas calcium and a function of age and exoskeleton calcium (Table 1).

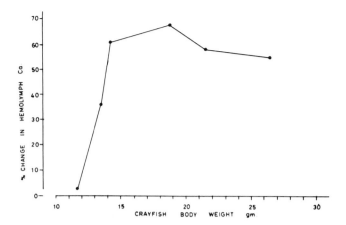

Figure 1.   The effect of body size (≡ age) on the response of
hemolymph calcium concentrations to cooling.   The percent change in
hemolymph calcium increases with age in young crayfish weighing less
than 18g.

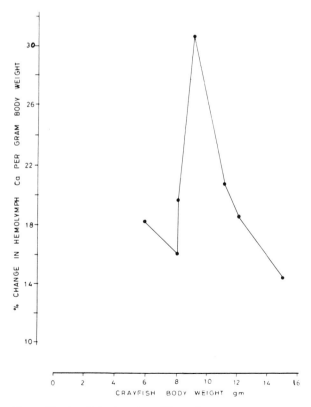

Figure 2.   The effect of body size (≡ age) on the response of hemolymph
calcium concentration to crustecdysone injection.   The effect falls off
with age in crayfish weighing above 10 g.

Table 1. The influence of age on the relationships between calcium concentrations in the tissues of the crayfish (10 crayfish at intermoult stage C_4).

		Correlation Coefficient	P Significance
(A)	Without Age Consideration		
	Hemolymph Ca vs Exoskeleton Ca	0.13	P>0.10
	Hemolymph Ca vs Hepatopancreas Ca	-0.14	P>0.10
	Hepatopnacreas Ca vs Exoskeleton Ca	-0.50	P>0.10
(B)	With Age plugged into the function		
	Hemolymph Ca vs Exoskeleton Ca (Age)	0.88	P<0.001
	Hepatopancreas Ca vs Exoskeleton (Age)	0.92	P<0.001

Table 2 - The influence of age on the relationship between the stage in the moulting cycle and hemolymph calcium concentration.

n = 147		
	Correlation Coefficient	P Significance
Length vs Hemolymph Ca	0.074	>0.10
Moulting vs Hemolymph Ca	0.609	<0.001
Moulting vs (Age) Hemolymph Ca	0.653	<0.001

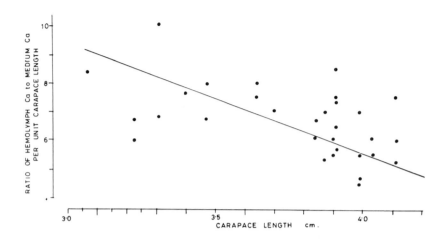

Figure 3. The effect of body size ($\equiv$ age) on the ratio of hemolymph calcium to medium calcium in the mid-intermoult stage C_1-C_2.

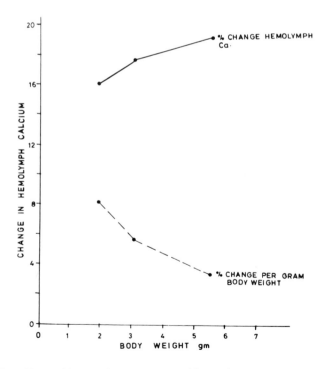

Figure 4. The effect of body size ($\equiv$ age) on the response of the crayfish to dehydration. The older the crayfish the greater the change in hemolymph calcium, and the less the change in body weight.

4. Age-dependent exchange of calcium between the crayfish and the medium:

In a study of the relationships between the body size, hemolymph Ca, medium Ca and the ratio of hemolymph Ca to medium Ca in _Procambarus acutus acutus_ (C_1-C_2), a direct relationship between hemolymph Ca and medium Ca ($R = + 0.581$, $0.05 < P > 0.02$) was observed. The relationship between age and hemolymph Ca was an indirect one ($R = -0.537$, $0.05 < P > 0.02$). The correlation between medium Ca and hemolymph Ca was improved when medium Ca was regressed against a function of age and hemolymph Ca ($R = 0.629$, $0.02 > P$). As shown in Figure 3, there was an indirect relationship between the age and the ratio of hemolymph Ca: medium Ca expressed as per cm standard length ($R = -0.748, P < 0.001$).

5. Age-dependent response to dehydration:

When the specimens of crayfish were kept out of the water for 1 hour, there was a general decrease in body weight coupled with an increase in hemolymph calcium concentration (Figure 4). The lowest percent change in hemolymph calcium concentration was 16.07% while the highest is 19.04%. When the increase in hemolymph calcium concentration was expressed as percent change per-gram-body-weight, the oldest crayfish had the lowest change in hemolymph calcium while the youngest crayfish experienced the highest change.

6. Age-dependent response to moulting process:

Table 1 shows the relationship between the stages in the moulting cycle and the concentration of calcium in the hemolymph of the crayfish. There was no correlation between the age of the crayfish and hemolymph Ca ($P > 0.05$). However, there was a highly significant relationship between the stages in the moulting cycle and the hemolymph Ca of the the 147 specimens of crayfish ($P < 0.001$). The correlation obtained for the relationship between the moulting stage and hemolymph Ca was improved by regressing the moulting stage against a function of age and hemolymph Ca ($R = 0.653; P < 0.001$).

DISCUSSION

The preliminary observations presented in this paper, though fragmentary in nature, do suggest that the accumulation, concentration and regulation of calcium ion concentrations in the tissues of the crayfish are all influenced by the size of the crayfish.

In this report, the age of the crayfish has been estimated from the body size. This is in order, because decapod crustaceans in particular, and arthropods, in general, are known to undergo increases in size during the period immediately following ecdysis. The number of ecdysis per specimen per year is also relatively constant for any given species of crustaceans. Hence, within limits, it is reasonable to believe that the body dimension is an index of age in decapod crustaceans. Ito (1970) is reported to have estimated the age and growth on the basis of the seasonal changes in the carapace width frequencies and carapace hardness of _Chionoectes opilia_. The life history of the crayfish _Procambarus acutus acutus_ has been well studied by Penn and Hobbs (1958), Penn (1956), Ham (1971), and Avault et al., (1970), and

131

observations of these workers suggest that within limits, the size of the crayfish is a reflection of its age.

When the crayfish are subjected to freezing conditions, the mechanisms for maintaining hemolymph Ca within the normal levels appear to have been disrupted, resulting in a general increase in hemolymph calcium concentration. For young crayfish specimens weighing between 11g and 18g, the increase in hemolymph Ca is related to the age of the crayfish: such that the older the crayfish, (up to 18g weight) the greater the change in hemolymph Ca. However, above 18g body weight, it appears that the older crayfish are capable of opposing the change in hemolymph Ca that is brought about by the decrease in ambient temperature. Thus, the older the crayfish, the more efficient is the calcium homeostatic mechanisms.

Although the injection of crustecdysone into the crayfish resulted in hypercalcemia (Adegboye, et al., 1974a, 1978c), the response of the crayfish is not uniform. For instance, young crayfish weighing 9g or less do not appear to be able to maintain their hemolymph calcium at the normal level following crustecdysone injection. On the other hand, older crayfish (of more than 9g body weight) are able to regulate their hemolymph Ca close to the normal level in face of the hypercalcemic activity of crustecdysone. The ability to oppose crustecdysone-mediated hypercalcemia appear to be age-dependent. Indeed, the oldest crayfish exhibits the least hypercalcemic effect of crustecdysone among the crayfish investigated. The frequency of moulting falls off with age as well (Passano, 1960).

The fact that the concentration of calcium in the hemolymph is directly related to the concentration of calcium in the environmental water indicates that there is a continuous calcium exchange between the crayfish and the medium, (see Greenaway, 1972). We have reported a 15% drop in hemolymph calcium in crayfish kept in tap-water for a period of 72 hours (Adegboye, et al., 1978b). However, it appeared that older crayfish experience greater dilution from the medium than the younger crayfish. Furthermore, the mechanism for calcium exchange between the medium and the hemolymph appears to be affected by age. The last statement might also imply that the loss of calcium from the hemolymph into the medium increases with the age of the crayfish. Figure 4 seems to support this idea.

The lack of correlation between exoskeleton Ca and hemolymph Ca, between exoskeleton Ca and hepatopancreas Ca and between hemolymph Ca and hepatopancreas Ca is interpreted to mean that the large amount of exoskeleton calcium is not available in ionic form to correct for calcium losses from the hemolymph at the late intermoult stage (Adegboye, et al., 1978b). As shown in Table 2, when the concentration of calcium in the hemolymph is regressed against a function of age and exoskeleton Ca, there is a highly significant relationship between the calcium concentrations of the two compartments. The last observation would indicate that the probability of exchange of calcium ions between the tissues increases with the age of the crayfish.

Dehydration of the crayfish resulted in decrease in body weight as the crayfish lost water and also in an increase in hemolymph calcium concentration. The younger crayfish is the most vulnerable to the hypercalcemic effects of dehydration. The older crayfish appears to be capable of at least reducing the adverse effect of dehydration on the hemolymph Ca.

The moulting process is the most important factor that affects the concentration of calcium in the hemolymph of the crayfish (Travis, 1955; McWhinnie, 1962; Adegboye, et al., 1974, 1975b, 1978b). The influence of the moulting process on the hemolymph Ca appears to increase with the age of the crayfish. While the age by itself is not statistically correlated with hemolymph Ca, the regression of moulting stages against a function of age and hemolymph Ca gave a correlation coefficient of 0.653 which was higher than the correlation of 0.609 obtained for the relationship between moulting stages and hemolymph Ca. of calcium appears to be more pronounced in the older crayfish than in the younger specimens.

Finally, age-dependent processes in the crayfish appear to influence the accumulation and concentration of calcium and phosphate in the tissues of the crayfish. As the crayfish grows older, the ability to increase the proportion of calcium in the tissues appear to improve with age.

Although our results perhaps demonstrate for the first time a direct indication of age-dependent metabolism of calcium in the crayfish, there are several other instances of age-related process in the decapod crustaceans. Harris and his co-workers (1972) have also reported an increase in egg number with the age (carapace length) of the shrimp Pandalopsis dispar. According to Kamiguchi (1971) younger specimens of Palaemon paucidens moulted more frequently than the older ones. Furthermore, the weight of the ovary is a direct function of the carapace length (age) of Palaemon sp. Prior to ecdysis, older crayfish generally constructed larger gastroliths than the younger specimens (McWhinnie, 1962; Adegboye, et al., 1978d). The attainment of sexual maturity in the Alaska tanner crab (Brown & Powell, 1972), and in the rock lobster, Jasus lalandii (Newman & Pollock, 1971) was dependent on the age of the crustacean. Lindquist (1970) has reported a relation-ship between the osmotic pressure and body size (age) of Porcellio scaber and Oniscus asellus.

These results support the idea of age-dependent metabolic processes in crustaceans. While the results reported in this paper might not be conclusive, they however, indicate that the age of the crayfish is an important factor in the physiology of the crayfish.

In order to confirm the age-dependent metabolism of calcium that has been indicated by our results, it would be necessary to hand-rear the crayfish in the laboratory in order to ascertain the chronoloical age of the crayfish. Although it is not possible at our present state of knowledge to give a definitive description of age-dependent calcium homeostasis in decapods, it is hoped that by reporting certain of our observations that indicate aging with respect to calcium metabolism in Procambarus acutus acutus, the concept will direct attention to this and other unanswered questions of calcium homeostasis in decapod crustaceans. Of particular interest is the role of hormones in the age-dependent physiological processes of the decapod crustaceans, as well as diet formulation for different ages of crayfish with particular reference to calcium intake.

ACKNOWLEDGMENTS

The author is grateful to Professor I.R. Hagadorn and P.F. Hirsch for their valuable suggestions; to Mr. Louis Fulkerson and Mr. Julius Oladipo for their help in statistical analyses. The help received from my students David Ogunleye, John Angwan and Yusufu Shibayan is gratefully acknowledged.

This study has been supported jointly by the USPHS Grant Am 10558 and by the Ahmadu Bello University Board of Research Grant 52073/JODA.

LITERATURE CITED

Adegboye, J.D. 1975. Regulation of hemolymph calcium in the crayfish, Procambarus acutus acutus Girard (Crustacea, Decapoda) Ph.D dissertation, University of North Carolina, Chapel Hill, N.C. 27514, U.S.A.

Adegboye, J.D., I.R. Hagadorn & P.F. Hirsch. 1975b. Variations in hemolymph calcium associated with the moulting cycle in the crayfish. 2nd Int'l Crayfish Symposium; Baton Rouge, La., U.S.A. Freshwater Crayfish (2):227-247.

Adegboye, J.D., I.R. Hagadorn and P.F. Hirsch. 1978a. Metabolism of calcium in the crayfish. 4th Int'l Crayfish Symposium, THonon, France. Freshwater Crayfish 4:25-34.

Adegboye, J.D., I.R. Hagadorn, and P.F. Hirsch. 1978b. Variations in hemolymph calcium associated with the mouting cycle in the crayfish. 2nd Int'l Crayfish Symposium; Baton Rouge, LA:U.S.A., Freshwater Crayfish (2):227-247.

Avault, J.W. Jr., L. de la Bretonne, Jr., and E. Jaspers. 1970. Culture of crawfish, Louisiana's crustacean king. The American Fish Farmer 1(10):8-14 and 27.

Brown, R.B. and G.C. Powell. 1972. Size at maturity in the male Alaskan tanner Crab, Chionecetes bairdi, as determined by chela allometry, reproductive tract weights, and size of precopulatory males J. Fish. Res. Bd. Canada 29:423-427.

Ennis, G.P. 1972. Growth per moult of tagged lobsters (Homarus americanus) in Bonavista Bay, Newfoundland. J. Fish. Res. Ed. Can. 29:143-148.

Greenaway, P. 1972. Calcium regulation in the freshwater crayfish Austropotamobium pallipes (Lereboullet). 1. Calcium balance in the intermoult animal. J. Exp. Biol. 57:471-487.

Greenaway, P. 1974a. Calcium balance at the pre-moult stage of the freshwater crayfish Austropolamobius pallipes (Lereboullet) J. Exp. Biol. 61:27-34.

Greenaway, P. 1974b. Calcium balance at the post-moult stage of the fresh water crayfish Austropotamobius pallipes (Lereboullet), J. Exp. Biol. 61:35-45.

Haefner, P.A. 1972. The biology of sand shrimp, Crangon septemspinosa, at Lamoine, Maine. J. Mitchell society 36:42.

Ham, G.B. 1971. Crayfish culture techniques. Amer. Fish Farmer and World Agriculture News. 2:5-6, 24.

Harris, C., K.K. Ches, and V. Price. 1972. Relation of Egg Number to Carapace Length of Sidestripe Shrimp (Pandalopsis dispar) from Dabon Bay, Washington, J. Fish. Res. Ed. Canada 29:464-465.

Ito, K. 1070. Ecological studies on the edible crab, Chionoectes opillio O. Fabricius in the Japan Sea. 3. Age and growth. Bull. Japan Sea Reg. Fish Res. Lab. 22:81-116. (English Summary).

Kamiguchi, Y. 1971. Studies on the moulting in the freshwater prawn, Palaemon paucidens I. Some endogenous and exogenous factors influencing the intermoult cycle J. Fac. Sci. Hokkaido Univ. Ser. VI. Zool. 18(1).

Lindquist, O.V. 1970. The blood osmotic pressure of the terrestrial isopods Porcellio scaber Latr. and Oniscus asellus L., with reference to the effect of temperature and body size. Comp. Biochem. Physiol. 37:503-510.

McWhinnie, M.A. 1962. Gastrolith growth and calcium shifts in the freshwater crayfish, Orconectes virilis. Comp. Biolchem. Physiol. 7:1-14.

Newman, G.G. and D.E. Pollock. 1971. Migration and availability of the rock lobster Jasus lalandii at Eland's Bay, S. Africa. Invest'l Rep. Div. Sea Fish. S. Afr. 94.

Orimo, H. and P.F. Hirsch. 1973. Thyrocalcitonin and Age. Endocrinology, 93(5):1206-1211.

Passano, L.M. 1960. Moulting and its control. In: the Physiology of Crustacea: Volume I. London: Academic Press.

Payne, J.F. 1971. Fecundity studies on the crayfish Procambarus hayi, Tulane studies in Zoology and Botany. 17(2):35-37.

Penn, G.H., and H.H. Hobbs, Jr. 1958. A contribution toward a knowledge of the crayfish Orconectes (Faxonella) Clypeatus (Hay). Tulane Stud. Zool. I(7):77-96.

Price, V.A and K.K. Chew. 1972. Laboratory rearing of spot shrimp larvae (Pandalus platyceros) and descriptions of stages. J. Fish. Res. Biol. Canada. 29:413-422.

Silbergbauer, B.I. 1971. The biology of the South African rock lobster Jasus lalandii (H. Milne Edwards) 1. Development. Investl Rep. Div. Sea Fish. S. Afr. 92:1070.

Speck, U. and K. Ueich. 1972. Nachweis einer Regulation der Glucosaminbildung bei dem FluBkrelos Orconectes limosus Zur Zeit der Hautung. Z. Vergl. Physiol. 76(3):341-346.

135

Stypinska, M. 1978. Individual variabilities in absolute fertility of crayfish occurring in the water of the Mazurian Lake District. Rocz. Nauk Roln H-98-3:177-203.

Travis, D.F. 1954. The moulting cycle of the spiny lobster *Panulirus argus* Latreille. 1. Moulting and growth in laboratory-maintained individuals. Biol. Bull. 107:433-450.

Travis, D.F. 1955. The moulting cycle of the spiny lobster, *Panulirus argus* Latreille. III. Physiological changes which occur in the blood and urine during the normal moulting cycle. Biological Bulletin Marine Biology Lab., Woods Hole, 109:484-503.

Travis, D.F. 1957. The moulting cycle of the spiny lobster, *Panulirus argus* Latreille. IV. Postecdysial histological, and histochemical changes in the hepatopancreas and integumental tissues. Biological Bulletin, 133(3):451-479.

CALCIUM HOMEOSTASIS IN THE CRAYFISH

Duro Adegboye
Department of Biological Sciences
Ahmadu Bello University
Zaria, Kaduna State, Nigeria

ABSTRACT

The major calcium compartments of the crayfish are the hemolymph, exoskeleton, hepatopancreas, new exoskeleton and the gastrolith. Calcium homeostasis in the crayfish has been monitored by determining the relationships between the concentration of calcium ions in the hemolymph and the calcium concentrations of the other tissues. Apart from the early premoult stage (D_1 - D_2), the concentration of calcium ions in anyone of the tissues does not significantly contribute to the concentration of calcium in the hemolymph. However, in the larger specimens of the crayfish, calcium ions in the other tissues do have a synergestic effect on hemolymph calcium during the period immediately preceeding moult. A model of calcium homeostasis which incorporates the existing ideas on calcium in decapod crustaceans is proposed.

INTRODUCTION

Calcium ion is perhaps the most important bioelement in the life of decapod crustaceans. The importance of calcium in these animals is borne out by the fact that the exoskeleton, which is the overruling single factor in all arthropods, is impregnated by calcium slats in all decapods, including the crayfish. Calcium homeostasis in the crayfish in particular, and in dacapod crustaceans in general, has not received as much attention as in vertebrates. The bulk of the data on calcium in the decapods is invariably concerned with the accumulation of calcium ions in the various tissues during the moulting cycle (Travis, 1955; McWhinnie, 1962; Mc Whinnie et al., 1969; Adegboye, 1975; Adegboye et al., 1975, 1978). The balance of calcium between the hemolymph and the environment (water) has been worked out by Greenway (1972, 1974), while the effect of dietary calcium on the growth rate of crayfish has been investigated by Westman (1972). Studies by Louisiana astacologists have clearly indicated that the addition of calcium salts to the ponds increased crayfish yields considerably (Avault, 1972; et al., 1970; de la Bretonne et al., 1969). The increase in crayfish production following liming is probably directly related to the increased accumulation of calcium ions in the tissues of the crayfish (Adegboye et al., 1978b).

While the effects of the moulting cycle, as well as those of the environmental calcium on the level of calcium in the hemolymph are well documented, very little is known about the influence of calcium in the tissues on the hemolymph calcium concentrations. The moulting process is perhaps the greatest single factor that influences the concentration of calcium in the hemolymph (McWhinnie, 1962; Adegboye et al., 1975b). However, the moulting process per se effects changes in hemolymph calcium concentrations only indirectly and also as a result of previous moult-related changes in the exoskeleton and other tissues. Thus, the tissues of the crayfish are the vehicles by which the moulting process influences the level of calcium in the hemolymph. Therefore, any discussion of calcium homeostasis in the crayfish or in any other animal for that matter, must include the interrelationships between the concentrations of calcium in the tissues.

Since calcium is an important ion in the growth development and production of the crayfish, any model that elucidates the homeostasis of calcium will go a long way to broaden our knowledge of the role of calcium in crayfish physiology and economics. This paper tackles the problems of calcium homeostasis in the crayfish by considering the interrelationships of calcium ions in the tissues of the crayfish. The observations made in this study are combined with the data from the literature to produce a model of calcium homeostasis in the crayfish.

MATERIAL AND METHODS

The specimens of crayfish used in this study were obtained from Carolina Biological Supply Company of Burlington, North Carolina, U.S.A. The stages in the moulting cycle of the crayfish were determined with the methods of Stevenson (1972, 1975). The male form I crayfish used in the analysis were identified with the criteria laid down by Hobbs (1973). Methods for determining the concentration of calcium ions in the tissues have been described elsewhere (Adegboye et al., 1975a, 1978b). Hemolymph calcium was determined by a microfluorometric method, while calcium in the ash of the hepatopancreas, exoskeleton, and gastrolith was determined by automated colorimetric spectrophotometry.

The Crayfish Constant was estimated by dividing the wet body weight by the cube of the carapace length (W/L^3). As explained in a previous work, the stages in the moulting cycle were assigned numerical units to facilitate statistical analysis: The period immediately following ecdysis (=A) was assigned "1"; the late postmoult state (=B) was given "2"; Early intermoult stage (=C_1-C_2), "3"; Late intermoult stage, "4"; Early premoult stage (=D_1-D_2), "5"; and the Late premoult stage (=D_3-D_4) was assigned the number "6' (Adegboye, et al., 1975b). The Physiological Condition Factor of the crayfish was estimated by multiplying the Crayfish Constant by the numerical equivalence of the stage in the moulting cycle.

Measurements of the physical parameters as well as chemical analyses were carried out in the Department of Zoology and Pharmacology of the University of North Carolina Chapel Hill, U.S.A. Statistical analyses of the data were carried out at the Computer Center, Ahmadu Bello University, Zaria, Kaduna State, Nigeria. Multiple regression analyses were performed on the data using SPSS version 6.50 of April 1, 1976 from the Northwestern University, U.S.A. Coefficient of Determination (=R^2) was used in the case of multiple regression analysis to test the level of significance of the various relationships examined in this study.

Body size, the moulting cycle, the availability of calcium in the environment, as well as the relationships between the concentrations of calcium ions in the tissues were considered in the construction of a model of calcium homeostasis in the crayfish.

138

RESULTS

As shown in Table 1, there were no significant relationships between the concentration of calcium in the hemolymph and the concentration of calcium in any of the other tissues of the crayfish during the late intermoult stage (C_4) and the late premoult stage (D_3-D_4). However, during the early premoult stage, the concentration of calcium in the hemolymph was highly correlated with the concentrations of calcium in the exoskeleton and hepatopancreas.

The synergistic effect of calcium ion concentrations in the tissues on the hemolymph of the crayfish during the late intermoult stage is summarized in Table 2. At this stage, there were no significant relationships between hemolymph calcium and a function of age and tissue calcium concentrations $(P > 0.05)$. However, during both the early and late premoult stages (D_1-D_4), the concentration of calcium in the hemolymph, was under the synergistic influence of the calcium ions in the other calcium compartments of the crayfish (Table 3 and 4). The hemolymph, exoskeleton and hepatopancreas were the major cacium compartments during the late intermoult and early premoult stages. However, the late premoult crayfish had five major stores of calcium, namely, the hemolymph, old skeleton, hepatopancreas, new exoskeleton and the gastrolith.

During the late premoult stage (D_3/D_4), the concentration of calcium in the hemolymph was under the synergistic influence of tissue calcium only in the larger specimens of crayfish (see Table 4, and compare R^2 of equaltion 1 and 2). As shown in Table 4 equation 1, the concentration of calcium in the hemolymph was not significantly related to the function of tissue calcium concentrations. The concentration of calcium in any of the other tissues (Exoskeleton, hepatopancreas, new exoskeleton or gastrolith) was directly related to the concentrations of calcium ions in the other tissues (Table 4 equations 3 to 10). The equations shown in Table 4 indicate that the body size played a role in the quantity of calcium ions contributed by the various tissues in calcium homeostasis. For instance, the bigger (or older) the crayfish, the larger was the contribution of calcium ions by the other tissues to the hemolymph, gastrolith and the new exoskeleton. However, during the same late premoult stage (D_3/D_4), the contribution of calcium ions from the other tissues to the old exoskeleton and hepatopancreas was drastically reduced in the larger specimens of crayfish (Table 4).

The contributions of calcium ions from the tissues of the crayfish to any particular tissue of the crayfish during the late premoult stage are summarized in Table 5. In the first instance, the higher the concentration of calcium in the new exoskeleton, the hepatopancreas and the gastrolith, the less was the concentration of calcium retained in the old exoskeleton. Furthermore, the higher the concentrations of calcium in the old exoskeleton, the new exoskeleton and the gastrolith, the less was the concentration of calcium in the hepatopancreas. Similarly, the higher the concentrations of calcium in the old exoskeleton, hepatopancreas, and gastrolith, the less was the concentration of calcium stored in the new exoskeleton. Finally, if less calcium was retained in the exoskeleton and less calcium was stored in the hepatopancreas and the new exoskeleton, the higher was the concentration of calcium in the gastrolith.

139

Table 1. Relationships between Hemolymph calcium ion concentrations of the various tissues of the crayfish:

ExoCa	-	Exoskeleton Ca mg % ash
HepCa	=	Hepatopancreas Ca mg % ash
NSCa	=	New Exoskeleton Ca mg % ash
GasCa	=	Gastrolith Ca mg % ash

Late Intermoult Stage C$_4$

	HemCa	ExoCa	HepCa
HemCa			
ExoCa	0.13		
HepCa	-0.14	-0.50	

Early PreMoult Stage D$_1$-D$_2$

	HempCa	ExoCa	HepCa
HemCa			
ExoCa	0.83		
HepCa	0.85	0.84	

Late Premoult Stage D$_3$-D$_4$

	HemCa	ExoCa	HepCa	NSCa	GasCa
HemCa					
ExoCa	0.19				
HepCa	-0.08	-0.93			
NSCa	-0.37	-0.06	-0.06		
GasCa	0.25	0.02	-0.11	-0.56	

140

Table 2. Relationship between Hemolymph Calcium and other parameters of the crayfish during the late intermoult stage (C_4).

A = Exoskeleton Ca	mg % ash
B = Hepatopancreas Ca	mg % ash
CL = Carapace Length cm	
BW = Body Weight gm	
K = Crayfish Constant gm/cm^3	
F = Physiological Condition	

n = 10

		R^2
Hemolymph Ca = 42.01 + 0.93"A"	- 0.37"B"	0.02
Hemolymph Ca = 7.94 - 15.98"CL"	+ 1.61"A" + 1.65"B"	0.38
Hemolymph Ca = -39.99 - 1.17"BW"	+ 1.67"A" + 2.65"B"	0.53
Hemoplymph Ca = 20.27 - 35.52"K"	+ 0.44"A" + 0.40"B"	0.22
Hemolymph CA = 29.27 - 8.88"F"	+ 0.45"A" + 0.40"B	0.23

Table 3. Relationships between the Hemolymph calcium concentration and other parameters of the crayfish during the early premoult stage (D_1-D_2).

N = 10

A = Exoskeleton Ca	mg % ash
B = Hepatopancreas Ca	mg % ash
CL = Carapace Length cm	
BW = Body Weight gm	
K = Crayfish Constant gm/cm^3	
F = Physiological Condition	

		R^2
1. Hemolymph Ca = 26.25 + 0.23"A" + 2.35"B"		0.77
2. Hemolymph Ca = -65.63 + 31.73"CL" - 0.15"A" - 0.032"B"		0.83
3. Hemolymph Ca = 14.50 + 1.51"BW" + 0.13"A" + 0.06"B"		0.78
4. Hemolymph Ca = 104.47 - -255.87"K" + 0.067"A" + 2.16"B"		0.95
5. Hemolymph Ca = 105.93 - 52.15"F" + 0.062"A" + 2.17"B"		0.95

Table 4. Relationships between the concentrations of calcium in the various compartment of the crayfish during the late premoult stage.

CL = Carapace Length   cm

A = Exoskeleton Ca   mg % ash

B = Gastrolith Ca   mg % ash

C = New Skeleton Ca   mg % ash

D = Hepatopancreas Ca   mg % ash

E = Hemolymph Ca   mg/100ml

n = 7

Equation

						R^2
1. Hemolymph Ca	= -18.49 + 1.09"A"	- 0.097"B"	- 1.06"C"	+ 5.15"D"		0.21
2. Hemolymph Ca	= -54.32 + 49.66"CL"	+ 0.21"A"	- 0.54"B"	- 16.13"C"	+ 7.57"D"	0.92
3. Exoskeleton Ca	= 71.15 + 0.065"E"	- 5.36"D"	- 0.074"B"	- 1.00"C"		0.91
4. Exoskeleton Ca	= 74.45 - 3.66"CL"	+ 0.033"B"	+ 0.13"E"	+ 5.84"D"	+ 0.19"C"	0.91
5. Hepatopancreas Ca	= 12.49 + 0.0096"E"	+ 0.17"A"	- 0.014"B"	- 0.19"C"		0.91
6. Hepatopancreas Ca	= 9.86 - 3.64"CL"	+ 0.034"B"	+ 0.071"E"	- 0.089"A"	+ 1.08"C"	0.96
7. New Skeleton Ca	= 21.71 - 0.015"E"	- 1.38"D"	- 0.046"B"	+ 0.23"A"		0.54
8. New Skeleton Ca	= -2.68 + 3.03"CL"	- 0.034"B"	- 0.055"E"	+ 0.0011"A"	+ 0.40"D"	0.99
9. Gastrolith Ca	= 304.66 + 0.27"E"	- 20.73"D"	- 9.33"C"	- 3.55"A"		0.51
10. Gastrolith Ca	= -58.90 + 82.24"CL"	- 1.47"E"	- 0.15"A"	- 27.40"C"	+ 9.81"D"	0.95

Table 5. The apparent competition between the various tissues for calcium ions in the crayfish during the late premoult stage D^3-D^4.

		Hem Ca	Exo Ca	Hep Ca	New ExoCa	GasCa
↑ = Rise in Ca concentration						
↓ = Fall in Ca concentration						

RISE IN:

	Hem Ca	Exo Ca	Hep Ca	New ExoCa	GasCa
Old Skeleton Ca ↑	↑		↓	↓	↓
Hepatopancreas Ca ↑	↑	↓		↓	↓
New Skeleton Ca ↑	↓	↓	↓		↓
Gastrolith Ca ↑	↑	↓	↓	↓	

Thus, as shown in Table 5, it appeared as if the old exoskeleton was "competing" with the hepatopancreas, new skeleton and gastrolith for calcium ions. Hepatopancreas was competing against the old exoskeleton, new skeleton, and gastrolith; while the new skeleton was competing against the hemolymph, the hepatopancreas and gastrolith. And lastly, the gastrolith competed against the old exoskeleton, hepatopancreas, and the new exoskeleton for calcium ions in the overall calcium homeostasis of the crayfish.

DISCUSSION

Studies carried out so far indicate that the cyclical fluctuation in in the levels of calcium ions in the hemolymph, exoskeleton, hepatopancreas and gastrolith are directly related to the moulting process of decapod crustaceans Travis, 1955; McWhinnie, 1962; McWhinnie et al., 1969; Adegboye et al., 1978a). Apart from the moulting process, the concentration of calcium ions in the environment (water) is also a major factor which influences the level of calcium ions in the tissues of the crayfish (Adegboye et al., 1978c). The influence of environmental calcium ions on the levels of calcium ions in the tissues has been given a practical boost in crayfish industry by the reports of the Louisiana group and Westman who found that the addition of calcium ions to the ponds and diets significantly increased the table size of the crayfish (de la Bretonne et al., 1971; Avault et al., 1970; Avault, 1972; Westman, 1972).

In previous studies, the conclusions reached on the relationships between the concentrations of calcium ions in the various tissues of the crayfish have always been based on circumstantial evidence. Although, it is generally agreed upon by the carcinologists that the increase in the concentration of calcium in the soft tissues is due to the reabsorption of calcium from the exoskeleton prior to ecdysis, there is really no actual experimental observation to support the idea. In this paper, however, we have been able to determine the extent of the interrelationship between the concentrations of calcium in the tissues by means of simple statistical analyses.

From the observations made in this study, it is clear that there is an apparent competition for calcium ions among the tissues of the crayfish. During the late intermoult stage (C_4) the major tissues of the crayfish which compete for the available calcium ions are the exoskeleton, hemolymph, and the hepatopancreas. The picture of calcium storage in the crayfish during the early premoult stage (D_1-D_2) is similar to that of the late intermoult stage (C_4). However, during the late premoult stage (D_3-D_4), the available calcium ions are distributed among five compartments, namely, the hemolymph, old exoskeleton, hepatopancreas, new exoskeleton and the gastrolith. The major structures involved in calcium homeostasis in the crayfish are schematically represented in Figures 1-5.

The two major sources of calcium ions for the decapods are the environment (Miyawaki et al., 1961; Travis, 1955; Digby, 1967; Adegboye et al., 1978b) and the ingested food (Digby, 1967; and Westman, 1972). The kidneys, membranous surfaces, as well as the gills are probably responsible for the uptake and losses of calcium ions from the body of the crayfish Bryan, 1960; Bergmiler and Bielawski, 1979; Kroghan,

Figure 1: The major calcium compartments of the crayfish during stage C_4, late intermoult stage. S = Exoskeleton, (EX = Exocuticle + ED = Endocuticle), EP = Epidermis, E = Ecdysial Gland, N = Neurosecretory Cells, K = "Kidney," HEM = Hemolymph, P = Hepatopancreas, MED = Water, PM = Permeable Membrane.

Figure 2: Calcium homeostasis in the crayfish. Stage D_1-D_2 (early premoult stage). The major calcium compartments are Exoskeleton ("S" = EX + ED); Hepatopancreas ("P") and Hemolymph ("HEM"). The epidermal layer (has been separated from the endocuticle ("ED").

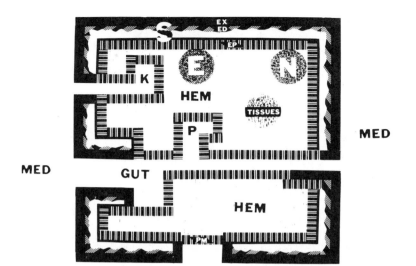

Figure 3: Calcium homeostasis in the crayfish. Stage D_2-D_3 (mid-premoult stage). The major calcium compartments are Exoskeleton ("S" = EX + ED); Hepatopancreas ("P") and Hemolymph ("HEM"). <u>The new exoskeleton is laid down on the distal part of the withdrawn epidermal layer. The endocuticle undergoes resorption while the gastrolith ("g" is being formed at this stage.</u>

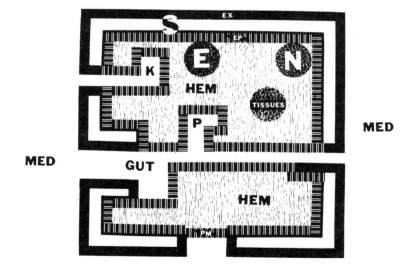

Figure 4: Calcium homeostasis in the crayfish. Stage D_3-D_4 (late premoult stage). The major calcium compartments are Exoskeleton ("S" = EX), Hepatopancreas ("P"), Hemolymph ("HEM"), new Exoskeleton (distal to epidermal layer "EP"), Gastrolith ("G").

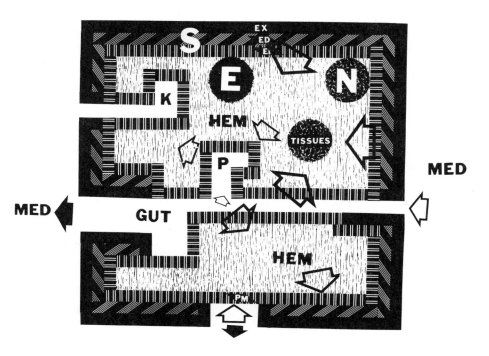

Figure 5: Model of calcium homeostasis in the crayfish. Arrows indicate the probable paths of calcium. Black solid arrows indicate losses into the environment.

1958a, b, c: Mantel, 1967; Krogh, 1939; Maluf, 1940, 1941; Shaw, 1960; Kamemoto and Ono, 1969; Ramamurthi and Scheer, 1967; Travis, 1954; Dandrifosse, 1966). The stores for the calcium ions in the crayfish have already been enumerated elsewhere in this report.

Calcium homeostasis during the late intermoult stage.

During the late intermoult stage (C_4), there is no direct relationship between the concentrations of calcium in the hemolymph and the exoskeleton; and between the hemolymph calcium and hepatopancreas calcium. It is obvious that at this stage, the calcium ions that are stored in both the hepatopancreas and the exoskeleton are not available for correcting the losses in calcium ions that the hemolymph experiences during the intermoult stage. The experimental observations of Greenaway (1972, 1974) and Adegboye et al., (1978a) lend support to the idea that the loss of calcium from the hemolymph into the environment (which is low in calcium ions) is not corrected for by the very large concentration of calcium ions in the exoskeleton. The exoskeleton is separated from the hemolymph by the epidermal layer of cells (Stevenson, 1975) and no reabsorption of calcium ions from the exoskeleton into the hemolymph is possible until after the moulting hormone-mediated apolysis or the withdrawal of the epidermal layer from the exoskeleton. However, during the late intermoult stage, no reabsorption of calcium ions from the exoskeleton has been reported. In their radioautographic studies of Ca^{45} uptake by the crayfish <u>Procambarus clarkii</u>, Miyawaki et al. (1961) have shown that Ca^{45} <u>absorption was mainly</u> via the hepatopancreas, and not via the abdominal muscles and gills. Thus, it appears that the accumulation of calcium in the hepatopancreas, might not be via the hemolymph. The quoted pieces of experimental evidence lend support to our statistical observation that the concentrations of calcium in the hepatopancreas or exoskeleton has no significance influence on the concentration of calcium in the hemolymph during the late intermoult stage (C_4).

Since calcium ion is a major factor in the physiology of the crayfish, the level of calcium in the hemolymph must be maintained at the level that the physiological role of calcium demands (Borle, 1967). As we have proved earlier on, the extracellular concentration of calcium ions cannot be maintained at a relatively constant level by either the calcium ions in the exoskeleton or the hepatopancreas. Therefore, it is suggested that the environment (de la Bretonne et al., 1969; Adegboye et al. 1978a) and the diet (Westmanm, 1972) are the only possible sources of calcium ions to regulate the concentration of calcium in the hemolymph during the late intermoult stage (C_4). Crayfish have been observed to supplement their calcium intake by consuming the calcareous exuvia during the postmoult period (unpublished reports).

Calcium homeostasis during the early premoult stage.

At the beginning of the premoult period (stage D_1-D_2) after the moulting hormone-mediated apolysis has taken place and the reabsorption of the old exoskeleton has commenced, there appears to be a relationship between the concentration of calcium ions in the hemolymph and the calcium ions concentrations in the hepatopancreas and the exoskeleton. The significant direct relationships between the hemolymph, hepatopancreas and the exoskeleton calcium would suggest that during the early

148

premoult stage, there is a substantial calcium mobilization between the tissues. The higher the concentration of calcium in the exoskeleton and the hepatopancreas, the higher is the concentration of calcium stored in the hemolymph during the early premoult stage (Table 3, equation 1). The bigger the crayfish, the bigger is the synergistic effect of the calcium ion concentrations in the exoskeleton and hepatopancreas on the hemolymph calcium (Table 3, equations 2 and 3).

Calcium homeostasis in the late premoult stage.

The problem of calcium homeostasis in the crayfish during the late premoult stage is complicated by two major events: 1) the excessive reabsorption of calcium ions from the exoskeleton and 2) the increase in the number of possible calcium compartments from the original three to five. One of the problems of calcium homeostasis during this period is the fact that the concentration of calcium in the hemolymph bears no significant relationships with the concentration of calcium in any of the tissues of the crayfish. Indeed, the hepatopancreas, gastrolith, new exoskeleton, an old exoskeleton do not exert any synergistic effect on hemolymph calcium concentration (Table 4, equation 1). Any irregularities observed in the distribution of calcium ions in the tissues of the crayfish prior to ecdysis might be explained by the fact that the quantity of calcium ions resorbed from the exoskeleton might not necessarily follow any pattern. However, in the bigger and older crayfish, the synergistic influence of all the (solid) tissues in regulating the hemolymph calcium becomes significant. This is understandable, since the larger the crayfish, the larger is the reabsorbable quantity of calcium in the exoskeleton. If the quantity of reabsorbed calcium ions is large enough, there will be significant increases in hemolymph calcium concentrations. The observations on calcium homeostasis during the late premoult stage further suggests that if a large quantity of calcium ion is retained in the old exoskeleton, the concentrations of calcium in the hepatopancreas, new exoskeleton and gastrolith will be low (Table 5). Conversely, a rise in calcium ion concentrations in the hepatopancreas, new exoskeleton and the gastrolith is as a result of extensive calcium reabsorption from the old exoskeleton. The bulk of the calcium ions that is mobilizable during the late premoult stage must have originated from the exoskeleton. Since the hemolymph is the extracellular fluid that bathes both the old exoskeleton, the hepatopancreas and the gastrolith, it is obvious that the higher the concentration of calcium is, in the hemolymph, the greater would be the quantities of calcium ions stored in hepatopancreas and the gastrolith prior to ecdysis (see Table 5). The new exoskeleton appears to benefit from the excessive load of calcium in the hemolymph during the late premoult stage: The more calcium ions are retained in hemolymph, the less the quantities of calcium ions that are deposited in the new exoskeleton and vice versa.

That the hepatopancreas, new exoskeleton and the gastrolith are the major stores of calcium has been confirmed by this report. It has also been shown that all of the three compartments are, as it were, in open competition for the calcium ions reabsorbed from the old exoskeleton. Therefore, any increase in calcium ions in one of the three compartments would mean reduction in the quantities of calcium stored in the remaining two compartments (Table 5).

Calcium homeostasis in the crayfish.

From the information presented, it is very obvious that the problem of calcium homeostasis in the crayfish is far more complex than it was at first imagined. While calcium homeostasis in vertebrates is concerned with the acquisition, absorption, transportation, storage, utilization and mobilization of calcium ions, calcium homeostasis in the crayfish is further complicated by the phenomenon of moulting, which is the price that all arthropods pay for their hard and protective exoskeleton. According to Langley (1965), the term homeostasis embraces the idea that within a living organism there are self-regulatory process that serve to maintain the internal environment relatively constant. Furthermore, when, and if the internal environment is disturbed, the self-regulatory processes are able to return the internal environment to near normal.

As indicated by this report, the classical picture of calcium homeostasis in vertebrates as suggested by Talmage (1969) appears to be none-existent in the crayfish at least not during the late intermoult period. In the vertebrate, the level of calcium ions in the blood is regulated by the dietary intake of calcium, kidney reabsorption as well as calcium mobilization between the bone tissue and the blood. The exoskeleton of the crayfish does not appear to participate (directly by means of Ca ions) in regulating the calcium levels during the late intermoult stage. In both the vertebrate and the crayfish, hormonal control of calcium mobilization exists. While the parathyroid hormone raises the concentration of calcium in the blood throughout the life of the normal vertebrate, the hormonal control of calcium homeostasis is not as direct in the crayfish as it is in the vertebrate. During the intermoult period the contribution of hormones to calcium homeostasis is probably restricted to the hormonal-control of membrane permeability to ions such as calcium. Hormone-mediated contribution of calcium ions by the exoskeleton to the hemolymph is restricted to the premoult period in the crayfish. Could the major difference between calcium homeostasis in vertebrates and the crayfish be due to the fact that the vertebrate possesses an internal calcareous skeleton from which calcium ions could be mobilized to regulate the calcium concentrations in the blood while the crayfish possesses a physiologically inaccessible store of calcium in an external skeleton? How is it that the tissues of the intermoult crayfish can tolerate high concentrations of calcium ions in the extracellular fluid (41 to 50 gm Ca/100ml) while the vertebrate physiology depends on a much lower level of calcium in the serum (9 to 12 mg Ca/100ml)? These and many other questions on calcium homeostasis in the crayfish demand our urgent attention. According to Huxley (1880):

"The crayfish may be regarded as a kind of chemical manufactory -- and the first physiological problem which offers itself to us is the mode of operation of the apparatus contained in this factory, and the extent to which the products of its activity are to be accounted for by reasoning from known physical and chemical principles."

ACKNOWLEDGMENTS

Much of the author's ideas about calcium homeostasis orginated from the lessons he received from Professors Paul L. Munson, Philip F. Hirsch, Cary W. Cooper, Kenney T. Gray, T-C Peng, Roy V. Talmage, Anderson and Professor Irvine Ray Hagadorn, all of the University of North Carolina, Chapel Hill, U.SA. Messrs. Simon Ebele and Steven Olorunju helped in statistical procedures. This study has been made possible by funds provided by the Ahmadu Bello University, Zaria, and the United States Agency: "The African-American Institute," New York, U.S.A. Mr. Halilu Usman was kind enough to type the manuscript.

LITERATURE CITED

Adegboye, J.D. 1975. Regulation of hemolymph calcium in the crayfish, Procambarus acutus acutus. Girard (Crustacea, Decapoda) PhD. dissertation, University of North Carolina, Chapel Hill, N.C. 27514, U.S.A.

Adegboye, J.D., I.R. Hagadorn and B.F. Hirsch. 1975a. Microfluorometric determination of Calcium in the hemolymph and other tissues of the Crayfish. 2nd Int'l. Crayfish Symposium; Baton Rouge, La; U.S.A. Freshwater Crayfish (2):211-225.

Adegboye, J.D., I.R. Hagadorn and P.F. Hirsch. 1975b. Variations in hemolymph calcium associated with the moulting cycle in the crayfish. 2nd Int'l Crayfish Symposium; Baton Rouge, La:U.S.A., Freshwater Crayfish (2):227-247.

Adegboye, J.D, I.R. Hagadorn and P.F. Hirsch. 1978a. Metabolism of calcium in the crayfish. 4th Int'l; Crayfish Symposium Thonon, France. Freshwater Crayfish 4:25-34.

Adegboye, J.D., I.R. Hagadorn and P.F. Hirsch. 1978b. Factors modifying calcium concentration in the hemolymph of the crayfish Procambarus acutus acutus Girard. 4th Int'l Crayfish Symposium. Thonon, France. Freshwater Crayfish, 4:1-14.

Adegboye, J.D., I.R. Hagadorn and P.F. Hirsch, 1978c. Hypercalcemic effect or crustecdysone in the crayfish Procambarus acutus acutus Girard. 4th Int'l Crayfish Symposium. Thonon, France. Freshwater Crayfish 4:15-24.

Avault, J.W. Jr., L. de la Bretonne, Jr., and E. Jaspers. 1970. Culture of crawfish, Louisiana's crustacean king. The American Fish Farmer 1(10):8-14 and 27.

Avault, J.W. Jr. 1972. Crayfish farming in the United States. Freshwater Crayfish 2:240-250.

Bergmiler, E., and J. Blelawski. 1979. Role of the gills in Osmotic regulation in the crayfish Astacus heptodactylyus. Esch. Comp. biochem. Physiol. 37:85-91.

Borle, A.B 1967. Correlation between morphological biochemical and biophysical effects of parathyroids hormone on cell membranes (monkey kidney, toad bladder). Calcified Tissues Proc. Europe Symp. 3:45-48.

Bryan, G.W. 1960. Sodium regulation in the crayfish Astacus fluriatilis. I, II, III J. Exp. Biol. 37:83-128.

de la Bretonne, L. Jr., J.W. Avault, Jr. and R.O. Smitherman. 1969. Effects of soil and water hardness on survival and growth of red swamp crawfish, Procambarus clarkii, in plastic ponds. Proceedings of the 23rd Annual Conference of Southeastern Association of Game and Fish Commissioners 23:626-633.

de la Bretonne, L. Jr. and J.W. Avault, Jr. 1971. Liming increases crawfish production. Louisiana Agriculture 15 (1):10.

Dandrifosse, G. 1966. Absorption d'eau au moment de la nue chez un crustace decapode: Maria Squinado Herbrt. Archs Int. Physiol. Biochem. 74:329-331.

Digby, P.S.B. 1967. Calcification and its mechanism in the shore-crab, Carcims maenas (L) Proce. Linn. Soc. Lond. 178:91-108.

Greenaway, P. 1972. Calcium regulation in the freshwater crayfish Austropotamobium pallipes (Lereboullet). 1. Calcium balance in the intermoult animal. J. Exp. Biol. 57:471-87.

Greenaway, P. 1974a. Calcium balance at the premoult stage of the freshwater crayfish Asutropolamobius pallipes (Lereboullet) J. Exp. Biol. 61:27-34.

Greenaway, P. 1974b. Calcium balance at the post-moult stage of the freshwater crayfish Austropotamobius pallipes (Lereboullet), J. Exp. Biol. 61:35-45.

Hobbs, H.H. Jr. 1973. Crayfishes (Astacidae) of North and Middle America Biota of Freshwater Ecosystems: Identification Manual No. 9. Environmental Protection Agency. Washington, D.C.

Huxley, T.H. 1880. The crayfish. Note II. on chap. I, pp. 29, 346-349. D. Appleton and Company, New York.

Kamemoto, F.I., and J.K. Ono. 1969. Neuroendocrine regulation of salt and water balance in the crayfish Procambarus Clarkii. Comp. Biochem. Physiol. 29(1):393-401.

Krogh, A. 1939. Osmotic Regulation in Aquatic Animals. Cambridge University Press, New York.

Kroghan, P.C. 1958a. The Osmotic and ionic regulation of Artemia Salina J. Exp. Biol. 35:219-233, 243-249.

Kroghan, P.C. 1958b. The mechanism of osmotic regulation in Artemia salina L. The physiology of the branchiae. J. Exp. Biol. 35:234-242.

Kroghan, P.C. 1958c. The survival of Artemia salina L. in various media. J. Exp. Biol. 35:213-218.

Maluf, N.S.R. 1940. The uptake of inorganic electrolytes by the crayfish. J. Gen. Physiol. 24:151-167.

Maluf, N.S.R. 1941. Secretion of Imilin, Xylose and dyes and its bearing on the manner of Urine formation by kindey of the crayfish. Biol. Bull., Woods Hole. 81:235-260.

Mantel, L.H. 1967. The foregut of Gecarciumus lateralis as an organ of water balance. Am. Zool 7:765.

McWhinnie, M.A. 1962. Gastrolith growth and calcium shifts in the freshwater crayfish, Orconectes virilis. Comp. Biochem. Physiol. 7:1-14.

McWhinnie, M.A. and P.N. Saller, Sr. 1969. An analysis of blood sugars in the crayfish, Orconectes virilis. Anat. Rec. 134(3):604.

Miyawaki, M. and N. Sasaki. 1961a. Ca^{45} by the hepatopancreas of crayfish Procambarus clarkii. Kumamoto J. Sci. 5:170-172.

Miyawaki, M., and N. Sasaki. 1961b. Histochemical studies on the hepatopancreas of the crayfish, Procambarus clarkii. Kumamoto J. Sci. 5(b), 2:161-169.

Ramamurthi, R and B.T. Scheer. 1967. A factor influencing sodium regulation in crustaceans. Life Sci. 7:2171-2175.

Shaw, J. 1960. The absorption of chloride ions by the crayfish Astacus pallipes Lereboullet. J. Exp. Biol. 37(3):557-572 (also 534-556).

Stevenson, J.R. 1972. Changing activities of the crustacean epidermis during the moulting cycle. American Zoologist 12:373-380.

Stevenson, J.R. 1975. The molting cycle in the crayfish: recognizing the molting stages, effects of ecdysone and changes during the cycle. Freshwater crayfish 4:255-269.

Talmage, R.V. 1969. Calcium homeostasis-calcium transport parathyroid action. Clin. Orthop. Related Res. 67:210-224.

Travis, D.F. 1954. The moulting cycle of the spiny lobster Panulirus argus Latreille. 1. Moulting and growth in laboratory-maintained individuals. Biol. Bull. 107:433-450.

Travis, D.F. 1955. The moulting cycle of the spiny lobster, Panulirus argus Latreille. III. Physiological changes which occur in the blood and urine during the normal moulting cycle. Biological Bulletin Marine Biology Lab., Woods Hole, 109:484-503.

Travis, D.F 1957. The moulting cycle of the spiny lobster, Panulirus argus Latreille. IV. Postecdysial histological, and histochemical changes in the hepatopancreas and integumental tissues. Biological Bulletin, 133(3):451-479.

Westman, L. 1972. Cultivation of the American crayfish Pacifastacus leniusculus. Freshwater Crayfish 1:212-220.

THE "CRAYFISH CONDITION FACTOR":
A TOOL IN CRAYFISH RESEARCH

Duro Adegboye
Department of Zoology
University of North Carolina
Chapel Hill, NC 27514, U.S.A.

ABSTRACT

The increased global interest in crayfish research demands an internationally acceptable means of standardizing the physiological conditions of the crayfish under investigation. A "Crayfish Constant, K", with a mean of $0.222 \pm$ S.E. 0.003 and a "Crayfish Condition Factor, F" with a value of $0.855 \pm$ S.E. 0.023 were calculated for 147 specimens of _Procambarus astacus astacus_. The "Crayfish Constant, K" was constant for all sizes and both sexes of the crayfish in all stages of moulting cycle $(P < 0.001)$. The need to introduce a "Crayfish Condition Factor" into astacology is discussed.

INTRODUCTION

One of the major problems confronting researchers in the interpretation of reports on the physiology of the crayfish is the fact that, at present, there are no ways of precisely stating the physiological status of the crayfish under investigation. Although, the size and the stage in the moulting cycle of the crayfish are often given in most reports, the interplay of these factors in the overall physiology of the crayfish is often not easily realized. In order to facilitate comparative study of the physiology of the various species of crayfish from different laboratories, it is desirable to introduce an internationally acceptable "condition factor" which could be used to standardize the physiological state of the crayfish. Such a "condition factor" could be used to compare individual specimens within an experimental group or it could be used to compare different crayfish populations.

Body size is one of the major factors which is correlated with certain physiological processes of crustaceans. Such processes include the metabolism, frequency and duration of moulting (Passano, 1960; Kamiguchi, 1971; Speck et al., 1972; McWhinnie et al., 1969; Stewart et al., 1972; Ennis, 1972; Travis, 1954; McWhinnie, 1962; Brown and Powell, 1972; Perkins, 1972), the onset of sexual maturity (Newman, and Pollock, 1971; Silbergbauer, 1971; Haefner, 1972; Harris et al., 1972; Brown and Powell, 1972) number of eggs on the pleopoed (Abrahamsson, 1972); the rate of embryonic development (Perkins, 1972) and the rate of larval development (Price and Chew, 1972) and the weight of the tail muscle of crayfish (Mikkola, 1978). The standard measure used in size determination of crustaceans include those of body weight (Siebers, 1972, Speck et al., 1972; McWhinnie et al., 1969; Stewart et al., 1972); total length Haefner, 1972; Price and Chew, 1972); carapace length (Armitage et al., 1973; Ennis, 1972, Harris, et al., 1972;

---

* Present Address: Department of Biological Sciences, Ahmadu Bello University, Zaria, Kaduna State, Nigeria.

Kamiguchi, 1971; Newman and Pollock, 1971; Haefner, 1972; Travis, 1954; McWhinnie, 1962); carapace width (Brown and Powell, 1972; Smith and Naylor, 1972; Spirito et al., 1972; Watson, 1972); and eye index (Perkins, 1972).

During the second international symposium on crayfish at Baton Rouge (Louisiana, U.S.A.), the need to standardize techniques on crayfish biometrics became obvious after several authors had presented reports in which different body dimensions were used as indices of size in the crayfish (Adegboye et al., 1975b; and others). A committee was therefore set up to look into the different techniques of body measurements in the crayfish. That committee (of which this author was a founding member) presented four linear measurements that could be used to estimate body size in the crayfish. These linear dimensions included total length, standard length, maximum carapace length, post orbital carapace length, and several dimensions of chela (Fitzpatrick, 1977).

Quite recently, Romaire, Forester and Avault (1977) and Fitzpatrick (1977) made detailed studies of interrelationships between different body dimensions of the crayfish. Previous studies on length-weight relationships include those of Adegboye et al. (1975b); Romaire et al. (1977) and Momot and Jones (1977). Romaire and co-workers have not only shown that growth is allometric "with weight increasing faster than the cube of the length," they have also provided regression equations for Form I male, Form II immature males and females of Procambarus acutus acutus and P. clarkii. According to Fitzpatrick (1977), a linear relationship exists between the standard carapace length, postorbital carapace length and maximum carapace length in the crayfish Procambarus vioscai. Previous studies on relative growth were carried out by using the various crustacean dimensions independently of each other (Travis, 1954, 1957; Watson, 1972; Tack, 1941; Hobbs et al., 1957; McWhinnie et al., 1964; Adegboye et al., 1978a, b, c). The present study was undertaken to examine several parameters of the crayfish with the view of estimating a "crayfish condition factor" which could be used to describe the physiological condition of the crayfish.

In selecting a "crayfish condition factor," the factor must meet the following criteria:

(i)   the factor must incorporate most, if not all, of the crayfish measurements already in use.

(ii)   it must be easy to calculate from the already known and easily determined parameters of the crayfish.

(iii)   it must be such that other parameters of the crayfish could be computed from it, and

(iv)   the factor must be directly correlated to, at least an aspect of the physiology of the crayfish.

The "crayfish condition factor" calculated for Procambarus acutus acutus met all the four criteria required for acceptability as set forth in the foregoing.

METHODS AND MATERIALS

(i) Crayfish parameters:

The specimens of Procambarus acutus acutus used in this study were obtained from Carolina Biological Supply Company of Burlington, North Carolina. The stages in the moulting cycle were determined by the method of Stevenson (1972). For statistical purposes, the early postmoult stage "A" crayfish were designated "stage 1," while stages B, C_{1-2}, C_{3-4}, D_{1-2}, and D_{3-4} were assigned 2, 3, 4, 5, and 6 respectively (Adegboye et al., 1974). Vernier calipers were used for all linear measurements, and a Mettler balance, Model 1200 (reading to nearest 0.01 mg) was used for determining crayfish weight. The total length of the crayfish was the rostrum-telson length, while the carapace length was the distance between the tip of the rostrum and posterior end of the carapace. The maximum width of the cephalothorax was taken as the carapace width of the crayfish.

(ii) Hemolymph Calacium

Crayfish hemolymph was obtained from the mid-dorsal portion of the cephalothorax, namely, the areola. The sample of hemolymph contained in a 5 μL disposable micropipette was rapidly and quantitatively transferred into the 5% weight/volume trichloroaetic acid solution contained in a Falcon tube. The tube was capped and inverted five times to mix thoroughly. By means of an automatic reagent dispenser, 3 ml of a calcein solution containing 1.2 mg calcein per 100 ml of potassium hydroxide solution (155g KOH in 2 litres of glass-distilled water) was added to the hemolymph-TCA mixture. Immediately after adding the calcein to each tube, the tube was capped and inverted five times before it was placed in a Turner Model 111 fluorometer. The flourescence of the calcium-calcein complex was measured at a wavelength between 425 mμ and 525 mμ (Adegboye et al., 1974).

(iii) The Crayfish Constant "K"

The "crayfish constant," was calculated from the ratio of body weight to the products of the linear measurements. The formula used in obtaining the "crayfish constant" for each specimens of crayfish was:

$$K = \frac{\text{Body Weight (fresh)}}{\text{Total Length X Carapace Length X Carapace Width}}$$

and the unit of "K" was gm/cm^3.

(iv) The crayfish Condition Factor "F"

The "Crayfish Condition Factor" "F" was derived from a product of the crayfish constant, "K", and the numerical equivalence of the stage in the moulting cycle, thus:

$$F = K \times \text{Moultling Stage}$$

The unit of "F" is $gm \ stage/cm^3$. Other forms of physiological condition factors were calculated for the crayfish, however, only the factor that had the highest correlation with the concentration of calcium in the hemolymph was selected.

(v) Statistical Analyses:

Statistical analyses were carried out at the Ahmadu Bello University Computer Centre. The programmes used included those for descriptive statistics (destat), correlation and regression as well as student's "t-test.". The correlation between any two parameters of the crayfish was determined by means of polynomial regression methods, and the level of significance in each statistical test was set at 0.05 or 95% confidence interval.

RESULTS

Physical Parameters of Procambarus acutus acutus:

Of the 147 specimens of crayfish investigated, only 21 were in the early postmoult stage A, 30 in the late postmoult stage B, 25 in the early intermoult stage C_{1-2}, 35 in the late intermoult stage D_{1-2} and 18 in the late premoult stage D_{3-4}. The specimens of crayfish measured between 35.00 and 98.00 mm total length, 15.00 and 45.00 mm carapace length, 6.20 and 21.60 mm carapace width. The crayfish weighed between 0.60 and 22.40 gm (Table 1). As shown in Table 2, the total length is directly related to the body weight, carapace length and carapace width respectively. ($P < 0.001$).

Hemolymph Calcium Concentrations:

The descriptive statistics of the hemolymph calcium concentration of the crayfish are show in Table 3. The mean value of hemolymph calcium is 43.22 mg per 100 ml with a standard error of $\pm$ 0.77. There is no significant correlation between any aspect of body size and the concentration of calcium in the hemolymph ($P < 0.05$). However, there is a direct relationship between the stage in the moulting cycle and the concentration of calcium in the hemolymph (Table 4). Although the hemolymph calcium is not directly related to any single linear measurement of the crayfish, several statitistical combinations of total length, carapace length and carapace width show positive correlation with the concentration of calcium in the hemolymph. The inclusion of body weight in the statistical combination of other physical parameters that were regressed against hemolymph calcium gives a higher correlation coefficient in each case. As shown in Table 5 and 6, the greatest direct correlation is obtained when the stage in the moulting cycle is combined statistically with the body weight, total length, carapace length, and carapace width and then regressed against hemolymph calcium concentrations.

The Crayfish Constant "K":

The mean value of "crayfish constant" for the 147 specimens of Procambarus acutus acutus is 0.222$\pm$ standard error of 0.003. Table 7 shows the "crayfish constant" of Procambarus at various stages of its moult cycle. The "crayfish constant" is relatively constant for all the crayfish irrespective of the stage in the moulting stage ($P < 0.05$). The "crayfish constant" of the female crayfish was not significantly different from that of the male crayfish (Table 8). As

Table 1. Body Size of Procambarus acutus acutus.

n = 147

	Total Length	Carapace Length	Carapace Width	Body Weight
Maximum	98.00	45.00	21.60	22.40
Minimum	35.00	15.20	6.20	0.60
Range	63.00	29.80	15.40	21.80
Mean	63.43	29.04	13.21	6.97
Variance	263.29	54.87	16.43	31.04
Std. Deviation	16.23	7.41	4.05	5.57
Std. Error	1.34	0.61	0.33	0.46
Mean Deviation	14.59	6.56	3.65	4.76
Median	64.00	29.40	13.60	5.20

Table 2. Relationships Between Total Length (mm) and other Measures of Body Size.

n = 147
T.L. = Total Length (mm)

Equation	Correlation	Test
TL = A + BX	R	P
TL = 44.39 + 2.73 (Body Weight bm)	0.939	<0.001
TL = 11.53 + 3.93 (Carapace Width mm)	0.981	<0.001
Tl = 0.81 + 2.16 (Carapace Length mm)	0.984	<0.001

Table 3. Hemolymph Calcium Concentration in _Procambarus_ _acutus_ _acutus_ (n = 147).

	mg Ca per100 ml
Maximum	80.00
Minimum	26.00
Range	54.00
Mean	43.22
Variance	83.94
Std. Deviation	9.16
Std. Error	0.77
Mean Deviation	6.68
Median	42.50

Table 4. Relationships Between Hemolymph Calcium Concentration Body Size of _Procambarus_ _acutus_ _acutus_.

Hem Ca = Hemolymph Ca mg/100 ml	Correlation F. Ratio	R	Test P
Hem Ca vs Total Length mm	0.40	0.054	>0.10
Hem Ca vs Carapace Length mm	0.77	0.074	>0.10
Hem Ca vs Body Weight	2.27	0.127	>0.10
Hem Ca vs Moulting Stage*	81.88	0.609	<0.001

*Indicates statistical significance

Table 5. Relationships between various aspects of body size and Hemolymph Calcium of _Procambarus acutus acutus._

Hemolymph Ca mg/100 ml. Vs Multivariates	F. Ratio	Correlation R	P
Total Length, Carapace Length	1.20	0.131	>0.10
Total Length, Width	0.32	0.109	>0.10
Carapace Length, Width	0.38	0.074	>0.10
Total Length, Weight*	3.76	0.227	<0.01
Carapace Length, Weight*	2.36	0.182	<0.05
Carapace Width, Weight*	2.85	0.199	<0.05
Total Length, Carapace Length, Width	0.95	0.143	>0.10
Total Length, Carapace Length, Weight*	0.72	0.237	<0.01
Total Length, Width, Weight*	2.49	0.227	<0.01
Total Length, Carapace Length, Width, Weight	2.03	0.238	<0.01
Total Length, Carapace Length, Width, Weight, Moulting Stage*	20.01	0.653	<0.001
(Hem. Ca)^2 Versus Total Length, Carapace Width, Moulting Stage*, Length, Weight	16.88	0.385	<0.001

n = 147

*Indicates statistical significance at 95% confidence interval or more.

Table 6. Relationship between Hemolymph Calcium and Multivariates.

G = Hemolymph Ca mg/100 ml
B = Total Length mm
C = Carapace Length mm
D = Carapace Width mm
E = Body Weight gm
F = Moulting Stage

G = 22.42 + 60.26"B" + 0.40"C" + 0.62"D" + 0.17"E" + 4.37"F"
r = 0.653
n = 147
P = <0.001

Table 7. The "Crayfish Constant K" of _Procambarus_ _acutus_ _acutus_.

Moulting Stage	n	Crayfish Constant K:	Condition Factor F
1	2	0.234 ± 0.002	0.234 ± 0.002
2	30	0.227 ± 0.004	0.454 ± 0.008
3	27	0.217 ± 0.005	0.651 ± 0.051
4	35	0.220 ± 0.003	0.880 ± 0.012
5	35	0.217 ± 0.005	1.085 ± 0.025
6	18	0.234 ± 0.008	1.404 ± 0.030

Table 8. Crayfish Constant of Male and Female Crayfish.

	Male	Female
	Mean $\pm$ S.E.	Mean $\pm$S.E.
Body Weight (gm)	7.50 ± 0.70	6.10 ± 0.50
Total Length (cm)	6.32 ± 0.20	6.40 ± 0.18
Carapace Length (cm)	2.91 ± 0.08	2.80 ± 0.08
Carapace Width (cm)	1.44 ± 0.05	1.42 ± 0.04
K =	0.253 ± 0.010	0.247 ± 0.012

Table 9. Lack of significance in the relationships between hemolymph calcium and products of linear dimensions and weight.

n = 147

Hem Ca = Hemolymph Ca mg/100 ml
C.L.   = C. Length = Carapace Length
C.W.   = C. Width = Carapace Width
T.L.   = Total Length

	Correlation	
	R	P
Hem Ca vs $\frac{Body\ Weight}{(Total\ Length)}$	0.165	>0.05
Hem Ca vs $\frac{Body\ Weight}{(Total\ Length) \times (C.\ Length) \times (C.\ Width)}$ *	0.126	>0.05
Hem Ca vs (Body Weight) x (T.L.) x (C.L.) x (C.W.)	0.140	>0.05

*Crayfish Constant

shown in Table 9 there is no significant relationship between the "crayfish constant" and the concentration of calcium in the hemolymph (P < 0.05).

The Crayfish Condition Factor, "F":

Table 10 shows the descriptive statistics of the "condition factor, F" of Procambarus acutus acutus. The mean value of "F" is 0.8543 with a standard error of ±0.023. The concentration of calcium in the hemolymph is directly related to the "crayfish condition factor" (Table 11). Table 12 shows the analysis of variance for relationship between hemolymph calcium concentration and the "crayfish condition factor." As shown in Table 13 there was no significant difference between the observed and calculated mean concentrations of calcium in the hemolymph of the crayfish.

DISCUSSION

Body Size as an Index of Physiological Condition:

Body weight, total length, carapace length and carapace width are the most frequently used dimensions in the study of crustaceans. Apart from a few studies (Takeda et al., 1970; Strel'nikova, 1970; Lasker et al., 1970; Belusova, 1970; Abolmasova, 1970; Kititsyna, 1970; and Chung, 1970) most carcinologists use just one dimension as the index of size. Body weight has been used by many workers as an index of size in decapod crustaceans (Siebers, 1972; Speck et al, 1972, Travis, 1954). In his study of the biology of the sand shrimp, Haefner (1972) used the total length as a measure of size. Price and Chew (1972) have also used total length measurements in the study of larval development of Pandalus platyceros.

Armitage et al. (1973) classified their specimens of Orconectes nais into large and small crayfish on the basis of carapace length measurements. In Homarus americanus, growth per moult was estimated by means of carapace length measurements (Ennis, 1972), while in the shrimp Pandalopsis dispar, a linear relationship existed between egg number and carapace length (Harris, et al., 1972). Kamiguchi (1971) represented the size of Palaemon paucidens as the carapace length. He found that the shorter specimens moulted more frequently than the longer ones. Moulting frequency was high in sexually immature specimens and this fequencey was dependent on the carapace length. Kamiguchi was also able to calculate an ovarian factor from the ratio of ovarian weight to the cube of the carapace length.

Kamiguchi's observation is a corollary of the one made by Harris, et al. (1972). This is because, since egg number is invariably related to ovarian weight, and egg number is related to carapace length (Harris et. al., 1972), the ovarian weight must be related to carapace length because:

1. Ovarian weight is related to body weight and in most cases

2. The body weight is related to carapace length (Adegboye et al., 1974b).

162

Table 10. Descriptive Statistics of "Condition Factor F".

Condition Factor = K. Moulting Stage.

Where K = $\dfrac{\text{Body Weight}}{\text{Total Length x Carapace Length x Carapace Width}}$ gm/cm^3

Units of Condition Factor = gm. Stage/cm^3

	gm. Stage/cm^3
Maximum	2.0357
Minimum	0.2326
Range	0.1804
Mean	0.8543
Median	0.8597
Std. Deviation	0.3410
Std. Error	0.023

Condition Factor = 0.8543 gm. stage/cm^3

Table 11. Relationships between Hemolymph Calcium and Products of Size and Moulting Stage.

$$n = 147$$

Hem Ca = Hemolymph Ca mg/100 ml
T.L.   = Total Length
C.L.   = Carapace Length
C.W.   = Carapace Width
B.W.   = Body Weight
M.S.   = Moulting Stage

	F. Ratio	R	P
Hem Ca vs (TL X CL X CW X BW)	12.74	0.290	<0.02
Hem Ca vs $\dfrac{\text{BW x MS}}{(\text{TL X CL X CW})}$	83.97	0.614	<0.001
Hem Ca vs $\dfrac{\text{BW}}{(\text{TL X CL X CW X MS})}$	17.58	0.335	<0.02

Table 12. The Relationship Between Hemolymph Calcium and "Condition Factor."

Hem. Ca = Hemolymph Ca mg/100 ml
    f = Condition Factor gm. stage/mm^3
Hem. Ca% = 29.1437 + 16.4724 (Condition Factor)
           Unit of Factor = (gm/cm^3) x stage.

Analysis of Variance

	Df	Sum of sq	Mean sq	F. Ratio
Regression	1	4425.74	4425.74	83.97
Residual	139	7325.62	52.70	
Total	140	11751.36		

         F   Probability = 0

Correlation Coefficient (R)

R = 0.6137

P < 0.001

Table 13. Back Calculations of Hemolymph Calcium Concentrations from the Equation Governing the Relationship between Hemolymph Calcium mg% and "Condition Factor."

Hemolymph Ca% = 29.1437 + 16.4724 (Condition Factor)
                   n = 30
              Hemolymph Ca mg 100 ml.

	Calculated	Observed
Maximum	53.39	60.00
Minimum	36.19	33.50
Range	17.20	26.50
Mean	45.05	45.67
Standard Error	±0.60	±0.62

Newman and Pollock (1971) were able to determine the state of sexual maturity in the female rock lobster Jasus lalandii by measuring the carapace length. They found that above 70 mm, over 70 percent of the lobsters were sexually mature, while all specimens above 85 mm were mature. Sexual maturity in broods of the sand shrimp were also indicated by carapace length measurements (Haefner, 1972). According to McWhinnie (1962), the ratio of gastrolith length to carapace length yielded consistent values at the time of moult for all sizes of crayfish and so it was possible to estimate the time of moulting by the use of the ratio obtained through radiography. According to Farmer (1973), the age of the lobster Nephrops norvegicus could be estimated from the carapace length of the lobster.

The size index of most crabs is based on carapace width (Brown et al., 1972; Smith and Naylor et al., 1972; Spirito et al., 1972). In commercial "fishing," the acceptable size of catch is also based on the carapace width measurements of the crab (Watson, 1972). Brown and Powell (1972) observed that the carapace width was related to chela width of the Alaskan tanner crab. In a plot of chela width against carapace width, they found that the break that occurred in chela width data at about 110 mm carapace width was an indication of average size at the moult of puberty of male crabs. The average size at which females attained maturity was at about 90 mm carapace width.

A linear relation was found between the carapace length and the number of eggs on the pleopod of Austropotamobius pallipes (Bowler and Brown, 1977), and Pacifastacus leniusculus (Mason, 1977). Similarly, there was positive correlation between body length and tail muscle weight in the crayfish Procambarus clarkii from Kenya (Mikkola, 1978).

Equations relating several measurement for body dimensions are useful in estimating relative physiological condition, growth rate, size, the onset of sexual maturity (Romaire et al., 1977) and reproductive efficiency (Harris et al., 1972; Mason, 1977). Such equations permit the comparison of individual specimens of crayfish within and between groups from the same or different localities (Romaire et al, 1977).

As far as linear dimensions are concerned, it has been shown by different authors that each technique of measurement used represents an equivalent measure since the correlation between any two dimensions was significantly high (P < 0.001) and because the regression equations governing any two linear dimension could easily be calculated (Adegboye et al., 1975b; Romaire et al., 1977; Fitzpatrick 1977). In view of the fact that there is no consensus on the dimension of the decapod crustaceans to be used as size index, and also because no single dimension truly and completely reflects the physiological condition of the crayfish, it would be desirable to introduce a "crayfish constant" into the study of crayfish.

The Crayfish Constant K

The "crayfish constant" calculated for Procambarus acutus acutus was derived from the ratio of body weight to the product of linear dimensions. In crustaceans, some attempts have been made to relate the body length to body weight (Belousova, 1970; Kasker et al., 1970; Abolmasova, 1970; Strel'nikova, 1970; and Kititsyna, 1970).

Of particular interest is the relationship obtained between ovarian weight and the cube of the carapace length (Harris et al., 1972). The "cubing" of the carapace length is reminiscent of the formula for calculating the "physiological condition" factor in teleost fishes. Lagler et al., (1962) gave the following formula for calculating the teleost's factor: Body Weight/ $(\text{Length})^3$. In the formula, the exponent of body length is 3 because fish growth, is not just a unidirectional or even two-dimensional phenomenon, but rather, growth represents increases in three dimension. While the idea of a three-dimensional growth has been retained in the "crayfish constant" (mg/cm^3), it was felt that neither the fish nor the crayfish was a cube (of equal sides). Therefore, rather than use Lagler's equation per se, we have developed a new formula in which the total length and carapace width play a part:

$$\text{Crayfish Constant} = \frac{\text{Wet Body Weight}}{\text{Total Length X carapace Length X Width}}$$

Unlike Lagler (1962), we do not refer to the ratio of body-weight-to-the-cube-of-length as <u>condition factor</u> in the study of crayfish because the most significant <u>physiological</u> factor in the crayfish is the moulting process (Adegboye et al., 1978a, 1978c). Hence the term Crayfish Constant was coined for the ratio of wet body weight to the product of the three linear measurments.

The fact that the Crayfish Constant was relatively unchanged for varying sizes of crayfish is an indication that within limits, increases in linear dimensions are accompanied by increases in body weight. The similarities in the values of crayfish constant for male and female specimens is borne out by the fact that there is no size distinction between male and female crayfish. The unchanging nature of the crayfish constant over the whole of the moulting cycle is expected because all sizes of crayfish undergo moulting under normal circumstances. Thus, while the hemolymph calcium increases with the approach of ecdysis (Adegboye et al., 1978c), the "crayfish constant," is unchanging, as the name implies.

The unchanging nature of the "crayfish constant" makes the constant a useful tool in selecting crayfish for controlled experiments. Only those crayfish specimens that have crayfish constant which is not significantly different from the mean "crayfish constant" obtained for a large population of crayfish (say 147 as in <u>Procambarus acutus acutus</u>) would be used for experimentation. Any specimen of crayfish with a "crayfish constant" which is significantly different from the "normal" crayfish constant would be discarded.

The "Crayfish Condition Factor F"

The most significant single factor in the physiology of decapod crustaceans is the moulting process (Passano, 1960). Hence, any derived "Crayfish Condition Factor" that is worthy of our consideration must be directly related to the process of moulting itself or to some monitorable index of the moulting process. The most popular physiological index of the moulting metabolism is the blood picture. According to Moses (BC), and as agreed upon by modern physiologists, "the life of the animal is in the animal's blood." The physiological picture of the blood of an animal is generally a reflection of the animal's general metabolism. Hemolymph calcium concentration occupies a central position in the physiology of decapod crustaceans, and the concentration of calcium ions in the extracellular fluids, in general, is a critical factor in many physiological processes of all living organisms (Borle, 1967; Lockwood, 1960). To a very large extent, the concentration of calcium in the hemolymph reflects the overall moulting metabolism in decapod crustaceans (Travis, 1955; McWhinnie, 1962; Adegboye et al., 1978a; Greenaway, 1972, 1974a, b). Therefore, any "factor" of the crayfish which is directly related to the level of calcium in the hemolymph is a factor that could be described as the "crayfish condition factor."

There was a highly significant relationship between the "Crayfish Condition Factor" and the concentration of calcium in the hemolymph of Procambarus acutus acutus (P < 0.001). The reasons for this high correlation between the "crayfish condition factor" and the hemolymph calcium are obvious as we consider the equation for deriving the crayfish condition factor:

$$F = \frac{\text{Body Weight X Moulting Stage}}{\text{Total Length X Carapace Length X Carapace Width}}$$

In the first place, under normal circumstances, the weight of an animal is an indication of its growth (or even its age). Secondly, while the total length alone indicates growth in only one direction, the product of total length, carapace length and carapace width is an indication of size increase in three dimension. Thirdly, and the major factor of importance is the "stage in the moulting cycle" of the crayfish which has been incorporated in the equation for deriving the Crayfish Condition Factor.

Of all the exogenous and endogenous factors that regulate the physiology of the crayfish in particular and decapod crustaceans in general, the mouting process is the most important (Adegboye et al., 1974). Thus, by incorporating the body weight, moulting stage, total length, carapace length and carapace width in the equation for calculating the crayfish condition factor, we have in effect stated in precise numerical terms, the major physiological and physical conditions of the crayfish. The fact that an index of crayfish physiology namely, the hemolymph calcium concentration, could be calculated from the "crayfish condition factor" (Table 13) indicates the importance of the "crayfish condition factors" as an indicator of the physiology of the crayfish.

Apart from being very highly correlated with the concentration of calcium in the hemolymph, the "crayfish condition factor" is also easily calculated from the simple equation:

F = K moulting stage

where "K" is the "crayfish constant" discussed in the previous sections. Most astacologists are already used to the idea of measuring the crayfish dimensions that constitute the "crayfish constant K," namely the total length, carapace length, carapace width and body weight.

The stage in the moulting cycle of the crayfish could easily be determined by examining the state of the epidermal layer in the pleopod as proposed by Stevenson (1972).

Applicability of Crayfish Condition Factor "F"

When, and if, the idea of a "crayfish condition factor" is universally acceptable to all the students of astacology, one could have introduced this report by stating that:

"The crayfish condition factor of the 147 specimens of crayfish used in this study was 0.85 gm stage/cm^3."

In the absence of such a crayfish condition factor, one would have had to say that:

"The mean total length, carapace length, carapace width and body weight of the 147 specimens of crayfish used in this study were 6.34 cm, 290 cm, 1.32 cm, and 6.97 gm respectively. 2 of the crayfish were in the early postmoult stage, 30 in the late postmoult stage, 27 in the early intermoult stage, 35 in the early premoult stage, and 18 in the late premoult stage."

Apart from the length of the second statement, the figures given for each index of body size do not in any way aid comparison of results from different astacology laboratories because different researchers could be working with specimens of crayfish which vary much in size. Since the suggested crayfish condition factor takes into account the proportionality of different aspects of body size plus the stage in the moulting cycle; the crayfish condition factor is considered an adequate description of the physiology of the crayfish.

Whether one is merely studying the activities in the neuromuscular junction of the crayfish or one is totally committed to the overall physiology of the crayfish, it is essential that one informs the reader of the physiological condition of the crayfish under investigation. No two specimens of crayfish are alike, but a simple crayfish condition factor such as the one suggested in this paper would go a long way in bringing the works of various astacologists into a state, where comparison of observations from various species of crayfish would be possible.

ACKNOWLEDGMENTS

The author is indebted to Drs. I.R. Hagadorn and P.F. Hirsch for their valuable advice; to Messrs. Louis Fulkerson and Julius Oladipo of the Ahmadu Bello University, Computer Centre for technical advice; to John Agwan, Segun Ogunleye, Yusufu Shibayan for technical assistance. Funds for this study came jointly from the American USPHS Grant AM 10558 and Ahmadu Bello University, Zaria, Research Board.

LITERATURE CITED

Abolmasova, G.I. 1970. Size-weight characteristics of some Decapoda of the Black Sea. Hydrobiol. J. 6(1):75-79.

Abrahamsson, S. 1972. The crayfish _Astacus astacus_ in Sweden and the introduction of the American Crayfish _Pacifastacus leniusculus_. Freshwater Crayfish 1:27-40.

Adegboye, J.D., I.R. Hagadorn and P.F. Hirsch. 1974. Regulation of Hemolymph Calcium in the Crayfish. American Zoologist 14(4).

Adegboye, J.D., I.R. Hagadorn and P.F. Hirsch. 1975b. Variations in hemolymph calcium associated with the moulting cycle in the crayfish. 2nd Int'l Crayfish Symposium; Baton Rouge, La: U.S.A. Freshwater Crayfish 2:211-225.

Adegboye, J.D., I.R. Hagadorn and P.F. Hirsch. 1978a. Metabolism of calcium in the crayfish. 4th Int'l Crayfish Symposium Thonon, France. Freshwater Crayfish 4:25-34.

Adegboye, J.D., I.R. Hagadorn and P.F. Hirsch. 1978b. Factors modifying calcium concentration in the hemolymph of the crayfish _Procambarus acutus acutus_ Girard. 4th Int'l Crayfish Symposium. Thonon, France. Freshwater Crayfish, 4:1-14.

Adegboye, J.D., I.R. Hagadorn and P.F. Hirsch. 1978c. Hypercalcemic effect or crustecdysone in the crayfish _Procambarus acutus acutus_ Girard. 4th Int'l Crayfish Symposium. Thonon, France. Freshwater Crayfish 4:15-24.

Armitage, K.B., A.L. Buikema and N.J. Willems. 1973. The effect of Photoperiod on Organic Constituents and Moulting of the Crayfish _Orconectes nais_ (Faxon) Comp. Biochem. Physiol. Vol. 44A:pp 431-456.

Belousova, S.P. 1970. Estimation of some plankton Crustacea weight according to body dimension. Ixv. tikhookean. Nauchno-issled. Inst. ryb. Khoz. Okeanogr. 73:122-126. (Russian).

Borle, A.B. 1967. Correlation between morphological biochemical and biophysical effects of parathyroids hormone on cell membranes (monkey kidney, toad bladder). Calcified Tissues Proc. Europe Symp. 3:45-48.

Bowler, K., D.J. Brown. 1977. Some aspects of growth in the British freshwater crayfish, _Austropotamobius pallipes pallipes_ (Lereboullet). Freshwater crayfish 3:295-308.

Brown, R.B. and G.C. Powell. 1972. Size at maturity in the male Alaskan tanner Crab, _Chionopecetes bairdi_, as determined by chela allometry, reproductive tract weights, and size of precopulatory males. J. Fish. Res. Bd. Canada 29:423-427.

Chung, K.S. 1970. Biological studies on the freshwater shrimps of Korea. 1. Relative growth of _Macrobrachium nipponensis_ (de Haan). Bull. Korean Fish. Soc. 3:71-76. (English Summary).

Ennis, G.P. 1972. Growth per moult of tagged lobsters (_Homarus americanus_) in Bonavista Bay, Newfoundland. J. Fish. Res. Ed. Can. 29:143-148.

Farmer, A.S. 1973. Age and growth in <u>Nephrops</u> <u>norvegicus</u> (Decapoda: Nephropidae). Marine Biology 23:315-325.

Fitzpatrick, J.F. 1977. The statistical relationships of different techniques of measurements in a crayfish species. Freshwater Crayfish 3:471-479.

Greenaway, P. 1972. Calcium regulation in the freshwater crayfish <u>Austropotamobium</u> <u>pallipes</u> (Lereboullet). 1. Calcium balance in the intermoult animal. J. Exp. Biol. 57:471-487.

Greenaway, P. 1974a. Calcium balance at the premoult stage of the freshwater crayfish <u>Austropolamobius</u> <u>pallipes</u> (Lereboullet) J. Exp. Biol. 61:27-34.

Greenaway, P. 1974b. Calcium balance at the post-moult stage of the freshwater crayfish <u>Austropotamobius</u> <u>pallipes</u> (Lereboullet) J. Exp. Biol. 61:35-45.

Haefner, P.A. 1972. The Biology of sand shrimp, <u>Crangon</u> <u>septemspinosa</u>, at Lamoine, Maine. J. Mitchell society 36:42.

Harris, C., K.K. Chew, and V. Price. 1972. Relation of Egg Nubmer to Carapace Length of Sidestripe Shrimp (<u>Pandalopsis</u> <u>dispar</u>) from Dabon Bay, Washington, J. Fish. Res. Ed. Canada 29:464-465.

Hobbs, H.H., Jr. and M. Walton. 1957. Three new crayfish from Alabama and Mississippi. Tulane Studies in Zoology 5(3):39-52.

Kamiguchi, Y. 1971. Studies on the moulting in the freshwater prawn, <u>Palaemon</u> <u>paucidens</u> I. Some endogenous and exogenous factors influencing the intermoult cycle J. Fac. Sci. Hokkaido Univ. Ser. VI. Zool. 18(1).

Kititsyna, L.A. 1970. Ratio of the weight and linear dimensions in population of <u>Pontogammarus</u> <u>robustoides</u> Grimm. Hydrobiol. J. 6(2):57-64.

Lagler, K.F., J.E. Bardach, and R.R Miller. 1962. <u>Icthyology</u>. Ann. Arbor. The Unviersity of Michigan. 545 pp.

Lasker, R., J.B. Wells, A.D. McIntyre. 1970. Growth reproduction, respiration and carbon utilization of the sand-dwelling harpacticoid copepod <u>Assellopsis</u> <u>intermedia</u>. J. Mar. Biol. Ass. U.K. 50:147-180.

Lockwood, A.P.M. 1960. "Ringen" solutions and some notes on the physiological basis of their ionic composition. Comp. Biochem. Physiol., 2:241-289.

Mason, J.C. 1977. Reproductive efficiency of <u>Pacifastacus</u> <u>leniusculus</u> (Dana) in culture. Freshwater Crayfish 3:101-117.

McWhinnie, M.A. 1962. Gastrolith growth and calcium shifts in the freshwater crayfish, <u>Orconectes</u> <u>virilis</u>. Comp. Biochem. Physiol. 7:1-14.

McWhinnie, M.A. and A. Chua. Eyestalk influence on carbohydrate metabolism in the crayfish, Orconectes virilis.

McWhinnie, M.A. and P.N. Saller, Sr. 1969. An analysis of blood sugars in the crayfish, Orconenctes Virilis. Anat. Rec. 134(3):604.

Mikkola, H. 1978. Ecological and Social problems in the use of the crayfish Procambarus clarkii in Kenya. Freshwater Crayfish 4:197-206.

Momot, W.T. and P.D. Jones. 1977. The relationship between biomass, growth rate and annual production in the crayfish, Orconectes virilis. Freshwater crayfish 3:3-31.

Moses. BC. The Holy Bible: Leviticus 17:11. B.B. Kirkbrioe, Bible Co., Inc., Indianapolis, Indiana. U.S.A. 1964.

Newman, G.G. and D.E. Pollock. 1971. Migration and availability of the rock lobster Jasus lalandii at Eland's Bay, S. Africa. Invest'l Rep. Div. Sea Fish. S. Afr. 94.

Passano, L.M. 1960. Moulting and its contol. In: the Physiology of Crustacea: Volume I. London: Academic Press.

Perkins, H.C. 1972. Developmental rates at various temperatures of embryos of the northern lobster (Homarus americanus). Milne-Edwards Fishery Bull, Fish. Wildl. Serv. 70(1):95-99.

Price, V.A. and K.K. Chew. 1972. Laboratory rearing of spot shrimp larvae (Pandalus platyceros) and descriptions of stages. J. Fish. Res. Biol. Canada. 29:413-422.

Romaire, R.P., J.S. Forester and J.W. Avault, Jr. 1977. Length-weight relationships of two commercially important crayfishes of the genus Procambarus. Freshwater crayfish 3:463-470.

Siebers, Dietrich. 1972. Mechanisms of Intracellular Isosmotic regulation of the amino-acid concentration in the Crayfish Orconectes limosus. Z VGL Physiol. 76(1):97-114.

Silbergbauer, B.I. 1971. The biology of the South African rock lobster Jasus lalandii (H. MILNE Edwards) 1. Development. Invest. Rep. Div. Sea Fish. S. Afr. 92:1-70.

Smith, G. and E. Naylor. 1972. The neurosecretory system of the eyestalk of Carcinus maenas (Crustacea: Decapoda) J. Zool. Proc. Zool. Soc. Lond. 166(3):313-321.

Speck, U. and K. Ueich. 1972. Nachweis einer Regulation der Glucosaminbildung bei dem FluBkrelos Orconectes limosus Zur Zeit der Hautung. Z. Vergl. Physiol. 76(3):341-346.

Spirito, Carlp, William H. Evoy and W. Jonathan, P. Barnes. 1972. Nervous Control of walking in the Crab, Cardosoma guanbumi: 1. Characteristics of resistance reflexes. Z. VGL. Physiol. 76(1):1-15.

Stevenson, J.R. 1972. Changing activities of the crustacean epidermis during the moulting cycle. American Zoologist 12:373-380.

Stewart, J.E., G.W. Horner and B. Arie. 1972. Effects of temperature, food and starvation on several physiological parameters of the lobster, Homarus americanus. J. Fish. Res. Biol. Canada 29:461-464

Strel'nikova, V.M. 1970. Size-weight characteristic of Idotea ochotensis and Cymodoce acuta (Isopoda) from the sea of Japan. Hydrobiol. J. 6(1):82-84.

Tack, P.I. 1941. The life history and ecology of the Crayfish Cambarus immunis Hagen. AM. Mild. Nat. 25:420-446.

Takeda, M. and S. Mitake. 1970. Crabs from the East China Sea. 4 Gymnopleura, Dromiacea and Oxystomata. J. Fac. Agric. Kyushu Univ. 16:191-236.

Travis, D.F. 1954. The moulting cycle of the spiny lobster Panulirus argus Latreille. 1. Moulting and growth in laboratory-maintained individuals. Biol. Bull. 107:433-450.

Travis, D.F 1955. The moulting cycle of the spiny lobster, Panulirus argus Latreille. III. Physiological changes which occur in the blood and urine during the normal moulting cycle. Biological Bulleting Marine Biology La., Woods Hole, 109:484-503.

Travis, D.F. 1957. The moulting cycle of the spiny lobster, Panulirus argus Latreille. IV. Postecdysial histological and histochemical changes in the hepatopancreas and integumental tissues. Biological Bulletin, 133(3):451-479.

Watson, J.E. 1972. Ecdysis of the snowcrab, Chionoecetes opilio. 1971 Can. J. Zool. 49:1025-1027.

THE RELATIONSHIP BETWEEN MEDIUM CALCIUM AND HEMOLYMPH CALCIUM CONCENTRATIONS OF THE CRAYFISH DURING THE MID-INTERMOULT STAGE

J.O.D. Adegboye
Department of Biological Sciences
Ahmadu Bello University
Zaria, Nigeria

ABSTRACT

The hemolymph picture of the crayfish Procambarus acutus acutus during the mid-intermoult stage was examined. The specific gravity of the hemolymph and the concentration of calcium in the hemolymph were (mean ± S.E.) $1.01 ± 0.01$ and $41.76 ± 1.41$ mg Ca/100 ml respectively. There was an indirect relationship between body size and hemolymph calcium concentration at the mid-intermoult stage $(0.05 > P > 0.02)$. However, a direct (positive) relation was found between hemolymph calcium and medium calcium concentrations $(0.05 > P > 0.02)$. The relationships between the hemolymph parameters appear to be size-dependent.

INTRODUCTION

Greenaway (1972) has shown that the crayfish in the mid-intermoult stage is characterized by a continuous exchange of calcium ions between the hemolymph and the environment. Studies on Orconectes virilis (McWhinnie, 1962), Austropotamobius pallipes (Greenaway, 1974), and Procambarus acutus acutus (Adegboye et al., 1974, 1978) have established the fact that the concentration of calcium in the hemolymph of the crayfish is conditioned by the concentration of calcium in the environment. As far back as 1880, Huxley observed that crayfish were scarce in areas in which the environment contained very little calcium. Quite recently, increase in the growth rate and production of crayfish have also been shown to accompany increase in the concentration of calcium in pond waters (Avault et al., 1970).

This study was carried out to determine the statistical relationships between the concentrations of calcium in both the hemolymph and the environment, and also to discover how the concentrations of calcium ions influence the weight and specific gravity of the hemolymph. The regression equations governing the significant relationships could be of value in predicting the concentration of hemolymph calcium in intact crayfish.

MATERIAL AND METHODS

Specimens of Procambarus acutus acutus were obtained from the Carolina Biological Supply Company, Burlington, North Carolina. Crayfish in the mid-intermoult stage were identified by means of the criteria reported by Stevenson (1972). The method described by (Hobbs, 1973) was used to identify the 15 male crayfish used in this study. The hemolymph was weighed in a corning 10 microlitre disposable micropipette on a Mettler balance weighing to 0.1 microgramme and the specific gravity of the hemolymph was obtained by dividing the volume of the hemolymph contained in a 20 microlitre disposable micropipette by the weight of the hemolymph. The specific gravity of the hemolymph was calculated from the weight of 20 microliter distilled water, while a fluorometric micro-method was used to determine the concentration of calcium in the hemolymph and the environment of the individual crayfish (Adegboye, 1975, Adegboye et al., 1975a). Carapace length was used as the size index of the crayfish and statistical analyses were carried out at the Ahmadu Bello University Computer Centre to detemine the descriptive statistics of each parameter and also to determine the relationships between the various parameters.

RESULTS

For the specimens of <u>Procambarus acutus acutus</u> measuring between 30.7 and 41.4 mm carapace length, the specific gravity and hemolymph calcium concentration were 1.01 ± 0.01 and 41.67 ± 1.41 mg Ca/100 ml (means ± standard error) respectively. As shown in Table 1, the environment of the crayfish contained 1.72 ± 0.07 mg Ca/100 ml water. Table 2 shows the correlation between different parameters of the hemolymph at the mid-intermoult stage when the calcification of the exoskeleton is in progress.

The only statistically significant relationships were those between carapace length and hemolymph calcium, and between hemolymph calcium and medium calcium of the individual crayfish. At the mid-intermoult stage, there was a negative (that is, indirect) relationship between the carapace length and the hemolymph calcium concentration (Table 2). The relationship between the concentration of calcium in the hemolymph and the concentration of calcium in the medium at the mid-intermoult stage was highly significant at the 98 percent ($P<0.02$) confidence interval (Table 2).

As shown in Table 2, the insertion of the body size into the function of hemolymph weight, hemolymph specific gravity, and medium calcium significantly improved the correlation between the hemolymph calcium concentration and other hemolymph parameters.

DISCUSSION

The hemolymph picture of the crayfish is a good indication of the overall metabolism of the crayfish particularly during the mid-intermoult stage, when the calcification of the new exoskeleton is still in progress. In terms of our present state of knowledge, the variations in the concentration of calcium in the tissues of the crayfish with respect to the moulting cycle appear quite straight forward (Mc Whinnie, 1962; Mc Whinnie et al., 1969; Greenaway, 1972; Adegboye et al., 1974; 1978a, b, c).

The mid-intermoult stage is characterized by the deposition of calcium salts in the new exoskeleton, and this process is continued until the late intermoult stage when the calcification of the exoskeleton is completed. The probable sources of calcium for the mineralization of the exoskeleton are the hemolymph, the hepatopancreas, the gastrolith, the environment and the exuvium. The quantity of calcium in the gastrolith was such that it could not possibly account for all the calcium ions required for the calcification of the exoskeleton

Table 1. Descriptive Statistics of the Hemolymph of the Crayfish at the mid-intermoult state C_{1-2}.

	n = 30				
	Carapace Length mm	Hemolymph Weight mg	Hemolymph Specific Gravity	Hemolymph Calcium mg/100 ml	Medium Calcium mg/100 ml
Maximum	41.1	20.81	1.04	50.00	2.3
Minimum	30.7	19.03	0.95	72.00	1.3
Range	10.4	1.78	0.09	23.00	1.00
Mean	36.92	20.23	1.01	41.67	1.71
Variance	10.38	0.204	0.0005	29.99	0.073
Std. Deviation	3.22	0.452	0.022	5.48	0.27
Std. Error	0.83	0.117	0.006	1.415	0.07
Mean Deviation	2.71	0.304	0.015	3.96	0.205
Median	38.4	20.32	1.01	42.00	1.80

Table 2. Relationships between Hemolymph Ca and other parameters of Crayfish.

Hem. Ca. = Hemolymph Ca mg/100 ml

W = Hemolymph weight mg

L = Carapace Length mm

G = Hemolymph specific gravity

M = Medium Ca mg/100 ml

Equation	R	P
Hem. Ca mg/100 ml = 75.3795 - 0.9131 Carapace length cm	-0.5372	<0.05
Hem. Ca mg/100 ml = 21.4264 + 11.7676 Medium Ca mg/100 ml	0.58116	<0.02
Hem. Ca mg/100 ml = 43.2414 - 0.9264 "L" + 1.63123 "W"	0.5534	<0.05
Hem. Ca mg/100 ml = 84.6236 - 0.9217 "L" - 8.8634 "G"	0.5383	<0.05
Hem. Ca mg/100 ml = 77.4259 - 1.1166 "L" + 7.5803 "W" - 146.8411 "G"	0.6255	<0.01
Hem. Ca mg/100 ml = 112.4501 7.666 "W" - 1.0463 "L" - 197.1654 "G" + 12.469 "M"	0.8455	<0.001

(Adegboye et al., studies on gastrolith, in prep.). The calcium contents of the hepatopancreas as well as in the hemolymph are so small that calcium ions must be obtained from other sources for the complete calcification of the exoskeleton. The exuvium is a ready source of exogenous calcium and Procambarus acutus acutus is known to consume its own exuvium following moult (Adegboye, notes on the moulting process, in prep.). The external environment of the crayfish appears to be the most reliable source of calcium ions for the crayfish, following ecdysis.

Several studies by Greenaway (1972, 1974a, b) have indicated that there is a constant exchange of calcium ions between the hemolymph and the environment. In this study, it was also observed that the concentration of calcium in the hemolymph is directly related to the concentration of calcium in the environment (P<0.02); and that the correlation between the two calcium compartments improves with the size of the crayfish (P<0.001). The negative (indirect) relationship between the carapace length and the hemolymph calcium concentration may be due to a combination of factors. According to Greenaway (1972), there is a (net) negative calcium balance in the hemolymph of the crayfish. We have also found that calcium losses into the environment from the hemolymph increases with size (Adegboye, age-dependent calcium metabolism in preparation). Therefore, the indirect linear relation-ship between hemolymph calcium and environmental calcium is probably due to the size-dependent losses of calcium into the environment. Another possible explanation could be that larger specimens of crayfish require relatively more calcium ions from the hemolymph for the calcification of exoskeleton than smaller specimens: and so, the depletion of calcium from the hemolymph increases with the size of the crayfish (Adegboye, 1975b).

The hemolymph is only slightly heavier than water (20.23 mg and 20.0 mg respectively), and the larger protein molecules of the hemolymph such as haemocyanin did not appear to contribute appreciably to the weight of the hemolymph.

The intermoult stage in Panulirus argus is characterized by a continuous increase in wet weight. This increase in weight is primarily due to water absorption from the environment (Travis, 1954). According to Travis' Table 1, the larger spiny lobster absorbed more water than the smaller specimens. Capen (1972) has also reported water uptake by euryhaline crabs. Perhaps, the inverse relationship between body size and hemolymph calcium concentration might also be due to the fact that crayfish suffer a greater amount of dilution during the intermoult period. Because of their large size relative to their molar concentration, blood proteins occupy a volume that is far out of proportion to their weight and water appears to contribute over 90 percent of the weight of the human plasma (Mountcastle, 1974). The specific gravity of hemolymph compares favourably with the specific gravity of whole blood of vertebrates and invertebrates (Table 3).

The lack of correlation between the hemolymph calcium (or even medium calcium) and hemolymph specific gravity is a confirmation of the general belief that the total mass of calcium ions in the hemolymph is so small that calcium ions do not contribute in any significant way to the weight of the hemolymph. It is estimated that with a hemolymph volume of about 5 ml (unpublished results), the total calcium ions

Table 3 - Specific gravity of whole blood of some animals (Handbook of Biological Data; National Academy of Sciences, National Research Council. W.B. Saunders Co., Philadelphia, PA U.S.A. 1961).

Water Invertebrates	1.000 Mean	Minimum/Maximum
Apix	1.045	
Bombyx	1.037	(1.032 - 1.041)
Calliphora	1.021	
Deilephila	1.031	
Dytiscus	1.026	
Gastrophilus	1.062	
Hydrophilus	1.012	
Periplanets	1.016	
Phormia	1.018	
Prodenia	1.032	
Procambarus acutus acutus	1.010	(0.950 - 1.040)*
Vertebrates		
Man	1.056	(1.052 - 1.061)
Cat	1.052	(1.045 - 1.057)
Cattle	1.052	(1.046 - 1.058)
Dog	1.052	
Goat	1.022	(1.046 - 1.059)
Horse	1.053	(1.046 - 1.059)
Mouse	1.057	(1.052 - 1.062)
Rabbit	1.050	(1.048 - 1.052)
Rat	1.051	(1.046 - 1.061)
Sheep	1.051	(1.041 - 1.061)
Swine	1.046	(1.046 - 1.054)
Chicken	1.056	(1.050 - 1.064)

* (this study)

would be about 0.009 mg per 20 microlitres of sample.

The data presented in this report indicate that there is a direct relationship between the concentration of calcium in the hemolymph and the concentration of calcium in the medium. Hence, the concentration of calcium in the environment is a major factor in the regulation of hemolymph calcium.

ACKNOWLEDGMENTS

The author is grateful to Drs. I.R. Hagadorn, P.F. Hirsch, Paul Munson and Cary Cooper for their useful advice, useful comments and constant encouragement. Messrs Louis Fulkerson and Julius Oladipo were most helpful in all the statistical analyses. This study was jointly supported by the American USPHS Grant AM 10559 and Ahmadu Bello University, Zaria (Nigeria) Board of Research.

LITERATURE CITED

Adegboye, J.D. 1975. Regulation of hemolymph calcium in the crayfish, Procambarus acutus acutus Girard (Crustacea, Decapoda). PhD. dissertation, University of North Carolina, Chapel Hill, N.C. 275414, U.S.A.

Adegboye, J.D., I.R. Hagadorn and P.F. Hirsch. 1974. Regulation of Hemolymph Calcium in the Crayfish. American Zoologist 14(4).

Adegboye, J.D., I.R. Hagadorn and P.F. Hirsch. 1975a. Microfluorometric determination of calcium in the hemolymph and other tissues of the crayfish. 2nd Int'l Crayfish Symposium; Baton Rouge, LA; U.S.A. Freshwater Crayfish (2):211-225.

Adegboye, J.D., I.R. Hagadorn and P.F. Hirsch. 1975b. Variations in hemolymph calcium associated with the moulting cycle in the crayfish. 2nd Int'l Crayfish Symposium; Baton Rouge, LA; U.S.A., Freshwater Crayfish (2):227-247.

Adegboye, J.D., I.R. Hagadorn and P.F. Hirsch. 1978a. Metabolism of calcium in the crayfish. 4th Int'l Crayfish Symposium, Thonon, France. Freshwater Crayfish 4:25-34.

Adegboye, J.D., I.R. Hagadorn and P.F. Hirsch. 1978b. Factors modifying calcium concentration in the hemolymph of the crayfish Procambarus acutus acutus Girard. 4th Int'l Crayfish Symposium. Thonon, France. Freshwater Crayfish, 4:1-14.

Adegboye, J.D., I.R. Hagadorn and P.F. Hirsch. 1978c. Hypercalcemic effect or crustecdysone in the crayfish Procambarus acutus acutus Girard. 4th Int'l Crayfish Symposium. Thonon, France. Freshwater Crayfish 4:15-24.

Avault, J.W. JR., L. De la Bretonne, Fr. and E. Jaspers. 1970. Culture of crawfish, Louisiana's crustacean king. The American Fish Farmer 1(10):8-14 and 27.

Capen, R.L. 1972. Studies of water uptake in the euryhaline crab, Ehinthropanopeus harris J. Exp. Zool., 182:307-320.

Greenaway, P. 1972. Calcium regulation in the freshwater crayfish Austropotamobius pallipes (Lereboullet). 1. Calcium balance in the intermoult animal. J. Exp. Biol. 57:471-487.

Greenaway, P. 1974a. Calcium balance at the pre-moult stage of the freshwater crayfish Austropotamobius pallipes (Lereboullet) J. Exp. Biol. 61:27-34.

Greenaway, P. 1974b. Calcium balance at the post-moult stage of the freshwater crayfish Austropotamobius pallipes (Lereboullet) J. Exp. Biol. 61:35-45.

Hobbs, H.H. Jr. 1973. Crayfishes (Astacidae) of North and Middle America Biota of Freshwater Ecosystems: Identification Manual No. 9. Environmental Protection Agency. Washington, D.C.

Huxley, T.H. 1880. The crayfish. Note II. on chap. I, pp. 29, 346-349. D. Appleton and Company, New York.

Mc Whinnie, M.A. 1962. Gastrolith growth and calcium shifts in the freshwater crayfish, Onconectes virilis. Comp. Biolchem. Physiol. 7:1-14.

Mc Whinnie, M.A., and P.M. Saller, Sr. 1969. An analysis of blood sugars in the crayfish, Onconectes virilis. Anat. Rec. 134(3):604.

Mountcastle, V.B. (Ed.) 1974. Medical Physiology Vol. 1, 13th Ed. C.V. Mosby Co. Saint Louis MO. U.S.A.

Stevenson, J.R. 1972. Changing activities of the crustacean epidermis during the moulting cycle. American Zoologist 12:373-380.

Travis, D.F. 1954. The moulting cycle of the spint lobster Panulirus argus Latreille. 1. Moulting and growth in laboratory-maintained individuals. Biol. Bull. 107:433-450.

IV
ECOLOGY OF CRAYFISH

STUDIES OF THE LIFE HISTORY AND ECOLOGY OF
ORCONECTES PALMERI PALMERI (FAXON)

James F. Payne and James O. Price
Memphis State University
Memphis, TN 38152

ABSTRACT

The life history and ecology of the crayfish Orconectes palmeri palmeri were studied in the Chickasaw Basin of western Tennessee. Prior to oviposition in February and March, females inhabited shallow burrows, especially within and around heavy debris. Juveniles appeared in open-water samples in April, and most specimens reached adult size (carapace length 18-25 mm) by late September and October of the same year. In November, 85% of all males were form I, and females had active cement glands; most form I males molted to form II during June. Frequency distribution graphs of population samples indicated that three adult age classes existed in open water during most of the year; these were: first year (18-25 mm), second year (26-35 mm), and third year (36-42 mm). The majority of reproduction occurred in the first year individuals with few specimens living longer than two years. Life history strategies and adaptations which minimize interspecific competition and interaction are discussed.

INTRODUCTION

Orconectes palmeri palmeri (Faxon), a member of the virilis section and palmeri group of this genus, inhabits a variety of streams within the lower Mississippi Valley in western Tennessee, Mississippi, and Louisiana and in eastern Missouri and Arkansas as noted by Hobbs (1972). Literature references to this species are few and provide little information regarding its life history. O. p. palmeri was first described by Faxon (1884), the type locality being a small stream in the vicinity of Reelfoot Lake in northwestern Tennessee. Hobbs and Marchand (1943) noted that this species was confined primarily to running waters in the Reelfoot Lake vicinity and provided the following statements concerning its life history: "First form males were taken in June and July, and in June one freshly moulted first form male was collected. No females with eggs were found." Penn (1952, 1959) described distribution of O. palmeri in Louisiana, and in 1957 published results of his studies on variation within the subspecies of O. palmeri.

The paucity of ecological information and the abundance of this species in western Tennessee prompted a study of its life history. The purpose of this investigation was to elucidate major features of the life cycle of this species and to compare these features with those of crayfish species associates to determine which factors minimize species interaction and competition in a single locality.

183

MATERIALS AND METHODS

Study Area

     The primary study site was Days Creek, a permanent stream which
flows northward into Nonconnah Creek, the latter being a major drainage
stream of the Chickasaw Basin in southwestern Tennessee flowing into
the Mississippi River via McKellar Lake (Fig. 1). The stream bed and
bank of Days Creek consist primarily of clay, and throughout most of
the stream's lower reaches, sand and coarse gravel constitute the major
substrate. The stream channel varies from 30 to 60 m in width and 15
to 20 m in depth; the channel is subject to torrential flows and annual
water levels range from 80 cm to 15 m. Accumulations of debris exist
at irregular intervals along the stream providing temporary refuges for
crayfish. Riparian vegetation consists primarily of grasses and of the
willow, Salix nigra since most of the stream runs through open fields.
No prominent aquatic macrophytes exist in the stream due to its broad
fluctuations in turbulence and depth. These physical features of the
stream are generally characteristic of numerous habitats of O. p.
palmeri in western Tennessee. Crayfish associates of O. p. palmeri in
Days Creek are Cambarus diogenes and Procambarus acutus acutus. P.
clarkii occurs with O. p. palmeri in some streams in the Chickasaw
Basin although this species was never taken in Days Creek.

Field Techniques

     Life history data were obtained from field and laboratory
observations. Periodic monthly field samples of crayfish were taken
from February 1978 through April 1979. Data from these samples were
compared with those from numerous other samples of O. p. palmeri taken
from various locations within the Chickasaw Basin. Seasonal and
temporal changes within the poulation were noted from study of 857
specimens taken from field samples in Days Creek. These specimens were
captured with 5 mm mesh nylon seines set across the stream channel.
Fine-mesh delta-ring dip nets were employed to insure capture of the
smallest members of the population during each sampling period. Also,
burrows along the stream bank were excavated and notes on the
relationships of O. p. palmeri and its crayfish associates were
maintained. Following capture, specimens were separated by sex and
placed in plastic containers until desired data were taken. Adult
males were recorded as form I or form II. Adult females were examined
for active cement glands (Stephens 1952) or for the presence of eggs
attached to the abdomen. Length of carapace (from tip of rostrum to
posteriomedian margin of cephalothorax) of each specimen was recorded
to the nearest 0.5 mm with a vernier caliper. Previous studies of
crayfish populations (van Deventer 1937, Prins 1968, Payne 1972) have
employed this measurement as a standard expression of body length since
it represents approximately one half of total body length. Length
measurements given hereafter are length of cephalothorax and should be
doubled for total body length.

     Most specimens were returned to the stream immediately after data
were taken and recorded. Crayfish too small for sex determination or
measurement in the field were returned to the laboratory, and this
information was obtained by using a binocular dissecting microscope
equipped with an ocular micrometer. Certain specimens were maintained
in laboratory aquaria for observations of amplexus, oviposition, and

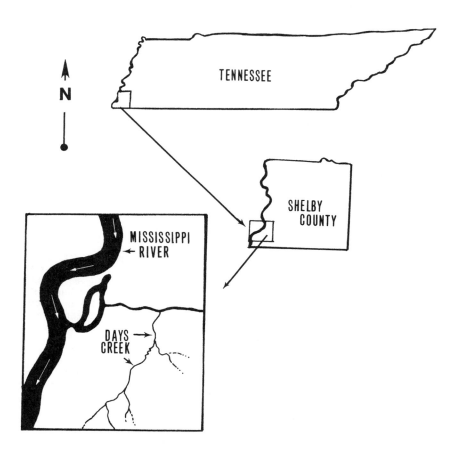

Figure 1. Map showing the relationship of the study site, Days Creek to the Mississippi River.

hatching; these specimens were fed fragments of Purina Cat Chow pellets along with pieces of dried leaves.

Separate analyses of juvenile and adult components from field samples were conducted; range, mean, standard deviation and standard error of the mean were determined for each component from each monthly sample. Ratios of adult males to females and ratios of form I and form II males were determined for each sample.

RESULTS AND DISCUSSION

Population Structure

The adult population structure of O. p. palmeri observed in this study is shown in Figure 2. Adult sex ratios ranged near 50% for each sample and we therefore conclude that our sampling techniques adequately reflect the total population structure. During February, March, and June, males constituted a slightly greater percentage of adults captured.

A major adult molt occurred from late September through November; within this period, form I males increased in number from 8% to 85% of the total adult male population. Form I males predominated from November through March (Fig. 2); this period is designated as the breeding season for O. p. palmeri at the latitude of Shelby County, Tennessee. During these months amplexus occurred in the stream; also during this period, form I males would frequently amplex females when both were confined in field or laboratory containers. A second major adult molt occurred from late March through April. During this period form I males reverted to form II and persisted in this form throughout the summer months. The mean body length of the adult population increased slightly but consistently through the summer months. This increase may result from adults molting one or more times during this growing period. By using tagging and capture-recapture techniques during the summer months, Price and Payne (1979) observed multiple summer molts (form II to form II) in adult male Orconectes neglectus chaenodactylus from the North Sylamore Creek in Stone County Arkansas. A similar phenomenon was reported for O. limosus in Poland by Kossakowski (1971) and may represent a strategy to canalize energy requirements for growth into the non-reproductive form. In the present study, field samples during June and July contained form II males which had recently molted. The consistent rise in mean adult length of O. p. palmeri during summer months may also result from growth in subadults from the lower end of the mean minimum size.

Oviposition

The cement glands of females became active in late October and remained so throughout the winter months; ovigerous females were taken from late February through April. Egg bearing females were seclusive, and only 12 ovigerous females were taken in stream sampling; 17 specimens became ovigerous in laboratory aquaria. This pattern of winter oviposition followed by recruitment in the spring fits the general pattern for most species of Orconectes studied with major differences apparently due primarily to latitude as noted by Payne (1978).

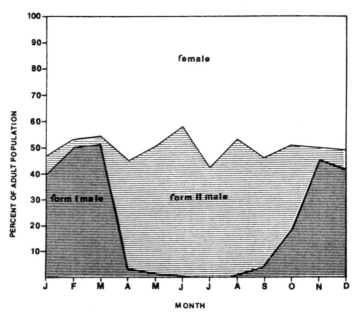

Figure 2. Annual composition of adult population of Orconectes palmeri palmeri in Days Creek, Shelby County, Tennessee.

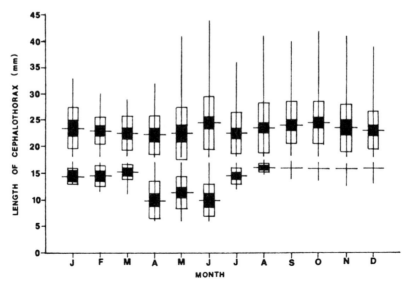

Figure 3. Summary of population size categories for Orconectes palmeri palmeri in Days Creek, Shelby County, Tennessee. (Horizontal lines are sample means; vertical lines the sample ranges; closed bar one standard error; open bar two standard deviations.

187

All ovigerous females of O. p. palmeri from field samples were taken from areas of heavy debris accumulation or from beneath stones. In such habitats females apparently construct shallow burrows; however, only two were taken from burrows along open portions of the stream bank.

Early Development

Early stages of postembryonic development of 428 hatchlings from 16 ovigerous females were studied in the laboratory. Abdominal eggs from these females had a mean diameter of 2.0 mm while first instar juveniles showed a cephalothorax length of 2.0 mm (range 1.8 to 2.2 mm) and were attached to the female pleopodia by the telson thread as described for other species. Second instars had a mean cephalothorax length of 2.5 mm (range 2.0 to 2.7) and clung tightly to the female pleopodia. Third instars showed a mean carapace length of 3.5 mm (range 3.1 to 4.2 mm) and occasionally left the female pleopodia to roam the laboratory containers. Fourth instars, mean carapace length of 5.1 mm (range 4.6 to 5.7 mm), were the first stage to behave truly independent of the females in laboratory confinement. Recruitment of these instars into the open stream occurred from April through June (Fig. 3). Juvenile growth was rapid during July and August, with the majority of the annual hatch reaching reproductive size by August. Juveniles were present in all monthly samples; however, the numbers of these from September through March were small (Fig. 3). These juveniles which remain we believe represent late hatchlings which overwinter in the juvenile condition and attain adult size in the spring.

Adult Size

Wenner et al. (1974) noted that the relationship between age and size in crustaceans is not as readily obtained as in other organisms (e.g. fish), and these investigators applied a method of obtaining mean minimum size of specimens at sexual maturity by plotting carapace length on probability paper. Stein et al. (1977) utilized this technique for Orconectes propinquus and determined that the mean minimum carapace length at the onset of sexual maturity of this species was 18.5 mm. For O. p. palmeri we selected a carapace length of 18 mm as a dividing point between juveniles and adults because this was the length of the smallest form I males (n=5) and the length of the smallest ovigerous females (n=2). We acknowledge the shortcomings of choosing one length increment to separate all specimens as either juveniles or adults, although such a point is most convenient for analyzing population dynamics.

Age Classes

Selected monthly samples are plotted as growth frequency distributions in Figure 4. Plots for the September through December samples were similar and are not shown. Estimates of the age class structure of the population are based on using the mean minimum size and natural breaks between peaks in these frequency distribution diagrams. Little change in poulation structure is apparent from January through March. First year animals are placed in the 18 to 25 mm category because of the mean minimum size noted earlier and because a consistent break in frequency distribution occurs following the 25 mm

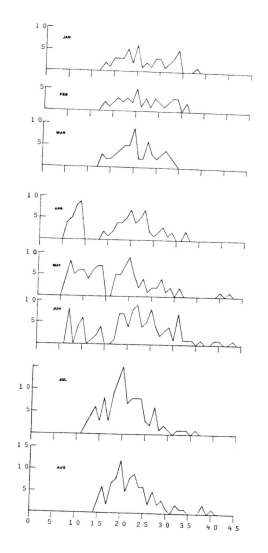

Figure 4.   Frequency distributions from monthly samples of <u>Orconectes</u> <u>palmeri</u> <u>palmeri</u> in Days Creek, Shelby County, Tennessee.   <u>Ordinate:</u> number of speciments; Abcissa: Length of cephalothorax in mm.

point. Following reproduction and the spring molt, the majority of adults had attained a carapace length of 18 to 25 mm. During June and July, this size category represented 66% and 81% of the total adult population respectively. Individuals within the 26 to 35 mm carapace length range apparently represent the second year specimens in Days Creek (Fig. 4). Percentage of second year specimens in samples was consistently small, indicating that the majority of specimens reproduce once and succumb before a second breeding season. Only 21 specimens with carapace lengths greater than 36 mm were taken during this study; these we believe represent third and perhaps fourth year individuals which apparently contribute very little to the reproductive pool. Boyd and Page (1978) also conlcuded that very few individuals of O. kentuckiensis lived longer than two years. The maximum size of an individual O. p. palmeri was a form I male (43.0 mm carapace length) taken on June 15, 1976; this specimen died attempting to molt two days following capture. Estimates of maximum ages of crayfishes as well as estimates of population age class structure are variable when the studies of several species are examined (Payne 1978).

Burrowing and Species Interaction

Hobbs (1942) distinguished between primary, secondary, and tertiary burrowing crayfish and related this classification to the degree of dependence upon and amount of the life cycle spent within a burrow. C. diogenes, a primary burrower, spends a major portion of its life cycle within burrows. Recruitment of young C. diogenes occurs primarily in early spring. However, juveniles are present in the open waters of the stream only briefly since they begin to establish their own burrows soon after recruitment. Thus interaction and competition with specimens of O. p. palmeri appears minimal. P. a. acutus, a secondary burrower, also constructs prominent burrows from April through early summer. Specimens inhabit burrows primarily during oviposition and breeding, with one male and one female per burrow, a feature noted for other species of Procambarus (Huner and Barr 1981, Payne 1972). Recruitment of P. a. acutus juveniles to the open water occurs primarily from September through November, with juveniles overwintering in the open water. Most P. a. acutus reach adult size by June of the following year. Due to this temporal separation of fall and spring recruitments, interspecific interaction and competition especially between juveniles is minimized.

ACKNOWLEDGMENTS

Many persons who assisted in field work deserve thanks, especially Christopher Payne, Lindsey Riley, Alberto Ciniglio, and Thomas Baker.

LITERATURE CITED

Boyd, J. A. and L. M. Page. 1978. The life history of the crayfish Orconectes kentuckiensis in Big Creek, Illinois. Am. Midl. Nat. 99:398-414.

Faxon, W. 1884. Descriptions of new species of Cambarus to which is added a synonymicallist of the known species of Cambarus and Astacus. Proc. Amer. Adac. Arts Sci. 20:107-158.

Hobbs, H. H., Jr. 1942. The crayfishes of Florida. University of Florida Publ. Biol. Sci. Ser. 3:1-179.

Hobbs, H. H., Jr. 1972. Crayfishes (Astacidae) of North and Middle America. Identification Manual 9: Biota of Freshwater Ecosystems. U.S. Environ. Prot. Agency, Water Poll. Res. Contr. Ser. 173 pp.

Hobbs, H. H., Jr. and L. J. Marchand. 1943. A contribution toward a knowledge of the crayfishes of the Reelfoot Lake area. J. Tennessee Acad. Sci. 18:6-35.

Huner, J. V. and J. E. Barr. 1981. Red swamp crawfish: biology and exploitation. Sea Grant Publ. LSU-T-001. Louisiana State Univ. Center for Wetland Resources. Baton Rouge, LA. 148 pp.

Kossakowski, J. 1971. Crayfish For. Fish. (Translations) International Activities Staff, National Marine Fisheries Service, N.O.A.A., U.S. Dept. Commerce, Washington, D.C. TT 70-55114. 155

Payne, J. F. 1972. The life history of Procambarus hayi. Am. Midl. Nat. 87:25-35.

Payne, J. F. 1978. Aspects of the life histories of selected species of North American crayfishes. Fisheries 3:5-8, 16-19.

Penn, G. H. 1952. The genus Orconectes in Louisiana. Am. Midl. Nat. 47:743-748.

Penn, G. H. 1957. Variation and subspecies of the crawfish Orconectes palmeri (Faxon) (Decapoda, Astacidae). Tulane Stud. Zool. 5:231-262.

Penn, G. H. 1959. An illustrated key to the crawfishes of Louisiana with a summary of their distribution within the state. Tulane Stud. Zool. 7:3-20.

Price, J. A. and J. F. Payne. 1979. Multiple summer molts in adult Orconectes neglectus chaenodactylus Williams. pp. 94-104. In: P. J. Laurent (ed.) Freshwater Crayfish IV, Thonon-les-Bains, France (1978). 473 pp.

Prins, R. 1968. Comparative ecology of the crayfishes Orconectes rusticus rusticus and Cambarus tenebrosus in Doe Run, Meade County, Kentucky. Int. Rev. Gesamten Hydrobiol. 53:667-714.

Stein, R. A., M. L. Murphy and J. J. Magnuson. 1977. Exernal morphological changes associated with sexual maturity in the crayfish (Orconectes propinquus). Am. Midl. Nat. 97:495-502.

Stephens, G. C. 1952. The control of cement gland development in the crayfish Cambarus. Biol. Bull. 103:242-258.

VanDeventer, W. C. 1937. Studies on the biology of the crayfish Cambarus propinquus Girard. Illinois Biol. Monogr. 15:1-67.

Wenner, A. M., C. Fusaro and A. Oaten. 1974. Size at onset of sexual maturity and growth rate in crustacean populations. Can. J. Zool. 52:1095-1106.

THE BIOENERGETICS OF ORCONECTES VIRILIS IN TWO POTHOLE LAKES[3]

Patricia D. Jones[1] and Walter T. Momot[2]
Ohio State University
Columbus, Ohio 43210

ABSTRACT

The lifetime energy expenditure per individual Orconectes virilis is calculated to be 54.8 K cal. Summations of energy estimated necessary to support crayfish populations in North Twin Lake varied from 12.7 to 19.6 K cal/ m^2 as compared to 32.1 to 52.8 K cal/m^2 in West Lost Lake. Estimated net growth efficiency (K_2) of individual crayfish was calculated at 23.5%. Individual crayfish in our lakes thus process three times as much energy for maintenance as for growth. Over a three-year period, the energy required to support surviving crayfish is 2.2 to 2.7 times greater in West Lost Lake. The greater carrying capacity for crayfish in West Lost Lake is likely to be related to the greater quantity of food available in that lake.

INTRODUCTION

Crayfish often dominate the annual production of the macrobenthos of low nutrient lakes (Momot et al. 1978). A comparison of two such lakes in Michigan has shown that West Lost consistently produces more crayfish than nearby North Twin (Momot and Gowing 1977a, b, c). Jones and Momot (1981) have suggested that the greater carrying capacity of West Lost Lake results from a greater availability of nursery area for the young-of-the-year recruits and a greater availability of allochthonous nutrients for the female brood stock. In this paper we calculate an approximate energy budget for Orconectes virilis and show that when it is applied to our previous production estimates that the West Lost population would require from two to three times the energy input of North Twin to sustain its larger populations.

The crayfish energy budget can be summarized by the equation:

$$Q_c - Q_w = Q_g + Q_s + Q_d + Q_a$$

where: Q_c = energy of food consumed

Q_w = energy of waste products

Q_g = energy converted to growth

Q_s = metabolic energy expeded by unfed and resting animals (standard metabolism)

Q_a = energy used for activity (Warren and Davis 1967, p. 180)

Q_d = energy released in the course of digestion (specific dynamic action), assimilation and storage of materials consumed

[1]Present Address: Ohio Department of Natural Resources, Columbus, Ohio.
[2]Present Address: Lakehead University, Thunder Bay, Ontario P7B 5E1.
[3]Research sponsored by NSF Research Grant BMS 71-01540-A02 to Dr. Walter T. Momot.

Each value is affected by temperature and the type of food consumed. Within the normal temperature range of ectotherms such as crayfish, metabolism, activity, and feeding are related to temperature (Warren 1971). The ability of crayfish to assimilate ingested foods will determine the energy available for standard metabolism, specific dynamic action, growth, and activity.

Ideally one must determine the energy values for all categories of caloric consumption and expenditure to construct an energy budget for $\underline{O}$. virilis, but time constraints and logistics made this impossible. We therefore concentrated on the essential measurements of standard metabolism, specific dynamic action, and assimilation. These measurements, growth data (Table 1), together with production estimates (Momot and Gowing 1977a, b, c) provided a first order estimate of the energy budget of the North Twin and West Lost crayfish populations.

MATERIALS AND METHODS

We calculated the energy expenditure for standard metabolism and specific dynamic action by measuring the oxygen consumption of young-of-year male and female and adult male crayfish. Oxygen consumption of starved $\underline{O}$. virilis (standard metabolism) was determined at 9, 14, 19 and 24°C. Oxygen comsumption of fed animals (standard metabolism plus specified dynamic action) was measured at 19°C.

Intermolt animals were used in all tests. Experiments were conducted inside a constant temperature room with a controlled photoperiod of 14L:10D. Tests were run during daylight and lasted 1 to 3 h depending on temperature, crayfish size, and amount of test water which varied between 180 ml for young-of-year and 750 ml for adults. Dissolved oxygen levels were determined according to the methods of Burke (1962). At the end of each experiment crayfish carapace length was measured to the nearest 0.11 mm and wet weight was recorded.

Crayfish were acclimated to the test temperature for 48 h prior to testing. Fed animals had fish available for consumption until the start of the experiment.

Respiratory chambers, 250 and 750 ml glass jars, were filled with measured amounts of oxygenated water. Stirring bars were placed in the bottom and a thermometer suspended in each chamger. Aluminum screening supported by glass rods separated the crayfish from the stirring bars. Crayfish were allowed to acclimate to the chambers for 15 min before the start of each test.

Three 10-ml water samples were withdrawn from each chamber to determine mean initial oxygen level. Chambers were then sealed against fluid evaporation with 1.5 cm of mineral oil. Magnetic stirrers were operated for 5 min at hourly intervals to ensure complete mixing of test water. Chamber temperature was recorded at the start, the end, and at hourly intervals during the test. At the conclusion of the experiment three water samples were extracted from each respiratory chamber. By using 10 ml syringes water was drawn through the layer of mineral oil without disturbing the oxygen concentrations. Oxygen readings at the end of the experiment were used to determine mean final oxygen concentration for each test chamber. The difference between the

Table 1. Mean daily growth rates and standard deviations calculated for young-of-year, _Orconectes virilis_. Laboratory growth rates for starved animals and crayfish maintained on eight diets were calculated during the summer of 1975. Growth of field animals in North Twin and West Lost Lakes, Otsego County, Michigan are shown for the summers of 1974 and 1975.

	Mean Growth Rate g wet weight/day x 10^{-4}
Diet	
P. resinosa	-2.3 ± 4.7
Fine Litter	-1.4 ± 12.9
Marl	-0.1 ± 3.6
P. grandidentata	1.9 ± 7.2
Q. borealis	2.7 ± 7.0
Starved	5.4 ± 5.4
Carex	7.7 ± 3.5
Extruded Alginate Diet 21-5/72A	20.6 ± 6.6
Fish	50.5 ± 23.3
Lakes	
North Twin 1974	101.3 ± 66.4
West Lost 1974	66.6 ± 36.7
North Twin 1975	76.1 ± 36.7
West Lost 1975	63.3 ± 41.3

mean initial and mean final oxygen concentrations represented the oxygen consumed during the experiment.

Six control experiments were run for every volume of water used at each temperature. The mean oxygen loss in these systems was subtracted from the oxygen consumed in the test chambers. These corrected oxygen values were used to calculate oxygen consumption of each crayfish (Burke 1962). The value 3.42 cal/mg O_2 (Warren 1971) permitted conversion of oxygen consumption to calories. We calculated the assimilation efficiency of crayfish supplied with fish as food in both the growth and oxygen consumption experiments (Table 1). Both fish and diet (21-5/72A) described by Huner et al. (1974) were fed to young-of-year $\underline{O}$. $\underline{virilis}$ measuring 12-14 mm carapace length. The alginate diet was also fed to yearling animals from 19-24 mm in carapace length. Only intermolt animals were used in these experiments.

Food was dried at 80°C for 48 h and weighed on a Mettler balance sensitive to 0.001 gm and was allotted to each crayfish. Food ration varied from 0.5-1.0 gm dry weight depending on the size of the crayfish. Preliminary experiments had established that these amounts were in excess of expected consumption.

Each ration was placed in a container of filtered water. Crayfish starved for 48 h were introduced, allowed to feed for 90 min, then placed in a separate container of filtered water. Tests were conducted at room temperature (19 ± 1°C).

Water containing uneaten food was passed through pre-weighed 0.45 μ millipore filters. The filter and trapped food were dried at 80°C for 48 h, then weighed. The dry weight of ingested food was calculated as the difference between the weight of the filter and uneaten food from the initial food ration. After 48 h crayfish were weighed and measured. Water containing fecal material was filtered and dried in the same manner as uneaten food. Assimilation was calculated as the difference between dry weight of ingested food and dry weight of fecal material.

RESULTS

Per gram body weight, small crayfish consumed more oxygen than adults (Table 2). For each 5°C increment young-of-year crayfish showed the greatest increase in standard metabolism from 9-14°C while for adults the greatest increase was from 14-19°C. The relationships between size and increase in metabolism was non linear (Table 2).

Calculation of mean standard metabolic rate (Cal/day) for young and adult crayfish showed a general increase in standard deviation at higher temperatures (Table 2). Once again, the 10-15 g crayfish at 24°C were the exception to this rule. The mean O_2 consumption of fed animals at 19°C (standard metabolism plus specific dynamic action) was only slightly higher than for starved crayfish at the same temperature. Considering the large standard deviations, especially for adult animals, the difference between means was insignificant.

195

Table 2. Daily caloric consumption (A) and hourly rates of oxygen consumption (B) per gram wet weight for three size groups of starved,[a] Orconectes virilis, at four temperatures. Data presented as mean ± S.D. al/day for A and mg O_2/hr/g for B.

Temp. °C	0.4-0.9[b] A	0.4-0.9[b] B	5-9[b] A	5-9[b] B	10-15[b] A	10-15[b] B
9	0.04 ± 0.12	2.4 ± 1.3 (10)[c]	0.03 ± 0.01	12.3 ± 3.5 (5)	0.02 ± 0.01	21.9 ± 5.6 (6)
14	0.12 ± 0.07	5.5 ± 2.7 (10)	0.03 ± 0.01	19.6 ± 4.3 (5)	0.03 ± 0.01	31.1 ± 5.1 (4)
19	0.14 ± 0.06	7.5 ± 3.0 (12)	0.07 ± 0.02	41.8 ± 13.4 (5)	0.08 ± 0.03	81.0 ± 21.6 (4)
24	0.23 ± 0.09	11.8 ± 4.1 (11)	0.12 ± 0.05	85.3 ± 36.7 (5)	0.08 ± 0.01	83.1 ± 9.8 (4)
19 (fed)		9.3 ± 2.6 (11)		52.4 ± 25.2 (5)		89.8 ± 33.6 (3)

a Except where noted
b Grams wet weight
c Sample size

The amount of food consumed varied with crayfish size (Fig. 1). Ingestion (I) per unit time was greater for yearling animals than for young-of-year. The regression line relating ingestion to body weight had a correlation coefficient (r) of 0.89 (Fig. 1). Ingested food was the sum of two components: egested (E) and assimilated (A) materials. A plot of food weight assimilated (I-E) showed that for the extruded alginate diet assimilation also increased with crayfish size (Fig. 1). The correlation coefficient for this relationship was 0.87 (Fig. 1). The difference between the regression lines for ingestion and assimilation represented the amount of food egested.

The parallel regression lines for ingestion and assimilation indicated that the fraction of food assimilated remained fairly constant for small and large crayfish. A plot of percent assimilation (A/I x 100) produced a relatively flat line reflecting an average assimilation efficiency of 69-73% (Fig. 1). The low correlation coefficient was a result of the slope of the line, and in this case did not indicate a lack of fit between the points and the regression line (Li 1968).

Time-limited feeding measurements showed that young-of-year crayfish ingested comparable amounts of both the fish and alginate diets. However, assimilation efficiency was significantly higher for the fish diet (P < 0.05, t-test) (Table 3). Similar ingestion values for both diets allayed initial concern that young-of-year crayfish might not readily accept the dried fish material and thus bias the results. Smaller egestion values and higher assimilation efficiency indicated that the fish was more thoroughly utilized than the alginate material.

DISCUSSION

As expected, large <u>Orconectes</u> <u>virilis</u> consumed more oxygen than small crayfish (Table 2) but per gram body weight, oxygen consumption was greater for small animals. Measurements of standard metabolic weight were temperature dependent (Table 2).

Despite these consistent results, variability increased at higher temperatures. Thus metabolic rates for starved animals were higher than true standard metabolism (Table 2). Part of this variability for adult crayfish was attributed to the size range of animals tested. By starving animals at 19°C and increasing body weight from 5 to 9 g, we raised the metabolic rate by 62%. At the same temperature, increasing weight from 10 to 15 g resulted in a 70% increase in metabolism. Thus, determination of mean oxygen consumption for the 5 to 6 g weight range introduced a source of variability into the calculations. This could only be reduced by increasing the number of equal sized replicates.

Activity influenced the measurement of standard metabolism. Although test animals were allotted time for acclimation to the respiratory chamber, once the test began only the chamber size physically restricted crayfish movement. Observations of animals during periodic temperature readings showed that though crayfish had calmed they were not totally immobile. Some minor movements and changes in position added to the greater variability in mean oxygen consumption of young-of-year and 5-9 g animals at 24°C, and also

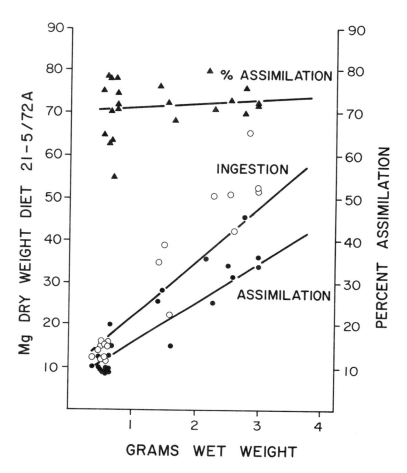

Figure 1. Relationship between ingestion (open circles), assimilation (closed circles), and assimilation efficiency (triangles) for an extruded alginate diet and weight of young-of-year and yearling Orconectes virilis. Calculations are based on a feeding period of 90 min. Equations describing the relationship between ingestion, assimilation, and assimilation efficiency of the extruded alginate diet 21-5/72A and wet weight of Orconectes virilis are of the form $Y = mx + b$, where Y = ingestion (I); assimilation (A), the difference between ingestion and egestion (I-E); assimilation efficiency (A/I x 100), and x = crayfish wet body weight. Number of observations (n) and correlation coefficients (r) are given for each equation. Values were as follows:

	m	b	r	n
Ingestion	0.013	0.009	0.89	22
Assimilation	0.009	0.067	0.87	22
Assimilation Efficiency	0.746	69.643	0.22	22

Table 3. Comparison of ingestion, egestion, assimilation, and assimilation efficiency for young-of-year, <u>Orconectes</u> <u>virilis</u>, fed fish and extruded alginate diet 21-5/72A. Physical data is shown for each test group. Means and standard deviations are listed for all feeding measurements.

	Alginate Diet 21-5/72A	Fish
Sample Size	12	6
Carapace Length (mm)	12 - 14	12 - 14
Wet Weight (g)	0.57 ± 0.08	0.43 ± 0.12
Ingestion (I) (g)	0.014 ± 0.002	0.012 ± 0.004
Egestion (E) (g)	0.004 ± 0.001	0.001 ± <0.001
Assimilation (I-E) (g)	0.010 ± 0.002	0.012 ± 0.004
Assimilation Efficiency (A/I x 100)	70.4 ± 7.5	95.0 ± 1.9

resulted in higher metabolic rates for 10-15 g animals at 19°C (Table 2). This increased activity was not a response to stress induced by oxygen depletion. The greatest oxygen loss recorded in any test was a decrease to 59% of saturation corresponding to 5.1 mg O_2/l which was far above levels limiting crayfish distribution in the Pigeon River lakes (Fast and Momot 1973, Gowing and Momot 1974).

Finally, temperature acclimation and degree of starvation may have influenced measurements of standard metabolism. McWhinnie and O'Connor (1967) showed standard metabolism determined for intermolt O. virilis varied with length of temperature acclimation and degree of starvation. After two weeks, animals tested at 5 and 18°C had a higher level of oxygem consumption than those tested at one week. They attributed this increase to a changeover in metabolic pathways within the crayfish hepatopancreas. In addition, animals acclimated and tested at 5°C had higher levels of oxygen consumption than animals acclimated at 18°C and tested at 5°C, but the metabolic rates of crayfish maintained at 5°C and tested at 18°C were not significantly different from those of animals acclimated and tested at 18°C. Other workers believed that an acclimation period of 1.5 days was sufficient to achieve stabilization of metabolic rate (Wiens and Armitage 1961).

Crayfish used in this study had no prior history of long term acclimation to warm temperatures. They were obtained from the field in mid-September and kept in chilled water (10-15°C) until testing started at the beginning of October. Other species of Orconectes (Table 6) have metabolic rates comparable to O. virilis (Eggleston 1973, Weins and Armitage 1961).

Though a longer period of temperature acclimation or greater control over activity may have provided metabolic rates closer to the definition of standard metabolism, our rates probably approximated the metabolism of non-feeding field animals. Crayfish probably never achieve a level of standard metabolism under field conditions. Even in winter, reduced levels of movement and feeding continue (Lorman 1975). In North Twin and West Lost Lakes, the bowl-shaped configuration of their basins restricts the size of the epilimnion, and many crayfish are located within the thermocline (Momot 1967, Fast and Momot 1973). Movements of only a few meters expose them to noticeable temperature changes. Thus, extended periods of metabolic adjustment in these crayfish would be a rare event in the field situation. For these reasons, we felt the measurements of metabolic rates obtained in this study were acceptable.

Measurements relating carapace length to age of field animals (Momot and Gowing 1976) and the value 3051 cal/g dry weight for O. virilis (Kelso 1973) were combined with respiratory data to calculate the caloric expenditure of an individual crayfish from hatching to 3.5 years of age (Table 4). Wet and dry weights for O. virilis (Jones 1979) are allometrically related.

In many ways, this summary greatly simplifies the energy expenditure of O. virilis. Some assumptions, such as the six month growing season and 19°C water temperature, are justifiable as average representations of field conditions (Momot 1967, Gowing and Momot 1974). Others such as 9°C used for winter temperature and the inclusion of specific dynamic action (Q_d) as a consistent factor in the

Table 4. Estimated lifetime energy expenditure for Orconectes virilis. Measurements were calculated for one individual living to 3.5 years. Yearly increase in size and calories converted to growth (Q_g) were derived from values reported by Momot and Gowing (1976) and Kelso (1973). Calculations of standard metabolism (Q_s) and specific dynamic action (Q_d) were determined from regression relationships between oxygen consumption and crayfish wet weight.

	Carapace (mm)	Wt. (g) wet/dry	Wt. Gain (g) wet/dry	Summer (6 mo – 19°C)		Winter (6 mo – 19°C)	Annual Total (Kcal)
				$Q_s + Q_d{}^a$ (Kcal)	Q_g (Kcal)	Q_s (Kcal)	
Age 0	13	0.6/0.1	0.6/0.1	2.3	0.3	0.5	3.1
Age I	28	6.2/1.4	5.6/1.3	6.9	4.0	2.1	13.0
Age II	33	10.2/2.3	4.0/1.0	10.4	3.1	3.3	16.8
Age III	40	18.4/4.1	8.2/1.8	16.4	5.5		21.9
TOTALS				36.0	12.9	5.9	54.8

[a]Values based on average size and wet weight attained during growing season: Age 0 – 6.5 mm, 0.07 g; Age I – 25.5 mm, 4.6 g; Age II – 30.5 mm, 8.0 g; Age III – 36.5 mm, 13.9 g.

201

summer energy budget, cause an overestimation of caloric expenditure. Despite these approximations the greatest discrepancy was the lack of data for activity (Q_a). This probably caused an overall underestimate of the total energy budget for O. virilis.

O. virilis is representative of the role of crayfish in the energy dynamics of a system. Values for the amounts of energy used by O. virilis were obtained by combining seasonal energy requirements (Table 4) and the end of summer population estimates derived for both North Twin and West Lost Lakes (Momot and Gowing 1976). The annual energy requirements of these populations are summarized for 1971, 1972, and 1973 (Table 5). In both lakes most of the energy is used to support young of the year and yearling animals (Ages 0 and 1). The difference between North Twin's and West Lost Lake's carrying capacity for crayfish is apparent when total K cal/m² were compared. For these years, energy which would be required to support the crayfish population is 2.2 to 2.7 times greater in West Lost Lake (Table 5). Because the growth of animals is not significantly different (Momot and Gowing 1974) and factors such as predation (Gowing and Momot 1979) and qualitative differences in potential food sources (Jones and Momot 1981) have been eliminated as controls of crayfish numbers, the amount of energy needed to support the West Lost Lake population indicates that the difference in carrying capacity may reflect the quantity of food available in each lake.

Our limited calculations in our energy budget suggest that individual crayfish in our small pothole lakes expand about three times as much energy for maintenance as for growth. Most other species also expend three to ten times as much energy for maintenance as for growth. An exception is Cherax destructor which apparently converts 50% of ingested energy into growth.

This discrepancy leads to apparent differences in calculated net growth efficiencies of K_2 (growth/assimilation). The K_2 values are essentially similar in several species including: Orconectes virilis (23.5%); Orconectes limosus (25.6%); Pacifastacus leniusculus (29.2%); and Astacus leptodactylus (21.1%) (Table 6). In contrast, the K_2 value for Cherax destructor is much higher (54%) (Table 6), nearly twice that calculated for other species.

The other reason for this difference is seen in a close examination of the estimated values for production and respiration in the energy budgets of the three species of Orconectes in comparison with Cherax destructor (Table 7). The ratios between production (P) and respiration (R) (Table 7) parallel the data for the K_2 values. For O. propinquus the ratio (P:R) is about 3:1; the same is true for O. virilis while in O. limosus it is about 10:1. For Cherax destructor the P:R ratio approximates 1:1, making this species far more efficient in converting detrital energy into crayfish biomass. The longer growing season in S. W. Australia must contribute substantially to the greater efficiency.

These animals play an important role in processing food materials within lake ecosystems. The assimilation experiments suggest the magnitude of this material transfer (Fig. 1, Table 4). In a 90 min feeding period, a 1 g animal consumes 0.022 g of alginate material. This corresponds to 2.2% of its body weight. In some experiments we

Table 5. Summation of energy supporting crayfish populations in North Twin and West Twin and West Lost Lakes, Otsego County, Michigan. Values are reported for end of summer population estimates calculated for 1971, 1972, and 1973 (Momot and Gowing 1975).

| | North Twin | | West Lost | |
	Total Kcal	Kcal/m^2	Total Kcal	Kcal/m^2
1971				
Age 0	56,246	14.7[a]	127,845	45.4[a]
Age I	53,010	2.8	52,592	3.5
Age II	36,738	1.9	54,304	3.6
Age III	4.133	0.2	4,864	0.3
Totals	150,127	19.6	239,605	52.8
1972				
Age 0	32,807	8.6[a]	65,224	23.1[a]
Age I	60,386	3.2	77,463	5.2
Age II	46,894	2.5	47,658	3.2
Age III	5,443	0.3	8,266	0.6
Totals	145,530	14.6	198,611	32.1
1973				
Age 0	27,368	7.1[a]	79,004	28.0[a]
Age I	62,951	3.3	46,580	3.1
Age II	35,630	1.9	40,264	2.7
Age III	7,157	0.4	2,747	0.2
Totals	133,106	12.7	168,595	34.0

[a]Calculated for estimated littoral area of North Twin (3834 m^2) and West Lost Lakes (2819 m^2) (Momot and Gowing 1976).

Table 6. Estimates of Gross Growth Efficiency (K1), Net Growth Efficiency (K2), Food Intake (I) as % body weight per day, Assimilation Efficiency (A/I) of various species of crayfish, all data expressed as %.

Species	Locality	K1	K2	A/I	I	Source
Orconectes virilis	Small Michigan Pothole Lakes		23.5	70-90	2.4-2.8	This study
Orconectes Timosus	1600 ha Polish Reservoir	7.3	25.6[e]	46	5a-7d	Kossakowski and Orzechowski (1975)
Pacifastacus Teniusculus	Small Creek in Oregon Lab Study	15.1	29.2	50	2.0[f]	Mason (1975)
				44-64		Moshiri and Goldman (1969)
Astacus Teptodactylus	Ponds, River Don U.S.S.R.	17[b] 60[a]	21[b] 75[a]		16[a] Ponds 9[a] Ponds 8.7[b] River 4.4[b] Don	Tcherkashina (1977)
Cherax destructor[g]	Australian Farm Ponds		54			Woodland (1969)

[a] Sub yearlings
[b] Yearlings
[c] Older subyearlings
[d] Adults
[e] Mean value
[f] Estimated as 0.6% dry wt./day
[g] Incorrectly given as Cherax albidus in the thesis

204

Table 7. Comparative estimates of production and respiration from several energy budgets of various crayfish species given as Kcal/m^2/yr.

Species	Production	Respiration	Source
Orconectes propinquus	32.5	100.7	Vannote (1963)
Orconectes limosus	28.9[a]	292[a]	Kossakowski and Orzechowski (1975)
Orconectes virilis	12.9[a]	36.0[a]	This study
Cherax destructor	158	132	Woodland (1969)

[a]Expressed as Kcal/individual crayfish/year.

obtained values as high as 2.8% per day. Actual daily consumption in nature is probably greater and is probably closer to the values of 5% recorded for O. limosus and 4.4% for yearling Astacus leptodactylus, and approximates the 2.0% estimated for P. leniusculus (Table 6).

Approximately 70% of the ingested alginate material was assimilated by O. virilis as was 90% of the fish. This compares favorably with values of 44%, 50% and 64% estimates for P. leniusculus fed on lettuce, leaves and chicken respectively and 46% for Orconectes limosus fed a variety of natural foods (Table 6).

For young O. virilis the alginate diet, judging from the mean growth rate (Table 1), served solely for maintenance. It did not promote weight increases comparable to that of field animals. Assuming an efficiency of between 50 and 70% for most natural food, crayfish in the field must either obtain more energy per weight of food assimilated or, more likely, they must be processing food at a faster rate than animals fed artificial diets or natural diets in the lab. A definitive answer to this problem can only come from a combination of bomb calorimetry studies determining energy values of ingested and egested foods and measurements of daily ingestion rates. However, the lower assimilation efficiencies (46% to 64%) (Table 6) for other species would suggest that crayfish in most habitats process considerable amounts of material in order to obtain the energy necessary for maintenance and growth. This is also alluded to from recent data on the role of the crayfish Cambarus bartoni on the Ca dynamics of small streams, which showed that crayfish often dominate the biomass and can account for as much as 53% of the total Ca ingestion in such streams (Webster and Patten 1979).

In most environments the availability of food resources may thus be the ultimate determining factor in regulating crayfish population size.

ACKNOWLEDGMENTS

We gratefully acknowledge the Michigan Department of Natural Resources for use of the Pigeon River Trout Research Station and the University of Michigan Biological Station for use of equipment and facilities. The following students assisted in field work: C. Cihra, W. Overholtz, W. Caine, M. Bigi, R. Rohrbaugh, P. Gerham, and C. Wencel. We also especially acknowledge Dr. T. Peterle and Dr. R. Stein, Ohio State University, for their considerable editorial assistance with Miss Jones' Ph.D. thesis, which contributes substantially to this paper.

LITERATURE CITED

Burke, J. D. 1962. Determination of oxygen in water using a 10-ml syringe. J. Elisha Mitchell Sci. Soc. 78(2):145-147.

Dean, J. L. 1969. Biology of the crayfish, Orconectes causeyi, and its use for control of aquatic weeds in trout lakes. U.S. Bur. Sp. Fish. Wildl. Tech. Paper No. 24, 15 pp.

Eggleston, P. M. 1975. The energy requirements of the crayfish, Orconectes rusticus, and its ability to utilize various species of algae as food. Ph.D. Thesis, Ohio State University, Columbus, 56 pp.

Fast, A. W. and W. T. Momot. 1973. The effects of artificial aeration on the depth distribution of the crayfish, Orconectes virilis (Hagen), in two Michigan Lakes. Amer. Midl. Natur. 89:89-102.

Gowing, H. and W. T. Momot. 1974. Population dynamics of trout, with crayfish as food, in three pothole lakes in Michigan. Mich. Dept. Nat. Res. Fish. Div. Lansing, Fish. Res. Rep. No. 1181, 69 pp.

Gowing, H. and W. T. Momot. 1979. Impact of brook trout (Salvelinus fontinalis) predation on the crayfish Orconectes virilis in three Michigan Lakes. J. Fish. Res. Board Can. 36:1191-1196.

Huner, J. V., S. P. Meyers and J. W. Availt, Jr. 1975. Response and growth of freshwater crayfish to an extruded, water-stable diet. pp. 149-157. In: J. W. Availt, Jr. (ed.) Freshwater Crayfish, Papers from the Second International Symposium on Freshwater Crayfish, Baton Rouge, Louisiana. 1974. Louisiana State University, Div. Cont. Educ., Baton Rouge, 676 pp.

Jones, P. D. 1979. An investigation of factors regulating the crayfish carrying capacity of two northern Michigan Lakes. Ph.D. Thesis, Ohio State University, Columbus, Ohio.

Jones, P. D. and W. T. Momot. 1981. Crayfish productivity allochthony and basin morphometry. Can. J. of Fisheries and Aquatic Science 38(2): 175-183.

Kelso, J. 1973. Seasonal energy changes in Walleye and their diet in West Blue Lake, Manitoba. Trans. Amer. Fish. Soc. 102:363-368.

Kossakowski, J. and B. Orzechowski. 1975. Crayfish, Orconectes limosus, in Poland. pp. 31-47. In: J. W. Availt, Jr. (ed.) Freshwater Crayfish, Papers from the Second International Symposium on Freshwater Crayfish, Baton Rouge, Louisiana. 1974. Louisiana State University, Div. Cont. Educ., Baton Rouge, 676 pp.

Li, J. C. R. 1968. Statistical Inference I. Edward Brothers, Inc., Ann Arbor. 658 pp.

Lorman, J. G. 1975. Feeding and activity of the crayfish, Orconectes rusticus, in a northern Wisconsin lake. Masters Thesis, University of Wisconsin, Madison. 56 pp.

Mason, J. C. 1975. Crayfish production in a small woodland stream. pp. 449-479. In: J. W. Avault, Jr. (ed.) Freshwater Crayfish, Papers from the Second International Symposium on Freshwater Crayfish, Baton Rouge, Louisiana. 1974. Louisiana State University, Div. Cont. Educ., Baton Rouge, 676 pp.

McWhinnie, M. A. and J. D. O'Connor. 1967. Metabolism and low temperature acclimation in the temperate crayfish, Orconectes virilis. Comp. Biochem. Physiol. 20:131-145.

Momot, W. T. 1967. Population dynamics and productivity of the crayfish, Orconectes virilis, in a marl lake. Amer. Midl. Natur. 78:55-81.

Momot, W. T. and H. Gowing. 1975. The cohort production and life cycle turnover ratio of the crayfish, Orconectes virilis, in three Michigan Lakes. pp. 489-511. In: J. W. Avault, Jr. (ed.) Freshwater Crayfish, Papers from the Second International Symposium on Freshwater Crayfish, Baton Rouge, Louisiana. 1974. Louisiana State University, Div. Cont. Educ., Baton Rouge, 676 pp.

Momot, W. T. and H. Gowing. 1976. Ricker equilibrium yield for unexploited poulations of crayfish. Mich. Dept. Nat. Res. Fish. Div. Lansing, Fish. Res. Rept. No. 1832. 43 pp.

Momot, W. T. and H. Gowing. 1977a. Response of the crayfish, Orconectes virilis, to exploitation. J. Fish. Res. Board Can. 34:1212-1219.

Momot, W. T. and H. Gowing. 1977b. Production and population dynamics of the crayfish, Orconectes virilis, in three Michigan lakes. J. Fish. Res. Board Can. 34:2041-2055.

Momot, W. T. and H. Gowing. 1977c. Results of an experimental fishery on the crayfish, Orconectes virilis. J. Fish Board Can. 34:2056-2066.

Momot, W. T., H. Gowing and P. D. Jones. 1978. The dynamics of crayfish and their role in ecosystems. Am. Midl. Natur. 99:10-35.

Moshiri, G. A. and C. R. Goldman. 1969. Estimation of assimilation efficiency in the crayfish, Pacifastacus leniusculus (Dana) (Crustacea: Decapoda). Arch. Hydrobiol. 66:298-306.

Tscherkashina. 1977. Survival, growth and feeding dynamics of juvenile crayfish (Astacus leptodactylus) in ponds and the river Don. pp. 95-100. In: Lindquist, O. V. (ed.). Papers from the Third International Crayfish Symposium, Kuoppio, Finland, University of Kuoppio, 504 pp.

Vannote, R. L. 1963. Community productivity and energy flow in an enriched warm water stream. Ph.D. Thesis. Michigan State University, East Lansing, Mich.

Warren, C. E. and G. E. Davis. 1967. Laboratory studies on the feeding, bioenergetics and growth of fishes. pp. 175-214. In: S. D. Gerking (ed.). The Biological Basis of Freshwater Fish Production. Blackwell Scientific Publications, Oxford. 495 pp.

Warren, C. E. 1971. Biology and Water Pollution Control. W. B. Saunders Co., Philadelphia. 434 pp.

Webster, J. R. and B. C. Patten. 1979. Effects of watershed perturbation on stream potassium and calcium dynamics. Ecol. Monogr. 49:51-72.

Wiens, A. W. and K. B. Armitage. 1961. The oxygen consumption of the crayfish, <u>Orconectes</u> <u>immunis</u>, and <u>Orconectes</u> <u>nais</u>, in response to temperature and to oxygen saturation. Physiol. Zool. 34:39-54.

Woodland, D. J. 1969. The population study of a freshwater crayfish, <u>Cherax</u> <u>albidus</u>* Clark. Ph.D. Thesis, Univ. of New England, New South Wales, Australia. 209 pp.

*<u>Cherax</u> <u>destructor</u> is the accepted name.

PACIFASTACUS LENIUSCULUS (DANA) PRODUCTION IN THE SACRAMENTO RIVER

Steven J. Shimizu and Charles R. Goldman
Division of Environmental Studies
University of California
Davis, Calfiornia, 95616 U.S.A.

ABSTRACT

A population of the crayfish, Pacifastacus leniusculus, was sampled for a two year period from a defined section of the Sacramento River. Estimates of production were calculated using the Allen and removal-summation methods from data collected on growth, reproduction, density and mortality. A comparison of Pacifastacus leniusculus production between the Sacramento River and Lake Tahoe populations showed production in the Sacramento River to be substantially higher and probably related to environmental differences.

INTRODUCTION

Pacifastacus leniusculus is a common freshwater crayfish inhabiting most of the coastal streams and drainages as well as several lakes in Washington, Oregon and northern California (Riegel, 1959; Miller, 1960). The Sacramento-San Joaquin delta area is perhaps the largest drainage system which affords habitat for this species. It is primarily a cold water species thriving in environments having water temperatures as low as 4-5°C as in Lake Tahoe, California. However, natural populations in the delta system thrive in water temperatures higher than most other environments in which it is found. Summer water temperatures can reach as high as 25°C which approach this species' lethal temperature limits (Becker, et al., 1975). In addition, high Pacifastacus densities in the delta and its increasing popularity as a food item have encouraged establishment of a commercial fishery (Nicola, 1971).

Little if any production or population dynamics research has been done on the Sacramento-San Joaquin delta Pacifastacus populations. Abrahamsson and Goldman (1970), Abrahamsson (1971), Flint (1975a), Flint and Goldman (1977), Goldman and Rundquist (1977) and Mason (1975) have published papers on natural Pacifastacus leniusculus populations. However, their populations were from colder unexploited environments.

The objective of this study was to outline details of Pacifastacus leniusculus growth, reproduction, density, and mortality. This information should make it possible to derive a production estimate for the population and establish factors which might influence or alter production. Importantly, results obtained here are used to help explain variation of this species' life history in different environmental situations.

STUDY AREA

The Sacramento-San Joaquin delta system is characterized by having relatively steep sloping banks with granite rock placed to reduced

210

erosion. Some portions, particularly in the slower moving areas, have been left unrocked and consist primarily of mud-clay bank with associated riparian vegetation. The river bottom topography is, for the most part, gently sloping reaching an average summer depth of about 9 meters. Heavy rains during the winter, however, can bring the Sacramento River up an additional 8 meters.

Studies of the Pacifastacus leniusculus population were conducted from May 1977 to February 1979. Field samples were taken from July 1977 to November 1978 within a 350 meter long by 105 meter wide segment of the Sacramento River. Location of the sample area was approximately 5.5 kilometers up-river from the town of Freeport. The site was chosen because it not only typified the general habitat occupied by Pacifastacus leniusculus but also included both types of river banks previously described. The west bank was made up entirely of granite rock while the east bank was characterized primarily by mud and clay. The bottom, sampled using an Eckman dredge, consisted primarily of coarse sand out from the submerged rock bank grading into silt or clay along the mud-clay bank.

METHODS

Sampling crayfish effectively proved to be a difficult task in this system as compared with oligotrophic Lake Tahoe (Flint, 1975a) or a small woodland stream (Mason, 1975). Turbid water conditions made in situ observations of crayfish extremely difficult. Secchi disc readings were less than .6 meters in the sample area as compared to readings of 40+ meters in Lake Tahoe (Goldman and Armstrong, 1969). In addition, the nature of the solid rocky substrate made sampling of juvenile crayfish almost impossible as they quickly evaded dip nets by scurrying deep into crevices. Consequently a variety of indirect methods involving determination of recruitment, growth and mortality were employed to effectively sample the total population.

Crayfish were captured by means of baited wire mesh traps, 61 centimeters long by 23 centimeters in diameter, with cone shaped entrances at both ends. Trapping was conducted on a weekly basis during the entire duration of field study (except for several weeks during the winter where high river conditions made it hazardous to sample).

Crayfish carapace lengths were measured from the posterior portion of the eye socket to the posterior margin of the carapace and recorded to the nearest .05 millimeter, using a Vernier caliper. The sex of each individual was noted and they were returned to the area from which they were caught. A small sample was retained for laboratory dissections of gonad maturation and gastrolith presence. A number of individuals at each sampling were cauterized with a numbered code similar to the method described by Abrahamsson (1965) and returned. Subsequent recapture revealed information on pre/post-molt growth, molting frequencies and possible dispersal patterns.

Density estimates of crayfish within the sample area were calculated using the Schnabel mark-recapture technique described in Ricker (1958). Momot (1967) indicated that reliable estimates of population size could be made using this method. Captured crayfish were tagged by tail punching over a one week period during which time recaptures were

211

counted and compared against marked crayfish.

Age determination is difficult due to the relatively long life span of Pacifastacus leniusculus. Statistical analysis, as described by Tanaka (1962), divided a polymodal length frequency distribution (Fig. 1) into its various age class components. A series of calculated parabolas was fitted to a semi-log length frequency distribution then redrawn on an arithmetic scale. Modes which were initially masked by overlapping can now be more easily distinguished (Fig. 2). Flint (1975b) used this method successfully in his growth studies of Pacifastacus leniusculus in Lake Tahoe. Basically the formula, $\text{Log } f(X) = -(0.217 \ X^2/\sigma^2 + \text{Log}(N/2\pi\sigma)$ where N = total number of animals captured and X represents the size class of crayfish (carapace length in millimeters), is used to obtain a series of parabolas by changing the standard deviation (σ) of mean molt increase by .12 increments (recommended by Tanaka, 1962).

Growth data collected in the laboratory and field also helped in the positioning of these modes. Pre- and post-molt growth data was obtained from recaptured, cauterized crayfish. Those that were observed to have molted were remeasured and compared with pre-molt lengths. Crayfish found nearing ecdysis (softening carapace) were observed in isolated laboratory aquaria as were juveniles less than 25 mm carapace length and newly hatched young.

Reproductive potential was estimated from pleopod egg counts for all size classes of females. The presence of cement glands on the underside of the female abdomen during the months of September and October as well as enlarged ovarian eggs indicated maturity. Maturity in males was determined by the presence or absence of sperm in the vas deferens during October before mating took place. The technique of Wenner (1974) was used to estimate the mean size of males and females at onset of sexual maturity.

Mortality was calculated using the relationship $\ln N_t = \ln N_0 - Zt$ to find instantaneous mortality rate (Z). Ideally Z is calculated using successive population estimates. However, questions arose as to the reliability of the Schnabel method application to this system. In addition, uneven representation of all age (size) classes tended to alter estimates. Consequently mortality was calculated using cohort length frequency data and estimates of recruitment.

RESULTS

Crayfish distribution within the sample site seemed to be influenced primarily by substrate. During one period of sampling the average weight of crayfish per trap on the rocky substrate was 2.27 kilogram per trap. The mud-clay substrate yielded an average of .18 kilogram per trap. A series of 22 traps were extended from one bank to the other at 5 meter intervals in an effort to plot crayfish distribution across a profile of the sample area. The rocky bottom was, by far, the preferred substrate. A total of 562 crayfish were caught, 87 percent of which were from 5 traps lying within the rocky area. The remaining 13 percent were scattered among those set on sand and mud-silt areas. Unlike many other crayfish, Pacifastacus leniusculus is a non-burrowing species and prefers areas which provide adequate cover.

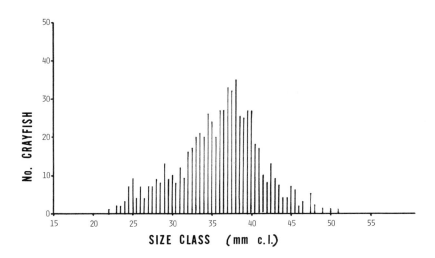

Figure 1. Sampled length frequency distribution of <u>Pacifastacus</u> <u>leniusculus</u> from the Sacramento River.

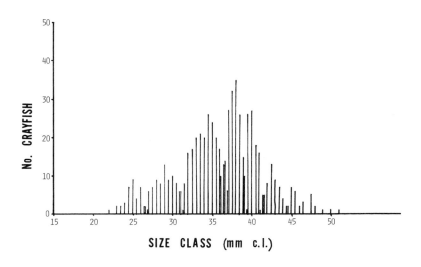

Figure 2. Adjusted length frequency distribution using the method of Tanaka (1962).

On several occasions crayfish tagged on one side of the river were recaptured on the opposite side. While some localized migration may occur, these movements are most likely a random dispersal phenomenon. Flint (1975a) mentions Pacifastacus leniusculus in Lake Tahoe migrating to deeper water in winter, however, he adds that this may be more of an escape from harsh winter conditions in the shallower areas.

Continuous trapping yielded information presented in the catch curve (Fig. 3). Presented along with it is a temperature curve taken at each sampling. The overall shape of the catch curve closely follows a seasonal dependence of crayfish activity to temperature. Major fluctuations in the curve during the months of July-October are due primarily to molting patterns and will be discussed shortly.

The Schnabel technique of capture-mark recapture (Ricker, 1958) is a useful method of population estimation especially for an area that is difficult to sample by more accurate or direct methods. However, it must be noted that population numbers may be underestimated if catch-ability of the population is variable during different times of the year. In Pacifastacus crayfish populations, attributes such as molt cycles, periods of dormancy, secluded habits of berried females must be taken into account. It is, therefore, important to first plot a curve similar to the catch curve in Figure 3 and conduct short duration population estimations using the Schnabel technique close to a time when a maximum number of animals are available. It is assumed by the model that enough traps are used to effectively attract all crayfish within the entire area. This assumption was not able to be tested since current flow altered trapping radius. Also murky water conditions made it impossible to determine actual crayfish densities in a way similar to Abrahamsson and Goldman (1970). If the total trapping radius did not cover 100 percent of the sample area, resulting population estimates would be low. Since there was no way of knowing whether or not the area was 100 percent covered, the number of crayfish calculated is a minimum with higher densities more likely the case. Unfortunately, due to the nature of the study site no other methods of estimation could be used and the capture-recapture method seemed the only one practicable. Tagging of crayfish by tail punching was conducted from August 2-6, 1977 and an estimated 66,381 crayfish greater than 25 millimeters carapace length were considered to be present in the 36.75 kilometer^2 sample area. This averages approximately 1.8 adult crayfish per meter^2. This minimum density projected for the entire study area, is somewhat misleading especially when one condsiders concentrations of crayfish within the area. As previously mentioned Pacifastacus leniusculus prefers rocky substrate almost exclusively. This substrate represents about a quarter of the total area into which almost all of the crayfish are concentrated. A more realistic population density in areas most likely occupied by crayfish would be 77.2 crayfish per meter^2 at the very minimum. However, for comparative purposes with other studies density over the entire area will be used. Traps sampled only those crayfish greater than 25 millimeters carapace length even though smaller mesh trap material was tested. Therefore, the number of crayfish per meter^2 was calculated for those having carapace lengths over 25 millimeters. Juvenile density was calculated using information collected on recruitment and mortality, which will be elaborated upon shortly. Using these methods an estimated 92.82 juveniles per meter^2 (less than 25 millimeters carapace length) was calculated for the entire area.

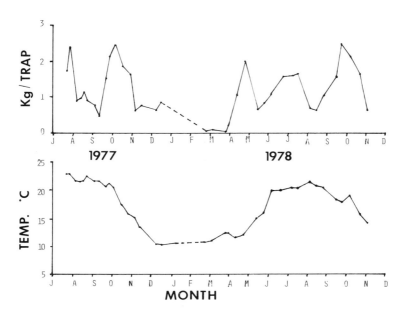

Figure 3. Crayfish catch and temperature curves from the Sacramento River over the duration of the field studies.

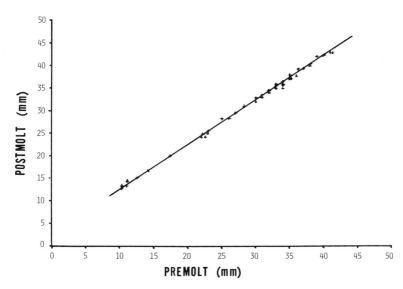

Figure 4. Linear regression of pre- and post-molt carapace length in Sacramento River crayfish. Post-molt length = 2.43 + .996 (pre-molt length). r = .99453.

Study of Lake Tahoe Pacifastacus populations and Mason's (1975) study of a woodland stream in Oregon supported evidence that Pacifastacus leniusculus is a long-lived crayfish. Flint (1975b) notes at least 9 years classes from Lake Tahoe. Length frequency analysis using the method of Tanaka (1962), described earlier, and growth information revealed about 6 year classes in the Sacramento River population.

Growth of crayfish is accomplished through a series of molts. Pre- and post-molt carapace length were recorded from aquaria specimens as well as from cauterized crayfish in the field that were recaptured and observed to have molted. Figure 4 presents the linear regression of pre- and post-molt carapace length for various sized crayfish. Using this data an increase of 2.3 millimeters carapace length was calculated at each molt. Growth is not only determined by an increase in size but also by the frequencey of molting. Gastrolith samples in Figure 5 indicate that almost all crayfish coming out of winter dormancy molt about March and most are through molting my mid-April. A little more than half of the population goes through a second molt in early June with a third molt occurring from mid-August to early September. Molt frequencey and time of molt between males and females are similar. Figure 6 presents gastrolith samples taken from crayfish greater than 40 millimeters carapace length. There appear to be only two molting periods and are extended over a broader period of time. All crayfish greater than 40 millimeters carapace length appear to have molted by July and a second molt, most likely in those crayfish closer to 40 millimeters carapace length, takes place in about 24 percent of the males and 15 percent of the females sampled during mid-August. The lag of females to males during the major spring molt is due primarily to the fact that most females greater than 40 millimeters carapace length are berried and don't molt until eggs hatch in April or May.

Fluctuation in catch, kilograms per trap (Fig. 3), are not only a result of temperature dependence but also a reflection on period of molting. During the 1978 season low catches in March and April correspond to high molt activity. The same is repeated in mid-May and mid-June and again from mid-August to about mid-September.

Using the pre/post-molt carapace increase of 2.3 millimeters, molting time and frequency of molting data, it is possible to construct an estimated growth profile for a Pacifastacus leniusculus cohort from the Sacramento River (Fig. 7). Age is superimposed on this growth profile and illustrates the number of molts each age class undergoes and the growth increment of each class from one year to the next. This data is also presented in tabular form (Table 1). Due to lack of adequate Age 0 samples, the number of molts is unknown. Flint (1975b) indicates about 11 molts for Age 0 crayfish in Lake Tahoe. Pacifastacus in the Sacramento River most likely has at least that many molts as indicated by studies of temperature effects on molting frequency (Hopkins, 1967). Growth can also be represented by the length-weight relationship in Figure 8. Crayfish with complete sets of appendages were weighed to the nearest .01 gram. No statistical difference was noted between males and females ($t_{.05}$ = .558; N = 372) and sexes were therefore combined.

Dissection of males and females just prior to mating yielded maturity information. Figures 9 and 10 present percent of mature males

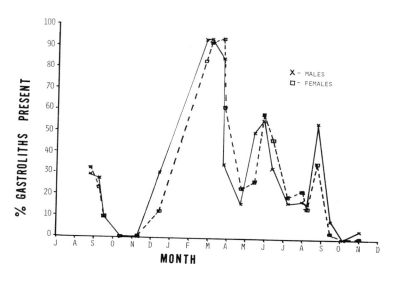

Figure 5. Percent gastroliths present in crayfish samples collected over the duration of the study for entire population. Graph represents all size classes greater than 25 mm carapace length.

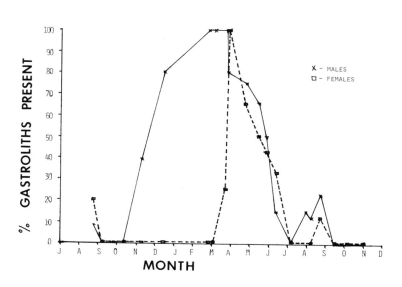

Figure 6. Percent gastroliths present in samples collected over duration of study. Graph represents crayfish greater than 40 mm carapace length.

Table 1. Comparisons of growth in <u>Pacifastacus leniusculus</u>. Molting frequency and size range of each <u>year class</u> are compared for the Sacramento River and Lake Tahoe Populations.

Year Class	Sacramento River		Lake Tahoe	
	# Molts	Carapace Length	# Molts	Carapace Length
0	?	3.5-18.65	11	3.2-12.95
1	4	20.95-27.85	4	15.36-22.35
3	3	30.15-34.75	3	24.64-29.18
4	2	37.05-38.2	2	31.43-33.66
4	1-2	40.5 -(42.8)	1	35.8
5	1	42.8 -(42.8)	1	38.08
6				

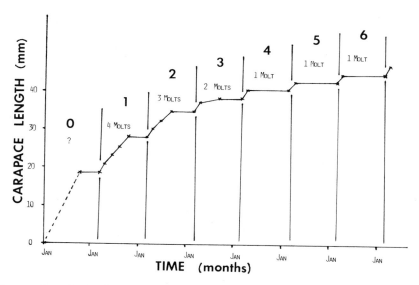

Figure 7. Probable growth profile for successive year classes of _Pacifastacus leniusculus_ from the Sacramento River. Time axis is marked at January of each year. Molting begins in March.

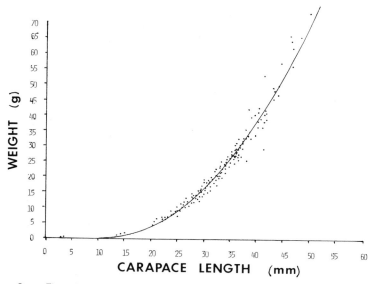

Figure 8. The length-weight regression of _Pacifastacus leniusculus_ from the Sacramento River. Log(weight) = -1.205 + 7.686 Log(carapace length). r^2 = .971.

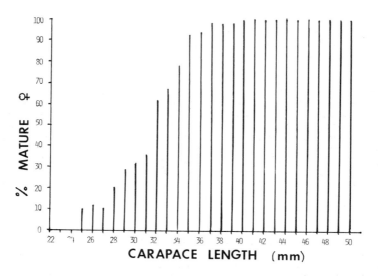

Figure 9. The percent of mature females for each size class collected prior to mating (Sept.-Oct.). N = 1041.

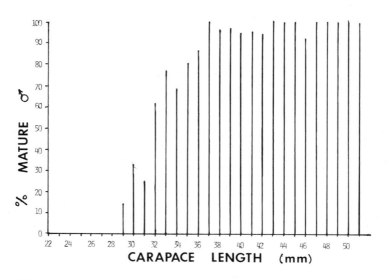

Figure 10. The percent of mature males for each size class collected prior to mating (Sept.-Oct.). N = 893.

220

and females as a function of size. The technique of Wenner et al. (1974) was used to determine the mean minimum size at the onset of sexual maturity. For females this was 30.5 millimeters carapace length and for males 32 millimeters carapace length was the mean. For both sexes this put the mature age of the population at 2 years old.

Crayfish mating was first detected on October 25, 1977 by the observance of spermatophores on the underside of the female cephalothorax. Subsequent berried females were observed in the trap catch a few days later. Based on this observation Sacramento River mating activity begins between mid- to late October. Pleopod egg counts were made on females sampled after mating and though there was a high degree of variability in number of attached eggs ($r^2 = .3711$, Fig. 11), an average of 190 eggs per female was estimated for the general population. Hatching in this system occurs when water temperatures rise to about 14°C. The first evidence of a female with attached juveniles was on March 24, 1978. Generally, hatching takes place between late March and April.

Estimates of crayfish mortaility were made using the relationship in $N_t = \ln N_0 - Zt$ and extrapolations from length frequency. Figure 12 represents a survival curve for the study site population. The instantaneous mortality rate (Z) was calculated to be .67. This represents all forms of mortality (predation, cannibalism, aging, etc.) as well as commercial fishing mortality.

Preliminary catch data indicated a sex ratio of about 1:1 and length frequency analysis of males and females did not indicate any differential mortality between sexes and were, for this reason, not segregated in mortality calculations.

Production

Ideally determinations of production for any population should be based on following the process of one or more year classes from birth to death. Unfortunately for a study of limited duration, following a population with a relatively long life span is difficult. However, by incorporating life history information from various portions of the population, production can be computed through the use of mathematical models. Each model has merits of its own based primarily on the assumptions that govern its use.

Several models are available that calculate production and are presented in Ricker (1958). Waters and Crawford (1973) present a comparison of production models using a stream mayfly population. Based on these results and discussions of each model, two were chosen.

The Allen (1951) method is graphical in nature and plots the number of individuals in a given year class (number per meter^2) against the average weight of an individual of that age taken from length-weight relationship (Fig. 8). Production is derived by measuring the area under the total curve. The Allen method is supposedly free of any assumptions.

The removal-summation method, as described by Waters and Crawford (1973), calculates production as the sum of losses, in weight, from one age class to the next. The model's assumptions of linear growth and

221

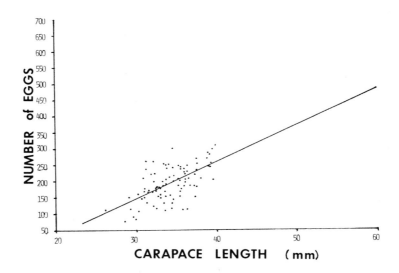

Figure 11. Pleopod egg counts for various sized female crayfish from the Sacramento River. r = .371.

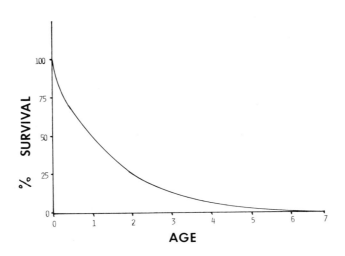

Figure 12. Survival curve of Sacramento River crayfish population. Mortality rate (z) = .67.

mortality were not at all critically based on computer simulation by Cushman et al. (1978). Cushman goes on to state that the removal-summation method was perhaps the single best method for calculating production. Because juveniles were not able to be adequately sampled, pleopod egg counts were used as Age 0 for production calculation. Original pleopod egg counts were made in the fall shortly after egg extrusion. However, losses of a percentage of these eggs occur during the incubation period as well as during the short period while juveniles are still attached to the female. Actual recruitment into the population may be considerably less than original pleopod egg counts predict. Mason (1970) observed that from 15-60 percent of the brood may be cannibalized by the mother or other crayfish. Flint (1975a) estimated that 15 percent of the original complement of eggs hatch and are successfully recruited into the population. For this reason recruitment into the population as Age 0 for production esti-mation was figured on this added pre-recruitment mortality. Berried female samples collected in early March from the Sacramento River, although few (N = 36), exhibited an extremely high egg loss. An average of about 20 percent of the total of eggs probably survive recruitment.

The Allen method resulted in an annual production estimate of 241.71 grams per meter^2 (Fig. 13). Area under the curve (production) was measured using a planimeter. The removal-summation method resulted in a surprisingly close production estimate of 235.8 grams per meter^2 (Table 2). The calculation of turnover, which is the ratio of produc-tion to biomass, is of particular importance in that, to an extent, it serves as another tool in which species or even environments can be compared. The turnover ratio for Pacifastacus leniusculus in the Sacramento River was .76.

Pacifastacus leniusculus occurs in both Lake Tahoe and the Sacramento River, however, environmental conditions to which they are subjected are very different. Lake Tahoe is a cold water ultra-oligotrophic lake system (Goldman and Armstrong, 1969). The lake has, in many cases, limited cover as is evidenced by relatively low adult crayfish density and the population is relatively unexploited. The Sacramento River, on the other hand, is a warm eutrophic system with extensive habitat particularly along the rocky levee banks. The population dynamics between Pacifastacus leniusculus in the Sacramento River sample area and Lake Tahoe is compared on Table 3. Bovbjerg (1970) points out that changes in certain environmental factors might prove important in altering certain biological characteristics such as rates of mortality, growth and reproduction.

Mortality in the Sacramento River population is over twice that in Lake Tahoe. This may in part be due to intra-specific interaction, a function of high densities, within the preferred rocky substrates of the study area (Svardson, 1949). Increased cannibalism during molt is a likely result. While it was not possible to distinguish fishing mortality from total mortality in this study it, nonetheless, probably constitutes a large percentage of the total. High mortality in a population may often be offset by a stimulation in fecundity (Giesel, 1976; Momot et al., 1978). Evidence of fecundity and presumed increased recruitment appears in Table 3 as the "average number of eggs per female." In addition the age at maturity is earlier in the Sacramento crayfish than Lake Tahoe crayfish.

223

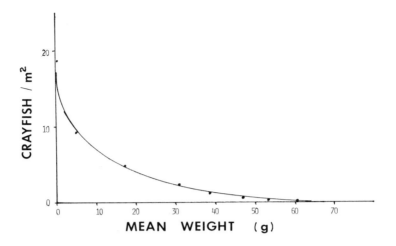

Figure 13 - Allen production curve for Sacramento River crayfish.   Area under curve (production) = 241.72 g/m^2.

Table 2.   Calculations for removal-summation production estimation.

Age	#/m^2	Mean Weight (g)	Standing Crop (g/m^2)	# loss/m^2	Weight Loss (g)	Weight Loss (g/m2)
0	18.56	.03	.556			
				9.08	2.5	22.7
1	9.48	5.0	47.4			
				4.62	11.25	51.97
2	4.86	17.5	85.05			
				2.41	24.25	58.44
3	2.45	31.0	75.95			
				1.19	34.75	41.35
4	1.26	38.5	48.51			
				.66	43.0	28.38
5	.6	47.5	28.5			
				.31	50.5	15.65
6	.29	53.5	15.1			
				.142	57.0	8.094
7	.148	60.5	8.9			
				.142	64.8*	9.20
			310.42 b/m^2			235.78 g/m^2

*Weight loss between 60.5 g (age 7) and observed maximum weight of 69.2 g

Turnover Ratio:
P/B = .759

Table 3. Comparison of differences between the Sacramento River and Lake Tahoe populations of *Pacifastacus leniusculus*.

	Production (g/m^2)	Biomass (g/m^2)	Turnover	Mortality	Min-Max H$_2$O Temp. C	Time of Hatching	Molt Increment	Mature Age	Ave. # Eggs
Sacramento River	235.8	310.42	.76	.67	9.9-25.0	Mar-Apr	2.3 mm	2	190
Lake Tahoe (Flint, 1975)	30.56	103.18	.29	.324	5.0-20.0	June-Jul	2.0 mm	3	120

225

Temperature is probably the single most influential enviromental factor contributing to the differences between Sacramento River and Lake Tahoe Pacifastacus leniusculus population dynamics. Mason (1974) and Westman (1973) present results indicating that increases in temperature have a positive effect on growth. Similarly, the differences in hatching time between eggs from Pacifastacus leniusculus in Lake Tahoe (Flint, 1975a) and the Oregon woodland stream (Mason, 1975) were attributed to warmer water temperatures in the stream. Time of hatching is indicated in Table 3 and is from 2-3 months earlier in the Sacramento River than in Lake Tahoe -- due primarily to the differences in water temperatures between the two systems. While an earlier hatch might suggest more growth through a greater number of molts, this does not seem to be the case here. The number of molts for each year class is basically the same for both environments (Table 1). Differences in growth rates between the two Pacifastacus leniusculus popultions is due primarily to increases in size at each molt. Svardson (1949) noted that higher temperatures increased the molt increment in Astacus astacus. In addition greater growth might be attributed to the abundance of nutrients and detritus in the river and concentration of it among the larger rocks where maximum density was observed.

The calculation of turnover ratios (production per biomass) are particularly valuable in the comparisions between populations and serve to give some indication of the energy flow within the popultion. While turnover ratios are primarily used to assist in assigning trophic levels, they can be used as an indicator of environmental influence on a population. In general, there is realtively high energy flow through a population living in a stressed or active environment and is usually characteristic in those having high production levels. Such is the case for Pacifastacus leniusculus in the Sacramento River. Turnover for the Sacramento River population was calculated to be .76 in contrast to a turnover of .29 for Lake Tahoe. While high densities and intra-specific interaction add stresses upon the population, total mortality (natural and fishing) may potentially stimulate production to higher levels. Temperature also increases production through its influences on growth and reproduction. Lake Tahoe on the other hand, though cold, is basically a stable environment. The crayfish population has few severe stresses placed upon it and due to the absence of fishing pressure, have relatively long life spans.

The major conclusion of this study is that environmental influences such as higher temperatures, ample habitat, and adequate food supply have served to enable Pacifastacus leniusculus in the Sacramento River to reach high relative densities. Stimulated by limited exploitation, production has been elevated to higher levels. In some cases the increase of one portion of an organism's population dynamics, such as mortality, often stimulates others such as fecundity. Warmer temperatures clearly play a major role influencing production as it increases not only growth but also feeding activity, reproductive development, and hatching. Basically this study serves two purposes. One is to describe the dynamics of a crayfish population in terms of production and secondly, to illustrate possible influences environment has on that production.

ACKNOWLEDGMENTS

We would like to thank the commercial fishermen of the Sacramento

River for their cooperation throughout the duration of this study. We are especially indebted to Ms. Darlene McGriff and the California Department of Fish and Game for providing laboratory facilities, equipment, and financial assistance. In addition valuable discussions with Ms. McGriff occurred throughout the duration of the research for which we are particularly appreciative.

LITERATURE CITED

Abrahamsson, S.A.A. 1965. A method of marking crayfish _Astacus astacus_ in population studies. Oikos 16:228-231.

Abrahamsson, S.A.A. 1971. Density, growth and reproduction in populations of _Astacus astacus_ and _Pacifastacus leniusculus_ in an isolated pond. Oikos 22:373-380.

Abrahamsson, S.A.A., and C.R. Goldman. 1970. Distribution, density and production of the crayfish _Pacifastacus leniusculus_ Dana in Lake Tahoe, Calfornia-Nevada. Oikos 21:83-91.

Allen, K.R. 1951. The Horokiwi stream, a study of a trout population. N.Z. Mar. Dp. Fish. Bull. 10:238 p.

Becker, C. Dale, R.G. Genoway and J.A. Merrill. 1975. Resistance of a Northwestern crayfish _Pacifastacus leniusculus_ (Dana), to elevated temperatures. Trans. Amer. Fish. Soc., 2:373-378.

Bovbjerg, R.V. 1970. Ecological isolation and competitive exclusion in two crayfish (_Orconectes virilis_ and _Orconectes immunis_). Ecology. 51:225-236.

Cushman, R.M. H.H. Shugart, S.G. Hildebrand and J.W. Elwood. 1978. The effect of growth curve and sampling regime on instantaneous-growth, removal-summation and Hynes/Hamilton estimates of aquatic insect production: A computer simulation. Limnol. Oceanogr. 23(1):184-189.

Flint, R.W. 1975a. The natural history, ecology and production of the crayfish, _Pacifastacus leniusculus_, in a subalpine lacustrine environment. Ph.D. Thesis, Univ. of Calif., Davis. 150 p.

Flint, R.W. 1975b. Growth in a population of the crayfish _Pacifastacus leniusculus_ from a subalpine lacustrine environment. J. Fish. Res. Bd. Can. 32(12):2433-2440.

Flint. R.W. and C.R. Goldman. 1977. Crayfish growth in Lake Tahoe: Effects of habitat variation. J. Fish. Res. Bd. Can. 34(1):155-159.

Giesel, J.T. 1976. Reproductive strategies as adaptations to life in temporally heterogeneous environments. Ann. Rev. Ecol. Syst. 7:57-79.

Goldman, C.R. and R. Armstrong. 1969. Primary productivity studies in Lake Tahoe, California. Verh. Int. Ver. Liminol. 17:49-71.

227

Goldman, C.R. and J.C. Rundquist. 1977. A comparative ecological study of the California crayfish, Pacifastacus leniusculus (Dana), from two sub-alpine lakes (Lake Tahoe and Lake Donner). In: Lindquist (ed.). Freshwater Crayfish, 51-80.

Hopkins, C.L. 1967. Growth rate in the freshwater crayfish, Paranephrops planifrons (White). N.Z. J. Sci. 9:50-56.

Mason, J.C. 1970. Maternal-offspring behavior of the crayfish, Pacifastacus trowbridgii (Stimpson). Am. Midl. Nat. 84:463-473.

Mason, J.C. 1974. Aquaculture potential of the freshwater crayfish (Pacifastacus). 1. Studies during 1970. Fish. Res. B. Can. Tech. Report 440. 43 p.

Mason, J.C. 1975. Crayfish production in a small woodland stream. In: J.W. Avault, Jr. (ed.). Freshwater Crayfish,. 449-479.

Miller, G.C. 1960. The taxonomy and certain biological aspects of the crayfish of Oregon and Washington. Thesis, Oregon State Coll., Corvallis. 216 p.

Momot, W.T. 1967. Population dynamics and productivity of the crayfish Orconectes virilis, in a marl lake. Am. Midl. Nat. 78(1):55-80.

Momot, W.T., H. Gowing and P.D. Jones. 1978. The dynamics of crayfish and their role in ecosystems. Am. Midl. Nat. 99(1):10-35.

Nicola, S.J. 1971. Report of a new crayfish fishery in the Sacramento River delta. Inland Fish. Admin. Report #71-7. Calif. Fish and Game. 21 p.

Ricker, W.E. 1958. Handbook of computations for biological statistics of fish populations. Bull. Fish. Res. Bd. Can. 119. 300 p.

Riegel, S.A. 1959. The systematics and distribution of crayfishes in California. Calif. Fish and Game. 45:29-50.

Svardson, G. 1949. Stunted crayfish populations in Sweden. Rep. Inst. Freshwater Res. Drottingholm. 29:135-145.

Tanaka, S. 1962. A method of analysing polymodal frequency distributions and its application to the length frequency distributions of the porgy Taius tumifrons. J. Fish. Res. Bd. Can. 19(6):1143-1159.

Waters, T.F. and G.W. Crawford. 1973. Annual production of a stream mayfly population; a comparison of methods. Limnol. Oceanogr. 18:286-296.

Wenner, A.M., C. Fusaro and A. Oaton. 1974. The mean size at onset of sexual maturity as an indication of growth rate in crustacean population. Can J. Zool. 51:1095-1106.

Westman, K. 1973. Cultivation of the American crayfish Pacifastacus leniusculus. In Abrahamsson (ed.). Freshwater Crayfish, 211-218.

228

POPULATION OF THE CRAYFISH AUSTROPOTAMOBIUS PALLIPES PALLIPES LEREB. IN A BROOK OF CORSICA FRANCE

Jacques C.V. Arrignon
Conseil Superieur de la Peche
10 Rue Péclet
75025 Paris

Bernard Roché
Service Regional d'Aménagement
des Eaux de Corse
20200 Bastia - France

ABSTRACT

The population of crayfishes (Austropotamobius pallipes pallipes Lereb.) was estimated using two statistical methods. Both results give populations of 7,000 females and 8,000 males per hectare.

Males are always more numerous than females, except in the size classes under 70/75 mm. The weight/size ratio is more pronounced for males than for females. The health of the population appears very good and there is no visible sign of plague, Thelohaniosis, or rust-disease. The number of animals reaching or exceeding the legal size of catching (90 mm) is also good (6.35%), and growth is apparently fast. The population is well balanced, with a large stock of juveniles.

INTRODUCTION

The island of Corsica has a system of rivers and hillbrooks of excellent hydrobiological quality, as evidenced by previous studies, that would be propitious to crayfish life. The aim of the present study has been to control the existence of the species Austropotamobius pallipes pallipes Lereb. or "white feet crayfish" in a brook of High Corsica where its presence has been reported by Laurent et Suscillon (1962) and confirmed thereafter by other reports. The control has been followed by a study of the quantitative importance of the crayfish stocks, with the aim of defining the dynamics of population in relation to its terrestrial and aquatic environment. The study will also examine growth and health state and, finally, the appropriateness of stocking this species in other brooks of the same type.

Description of the Study Site

The brook under investigation is the Lutina, a tributary of the Casabiance, itself a tributary of the Porta, that falls in a short stream off the eastern coast of Corsica: the Fium´ alto River (Figure 1). The brook of Lutina rises to about 800 m elevation and the study site is located at about 450 m. General characteristics are the following: Length is 150 m - mean breadth, currents: 2 m - Mean depth: currents : 0.30 m
          basins: 4 m   basins: 0.60 m

Mean slope: 12.5% (125o/oo)

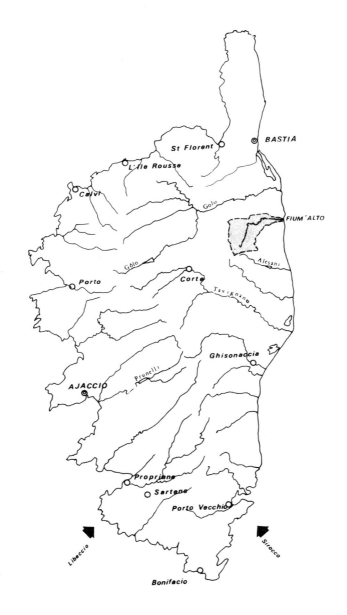

Figure 1. Catchment of Fium'alto River, in Corsica (France)

The summer flow is 17 liter/sec. The speed of the water averages 30 cm/sec. The rainfall can be divided into two periods of precipitation: one in February/March, the second, more important, in October/November, continuing until December. Fall and winter are the two dampest seasons; the summer is very dry. The study site is located in a catchment area receiving between 1000 and 1200 mm of water per year.

The specific study site is located in "Castagniccia" a region of woods of very large chestnut trees. In the Mediterranean climate, with strong montaneous influence, there is an extreme seasonal contrast of a long, cold winter and hot summer. In August, at the study site, air temperature was 27° C and water temperature 14.5° C at 10 a.m. Clear and sunny weather without wind prevailed. It should be noted that there is probably an influence from a dominant and sometimes violent wind which locally is called "Libeccio."

The valley, is of a 1A type upstream and 1B downstream (Arrignon, 1970) and is oriented ESE/WNW. The watershed consists of a pure stand of chestnut trees on the right bank, and on the left bank are chestnut trees bordered by ancient terrace cultures. Borders consists of open bushes composed of Alnus sp, brambles, Viburnum, fig trees, ivies in the upstream portion. Sixty percent of the bank is composed of tree and bush roots of Geraniaceae and wild mint. Forty percent of the stones are covered with moss and Marchantia. The brook flows over a schistous substrata. The stream bottom is arenitic, being composed of blocks and cobbles of schist and patches of sand, with alternatively quiet and fast waters.

Hydrobiological situation

After interpreting the physico-chemical and fauna data it was concluded that the brook is oligotrophic without important evolution from upstream to downstream (Tables 1 and 2).

Fishing

The dominant species of fishes is found to be the Brown Trout (Salmo trutta fario) which is, however, of small size and not abundant in the stream. Only occasional fishing occurs on the stream. Crayfish (Austropotamobius pallipes pallipes) do not appear to be harvested since the local people do not consider it as a native of the region, but rather a species that has been introduced from the continent, some decades ago.

METHODS AND MATERIALS

The most probable estimate of the crayfish population has been based on the two statistical methods commonly used in fish inventory (Arrignon, 1970-1976). These are the successive captures method (De Lury) and the capture/recapture method (Petersen). Marking has been made with spots of a coloured varnish (blue for males and red for females). Limits of confidence were calculated for the population estimates obtained by each method. Two techniques of capture were employed: Scandinavian traps and night fishing with lights.

Table 1. Chemical characteristics (Roche, 1974).

pH (u)	7,6
Conduct. 10_{-6} S/cm	160
D.B.O. 5 j.	0,6
O_2	10,4
HCO_3^-	146
HN_4^+	0
NO_2^-	0
NO_3^-	0,6
Cl^-	8
PO_4^{3-}	0
Ca^{2+}	39,5
Mg^{2+}	8
SO_4^{--}	2

Table 2. Aquatic invertebrates (Roche, 1974).

Famille	Genre
PLECOPTERES	
Perlodidae	Isoperla
Neumridae	Protonemura
Chloroperlidae	Chloroperla
TRICHOPTERES	
Rhyacophilidae	Rhyacophila
Hydropsychidae	Hydropsyche
EPHEMEROPTERES	
Baetidae	Baetis
Ecdyonuridae	Ecdyonurus
Caenidae	Caenis
COLEOPTERES	
Dryopidae	Elminae
DIPTERES	
Chironomidae	
Simulidae	

The trapping was carried out in a selected stream portion of about 250 m² area in a typical zone having successive quiet waters (14% length and 35% of area) and quick waters (86% length and 65% area). Each characteristic portion has been measured in its length, breadth, depth, and habitat description.

The traps have been set at 12 p.m. and raised the day after in the morning at 8 a.m. (10 cylindric, foldable net traps of the Swedish type). The night fishing went from 10:30 p.m. until 11:45 p.m. The operation took place 7 and 8 August 1978.

RESULTS

Numerical population estimation

The Petersen evaluation was based on 150 m length of stream averaging 1.70 m width and including 250 m² of area. The conversion for number of individuals per hectare is x 40 (Table 3).

The De Lury evaluation was also based on a 150 m length of stream averaging, 1.70 m width and given 250 m² of total area. The conversion of crayfish per hectare is given by multiplying x 40 (Table 4).

Table 5 gives the final and best evaluation (based on the DeLury method) for the estimation of the crayfish population.

Size classes and sex ratio

The data includes only those crayfish which exceeded 25 mm. The average weights have been established by size fractions of 5 mm, from 25 mm to 110 mm.

The size classes and the sex ratio appear in Table 6. The sex ratio always favors the males (53.4%) except in the size classes lower than 70 to 75 mm (47.9%). In the size classes above 70 to 75 mm, the difference is even greater (75%). This disproportion has been observed during a similar study of a brook of Lozere (Arrignon and Magne 1978) (see Table 7).

Health condition of population

Rust disease is not frequent and was observed on only three animals (1.3%). No mycose attack has been observed. Mutilations of animals are very frequent (18.6% of the caught stock) with most occurring on males (23% of the male stock). Moulting and post-moulting was observed only in the females (12.7% of the female stock).

DISCUSSION

On the choice of the surroundings

Brooks of the same size and of the same hydrobiological status exist in the natural region considered. It is not certain that the brook was really virgin since it is possible that the existing crayfish stock have been occasionally fished.

233

Table 3. Petersen evaluation.

Passage	Femelles	Mâles	Total
1er (m)	68	75	143
2ème			
(r) recaptures marquées	12	8	20
(u) recaptures non marquées	38	43	81
Total (m+u)	106	118	224
Total (u+r)	50	51	101
Peuplement /P/ le plus probable	283	478	722
Limites de confiance	(179-618)	(258-1,000)	(484-1,144)
Conversion à l'hectare	15,722	26,555	40,110
Conversion au kilomètre de			
cours d'eau	1,886	3,186	4,812

Table 4. DeLury evaluation.

Passage	Femelles	Mâles	Total
1er (m)	72	83	155
2 ème (n)	38	43	81
Total (m+n)	110	126	236
Total (m-n)	34	40	74
Peuplement /P/ le plus probable	152	172	324
Limites de confiance ± 2 Var P	±5	±5	±7
Conversion à l'hectare	6,080	6,880	12,960
Conversion au kilomètre de			
cours d'eau	1,013	1,147	2,160

Table 5. Final evaluation.

	Femelles	Mâles	Ensemble
Peuplement le plus probable			
dans le secteur	160	210	370
Conversion à l'hectare	6,400	8,400	14,800
Conversion au kilomètre			
de course d'eau	1,066	1,399	2,465

Table 6. Weight/Size Ratio.

Tallies mm	Mâles			Femelles			Ensemble		
	Nbre u	Pds total gr	Pds moyen gr	Nbre u	Pds total gr	Pds moyen gr	Nbre u	Pds total gr	Pds moyen gr
20 - 25	1			1			2		
25 - 30	4			7			11		
30 - 35	17	92	1,56	22	76	1,41	39	168	1,49
35 - 40	25			18			43		
40 - 45	9			2			11		
45 - 50	3			4			7		
50 - 55	6	26	4,3	12	50	4,2	18	76	4,2
55 - 60	2	11	5,5	9	50	5,6	11	61	5,5
60 - 65	20	152	7,6	14	105	7,5	34	257	7,6
65 - 70	3	27	9,0	9	81	9,0	12	108	9,0
70 - 75	3	38	12,7	7	74	10,6	10	112	11,2
75 - 80	3	56	18,7	3	44	14,7	6	100	16,7
80 - 85	9	210	23,4	1	16	16	10	226	22,6
85 - 90	6	159	26,5	1	20	20	7	179	25,6
90 - 95	8	237	29,6	/			8	237	29,6
95 -100	3	100	33,3	/			3	100	33,3
100-105	2	88	44,0	/			2	88	44,0
105-110	2	106	53,0	/			2	106	53,0
Total	126	1,302	10,3	110	516	4,7	236	1,818	7,7

Table 7. Biomass caught and biomass probable.

Secteur	Mâles		Femelles		Ensemble	
	Nbre (u)	Pds (gr)	Nbre (u)	Pds (gr)	Nbre (u)	Pds (gr)
Stock pêché	126	1,302	110	516	236	1,818
estimé	210	2,163	160	752	370	2,915
Biomasse Probable sur 100 m		1,427		496		1,924
Conversion à l' hectare	54,200 57,000 60,000		18,800 19,800 20,800		73,100 77,000 80,000	
au m²	5,7		2		7,7	

Concerning the choice of methods

The most favourable time for this investigation is summertime. During this period crayfish undergo intense activity. Before beginning an inventory study, it is necessary to first sample the population in order to avoid the ecdysis period.

The comparative use of the statistical methods of Petersen and De Lury does not give homogenous and analyzable results. The treated samples were too limited to permit a mathematical analysis (Laurent et Lamarque, 1974). It is confirmed that the De Lury method of successive captures (which allows more than 3 successive captures made under identical experimental conditions) is better than the Petersen method of capture and recapture. In the present case, different atmospheric conditions such as the violent wind which occurred during the second day of the investigation, probably stressed the animals which were re-introduced for the second time to their environment. Some of them died, the others hid in their holes. This explains the low rate of marked animals that were recaptured.

It was also found that night handfishing with a light will give a more definitive and diversified collection than is achieved with traps. This method however, can only be applied to the brooks having very clear water. Thus it is possible to capture small crayfishes which are generally not adequately trapped by ordinary means. The successive captures method involving more than three repeats each night, at the same hour, under the same collecting conditions (including meteorological conditions) leads to the most precise results.

Quantitative aspects

Basing the population estimation on the only credible method (De Lury), we can evaluate, as follows, the population of the investigated section of the water course; (Table 5).

```
        Females  - 160
        Males    - 210
TOTAL            370
```

We may consider as probable the presence of about 3 animals per square meter of the water course, or about 8 gr/m^2. This gives a biomass of 73 to 80/kg/ha of crayfish.

Qualitative aspects

The very important presence of young year classes contributes only a small part to the biomass estimation (168 gr per 1818: or 9.2%), but as the catchability in general is very weak, it probably represents a strong base for the demographic pyramid. The absence of credible numerical data, however, leads to a large uncertainty as to the exact interpretation of the structure of this crayfish population.

This uncertainty is increased by the facts that it is impossible to determine the animals' age and that the study of the crayfish as outlined is only based on the following assumptions:

14 months juveniles, measure about	35 mm
2 year + crayfishes, measure about	65 mm
3 years + crayfishes, measure about	80 mm
4 years + crayfishes, measure about	90 mm

Large animals are not rare; the crayfishes number reaching or exceeding the legal capture size (90 mm) is 15 per 236 subjects, or 6.35% of the population samples. The stock of this brook appears to be very healthy.

CONCLUSIONS

It may be said that crayfish population of the Lutina brook is abundant, in good balance, and that growth is apparently fast for a species that usually has a slow growth rate. In general, it could be proposed that brooks of the same type, flowing in Castagniccia region should be inventoried for crayfish and stocked and restocked with the Austropotamobius pallipes pallipes species which appear to be doing very well in Lutina brook.

LITERATURE CITED

Arrignon, J. 1970 Aménagment pisciocle des eaux intérieures, p. 302-314 Ed. S.E.D.E.T.E.C. A.S., Paris. 643 p.

Arrignon, J. 1976. Aménagement ecologique et piscicole des eaux douces, p. 36-43 Ed. Gauthier Villars, Paris. 322 p.

Arrignon, J. and P. Magne. 1979. Population de'écrevisses (Atlantoastacus pallipes pallipes Lereboullet) d'un ruisseau de Lozère - France, p. 131-140. In P.J. Laurent (ed) Freshwater Crayfish IV, Thonon-les-Bains, France (1978). 473 p.

Laurent, M. and P. Lamarque. 1974. Utilisation de la méthode De Lury pour l'évaluation des peuplements piscicoles. Ann. Hydrobiol. 5(2) 121-122. Paris.

Laurent, P.J. and M. Suscillon. 1962. Les écrevisses en France. Ann. Stat. Centr. Hydrobiol. Appl. Paris. p 336-395.

Roch, B. 1974. Étude de la rivière Fium'alto; détermination de la qualité biologique des eux. Doc S.R.A.E. Corse. Bastia.

TRAPPABILITY, LOCOMOTION, AND DIEL PATTERN OF ACTIVITY OF THE CRAYFISH ASTACUS ASTACUS AND PACIFASTACUS LENIUSCULUS DANA

Sture Abrahamsson
Department of Animal Ecology
University of Lund
S-223 62 Lund, Sweden

INTRODUCTION

The purpose of this investigation was to study the trappability in relation to the activity of the native European crayfish Astacus astacus Linné and the American crayfish Pacifastacus leniusculus Dana, introduced into Sweden in 1960, Finland in 1967, and various other European countries in the 1970s. Differences in activity patterns may be of significance in terms of interspecific interference and resource partitioning. Whether the two species may in the long run co-exist in European water systems is doubtful, because the crayfish plague fungus Aphanomyces is endemic in Pacifastacus.

Study areas

Field studies of Pacifastacus leniusculus were mainly carried out in Lake Tahoe (39° 10' N, 120° 7' W), located in the Sierra Nevada Mountains on the California-Nevada border (for description of the area see Abrahamsson and Goldman 1970). Some investigations of Pacifastacus were also made in Lake Natoma (38° 39' N, 121° 11' W) a man-made lake below Folsom Dam in the American River, California. Lake Natoma is a reservoir for the Folsom Dam power station, the water level of which fluctuates by about 2 m each day. The underlying substrate of the American River in the experimental area consists mostly of boulders on top of gravel providing an excellent habitat for crayfish.

Field studies of Astacus astacus populations in Sweden were mainly performed in the ponds at Rogle (55° 42' N, 13° 19' E), an area described by Abrahamsson 1966, and in Lake Jogen (57° 53, N, 13° 88' E). The substratum in the littoral zone of the oligotrophic Lake Jogen consists mainly of boulders on a layer of gravel and sand, which create suitable refuges for crayfish. Astacus populations in Sweden were also studied in the River Snytean (60° 0' N, 16° 2' E), the River Iskan (60° 47' N, 15° 51' E) and the River Ljungan (62° 30' N, 15° 33' E).

METHODS AND MATERIALS

Locomotory activity

In the laboratory the method of Muller and Schreiber (1967) was used to monitor activity. The method employed a circular aquarium with an infra-red photocell passage connected to a print-counter recording the hourly number of passages. The bottom substrate was gravel and stones. Five animals were placed in each aquarium. They were provided with a surplus of food (carrots). Males and females were studied separately in order to examine sexual differences.

The aquarium was placed close to a window in the laboratory. The interval between sunrise and sunset was used as an expression of daylength. The monthly average water temperature was calculated from daily readings at 08^{00}. The waterflow (tap water) through the aquarium was about 80 litre hour^{-1}. The chemical composition of the tap water changed little during the year.

The effect of temperature on the activity level was investigated by recording the locomotory activity in water of 8^{0} C, after which the temperature was raised to 15^{0} C. A 24-hour acclimatization was allowed after change of experimental conditions.

The locomotory activity of Pacifastacus leniusculus and Astacus astacus was studied in the field by means of SCUBA equipment (Self-Contained Underwater Breathing Apparatus). Crayfish of about 50 mm body length and larger which exposed their whole bodies above the substrate were considered "active" and counted every four hours during day and night. The amount of solar energy was measured at the surface of Lake Tahoe with a pyrheliometer.

Trappability

Seasonal variation of the trappability of Pacifastacus leniusculus in Lake Tahoe was investigated by setting out 80 traps once a month at 5 m intervals at certain depths. The crayfish traps had funnel entrances at both ends and were baited with fresh meat. The traps were emptied after 24 hours.

Differences in trappability during day and night were also investigated in Lake Tahoe by emptying the traps at 12-hour intervals. Traps were set at about 06^{00} and emptied at 18^{00}.

The seasonal variation in trappability of Astacus astacus was studied in Sweden in the ponds at Rogle where 10 traps were regularly set four times a month. The sex ratio and density of the population of A. astacus in Rogle was studied by Abrahamsson (1966). Sampling with 75 traps once a month was also done in the littoral zone of Lake Jogen at 2 m depth.

Diel variation in trappability was studied in Lake Jogen and the outlet of River Snytean. Ten crayfish traps were examined, rebaited and moved over a homogeneous habitat every four hours.

Dispersal

Dispersal of adult Pacifastacus leniusculus was investigated in Lake Natoma and Lake Tahoe.

In Lake Natoma, 1,273 males and 1,097 females were marked by cauterization (Abrahamsson 1965) and released at one place in the same area they were caught. The dispersal of marked crayfish was followed by intense trapping in the experimental area at certain time intervals.

1,000 Pacifastacus males were marked and released in Star Harbour at the border of the collection area in Lake Tahoe. Recaptures were made about 2 1/2 months later.

The dispersal of adult <u>Astacus astacus</u> was also studied by the trapping of marked specimens. 265, 1,279, and 477 marked crayfish were released in the Rivers Snytean, Iskan, and Ljungan, respectively. Recaptures were made after about 1 1/2 months and the positions of recaptured marked crayfish were related to the points of release.

RESULTS AND DISCUSSION

Diel activity and its relation to light and temperature

Fig. 1 shows the annual diel activity of <u>Pacifastacus</u> males and females in the laboratory. The crayfish were dark-active throughout the year. The reduction of light at the end of the day seems to act as an activity releaser ("Zeitgeger"). The number of passages counted in 1970 was 52,080 for the males compared with 36,249 for the females. The level of activity was influenced by the temperature. The seasonal pattern of diel activity is summarized in Table 1. Light was also found to regulate the timing of the active period of <u>Astacus</u>.

In a study of the diel activity of <u>Pacifastacus</u> by SCUBA diving in Lake Tahoe, the activity in an unshaded area did not begin until the light intensity at the surface was low (Fig. 2). From 20^{00} there was a considerable increase in activity, with a maximum around 24^{00}, at which time 214 crayfish per 100 m^2 were active. Later in the night there was a successive decrease in activity, and at dawn almost all crayfish were inactive. In a shaded area with fewer total hours of exposure to light, activity reached its maximum around 20^{00} and continued at about the same intensity to 04^{00}, at which time 210 crayfish were active per 100 m^2. Thus the activity of crayfish is related to the length of the dark period, and even a small amount of light can reduce the time of activity of the population.

Counting exposed <u>Astacus</u> by SCUBA every four hours in Lake Jogen, 16-17 July, 1968, demonstrated that in daylight (04^{00}-20^{00}) ≤ 8 individuals per 100 m^2 were exposed at each observation. The active period was mainly between sunset and sunrise with a peak around midnight. 33 crayfish per 100 m^2 were exposed at 24^{00}.

Trappability in relation to light intensity and population density

The influence of light and population density on the trappability of <u>Pacifastacus</u> was investigated in August-November 1967 at different depths off the Coast Guard Station in Lake Tahoe: 2,820 crayfish were caught (Fig. 3). The distribution of the 24-hr crayfish catch in relation to the depth shows a maximum density at 10 m and a rapid density decline below 40 m depth. The high proportion of the crayfish population trapped during the day shows that light was not the only factor controlling trappability. As a result of a high population density, competition apparently stimulated feeding activity even under daylight conditions. In populations where there was little competition for food and shelter, weak illumination almost eliminated feeding (cf. Smolian 1926).

The trappability of <u>Astacus</u> expressed by the number of crayfish entering traps at four hour intervals was studied in Lake Jogen (Fig. 4) and the River Snytean. The previous observation of night activity was confirmed.

241

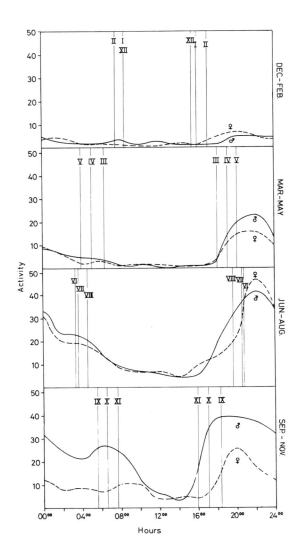

Figure 1. Diel activity pattern of Pacifastacus leniusculus males and females in different seasons, 1970. Activity is given as an average of passages per two hours and five individuals. Vertical lines denote sunrise and sunset in the middle of each month.

Table 1. The seasonal pattern of diel activity of P. leniusculus males and females in 1970 given as a percentage of the total number of passages for 5 individuals of each sex.

	Dec-Feb	Mar-May	Jun-Aug	Sep-Nov
Males	4.3	12.6	38.1	45.0
Females	3.7	11.5	49.7	35.1

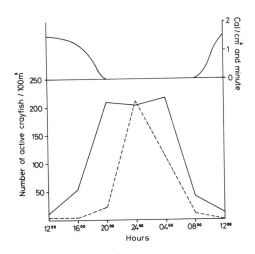

Figure 2. Diel variation in the activity of P. leniusculus as studied by SCUBA diving at a depth of about 2.5 m off Tahoe City State Park in Lake Tahoe, 29-30 July, 1967. Continuous line = activity in unshaded area.

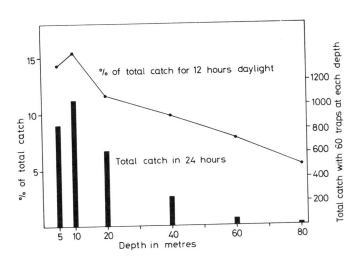

Figure 3. Trappability of P. leniusculus at different depths off the Coast Guard Station in Lake Tahoe, August-November 1967.

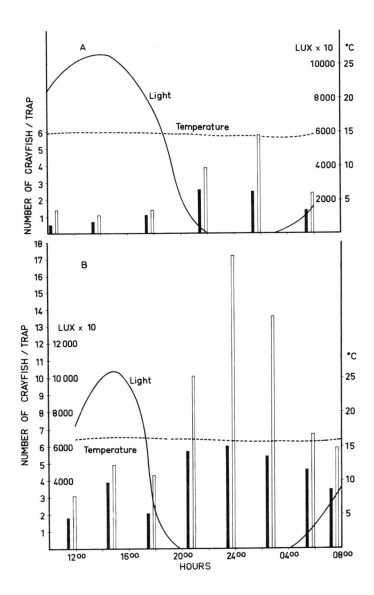

Figure 4. Diel variation in the trappability of <u>A. astacus</u> in Lake Jogen 5-6 July 1965 (A) and 30-31 August 1965 (B). The light and water temperature were measured at the surface. Black bars = females, white bars = males.

During moulting crayfish are incapable of rapid locomotion for about a week, and the chilipeds are incapable of defense. The high yield of the traps in Lake Jogen on 30 August 1965 compared with 5 July the same year demonstrates the increased feeding activity after the moulting period.

Seasonal variation in locomotory activity and its relation to envirormental factors

The seasonal variations in locomotory activity for _Pacifastacus_ males and females are shown in Fig. 5, which shows the monthly averages of passages per two hours in the photocell aquarium in natural LD regimen. The variation of the water temperature explains 75% of the variation in the level of the locomotory activity in both males and females. The variation of the length of day explains 50% of the variation in females but only 1% in males. In the aquarium, moulting crayfish were generally eaten by other crayfish, so they were replaced with recently moulted individuals with a semi-hard exoskeleton. When the exoskeleton of post-moulted crayfish became calcified, the locomotory activity level considerably increased, which explains the high level of activity at the end of August and the beginning of September.

The average number of passages per 24 hours for _Astacus_ males and females in the photocell aquaria between 1 February and 15 April 1971 was 39 and 38, respectively. The corresponding figures for _Pacifastacus_ males and females during the same period in 1970 were 51 and 53. The result indicates a difference between the males. There were differences in the physiological condition among the _Pacifastacus_ females, as some of them had extruded eggs on their pleopods during part of the period, probably resulting in a decrease of activity. Therefore, the results for the females of the two species are not comparable.

During the above mentioned period 80.6% of the activity of _Pacifastacus_ males and 85.1% of _Astacus_ males occurred during the night. The corresponding figures for females of _Pacifastacus_ was 72.7 and for _Astacus_ females 83.4.

The rate of locomotion of _Pacifastacus_ males in a photocell aquarium (Fig. 6) increased by abut 70% after 7° C rise in water temperature.

seasonal variation of trappability and its relation to environmental factors

Seasonal trappings of _Pacifastacus_ in Lake Tahoe (February 1967-February 1968) were based on 40 traps set monthly at each of two depth ranges, 5-40 and 60-150 m. Total catch was 5,344 and 196 for respective depth ranges. A seasonal pattern was found in the sex composition of crayfish captured (Fig. 7). In general, trap catches during mid-July from 5-40 m showed a predominance of females. This predominance remained until mid-October and the start of the breeding season. The proportion of females seemed to be lower at the 60-150 m depth.

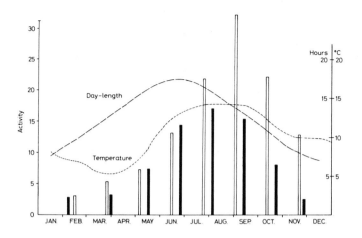

Figure 5. Seasonal variation in activity of <u>P. leniusculus</u> males and females in 1970. Activity is given as a monthly average of the number of registrations per 2-hr period for five individuals. Black bars = females, white bars = males.

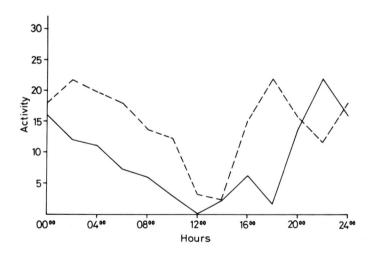

Figure 6. Locomotory activity of <u>P. leniusculus</u> males in water of 8° and 15° C. Activity is given as the average number of registrations per 2-hr period. Dashed line = 15° C, continuous line = 8° C.

247

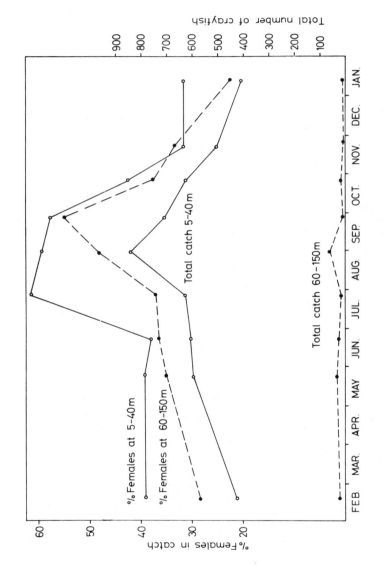

Figure 7. Seasonal variation in the trappability of P. leniusculus off the Coast Guard station in Lake Tahoe February 1967–January 1968, at 5–40 and 60–150 m depths. 5,344 and 196 crayfish were caught with 40 traps at two depths.

248

The forty crayfish traps used between 5 and 40 m depth yielded an average of 600 crayfish per night. The same number of traps, however, averaged only 20 animals at depths between 60 and 150 m. The greatest numbers of crayfish were caught at the end of August. This probably reflects an increase in the intensity of feeding activity after moulting. With the drop in temperature of the water during the autumn and winter months there was a continuous decrease in the number caught at the 5-40 m depths. During the same period but at depths ranging from 60-150 m no further decrease of activity was observed which is attributed to the uniformity of temperature at these depths (Abrahamsson and Goldman 1970).

When the yield and sex ratio of adult _Astacus_ in the trap catches are compared throughout the year, a seasonal pattern emerges (Figs. 8 and 9). Owing to the dominating behaviour of large males in the unexploited _Astacus_ population at Rogle (Abrahamsson 1966) the percentage of females caught was low and egg-bearing females did not enter the traps. The female catch increased at 10.7% in the middle of the summer and dropped almost to zero in the winter.

In the oligotrophic Lake Jogen the population of _Astacus_ was regularly harvested in the open season. Consequently there were not many large dominant males in the population. The density was high, resulting in high yields after moulting in August. From early November to late February the catch of females was below 10% being non-ovigerous and 25% ovigerous. In June and July, females caught during this period were primarily those that had recently dropped their young and were voraciously feeding prior to moulting. In the middle of the summer the female catch decreased because of moulting, but reached about 45% in late September. In August and September the proportion of females was 40-50%, but dropped to below 10% after the mating period in late autumn. The curves for sex ratio in trap catches reflect behavioural differences. Presumable the primary sex ratio is 1:1. However, in the ponds at Rogle, Abrahamsson (1966) showed that the adult crayfish population consisted of 65.3% males and 34.7% females.

Dispersal

The dispersal of marked _Pacifastacus_ in American River was studied by trapping in a bay (28,000 m^2) at certain time intervals after the release of marked crayfish. The percentage of marked crayfish in a sample of 2,000 individuals 5, 12, and 71 days after the release was 6.7%, 3.3%, and 1.7%. On the three occasions the percentage of females in the catches was 48.0%, 48.7%, and 45.3% proving that both sexes were available for trapping.

The dispersal in the bay of Lake Natoma in American River was studied by the capture-recapture method. One hundred fifty five of 2,370 released marked _Pacifastacus_ were recaptured. The total catch of _Pacifastacus_ at recapture was 1,949. One hundred thirty-five traps in the zone where marked crayfish were released yielded 1,145 crayfish, 45 traps in the middle zone collected 393 crayfish, and 45 traps in the most distant zone yielded 411 crayfish. The traps set at 5 m intervals in the three different zones yielded an average of 8.5, 8.8, and 9.1 crayfish/trap indicating that the population of _Pacifastacus_ was uniformly distributed in the bay.

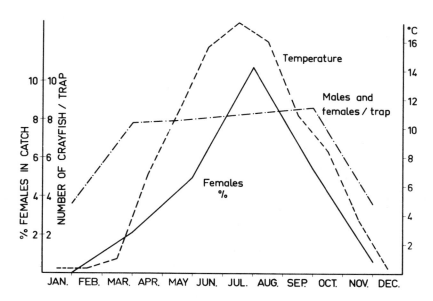

Figure 8. Annual variation in the trappability of adult _A. astacus_ in the ponds at Rogle in 1962. n = 5,990.

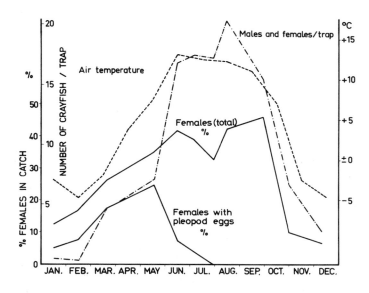

Figure 9. Annual variation in the trappability of adult _A. astacus_ in Lake Jogen in 1965. n = 16,486.

A similar experiment in Sweden was performed in the isolated pond No. 4 at Rogle (Abrahamsson 1971), where 1,050 Pacifastacus were marked. About 27% of the marked crayfish in a sample of 2,000 were collected about 30 days after the release. In contrast to the isolated pond in Sweden the crayfish in the American River could disperse over vast areas. Trapping on the opposite side of the river 12 days after the release of marked Pacifastacus confirmed that some marked crayfish had crossed the river. The distance traveled was about 700 m and the river is about 20 m at its maximum depth. The crayfish popultion was rather uniformly distributed in the river.

Trapping in Lake Tahoe 22 July-7 August 1970 yielded further information on the dispersal capacity of Pacifastacus. 450 traps per night yielded 58,675 males and 44,202 females in a 3.3 million m^2 area. Of 1,000 marked Pacifastacus released in Star Harbour on 15 May 1970, 46 were recaptured, but only 14 caught in the harbour (2,000 m^2). all other recaptures were scattered over the entire trapping field, which was part of a productive area where crayfish occur in great numbers down to 50 m depth.

A high rate of mortality in the upper littoral zone resulting from heavy wave action was observed in Lake Tahoe during autumn and winter storms. The low water temperature caused the animals to become lethargic and thousands, apparently unable to maintain their position, were washed up on the shore and killed.

Similar investigations, based on release and recapture, were carried out with marked Astacus in Sweden. In the River Iskan the average distance traveled in about 1 1/2 months was about 250 m. The maximum known distance traveled by a crayfish from the site of release was about 2,000 m. 17.7% of the marked 1,279 individuals were recaptured.

In the River Ljungan 31.8% of 477 marked individuals were recaptured. The crayfish spread both upstream and downstream, but most individuals were recaptured in the bay protected from the current where the crayfish were released. Few crayfish passed the main channel where the current was strong. The average distance traveled from the end of June to the middle of August was about 100 m.

In both Rivers Iskan and Ljungan recaptures were made after the moulting period.

At the outlet of the River Snytean the recapture of 265 marked Astacus was made immediately after the moulting period. The dispersal was not extensive, the average distance traversed by the recaptured 86 crayfish being only about 25 m.

Concluding remarks

Locomotory activity is influenced by a number of factors, such as density, amounts and availability of food and shelter, reproductive behaviour and dispersal capacity. Some of these factors vary with the season. As there seems to be a close relationship between diel locomotory activity and food-searching of Astacus and Pacifastacus, the trappability is mainly related to food searching activity, but might fluctuate with changes of other factors. During daylight crayfish hide in crevices and holes, but in darkness the movements are extensive. Changes in illumination regulate the locomotory activity of both species. Kalmus (1938) found that crayfish are activated at night by a substance released from the eyestalks.

The nocturnal activity of Astacus and Pacifastacus is controlled by light changes and is synchronized by sunset and sunrise. The gradual disappearance of this circadian rhythm in uniform environment indicates that it is produced through environmental induction. As with Pacifastacus a persistance of circadian rhythm during several days in uniform environmental conditions was demonstrated for Orconectes virilis by Roberts (1944) and Penaeus duorarum (Burkenroad) by Hughes (1968).

Feeding occurs mainly at night, but in overcrowded populations, crayfish may leave their retreats in daylight. The disposition of both Astacus and Pacifastacus to restrict their movements outside their hiding places to darkness is probably an adaptation of survival value in the presence of day-active predators.

Temperature was found generally to influence the locomotory activity. For the crayfish Cambarus affinis (Rafinesque) migration and habitat selection were reported to be regulated by temperature (cf. Bell 1906). In Lake Tahoe there was some evidence of a migration of Pacifastacus to deeper water in the autumn, probably caused by a lowering of the water temperature.

The slow rate of dispersal established for A. astacus is in accordance with the findings at Lake Loby, Poland (Kossakowski 1965). When the dispersal rates of A. astacus and A. leptodactylus (Eschsholz) were compared it was found that A. leptodactylus had a higher ability to invade new areas. During the present century A. leptodactylus, originating from the Caspian Sea area, has expanded to the west and forced out A. astacus from many habitats (Cukerzis 1958 and 1968).

In view of apparent lower dispersal intensity, lower egg production, later maturity, and other disadvantages of the native A. astacus compared to the introduced P. leniusculus (Abrahamsson 1971), it is likely that the introduced crayfish may have an adverse effect on A. astacus populations when both species occur in the same biotope.

ACKNOWLEDGMENTS

The present work was carried out at the Department of Animal Ecology, University of Lund and at the Institute of Ecology, University of California, Davis.

The investigations were supported financially by the Swedish Fishery Board and the Swedish Nature Conservancy Office.

LITERATURE CITED

Abrahamsson. S. 1965. A method of marking crayfish Astacus astacus Linne in population studies. Oikos 16:228-231.

----1966. Dynamics of an isolated population of the crayfish Astacus astacus Linne. Oikos 17:96-107.

---- and Goldman, C.R. 1970. The distribution, density, and production of the crayfish <u>Pasifastacus leniusculus</u> (Dana) in Lake Tahoe, California-Nevada. Oikos 21:83-91.

---- 1971. Density, growth, and reproduction in populations of <u>Astacus astacus</u> and <u>Pacifastacus leniusculus</u> in an isolated pond. Oikos 22:373-380.

Bell, J.C. 1906. The reactions of the crayfish. Harv. Psychol. Stud. 2:615-644.

Cukerzis, J.M. 1958. On the problem of ousting of <u>Astacus astacus</u> L. by <u>Astacus leptodactylus</u> (Esch.) in the lakes of Eastern Lithuania. (In Russian). Tr. AN LitSSR. Ser. B, vol. 4, p. 249-260.

---- 1968. Interspecific relations between <u>Astacus astacus</u> and <u>A. leptodactylus</u> (Esch.) Ekologia Polska-Seria A. 31:629-636.

Hughes, D.A. 1968. Factors controlling emergence of pink shrimp (<u>Penaeus duorarum</u>) from the substrate. Biol. Bull. 134:48-59.

Kalmus, H. 1938. Das Aktogramm des Flusskrebses und seine Beeinflussung durch Organextrakte. Zeitschr. f. vergl. Physiol. 25:798-802.

Kossakowski, J. 1965. Crayfish <u>Astacus astacus</u> (L.) and <u>Astacus leptodactylus</u> (Esch.) migrations in Lake Loby, Poland. Ekologia Polska - Seria A. 26:1-10.

Muller, K. und Schreiber, K. 1967. Eine Methode zur Messung der lokomotorischen Aktivitat von Susswasserfischen. Oikos 18:135-136.

Roberts, T.W. 1944. Light, eyestalk chemical, and certain other factors as regulators of community activity for the crayfish, <u>Cambarus virilis</u> (Hagen.) Ecological Monographs 14:360-392.

Smolian, K. 1926. Der Flusskrebs, seine Verwandten und die Krebsgewasser. Handb. d. Binnenfischerei Mitteleuropas. 5:423-524.

SUBSTRATE SELECTION BEHAVIOR OF THE CRAYFISH
PACIFASTACUS LENIUSCULUS

B.J. Klosterman and Charles R. Goldman
Division of Environmental Studies
University of California, Davis
Davis, California U.S.A.

ABSTRACT

Adult Sacramento River Pacifastacus leniusculus of two size classes were tested for nocturnal substrate selection behavior. After acclimation to laboratory conditions in flowthrough water systems, crayfish were individually exposed to three simple binomial choices (treatments) between substrate materials: (1) both inlet and outlet halves of tank lined with small gravel; (2) inlet half of tank small gravel and mixed-size rock, outlet half small gravel; (3) the reverse of treatment 2.

Comparison of mean proportions of time spent on the inlet half of tank revealed that both size classes displayed substrate selection preferences. However, non-substrate environmental factors such as current direction and speed appear to influence substrate choice. Crayfish spent more time on the mixed rock than on small gravel when the mixed rock substrate was closer to the water inlet (treatment 2) than when the reverse was true (treatment 3).

Between size classes, there was one apparent difference in substrate selection behavior; larger adults did not show the general tendency of slight inlet directional movement demonstrated by the smaller adults.

INTRODUCTION

Environmental factors influence an organism's development, survival and reproduction. Therefore, in terms of evolutionary fitness, it may be to the organism's advantage if it can respond to surrounding conditions by actively altering or choosing its environment.

The state of California has a variety of both indigenous and introduced crayfish species. Genera represented are Orconectes, Procambarus and Pacifastacus. Along the Sacramento River, a commercial crayfishery has been developed for Pacifastacus leniusculus.

Commercial crayfish catches may vary greatly, both in animal sizes and numbers. Catch size and composition often changes from inner to outer bend, one bend to the next, as well as from one part of the trapping season to another. Such variation suggests a nonrandom distribution of crayfish both through space and time. It appears likely that environmental conditions are excluding or reducing some crayfishes' ability to exploit particular habitats and/or crayfish are actively choosing their habitats.

The Sacramento River channel is characterized by silts, sands and gravels (NOAA). Substantial portions of the river's banks have been

254

lined with a mixture of large stones to prevent further current and wave-caused erosion (Scott, pers. comm.). Many commercial crayfishermen operating in these areas believe these rocks provide favorable habitat for Pacifastacus, helping to attract and support large numbers of the animals.

The relationship between substrate and crayfish distribution had been largely limited to informal observation. Recently, however, more rigorous attention has been applied to observing the relationship between substrate and organisms.

Abrahamsson and Goldman (1970) in their study of the distibution, density and production of Pacifastacus leniusculus in Lake Tahoe found that medium sized rocks, creating stony substrates, supported the largest populations of crayfish. Such rocks provided more crevices for crayfish to occupy than gravel or large boulders, thus affording inhabitants better protection from intraspecific cannablism and interspecific predation. However, the high density of animals in these stony areas resulted in food shortages and stunted crayfish.

Similarly, Flint and Goldman (1977) in their later study of the effects of habitat variation on the same species in Lake Tahoe revealed that the size and area of stone cover were factors affecting the size and number of adult crayfish present in a given area. One of three transects studied provided extensive stone cover and greater food availability. This transect contained (and perhaps attracted) higher numbers of crayfish than the other two.

The potential importance of animal response to substrate materials has been recognized in crayfish aquaculture studies as well as in fieldwork. Mason (1978) studying juvenile Pacifastacus leniusculus under different dietary and animal-density conditions, found that the availability of pebbles or burrows consistently increased crayfish survival, and to some extent biomass accumulation, over that on bare tank floors. Working with a different age class and species, in contrast to the above, Nelson and Dendy (1978), in their studies of adult Procambarus clarkii, found that the availability of burrows in ratios of 0, 0.5, 1, and 2 per animal had no significant effect on survival. Bovbjerg and Stephen (1975) found that the behavior of Orconectes virilis shifted from aggressive activities to the formation of passive aggregates when tank corners or crevices were in short supply.

This paper presents the results of an experiment designed to test whether crayfish actively choose their substratum conditions, that is, show substrate selection preferences.

MATERIALS AND METHODS

From an October week's commercial crayfish catch of Sacramento River Pacifastacus leniusculus, sexually mature males and females from two size classes (those under and those over the legal catch length of 3.43 cm. carapace length (Shimizu, 1981)) were chosen at random. Animals were transported to the laboratory and there housed in the following flow-through water system: three tiers of four tanks each, bottom and top tiers used for holding animals; middle tier, for experimental treatments. Tanks were of an opaque material. Tank dimensions were 1.5 x .3 x .2 meters. Each size class of crayfish was randomly assigned to four of the holding tanks, individual holding tanks containing only one animal size class (Figure 1).

Holding tanks were fitted with perforated partitions to provide housing for small groups of individuals while simultaneously allowing free flow of water (Figure 2). For individual housing, tanks were divided by the partitions into ten spaces of equal size. Within animal size classes, individuals were randomly assigned to a space within one of the four holding tanks assigned to that size class. Although perforated partitions greatly reduced cannibalism, they did not entirely isolate individuals from their neighbors. Partition holes allowed some visual, tactile, and possibly chemical communication among animals within a tank.

Every tank was fed at one of its ends with oxygen-saturated lake water through a common delivery system. As precise temperature and incoming flow volume for the flow-through system was not feasible, water temperature was kept within the range of 15.0-19.0°C for holding periods and 16.0-18.5°C for experimental periods. As water temperature increased from 18.7 to 19.0°C, many of the crayfish became "listless." That is, their responsiveness during the day to the visual stimulus of the experimenter, feeding and general activity levels decreased. When water temperature again dropped, animals rapidly resumed their previous higher levels of activity. Although water temperature could not always be kept below 18.7°C, experiments were not run during those few nights when water temperature had been measured to be above 18.5°C in the previous twenty-four hours. The flow stream into and out of each tank was held to be approximately the diameter of a standard width pencil. The total volume within each tank remained constant.

Each animal had neighbors to both the inlet and outlet sides of the tank except for animals in holding spaces 1 and 10. Animals randomly assigned to spaces 1 and 10 were not used for experiments. Individual holding spaces were large enough to allow for relatively unimpaired turning, forward and backward movements, as well as for grooming, digging, and feeding behaviors.

Photoperiod, set via a room switch timing device, ranged from 12 to 16 hours of light. During light hours, when Sacramento River Pacifastacus appear to be the least active, all holding tanks were covered with thin masonite sheets perforated with 5 mm diameter holes at 20 mm intervals. This allowed not only free exchange between tank water and air, but also provided crayfish with protection from direct bright light (to which they displayed a negative phototaxic response).

All holding tanks were lined at the bottom with a ten dry quarts measure of thoroughly rinsed grey pea gravel -- forming a layer of approximate 25 mm. thickness. Gravel was thoroughly cleaned of decaying food and animal wastes and tank water fully replaced every six weeks.

Animals were provided with a constant supply of fresh spinach leaves. Spinach leaves were sunk into place within each animal's holding space by attachment to one piece of pea gravel secured with a tiny orthodontic brace rubber band. Additionally, animals were fed

Figure 1. The flow through water system used in testing for crayfish substrate selection contained four tanks each top and bottom for housing animals and four tanks in the middle tier for experimental trials. Holding tanks were fitted with partitions--which are clearly visible in the uncovered tanks (lower right).

Figure 2. A closer view of part of one of the holding tanks shows the individual housing spaces formed by use of partitions. Each partition, which fits loosely into the grooved tank sides, has three perforations in its upper surface and extensive perforation along its lower part. These lower perforations, being less visible under water, are shown on the sample sheet (upper middle).

once every three days with a variety of fresh fruits and vegetables, predominately bananas and tomatoes, and either chopped whole smelt or one pellet per individual of a sinking feed high in protein and calcium.[1]

After all animals were acclimated to holding conditions for several weeks, twenty individuals from each of the two size groups (five subjects from each holding tank, excluding animals in spaces 1 and 10) were chosen at random as experimental units.

Experimental animals were provided individually with a series of three simple binomial choices concerning type of substrate. Experimental tanks were marked on the outside, visually dividing them into two equal areas, inlet and outlet half. A substrate material was then placed to completely cover the tank bottom of each half. For each experimental tank, the half of the tank set up to be the inlet half was chosen at random. Thereafter, the inlet and outlet halves (direction of overall water flow) were held constant for each animal tested in that tank.

In treatment 1, for each of the experimental tanks, each half of the tank was covered in 2 and 1/2 dry quarts measure of thorougly rinsed grey pea gravel. In treatments 2 and 3, one half also received mixed river rock consisting of one dry pint volume of washed large gravel (commercial size range 1.9-3.8 cm), six small cobbles (7.7 cm mean longest axis, 4.5-5.7 cm total range of average length of three axes for each rock), four medium cobbles (12.3 cm, 7.8-11.0 cm), and two large cobbles (27.0 cm, 16.3-19.0 cm) (Figure 3 and Figure 4).

Each animal was released individually into its experimental tank four or more hours before a test.[2] Release was along the division between the two tanks halves, the animal being oriented toward one of the tank side walls rather than toward either inlet or outlet half. At least three hours of darkness were provided in which the animal could explore the tank before readings were taken. (Preliminary observations had shown that actively moving crayfish of either size class could readily travel the entire length of an experimental tank several times within a five minute period.) Then, for ten readings, at half hour intervals from 12:30 a.m. to 5:00 a.m., the position of the animal relative to the inlet half of the tank was recorded. During the dark period, all readings were taken under red light to prevent any crayfish-visible light disturbances of the animal's behavior. Crayfish were not only light-sensitive, but also somewhat noise sensitive. However, the constant level of background noises, caused by water outflow drain disposal and laboratory machinery, masked irregular sounds; this probably prevented any noise induced alteration of subject behavior.[3]

Whenever an animal was placed in a test tank, any leftover food remaining in its holding space was transferred using a rubber-gloved hand to the center of the imaginary middle line which separated the two substrate covered tank halves. If no food remained in the test tank when the first reading for the night was taken, one pellet of sinking feed was dropped from the water surface to the approximate center of the tank middle line below.[4]

Figure 3. The substrate materials that were used are shown with an individual crayfish from each size class for perspective. Counterclockwise from upper left: small pea gravel; and the large gravel, small cobbles, medium cobbles and large cobbles used for the "mixed rock" substrate.

Inlet	Substrate Inlet Half	⋮	Substrate Outlet Half	Outlet

Experimental Tank

Experimental Treatments:	Substrate Inlet Half	Substrate Outlet Half
Treatment 1, (SG/SG):	Small Gravel	Small Gravel
Treatment 2, (MR/SG):	Mixed Rock	Small Gravel
Treatment 3, (SG/MR):	Small Gravel	Mixed Rock

Figure 4.

260

Between test runs, the experimental animal was returned to its holding space, and its experimental tank was drained, cleaned and refilled. Experimental substrates were removed, thoroughly rinsed and replaced in as close to their previous arrangements as possible. Originally, the rock arrangement for treatment 2 in each experimental tank was determined randomly, with the exception that the two large cobbles were purposefully not clustered and were placed far enough from tank walls to prevent subjects from escaping. Thereafter, the mixed rock arrangement was kept the same. The arrangement of mixed rock in treatment 3 for each tank was the stereo image (180 degree rotation of substrate) of the mixed rock arrangement in treatment 2. Each animal was housed in its holding space for approximately eight days between any two treatments.

Although all animals molted at some time during the total holding-and-experimental period, no individual crossed over from one size class to the next.

Smaller Units:

before: 2.73 cm mean carapace length, total range 2.44-2.97 cm

after: 3.10 cm mean carapace length, total range 2.89-3.38 cm

Legal Minimal Catch Size: 3.43 cm carapace length.

Larger Units:

before: 3.74 cm mean carapace length, total range 3.49-4.00 cm

after: 4.03 cm mean carapace length, total range 3.75-4.30 cm

No crayfish was tested within three days of molting. Within that time period before and after molting, the crayfish exoskeleton was soft in places and crayfish behavior was altered. Until the exoskeleton again hardened, pre- and post-molting animals displayed a notable decrease in movements, feeding and exploratory or roaming activities.

In the event that an animal died or escaped part way through the three treatment series, (3 animals of 40), its holding space was refilled using an individual chosen randomly from the spare animals in that holding tank. The new experimental animal received all three treatments as described above.

RESULTS

In interpreting results we would expect that if substrate were "chosen" randomly, crayfish would show no clear preferences between the two substrates presented in each treatment. The mean proportion of time spent on the inlet half of the tank would be approximately equal to the mean proportion of time spent on the outlet half.[5]

Smaller Units: Table 1

Treatment 1, all small gravel (SG): When presented with small gravel spread uniformly across the experimental tank bottom, smaller adults displayed a slight but highly significant preference for the inlet half of the tank. As the only readily discernable difference between tank halves was the in or outflow of water, this towards-inlet movement of crayfish suggests that either the water itself entering the tank or the slight directional movement of tank water induced animals to orient or move toward the inlet side. Thus, we might consider the inlet and outlet sides analogous to the upstream and downstream directions of a lotic water system.

In treatment 2 (MR/SG), where mixed rock (MR) substrate was present on the inlet half of the tank, but only small gravel was available on the outlet half, smaller adults displayed a clear highly significant preference for the mixed rock half of the tank.

When results of treatments 1 and 2 are compared (Table 1, HO_1), the highly significant difference between animals' responses under the two treatments suggests either a strong preference for mixed rock or a synergistic effect; i.e., "inlet mixed rock" is more desirable substrate than "outlet small gravel."

To further ascertain whether such an interaction effect exists, we look to the results of treatment 3 (SG/MR). Here, when mixed rock substrate was available on the outlet half of the tank, mixed rock appears to be a notably less desirable substrate than when it was present on the inlet half of the tank in treatment 2. Although results suggest that the presence of mixed rock influences substrate choice, when small gravel was on the inlet half, crayfish displayed no statistically significant difference in the proportion of time they chose outlet mixed rock versus outlet small gravel (Table 1, HO_2).

Larger Units: Table 2

Treatment 1, all small gravel (SG): When presented with small gravel spread uniformly across the experimental tank bottom, larger adults displayed no significant preference for one half of the tank over the other.

In treatment 2 (MR/SG), however, larger animals displayed a highly significant preference for mixed rock half of the tank. As with the results of tests with the smaller adults, this might suggest a synergistic effect of inlet-mixed rock rather than preference for mixed rock per se.

In treatment 3 (SG/MR), where mixed rock was available on the outlet half, but only small gravel on the inlet half of the tank, larger animals displayed a less strong, though still statistically significant, preference for mixed rock over small gravel. While the location of mixed rock relative to the inlet half of the tank influenced substrate choice, the overall mean choice level displayed for outlet mixed rock relative to inlet small gravel in treatment 3 was not statistically different from crayfish response to the all small gravel tank bottom of treatment 1 (Table 2, HO_2).

Table 1. Crayfish Size Group A, Smaller Units.

MEAN PROPORTION OF TIME SPENT ON INLET SIDE OF TANK, 95% CI:	
TREATMENT 1, ALL SG (N = 208)	0.639 ± 0.0652
TREATMENT 2, MR/SG (N = 198)	0.871 ± 0.0466
TREATMENT 3, SG/MR (N = 198)	0.545 ± 0.0694

HO_1: SG = MR/SG	Z = 5.66	P< 0.001
HO_2: SG = SG/MR	Z = 1.93	0.05<P<0.10

Table 2. Crayfish size Group B, Bigger Units.

MEAN PROPORTION OF TIME SPENT ON INLET SIDE OF TANK, 95% CI:	
TREATMENT 1, ALL SG (N = 216)	0.525 ± 0.0666
TREATMENT 2, MR/SG (N = 194)	0.778 ± 0.0584
TREATMENT 3, SG/MR (N = 208)	0.462 ± 0.0346

HO_1: SG = MR/SG	A = 5.59	P<0.001
HO_2: SG = SG/MR	Z = 1.32	0.10<P<0.20

Table 3.

MEAN PROPORTION OF TIME SPENT ON INLET SIDE OF TANK		
	Size A	Size B
TREATMENT 1 (all SG)	0.639	0.525
TREATMENT 2 (MR/SG)	0.871	0.778
TREATMENT 3 (SG/MR)	0.545	0.462
HO_1: A SG = B SG	Z = 2.40	$0.01 < P < 0.05$
HO_2: A MR/SG = B MR/SG	Z = 2.43	$0.01 < P < 0.05$
INTENSITY OF RESPONSE		
HO_4: (A SG - A MR/SG) = (B SG - B MR/SG)	Z = 0.346	$0.50 < P < 0.75$
HO_5: (A SG - A SG/MR) = (B SG - B SG/MR)	Z = 0.436	$0.50 < P < 0.75$

Responses of the Two Size Class Groups Compared: Table 3.

In comparing different mean group responses to treatment 1, we find that among the animals tested, larger animals did not show the general slight inlet directional movement characteristic of the smaller adults (Table 3, HO_1).[6]

This difference in response remains fairly constant as treatments were changed (Table 3, Probability for HO_2 compared to that for HO_1). The intensity of response change from one treatment to the next was statistically similar for both animal size class groups (Table 3, HO_4, HO_5). Thus it can be concluded that the differences in intergroup responses appearing under the three treatment conditions may consist entirely of the difference in the two groups' responses to the inlet versus outlet sides of the tank rather than to the substrate material.

DISCUSSION

The results of this experiment have important implications for further research because they suggest that under laboratory conditions adult <u>Pacifastacus leniusculus</u>, when active, may display definite preferences by choosing substrate materials in a nonrandom fashion. The results of this experiment also suggest that such preferences are not necessarily based on substrate availability alone, but may also be influenced by other nonsubstrate environmental factors, such as current direction and speed.

The study readily lends itself to amendment. Using whatever natural or human-made substrate materials that are of interest to the experimenter, the substrate preferences of crayfish of different species or size classes (e.g., juvenile versus adult) can be tested.[7]

Because nonsubstrate factors such as current can influence substrate selection behaviors, in further research it might be interesting to compare crayfish response to sets of substrates under different laboratory conditions (for example, stream-like versus pond-like experimental systems).

Since the successful production of high density crustacean populations is an important goal in aquaculture development, it appears likely that substrate testing may provide some of the knowledge necessary to ensure optimum culturing conditions.

FOOTNOTES

[1]Moist Meals® cat food was chosen as the sinking feed.

[2]All handling of crayfish was done with clean rubber-gloved hands.

[3]Soft-soled shoes were worn by the researcher as a precaution against making noise.

[4]This feeding pattern was done for two reasons. First, in preliminary tests, lack of food availability resulted in extreme interanimal variations in behavior, increasing searching or exploratory behavior in some crayfish while reducing apparent activity levels in others. Second, the influx of lake water carrying small quantities of organic materials might attract crayfish to the inlet half of the tank. Such an attraction would be an undesirable artifact not directly related to the substrates being tested.

[5]This analysis of the data readings obtained was based upon the assumption that there was little intragroup interanimal variation. Later computer analysis, using the BMDP2V repeated measures, interdependent treatments program, showed this assumption to be basically sound.

Preparation for the computer analysis was performed in a rigorous manner. Data for animals for which there were not ten complete readings for all three treatments were discarded. For the remaining data, each of the sets of ten readings were reduced to a single proportion. Thus the number of animals was reduced to twenty-eight and the data to eighty-four figures. With resultant larger standard deviations and with inter-animal variation separated out, treatment effects were still very significant.

[6]A discrepancy exists between the results of this analysis and those of the analysis run by computer as described above in footnote 5. The first analysis suggests significant differences in the extent of inlet directional movement by members of the two crayfish size classes. The later computer analysis suggests that the differences seen may have been due to chance (tail probability = .134).

[7]Persons working with animals that are active during daylight may wish to take into consideration substrate color as well as substrate shape, texture and size.

LITERATURE CITED

Abrahamsson, S.A.A. and C. R. Goldman. 1970. Distribution, density and production of the crayfish Pacifastacus leniusculus Dana in Lake Tahoe, California-Nevada. OIKOS 21:83-91

Bovbjerg, R.V. and S.L. Stephen. 1975. Behavioral changes with increasd density in the crayfish Orconectes virilis. In Freshwater Crayfish 2:429-441.

Flint, W.R. and C.R. Goldman. 1977. Crayfish growth in Lake Tahoe: effects of habitat variation. Journal of the Fisheries Research Board of Canada. 34(1):155-159.

Mason, J.C. 1978. Effects of temperature, photoperiod, substrate, and shelter on survival, growth, and biomass accumulation of juvenile Pacifastacus leniusculus in culture. In Freshwater Crayfish 4:73-82.

NOAA (National Oceanic and Atmospheric Administration), U.S. Department of Commerce, Nautical Chart, numbers 18661 (1978), 18662 (1978), 18664 (1976).

Nelson, R.G. and J.S. Dendy. 1978. Effects of various culture conditions on survival and reproduction of red swamp cayfish (<u>Procambarus</u> <u>clarkii</u>). <u>In</u> Freshwater Crayfish 4:305-312.

Scott, Paul. 1981. State of California, Chief of Maintenance, Sacramento River Bank Protection Project. Personal Communication.

Shimizu, S. 1981. Unpublished data on Sacramento River <u>Pacifastacus</u> <u>leniusculus</u>.

CLIMATE AND STREAM AS LIMITING FACTORS IN THE DISTRIBUTION OF ASTACUS ASTACUS L.

M. Furst and B. Eriksson
Institute of Freshwater Research
S-170 11 Drottningholm, Sweden

There are several important factors limiting the distribution of crayfish in Sweden, which include the occurrence of eel (Anguilla anguilla L.) and the acidification during recent years. There is also a climatic barrier that limits the crayfish populations to the southern part of the country. A great many attempts have been made to extend the distribution by means of stocking new lakes and streams. Most of the attempts have failed but some, especially in running waters, have succeeded. In lakes just above the border hatching occurs too late in the summer for the larvae to survive. At the same time they obviously survive in the running waters even farther upstream. The survival of the young of the year seems to be directly related to the positive factors which are involved in flowing waters. Ninety-eight percent of the water that runs through a formerly rich crayfish river has been drawn off through an electrical power plant. The decreasing reproduction in the pooled river bed is recorded successively. Attempts are made to compensate for part of the natural reproduction. This is done by stocking the former river bed with young of the year that are hatched earlier than normal. This may also be a method for extending the distribution of the crayfish now limited by climate.

PREDATION BY <u>ANAX JUNIUS</u> (ODONATA: AESCHNIDAE) NAIADS ON YOUNG CRAYFISH

J. F. Witzig,* J. W. Avault, Jr., and J. V. Huner
Fisheries Section
School of Forestry and Wildlife Management
Louisiana State University
Baton Rouge, Louisiana 70803, U.S.A.

Biology
Southern University
Baton Rouge, Louisiana 79813, U.S.A.

<u>A. junius</u> naiad predation on young crayfish, <u>Procambarus clarkii</u>, was measured in the lab. Predation rates (0.066-1.16 crayfish/day) were estimated at 3 temperatures (25, 15, 5 C) and for 2 size classes of crayfish (11-20 and 12-30 mm). Both factors had significant effects on predation rates. At 5 C, naiads did not significantly affect survival. At 15 and 25 C, predation rates were significantly affected by temperature and prey size. Concurrent with lab studies, relative naiad abundance in a 0.8 ha crawfish pond was estimated biweekly from Sept. 1978May 1979. Large naiads (head width> 6.5 mm; total length > 35 mm) capable of preying on young crayfish were not present until well after most crayfish recruitment ended. <u>A. junius</u> naiads probably did not have a serious affect on crayfish production in that predaceous water beetle, <u>Cybister fibriolatus</u>, showed that the water bug did prey on young crayfish but adult water beetles did not. <u>B. lutarium</u> was not present in sufficient numbers in the pond to affect crayfish production.

---

*Mr. Witzig's present address is Zoology, North Carolina State Univ., Raleigh, NC 27067, U.S.A.

A PORTABLE HYDRAULIC DIVER-OPERATED DREDGE-SIEVE FOR SAMPLING JUVENILE CRAYFISH. DESCRIPTION AND EXPERIENCES.

Tommy Odelstrom
Institute of Limnology
University of Uppsala
P.O. Box 557
S-751 22 Uppsala, Sweden

ABSTRACT

Juveniles of <u>Astacus astacus</u> L. and <u>Pacifastacus leniusculus</u> Dana in Swedish waters live in the littoral zone, where the substrate mainly consists of rocks and gravel. In order to sample the juveniles a hydraulic diver-operated dredge-sieve has been constructed. A pump injects a jet of water through a hose into a L-shaped plastic tube, causing suction in the tube. The outflow is passed through a sieve with 1 or 3 mm mesh size. The device is effective in sampling age-classes 0 and 1 in the littoral zone at depths of 0.2-0.8 m. Older specimens, when located in that depth-interval, are also sampled together with other benthic animals.

INTRODUCTION

The north-American crayfish species <u>Pacifastacus leniusculus</u> Dana was introduced into Lake Erken, central Sweden, in 1966-1969. Those introductions were made to replace a dense population of the native <u>Astacus astacus</u> L. that was eliminated by the crayfish plague in 1930-1931.

The bottom substrate in the littoral zone of L. Erken consists of thick layers of rocks and gravel, that provide very good shelter for crayfish of different ages. Especially young-of-the-year are rarely seen in this biotope. In fact, they have never been observed out of shelter in any time of the day. Their way of living makes sampling very difficult.

Several sampling techniques have been considered or tested. Seining, as described by Momot (1967) is not useful in L. Erken because the young are not exposed. Capelli and Magnuson (1974) used hand-collection, and Flint (1975) used dip-nets, on both occasions carried out by divers, to collect juvenile crayfish. In L. Erken the space between rocks is filled up with sediment deposits, and when the rocks are moved, sediment is washed out into the water causing deterioration of the visability. In the muddy water juveniles can be well hidden, and thus avoid being caught.

Given these factors, a sampling technique based on the aspirator principle, like a "gold sucker," seemed to be a possible solution. A diver-operated dredge-sieve, working like such an apparatus, was described by Brett (1964). His sampling equipment was designed for marine sampling purposes, and in the present paper a somewhat modified dredge-sieve is described, usable for crayfish sampling by divers.

270

Description of the apparatus

The dredge-sieve is constructed of units available in Sweden. The four basic parts are: (1) a portable gasoline-powered water pump, (2) a PVC hose, (3) a L-shaped plastic tube, and (4) a sampling frame (Figure 1).

The water pump assembly used by the author is a LÄNS-MAN 40. It consists of a 3 h.p. air-cooled Briggs & Stratton engine coupled to a LANS-MAN impeller pump unit, which discharges 450 litres per minute (maximum).

The hose used in this investigation is 15 m long, and made of textile-armoured, transparent PVC plastic. The inner diameter is 32 mm.

The L-shaped plastic tube is constructed of commercially available PVC units of 9 cm diameter (Figure 2). Jets of water are injected through a brass ring with four nozzles placed at the suction end of the tube. A ring-formed handle is situated at the middle of the tube. By placing an arm through this ring, it is easy to direct the tube during sampling, while the other arm is free to move rocks, etc. in the sampling area. At the discharge end a sieve (net bag) is fastened, with mesh sizes 1 mm for newly hatched crayfish (June-July), and 3 mm for larger juveniles (August-November).

The sampling frame is made of a 2.5 cm broad steel band, formed like a circle. The frame covers an area of 0.25 m².

Dredge-sieve sampling experiences

After preparation the dredge-sieve was tested in L. Erken by the author, in two small forest lakes - Lake Holmsjön and Lake Västra Stensjön, also in central Sweden - by Magnus Appelberg, and in River Ljungan in northern Sweden by Dr. Magnus Fürst.

Sampling was performed at the depth-interval 0.2-0.8 m. The main reason for this limitation in sampling depth was a change in rock size to larger rocks with an increasing water depth. At sampling the frame was dropped at random in the vicinity of the diver, and all rocks inside the frame was turned and removed. Sediment, sand, and gravel was simultaneously sucked up with the dredge-sieve together with potential crayfish.

The number of crayfish obtained with the dredge-sieve in the three lakes - L. Erken, L. Holmsjön, and L. Västra Stensjön - in 1980 and in July 1981 were on most occasions low (Table 1).

The age-class 0 crayfish dominated the catches at most of the sampling events. In L. Erken they first appeared in shallow waters about July 24, 1980, and they were represented in the catches during the rest of the sampling season. Also in July 1981 y-o-y from 1980 were found.

Juveniles of age-class 1 were collected from July 7 to August 6, 1980. In July-August one-year old _Pacifastacus_ in L. Erken start moving out to deeper areas.

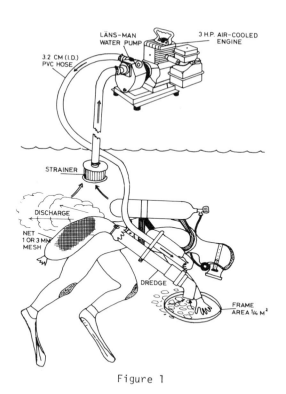

Figure 1

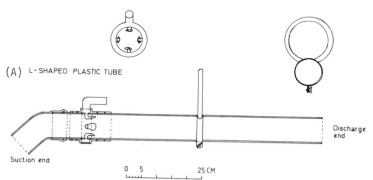

Figure 2. (A) L-shaped plastic tube, (B) cross-section of brass part
with four nozzles, (C) cross-section of brass handle.

272

Table 1. Results from dredge-sieve sampling in three Swedish lakes. The total number of collected crayfish and the number per m^2 are given for different age-classes at each sampling event.

Lake and Sampling area date	Sampled m^2	Age-class 0		Age-class 1		Age-class 2	
		number sampled	number per m^2	number sampled	number per m^2	number sampled	number per m^2
Lake Erken (Pacifastacus)							
1980							
July 7	5.0	0	-	1	0.2	0	-
July 24	2.0	9	4.5	1	0.5	2	1
August 2-6	4.5	17	3.8	2	0.4	0	-
August 15	2.5	6	2.4	0	-	0	-
Sept. 20	5.0	21	4.2	0	-	0	-
1981							
July 21	2.0	2	1.0	0	-	0	-
L. Holmsjon (Astacus)							
1980							
July 9	2.0	0	-	1	0.5	0	-
August 7	2.0	0	-	0	-	0	-
Sept. 10	4.5	6	1.3	0	-	0	-
Sept. 30	2.2	2	0.9	3	1.4	0	-
1981							
July 13	3.75	0	-	3	0.8	0	-
L. V:a Stensjon (Astacus)							
1980							
August 7	2.0	0	-	1	0.5	0	-

Older specimens were only found on July 24, 1980. In July age-class 2 and adult _Pacifastacus_ are molting, and they can be found close to the shoreline during their molting period. However, it should be mentioned that no efforts actually were made to collect adult crayfish.

In L. Holmsjön and L. Västra Stensjön the total numbers of collected crayfish were very low at all sampling events. Those two _Astacus_ populations are sparse, due to an increasing acidification of the lakes. The littoral zones in both lakes are steep, and a big portion of the population live in cavities at the shorelines. Dredge-sieve sampling of juveniles has shown to be very successful compared to normal hand-gathering by divers in this type of lakes.

In River Ljungan the number of juvenile _Astacus_ collected with the dredge-sieve was large (Dr. Fürst, personal communication).

At the dredge-sieve sampling other benthic animals are very common in the dredge remainings. Gammarids, _Asellus_ _aqualticus_, Ephemeroptera, and Odonata larvae are collected in large numbers.

The sampling procedure with the dredge-sieve have been successful. The capacity of the LÄNS-MAN water pump has been quite satisfactory, causing desirable suction in the tube. At sampling the diver has been able to move around rather freely, despite some resistence from the PVC hose.

ACKNOWLEDGMENTS

I wish to thank Prof. Birger Pejler, University of Uppsala, for reviewing my manuscript. Thanks also to Magnus Appelberg for _Astcus_-results from his forest lakes. Finanacial support was given by the Malmén Foundation at the University of Uppsala.

LITERATURE CITED

Brett, C.E. 1964. A portable hydraulic diver-operated dredge-sieve for sampling subtidal macrofauna. J. Mar. Res. 22:205-209.

Capelli, G.M. and J.J. Magnuson. 1974. Reproduction, molting, and distribution of _Orconectes_ _propinquus_ (Girard) in relation to temperature in a northern mesotrophic lake. In J.W. Avault (ed.) Freshwater Crayfish II, Louisiana, USA (1976). pp. 415-427.

Flint, R.W. 1975. Growth in a population of the crayfish _Pacifastacus_ _leniusculus_ from a subalpine lacustrine environment. J. Fish. Res. Board. Can. 32:2433-2440.

Momot, W.T. 1976. Population dynamics and productivity of the crayfish, _Orconectes_ _virilis_, in a marl lake. Am. Midl. Nat. 78:55-81.

V
PATHOLOGY AND
TOXICOLOGY OF CRAYFISH

FREQUENCY OF VISIBLE SYMPTOMS OF THE CRAYFISH PLAGUE FUNGUS (APHANOMYCES ASTACI) ON THE AMERICAN CRAYFISH (PACIFASTACUS LENIUSCULUS) IN NATURAL POPULATIONS IN FINLAND

Viljo Nylund and Kai Westman
Finnish Game and Fisheries Research Institute,
Fisheries Division,
P.O. Box 193, SF-00131 Helsinki 13, Finland

ABSTRACT

The occurrence of the brown spots indicative of crayfish plague was investigated in three lakes inhabited by self-propagating populations of the American Crayfish. The 236 crayfish examined came from catches made with crayfish traps. In two of the lakes, in which the crayfish populations originated from adults imported in 1969 from Lake Tahoe, Nevada, USA, the frequency of infected crayfish was high - 47% and 52%. In one lake, in which the population had developed from juveniles imported from Sweden in 1971, the crayfish showed no brown spots at all. The absence of crayfish plague from this small lake is also indicated by the fact that it also contains a population of the native crayfish (Astacus astacus). The spots caused by the plague most commonly occur on the walking legs (34% of spots), the upper and lower parts of the abdomen (33% of spots), and the chelae (27%). No data are available on the extent to which the stocks of American crayfish in these lakes have been damaged by the plague.

INTRODUCTION

Since 1967, the Finnish Game and Fisheries Research Institute has carried out research and experimental stockings with the American crayfish (Pacifastacus leniusculus), in order to introduce it into waters chronically infected with crayfish plague (Aphanomyces astaci). So far stockings have been restricted to small lakes. In certain cases the results have been promising (Westman 1973, Westman and Pursiainen 1979).

When trial stockings began, it was known that American crayfish may carry crayfish plague and thus might infect the native crayfish (Astacus astacus) species (Svärdsön 1968, Unestam 1969, 1973). The hyphae of the plague fungus occurs in the cuticle of infected Pacifastacus, but its growth is prevented by the natural resistance of the host. An arrested plague infection appears as dark brown spots on the crayfish exoskeleton. Although all American crayfish species have good natural resistance to the plague, this immunity is lacking e.g. in the European crayfish species (Unestam 1969). It has been suggested that chronic infection with crayfish plague may even be essential to the preservation of resistance in populations of the American crayfish (Fürst and Boström 1978).

Later observations have shown that chronic infection can become acute, leading to the rapid death of Pacifastacus, if its resistance is in some way reduced. The plague can also be fatal if it occurs during moulting (Söderhäll et al. 1981). The occurrence of visible symptoms of the plague was examined by Fürst and Boström (1978, 1979) in nine

277

Swedish lakes and three American lakes with Pacifastacus populations. With one exception, the frequency of infection was low in the Swedish lakes. While no serious damage to Pacifastacus populations in natural waters was observed, the crayfish plague may cause trouble in crayfish cultivation.

The aim of the present study was to investigate the occurrence of the dark brown spots indicative of crayfish plague, and possible harmful effects of the plague, in self-propagating Pacifastacus populations of different origin in three Finnish lakes. Such information is needed for planning future stocking with Pacifastacus, and for combatting the plague and protecting the native crayfish populations in Finland.

MATERIAL AND METHODS

The crayfish examined for occurrence of brown spots were caught on 6 August-5 September 1979 with baited crayfish traps in three Finnish lakes with self-progagating Pacifastacus leniusculus populations. In two of the lakes, Majajärvi and Karisjärvi in Central Finland, the stocks originated from adult crayfish imported in 1969 from Lake Tahoe, USA (Westman 1973). In the third lake, Slickolampi in southern Finland, the Pacifastacus stock had developed from juveniles imported from Sweden in 1971 (Simontorp, Blentarp). This lake also contains a self-propagating population of native crayfish (Astacus astacus) (Westman and Pursiainen 1979). In all three lakes crayfish are caught only a few times each year and only for purposes of research. All three are small (4-12 ha) unpolluted forest lakes in a completely natural state.

The mean length recorded for the samples were: Majajärvi 111.8 mm (66-132 mm), Karisjärvi 113.4 mm (64-164 mm), Slickolampi 87.5 mm (60-134 mm). The mean lengths of the males were: Majajärvi 117.3 mm, Karisjärvi 109.8 mm, Slickolampi 87.3 mm; the mean lengths of the females were: Majajärvi 106.3 mm, Karisjärvi 117.1 mm, Slickolampi 87.8 mm.

All crayfish trapped were checked for the occurrence of brown spots; in Majajärvi the catch was 83 crayfish, in Karisjärvi 29 and in Slickolampi 124. These numbers represent in Majajärvi ca. 23%, in Karisjärvi ca. 10% and in Slickolampi ca. 44% of the total stock of over 7 cm long crayfish estimated to be in the lakes in 1979 (Westman and Pursiainen 1979, 1981).

In the field after the crayfish were caught, records were made of the frequency of the dark brown plague spots, the position of the spots on the crayfish, and the size of the spots. To confirm the occurrence of the plague fungus, four of the spotted crayfish were taken at random (Majajärvi), and samples of the cuticle, cut from the sites of the spots, were examined under the microscope.

RESULTS AND DISCUSSION

Occurrence of the spots

Microscopic examination of the cuticle samples revealed the fungal hyphae typical of crayfish plague. Since the brown spots on the other

278

<u>Pacifastacus</u> were the same in shape and appearance as the spots on the specimens taken for microscopic examination, it is probable that all the brown spots were caused by crayfish plague. Several other species of fungi may occur in brown spots on <u>Pacifastacus</u>, however Söderhäll et al. (1981) has observed in Sweden that the majority of the brown spots are most likely caused by the crayfish plague.

In the two lakes originally stocked with adult Lake Tahoe crayfish, the proportion of infected crayfish was high, in Majajärvi 47% and in Karisjärvi 51.7% (Table 1). The number of infected crayfish and the frequency of the brown spots may actually be higher than appears in Table 1, since large specimens, which often are heavily infected, may be less active and therefore less easily caught.

The frequency of crayfish plague was only 3.0-11.2% in seven lakes investigated in Sweden using the same methods. In one lake it was 60.5%, however this high value was probably due in part to discharge of heavy metals (Fürst and Boström 1978). The frequencies reported by Söderhäll et al. (1981) from their two study lakes in Sweden were 30% and 40%.

In North America, the proportions of infected <u>Pacifastacus</u> in the two lakes from which the Swedish crayfish stocks originiated were 27.0-62.9% (Fürst and Boström 1978).

One reason for the rather high frequencies recorded in this study may be possible injuries incurred by the crayfish during previous test trappings. In a lake in Sweden, however, no differences in the frequency of infections were found between a stock subjected to repeated trapping and a stock that was fished very seldom (Fürst and Boström 1978).

According to Fürst and Boström (1978), the frequency of infection and the number of spots per specimen are greater in the females than in the males, especially among the large crayfish.

At Majajärvi the frequency of infection was somewhat greater in the females than in the males, but at Karisjärvi the proportion of infected individuals was clearly smaller among the females than the males (Table 1). The results for Karisjärvi may have been influenced by the small size of the material.

The influence of the size of the crayfish was not examined because of the small number of crayfish in different size groups.

No plague spots were observed in the sample from Slickolampi, whose crayfish stock has developed from juveniles imported from Simontorp, Sweden. Nor have any spots been recorded earlier. The absence of crayfish plague from this lake is also indicated by the fact that it is inhabited by a self-propagating stock of the native crayfish (<u>Astacus astacus</u>) (Westman and Pursianinen 1979). Fürst and Boström (1978) have reported plague-free populations of <u>Pacifastacus</u> in two lakes stocked with juvenile crayfish. In two other lakes, however, also stocked solely with juveniles, plague spots were found on the crayfish. The spots are generally absent from juveniles measuring less than 6 cm, but in crayfish farming small spots may appear in individuals over 2 cm in length (Fürst and Boströn 1978). Observations made

279

Table 1. Frequency of the brown spots caused by crayfish plague in _Pacifastacus leniusculus_ samples from two Finnish lakes.

Lake	Date of trapping	N	Healthy		Infected	
			No. of crayfish	%	No. of crayfish	%
Majajärvi						
Females	31.8 (1979)	50	25	50.0	25	50.0
Males		33	19	57.6	14	42.4
Total		83	44	53.0	39	47.0
Karisjärvi						
Females	5.9. 1979	13	8	61.5	5	38.5
Males		16	6	37.5	10	62.5
Total		29	14	48.3	11	51.7

Table 2. Size distribution of spots (n = 101) caused by crayfish plague in _Pacifastacus leniusculus_ (n = 39) from Majajärvi.

Diameter of spot mm	Males		Females		Total	
	Number of spots	%	Number of spots	%	Number of spots	%
2	3	7.0	10	17.2	13	12.9
2-3	19	44.2	21	36.2	40	39.6
3	21	48.8	27	46.6	48	47.5
	43	100.0	58	100.0	101	100.0

both in Sweden and Finland indicate that juveniles and the populations developed from them may in some cases be free from crayfish plague. This is of great importance in preventing the spread of the plague along with the stocking of P. leniusculus. But from the point of view of crayfish management, it may be harmful to attempt to keep stocks of the American crayfish completely free from the plague, since this may gradually reduce the natural resistance of the crayfish to new infections of the disease (Fürst and Boström 1978).

Size and occurrence of the spots

In the infected crayfish from Majajärvi (39 ind.), a total of 101 spots were counted (average 2.6/crayfish; range 1-9). Only one spot was found on 46.0% of the crayfish; two on 10.3% and three on 23.1%. The infected animals from Karisjärvi (15 ind.) were found to have 21 spots. One spot was counted on 73.4% of the crayfish, two on 13.3% and three on 13.3%.

In Majajärvi, the diameter of the spots varied from 1 to 5 mm. The most common size class was over 3 mm (47.5% of spots). The size of the spots did not vary significantly with the sex of the crayfish (Table 2).

The size and number of the spots indicate the strength of the infection and are also of aesthetic importance. The plague spots occurred most commonly on the walking legs (34% of spots), the dorsal and ventral parts of the abdomen (33%) and the chelae (27%). In the males the spots were concentrated on the walking legs and the chelae, whereas in the females they frequently occurred on the abdomen as well (Table 3). Of the 101 spots, 24.8% were located at the ends of broken chelae and walking legs (20.9% in males, 27.6% in females). The appendages are most probably broken during fighting between the crayfish, and the damaged parts are then easily infected with crayfish plague. The results thus suggest that the plague spots seem to concentrate on the most easily damaged parts of the crayfish, but the authors do not know whether all brown spots arose as the result of injuries.

Chronic infection with crayfish plague did not appear to harm the Pacifastacus populations in the lakes in this study. Söderhäll et al. (1981) indicated, however, that when the resistance of an individual of the Pacifastacus species is weakened in some way, a chronic infection can develop in a few hours into an acute and fatal form of the disease. In such cases Pacifastacus die much more quickly than the native Astacus, and it may be difficult to find the dying individuals. But, provided the resistance of the whole Pacifastacus population is not reduced, only a few individuals will die. This differs clearly from the case with the native crayfish, where infection with crayfish plague results in the destruction of the whole population (Westman and Nylund 1979).

LITERATURE CITED

Fürst, M. and U. Boström. 1978. Frekvens av en skalsvamp (kräftpest) på signalkräftor. (Summary: Frequency of visible symptoms of crayfish plague in populations of Pacifastacus leniusculus Dana). Information från Sötvattenslaboratoriet, Drottningholm. No. 1, 24 pp.

Table 3. Location of spots caused by crayfish plague on carapace of <u>Pacifastacus</u> <u>leniusculus</u> (n = 39) from Majajärvi.

	Males		Females		Total	
	Number of spots	%	Number of spots	%	Number of spots	%
Chelae	14	31.8	13	22.8	27	26.7
Mouth parts	3	6.8	4	7.0	7	6.9
Walking legs	16	36.4	18	31.6	34	33.7
Pleopods	4	9.1	3	5.3	7	6.9
Abdomen (dors.)	3	6.8	9	15.8	12	11.9
Abdomen (ventr.)	3	6.8	8	14.0	11	10.9
Uropod	1	2.3	2	3.5	3	3.0
	44	100.0	57	100.0	101	100.0

Fürst, M. and U. Boström. 1979. Frequency of symptoms of Aphanomyces on Pacifastacus leniusculus. The Second Scandinavian Symposium on Freshwater Crayfish, Lammi, Finland, 1979. (Manuscript).

Svärdson, G. 1968. Tio år med signalkräftan. - Svenskt Fiske (12): 377-379. (In Swedish).

Söderhäll, K., R. Ajaxon and M. Persson. 1981. Amerikanska kräftor och kräftpest. - (Mimeo). 9 pp. (Inst. Fys. Botany, Univ. of Uppsala, Box 540, 75121 Uppsala, Sweden).

Unestam, T. 1969. Resistance to the crayfish plague in some American, Japanese and European crayfishes. Rep. Inst. Freshw. Res. Drottningholm, 49:202-209.

-- 1973. Significance of diseases on freshwater crayfish, p. 135-150. In S. Abrahamsson (ed.) Freshwater crayfish I. Austria (1972). 252 p.

Westman, K. 1973. The population of the crayfish Astacus astacus in Finland and introduction of the American crayfish Pacifastacus leniusculus Dana, p. 41-55. In S. Abrahamsson (ed.) Freshwater Crayfish I. Austria (1972). 252 p.

Westman, K. and V. Nylund. 1979. Crayfish plague, Aphanomyces astaci, observed in the European crayfish, Astacus astacus, in Pihlajavesi waterway in Finland. A case study on the spread of the plague fungus, p. 419-426. In P.J. Laurent (ed.) Freshwater crayfish IV, Thonon-les-Bains, France (1978). 473 p.

Westman, K. and M. Pursiainen. 1979. Development of the European crayfish Astacus astacus (L.) and the American crayfish Pacifastacus leniusculus (Dana) populations in a small Finnish lake, p. 243-250. In P.J. Laurent (ed.) Freshwater crayfish IV, Thonon-les-Bains, France (1978). 473 p.

Westman, K. and M. Pursiainen. 1981. Size and structure of crayfish (Astacus astacus) populations of different habitats in Finland. Hydrobiologia, 86:67-72.

SUSCEPTIBILITY OF ORCONECTES LIMOSUS RAFF. TO THE CRAYFISH PLAGUE, APHANOMYCES ASTACI SCHIKORA

Alain Vey
Station de Recherches de Pathologie Comparée
30380 Saint-Christol-les-Alès, France

Kenneth Söderhäll, Ragnar Ajaxon[1]
Institute of Physiological Botany
University of Uppsala
Box 540
S-751 21 Uppsala
Sweden

ABSTRACT

Orconectes limosus a North American crayfish species was found to harbour the crayfish parasitic fungus, Aphanomyces astaci (Schikora), as a resting (chronic) infection in the cuticle. This crayfish could also function as a vector for the disease under laboratory conditions and could also under certain conditions die from A. astaci infection.

INTRODUCTION

The causal agent of the crayfish plague is a fungus, Aphanomyces astaci Schikora, belonging to Oomycetes (Unestam, 1969). This fungal parasite has devastated several populations of European crayfishes (Unestam et al., 1977). Orconectes limosus (Raff.) were introduced into Europe in 1890 and is now spread in different areas of middle and south Europe (Geelen, 1975, 1978; Laurent and Suscuillon, 1962). Schäperclaus (1935) reported that it was resistant to A. astaci infection but little experimental data were shown to support this. The purpose of this investigation was to investigate more carefully the resistance of this particular species to the crayfish plague parasite, A. astaci and also try to find evidence as to whether this species can act as a vector for A. astaci in the same way as one other North-American species, Pacifastacus leniusculus Dana (Unestam et al., 1977). This is important to know in Europe, since populations of O. limosus exist together with or in close association with different crayfish species (A. astacus, A. Pallipes and A. leptodactylus) which are susceptible to A. astaci (Unestam, 1969).

MATERIALS AND METHODS

CRAYFISH: Astacus astacus L. were obtained from Lake Vallsjön, Småland; Pacifastacus leniusculus Dana with chronic infections of A. astaci (Unestam et al., 1977) were a gift from Stellan Karlsson, Simontorps akvatiska avelslaboratorium, Blentarp, Sweden, and Orconectes limosus Raff. were kindly provided by Dr. P. Laurent, Thonon-les-Bains, France. All crayfish were kept in aerated aquaria at 13°C and one week before the experiments all crayfishes were fed with fish. Only intermolt animals were used in the experiments.

---

[1]Technical assistance.

Fungus: Aphanomyces astaci (Schik.) strain Si was obtained from our stock culture collection. Production of zoospores for infection experiments was done according to Unestam (1969).

Infection experiments

In aquaria. Aquaria containing 6 l of lake water and distilled water (2/1; v/v) were used for infection with swimming zoospores of A. astaci. Zoospores (10^4 ml^{-1}) were added to one aquaria containing 3 crayfish.

Consecutive infections were also make in aquaria at 13°C by adding 10^3 zoospores ml^{-1} once a week for 5 weeks. The water was replaced in each aquaria before addition of the zoospores. Controls were always run in which Astacus astacus L. were infected with A. astaci (10^3 or 10^4 zoospores per ml).

In some experiments O. limosus were slightly scratched on the epicuticle on the ventral soft abdominal cuticle. In one experiment 9 crayfish were scratched and then 6 crayfish were transferred to aquaria (3 crayfish in each) and immediately zoospores of A. astaci to a final concentration of 10^4 spores ml^{-1} were added. The 3 other crayfish were used as controls. This experiment was repeated once.

In all infection experiments the virulence of A. astaci zoospores was always checked with susceptible crayfish A. astacus. Mean time of death for A. astacus infected with zoospores of Aphanomyces astaci (Schik.) was 6 days.

In a closed system

Attempts were also made to infect A. astacus from chronically A. astaci infected O. limosus. These experiments were made in a closed aquaria system which allowed incubation of crayfish for at least 2 months (Fig. 1). This closed aquaria system has been designed and constructed by Ebbe Svensson, Kenneth Söderhäll, Torgny Unestam, Mikael Persson and Ragnar Ajaxon.

RESULTS AND DISCUSSION

When a concentration of 10,000 zoospores ml^{-1} of A. astaci was used to establish infection in O. limosus, only two of ten tested crayfish were infected (Table I). In the soft ventral abdominal cuticle of these animals heavily melanized hyphae were found. Such a melanotic deposit only appears when the fungus grows slowly and poorly in the integument, and when the host is able to develop a relatively efficient reaction. This observation led us to the hypothesis that our population of O. limosus already had an infection which developed very slowly, with limited symptoms, which could be called "chronic infection." Such chronic infections with A. astaci are known to occur in P. leniusculus where this kind of symptoms (melanized areas in the integument) in different populations vary from 5 up to 40 percent (Fürst and Boström, 1978; Söderhäll and Persson, unpublished).

This hypothesis was confirmed by the results of different other studies. Examination of 40 O. limnosus showed that 65 percent of these

Table 1. Infection of O. limosus and A. astacus with zoospores of Aphanomyces astaci strain Si in aquaria at 13 °C.

	Infection with A. astaci Strain Si (zoospores ml^{-1})	Animals dead/total
O. limosus	10,000	2/10[1]
A. astacus	10,000	8/8 [2]

[1] Dead in crayfish plague after approximately 2 months.

[2] Mean time of death 6 days.

Table 2. Infection of slightly wounded O. limosus with zoospores of A. astaci.

Crayfish (n)	A. astaci (zoospores/ml)	Dead in crayfish plague (dead/total)
10	10,000	7/10*

* Mean time of death 36 days.

Table 3. Turnover of chronic A. astaci infection in O. limosus to lethal infection in a closed aquaria system.

O. limosus (n)	Dead in crayfish plague (dead/total)

* Mean time of death 11 days.

286

animals had heavily melanized spots, mostly found on the walking legs, in which hyphae resembling A. astaci could be seen.

From these crayfish apparently carrying the disease in melanized wounds we succeeded in transferring the disease (A. astaci) to highly susceptible crayfish (Astacus astacus) and we have isolated a fungus which is an Aphanomyces sp. As this isolate produces only a small amount of spores in liquid culture (Unestam, 1969), the identification of this phycomycete requires pathogenicity tests with highly susceptible animals.

In some experiments we tried to establish infections by A. astaci on Orconectes which had been scratched on the soft abdominal epicuticle. As a control other O. limosus were also scratched, but no spores were added. Among the wounded and infected crayfish 70% died due to crayfish plague infection (Table 2). After 5 days one of the control animals was also found dead, and after examination in light microscope we discovered that the soft abdominal cuticle has a massive hyphal net-work of A. astaci.

This crayfish also showed a brownish area of chronic infection in the base of one "copulation leg" nearby the zone invaded by hyphae. The other control crayfish also died from A. astaci infection and they had chronic infections with Aphanomyces too. The same pattern has been observed in P. leniusculus (Söderhäll and Persson, unpublished) i.e. decreasing the host resistance by wounding enables the parasite to develop in the cuticle and eventually produce zoospores.

Melanization in crayfish and insect cuticle is probably a defense reaction against invading parasites (Aoki and Yanase, 1971; Götz and Vey, 1974; Söderhäll et al., 1979; Söderhäll and Ajaxon, 1981; Nyhlén and Unestam, 1980). Therefore if the host defense system is decreased by wounding it is possible that A. astaci in a heavily melanized area will have a chance to expand and spread in the cuticle. Then the balance between the parasite and host is shifted towards an advantage for the parasite. Whether this balance will turn out to become an acute infection, i.e. the crayfish dies, is probably dependent upon the physiological status of the individual host and the locatization of the chronic infection on the animal.

For example we note that when chronically A. astaci infected O. limosus were incubated in our closed aquaria system (Fig. 1) a high percentage of these crayfish died from lethal A. astaci infection within 21 days (Table 3). Apparently this aquaria system with running water enables A. astaci present in melanized areas of the cuticle (chronic infections) to grow outside the melanized area and will lead to extensive growth mainly within the soft cuticle on the abdomen. This will cause a lethal A. astaci infection and O. limosus die rather rapidly (6-21 days).

We cannot completely explain this phenomenon, but it agrees well with our infection experiments with Pacifastacus leniusculus (Söderhäll and Persson, unpublished).

Attempts to infect O. limosus with a lower, but multiple dose of A. astaci zoospores (10^3 ml^{-1}) at weekly intervals for 5 weeks in the small aquarias showed that 50% of the animals died due to crayfish plague. The mean time of death in these experiments was 77 days compared to 36 days for infection of slightly wounded O. limosus with 10,000 zoospores ml^{-1} (Table 4). The reason for the delay in establishing a lethal A. astaci infection when a low dose of spores was used is unknown. Considering that 65% of the tested O. limosus appeared to have chronic infections with A. astaci we cannot say whether these crayfish died from the added zoospores or if the chronic infections were the major cause of death.

Transfer of A. astaci from chronically infected O. limosus or P. leniusculus to susceptible Astacus astacus L. was only possible in the closed aquaria system (Fig. 1) and in three separate experiments we succeeded in transferring the disease from O. limosus to A. astaci. Furthermore, we have made consecutive transfers from A. astacus infected from O. limosus to other A. astacus (Fig. 2). The time needed for these consecutive transfers was approximately 12 h of incubation. The great difficulty in transferring A. astaci from O. limosus(this work) or P. leniusculus (Söderhäll and Persson, unpublished) in small aquaria (6 liter) to susceptible Astacus astacus L. is probably due to more heavy contamination of bacteria and other fungi under these experimental conditions than in the closed aquaria system (Fig. 1), where evidently "optimal" conditions for disease transfer are at hand.

Thus, we can conclude that the resistance of O. limosus to the crayfish plague is approximately the same as for other Orconectes species as well as other North-American species (Unestam, 1969), but it is important to emphasize also that this North-American species bears the crayfish plague parasite and under certain conditions (stress) these crayfish can die of A. astaci infection. O. limosus infected with A. astaci could also transfer the disease to A. astaci as was the case with P. leniusculus. It is noteworthy that the chronic infections both in O. limosus and P. leniusculus (unpublished) harbor many other water molds such as Phythium spp. and Saprolegnia spp., the importance of which in pathogenesis is not yet known.

Microscopic and microbiological studies of new samples of O. limosus, carried on immediately after fishing, will be necessary to confirm that this crayfish species bears A. astaci in its habitat in France. Additionaly experiments will also be needed to demonstrate that Orconectes can act not only as a vector under laboratory conditions, but also is able to transfer the disease to the native susceptible crayfish populations.

ACKNOWLEDGMENTS

This work was supported by the Fishery Board of Sweden, the Bergvalls' Foundation, and the French Conseil Superieur de la Pêche.

LITERATURE CITED

Aoki, j. and Yanase, K. 1970. Phenol oxidase activity in the integument of the silkworm Bombyx mori infected with Beauveria bassiana and Spicaria fumoso-rosea. J. Invertebr. Pathology 16, 459-464.

Table 4. Infection of O. limosus and A. astacus with 1000 zoospores ml^{-1} of A. astaci at weekly intervals in aquaria at 13°C.

Experiment 1.

Date of infection	O. limosus (n = 5)	A. astacus (n = 3)
801118	1000 zoospores ml^{-1}	1000 zoospores ml^{-1}
801125	"_	
801202	"_	
801209	"_	
801216	"_	
	One crayfish dead in crayfish plague 810323 and one dead 810408	Two dead in plague 801127 and one dead in plague 801128

Experiment 2.

Date of infection	O. limosus (n = 5)	A. astacus (n = 2)
810113	1000 zoospores ml^{-1}	1000 zoospores ml^{-1}
810120	"_	
810127	"_	
810203	"_	
810203	"_	
810216	"_	
	Three crayfish dead in crayfish plague 810427, 810504 and 810518	Two dead in plague 810122

289

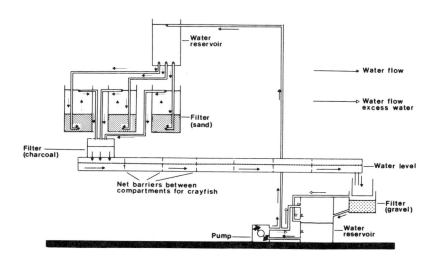

Figure 1. The closed aquaria system used for some of the infection experiments.

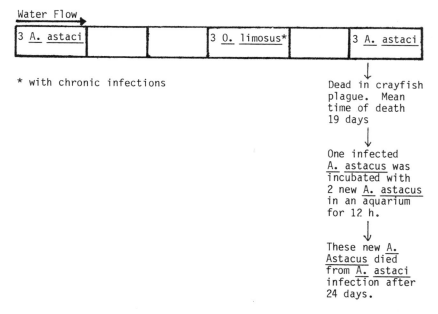

Figure 2. Transfer of A. astaci from chronically infected O. limosus to Astacus astacus in a closed aquaria system.

Fürst, M. and Boström, U. 1978. Frekvens av en skalsvamp (kräftpest) på signalkräftor. Inf. Från Sötvattenslab. Drottninghoml 1, 1-24. (Mimeographed in Swedish)

Geelen, J.F.M. 1978. The distribution of the crayfishes Orconectes limosus Raff. and Astacus astacus L. (Crustacea, Decapoda) in the Netherlands. Zool. Bijdragen 23, 4-19.

Götz, P. and Vey, A. 1974. Humoral encapsulation in Diptera (Insecta): defence reactions of Chironomus larvae against fungi. Parasitology 68, 193-205.

Laurent, P.J. and Suscillon, M. 1962. Les écrevisses en France. Ann. Station Centrale d'Hydrobiologie Appliquée 9, 333-395.

Nyhlén, L. and Unestam, T. 1980. Wound reactions and Aphanomyces astaci growth in crayfish cuticle. J. Invertebr. Pathology 36, 187-197.

Schäperclaus, W. 1935. Die Ursache der pestartigen Krebssterben. A. Fisch. 33, 343-366.

Söderhäll, K. and Ajaxon, R. 1981. Effect of quinones and melanin on mycelial growth of Aphanomyces spp. and extracellular protease of Aphanomyces astaci, a parasite on crayfish. J. Invertebr. Pathology (in press).

Söderhäll, K., Häll, L., Unestam, T., and Nyhlén, L. 1979. Attachment of phenoloxidase to fungal cell walls in arthropod immunity. J. Invertebr. Pathology 34, 285-294.

Unetam, K. 1969., On the physiology of zoospore production in Aphanomyces astaci. Physiol. Plant. 22, 236-245.

Unestam, T., Söderhäll, K., Nyhlén, L., Svensson, E., and Ajaxon, R. 1977. Specialization in crayfish defence and fungal aggressiveness upon crayfish plague infection. In: p. 321-331. O.V. Lindqvist (ed.) Freshwater Crayfish III, Kuopio, Finland (1976).

PACIFASTACUS LENIUSCULUS DANA AND ITS RESISTANCE TO THE PARASITIC FUNGUS APHANOMYCES ASTACI SCHIKORA

Mikael Persson and Kenneth Söderhäll
Institute of Physiological Botany
University of Uppsala
Box 540
S-751 21 Uppsala, Sweden

ABSTRACT

The resistance of Pacifastacus leniusculus towards the crayfish plague fungus Aphanomyces astaci can be reduced by injecting "non-self" particles into the crayfish. Latent "chronic" infections can then easily be converted to lethal "acute" infections resulting in death of the animal.

The resistance of P. leniusculus towards Aphanomyces astaci zoospores seems to have decreased.

INTRODUCTION

The American crayfish Pacifastacus leniusculus Dana is more resistant towards attack by the parasite Aphanomyces astaci Schikora than European crayfish species (Unestam et al., 1977). P. leniusculus harbors the parasite in the cuticle as a latent ("chronic") infection and functions as a vector for this disease (Unestam et al., 1977). In crayfish with such latent infections fungal growth apparently is restricted by the host defense reactions, which can be seen as melanized spots in the cuticle. This does not necessarily mean that all melanized spots (brown to dark brown) in the cuticle of P. leniusculus harbor the parasitic fungus, A. astaci. A wound or presence of other parasites will also induce melanization in the cuticle of most crustaceans (Söderhäll, 1982 a). The melanization reaction might constitute a defense system of the cuticle (Aoki and Yanase, 1971; Götz and Vey, 1974; Söderhäll et al., 1979; Nyhlén and Unestam, 1980). In a recent report Söderhäll and Ajaxon (1982) showed that melanin and precursors to melanin were fungistatic towards three Aphanomyces species including A. astaci. Furthermore, $\beta 1$, 3-glucans, carbohydrates from fungal cell walls, are specifically activating hemolymph prophenoloxidase in several arthropods such as freshwater crayfish (Unestam and Söderhäll, 1977; Söderhäll and Unestam, 1979), marine crustaceans (Smith and Söderhäll, to be published), and insects (Ashida, 1981; Söderhäll, unpublished). Therefore, in the hemolymph the prophenoloxidase activating system may constitute a recognition reaction towards not-self in arthropods (Söderhäll, 1981; 1982 a, b).

MATERIALS AND METHODS

Crayfish: Pacifastacus leniusculus Dana with latent ("chronic") infections and "healthy" crayfish were obtained from Lake Halmsjön, Uppland and kindly provided by Simontorps akvatiska avelslaboratorium, Blentarp, Sweden. Astacus astacus L. were obtained from Lake Vallsjön, Småland. Prior to the experiments all crayfish were fed with fish and

kept under the experimental conditions 1-2 weeks. Only intermolt
animals were used in the experiments.

Injection of purified fungal cell walls in crayfish

Purified fungal cell walls (Zymosan A, Sigma) were suspended in
0.85% NaCl (0.1 mg ml^{-1}) and then sterilized by autoclaving. The
Zymosan-suspension (0.05 ml) was injected using a sterile syringe
(needle, 0.6 × 25 mm) in the base of a walking leg.

The Zymosan suspension was injected in P. leniusculus with or
without latent infections of A. astaci and in Astacus astacus. As
controls P. leniusculus and A. astacus were given with 0.05 ml 0.85%
NaCl. After injection, the crayfish were kept in a recirculating
aquaria system (Fig. 1) during the course of experiments (usually 2-3
months).

Infection experiments with A. astaci

Fungus: Aphanomyces astaci strain Si was obtained from our stock
culture collection. Production of zoospores was according to Unestam
(1969). Infection experiments were performed as described in Unestam
(1975).

RESULTS AND DISCUSSION

The latent infections in the cuticle of P. leniusculus could be
converted into lethal ("acute") infections. The amount of circulating
hemocytes decreased rapidly after injection with purified fungal cell
walls (Söderhäll, unpublished), which probably affected the overall
resistance of the host, allowing growth of the parasite out from the
melanized area i.e. the latent infection developed into a lethal (Fig.
2).

A higher mortality was always found in Zymosan-injected P.
leniusculus with latent infections, than Zymosan-injected "healthy" P.
leniusculus (Fig. 3). Lethal infections of A. astaci could only be
found in Zymosan-injected P. leniusculus, which already had the parasite
in a latent form (Table 1). Due to the difficulty in examining all
tissues in each crayfish we restricted our investigations to look for A.
astaci hyphae in the soft integument (cuticle) of all dead crayfish.
Non-melanized hyphae of A. astaci (lethal infection) were only found in
40% of the dead P. leniusculus (Table 1), where this method was used.
In the present investigation a lethal infection is defined as hyphae of
A. astaci protruding out from a melanized area, which means that we
always checked that the hyphae originated from a latent infection (Fig.
4). Occassionally, we observed hyphae in parts of the cuticle where no
latent infections could be found. This suggests that when optimal
conditions for fungal growth in the cuticle occurred the hyphae
penetrated out through the cuticle, sporulated and produced new
infection units (zoospores). Probably the zoospores attached on new
areas of the cuticle, penetrated and a new infection was established.
This is how P. leniusculus functions as a vector for this disease
(Unestam et al., 1977). Bearing this in mind, it can be assumed that
more crayfish actually died in crayfish plague, a possibility we have
not verified, since only the soft cuticle was examined. Other

293

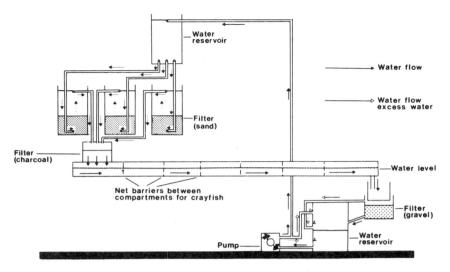

Figure 1: The recirculating aquaria system used for the Zymosan-injection experiments.

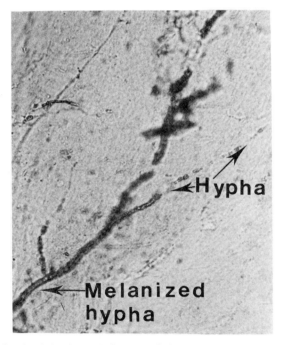

Figure 2: Melanized hyphae of _A. astaci_ in the cuticle of _P. leniusculus_. Development of a latent infection to a lethal infection ("acute") after Zymosan-injection, visualized by lack of melanin deposits on the hyphae (arrows).

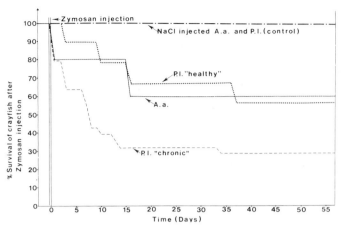

Figure 3: Survival of crayfish after injection with purified fungal cell walls
(Zymosan): 5 A. astacus and 5 P. leniusculus injected with 0.05 ml saline
(0.85% NaCl), control. 9 P. leniusculus without visible A. astaci infections
("healthy") injected with 0.05 ml Zymosan suspension (0.1 mg ml^{-1}). 5 A.
astacus injected with 0.05 ml Zymosan suspension (0.1 mg ml^{-1}). 28 P.
leniusculus with latent ("chronic") A. astaci infections injected with 0.05 ml
Zymosan suspension (0.1 mg ml^{-1}).

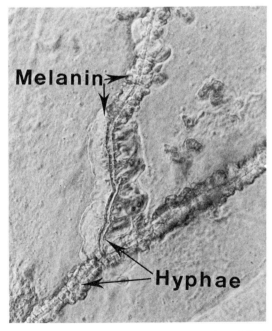

Figure 4: Melanized hyphae of A. astaci in the cuticle of P. leniusculus. The
hyphae are heavily melanized as a result of cuticle phenoloxidase activity
(latent infection).

Table 1. P. leniusculus dead in crayfish plague (A. astaci) after
Zymosan injection.

Crayfish	Zymosan injection	Dead in crayfish plague/total dead (%)
P. leniusculus (with latent infections)	yes	40 (20) [1]
P. leniusculus ("healthy")	yes	0 (4) [1]

[1]The number of crayfish in the experiments is shown in parentheses.

possibilities such as bleeding or bacterial/fungal infection in the wound made by the injection needle should also be considered as cause of death.

The resistance of P. leniusculus towards the crayfish parasite, A. astaci, seems to have decreased. Unestam (1975) exposed 25 P. leniusculus from Lake Tahoe, California, to A. astaci zoospores (10^4 spores ml^{-1}) in aquaria with 6 l water. No crayfish died. When we repeated the experiment with P. leniusculus from a lake in Sweden and from Simontorp 20 of 22 crayfish died due to A. astaci infection. In the cuticle of the infected animals non-melanized hyphae of A. astaci could be seen. We could not observe any difference in the amount of dead crayfish or mean time of death (16 days) between crayfish with latent infections and "healthy" crayfish. Before we can conclude that the resistance of P. leniusculus in Sweden towards A. astaci infection has drastically decreased it is necessary to test other populations and to more carefully determine the present level of resistance.

In summary the results in this paper and in one adjacent paper (Vey et al., 1982) conclusively show that the American crayfishes, Orconectes limosus and P. leniusculus harbor the crayfish parasitic fungus, A. astaci, as latent infections and function as vectors for the disease. Also, a latent infection in crayfish can be converted to a lethal by injecting foreign particles (purified fungal cell walls) in the hemolymph, imitating a sublethal fungal infection. P. leniusculus dies if infected with zoospores of A. astaci (crayfish plague).

Although our work has been carried out under laboratory conditions and one can argue that the crayfish are stressed during the experiments, we believe that these results should be taken into consideration when introducing P. leniusculus into a lake. This is especially valid since "chronically" infected P. leniusculus very seldom die from lethal infections in our aquaria. We would not like to imply that our experiments could be directly transferred to a natural situation, but some of the results which we have shown in this paper might as well happen in nature.

ACKNOWLEDGMENTS

This work was supported by the Fishery Board of Sweden. We are grateful to Stellan Karlsson, Simontorps Akvatiska Avelslaboratorium, and Magnus Fürst, Freshwater Research Institute, for delivery of Pacifastacus leniusculus. We thank Torgny Unestam for his interest in this study.

REFERENCES

Aoki, J. and Yanase, K. 1970. Phenoloxidase activity in the integument of the silkworm Bombyx mori infected with Beauveria bassiana and Spicaria fumoso-rosea. J. Invertebr. Pathology 16, 459-464.

Ashida, M. 1981. A cane sugar factor suppressing activation of prophenoloxidase in haemolymph of the silkworm, Bombyx mori. Insect Biochem. 11, 57-65.

Götz, P. and Vey, A. 1974. Humoral encapsulation in Diptera (Insecta): defence reactions of Chironomus larvae against fungi. Parasitology 68, 193-205.

Nyhlén, L. and Unestam, T. 1980. Wound reactions and Aphanomyces astaci growth in crayfish cuticle, J. Invertebr. Pathology 36, 187-197.

Söderhäll, K., Hall, L., Unestam, T., and Nyhlén, L. 1979. Attachment of J. Invertebr. Pathology 34, 285-294

Söderhäll, K. and Unestam, T. 1979. Activation of serum prophenoloxidase in arthropod immunity. The specificity of cell wall glucan activation and activation by purified fungal glycoproteins of crayfish phenoloxidase. Can. J. Microbiol. 25, 406-414.

Söderhäll, K. and Ajaxon, R. 1982. Effect of quinones and melanin on mycelial growth of Aphanomyces spp. and extracellular protease of Aphanomyces astaci, a parasite on crayfish. J. Invertebr. Pathology 39, 105-109.

Söderhäll, K. 1981. Fungal cell wall β1, 3-glucans induce clotting and phenoloxidase attachment ot foreign surfaces of crayfish hemocyte lysate. Developmental Comparative Immunology 5, 565-573.

Söderhäll, K. 1982 a. Prophenoloxidase activating system and melanization - a recognition mechanism of arthropods? A review. Developmental Comparative Immunology 6, x-xx.

Söderhäll, K. 1982 b. β1, 3-glucan enhancement of serine protease activity in crayfish hemoctye lysate Comp. Biochem. Physiol. B. (in press).

Unestam, T. 1969. On the physiology of zoospore production in Aphanomyces astaci. Physiol. Plant. 22, 236-245.

Unestam. T. 1975. Defence reactions in and susceptibility of Austraian and New Guinean freshwater crayfish to European-Crayfish-Plague fungus. Austr. J. Exp. Biol. Med. Sci. 53, 349-359.

Unestam, T., Söderhäll, K., Nyhlén, L., Svensson, E., and Ajoxon, R. 1977. Specialization in crayfish defence and fungal aggressiveness upon crayfish plague infection. In: Freshwater crayfish III, O.V. Lindqvist (ed.) pp. 321-331, Kuopio, Finland (1976).

Unestam, T. and Söderhäll, K. 1977. Soluble fragments from fungal cell walls elicit defence reactions in crayfish. Nature, 267, 45-46.

Vey, A., Söderhäll, K. and Ajaxon, R. 1982. Susceptibility of Orconectes limosus to the crayfish plague, Aphanomyces astaci. In: Freshwater crayfish V. C.R. Goldman (ed.) Davis, California (1981) (in press).

THE OCCURRENCE OF CRAYFISH DISEASES AND THEIR SIGNIFICANCE IN IRELAND

C. O'Keeffe and Julian D. Reynolds
Deparment of Zoology
University of Dublin
Trinity College
Dublin 2, Ireland

ABSTRACT

Populations of the crayfish Austropotamobius pallipes (Lereb.) were examined in 1979 and 1980 for evidence of parasitic disease. Only 0.7% of the White Lake crayfish and 1.2% of Brittas River specimens had porcelain disease caused by the microsporidian Thelohania, a parasite also noted in three other major rivers systems. Microscopic examination revealed that a 1.3% rate of infection went undiscovered in macroscopic screening. The disease was only seen in crayfish over three years old, and it is suggested that feeding behaviour of different life stages may be responsible for the restriction of the disease to larger forms. Diseased crayfish may live in deeper, less favourable lake habitats. Methods of capture of crayfish may also affect the occurrence of infected crayfish in a sample. "Burn-spot" fungal disease was noted on 3% of lake crayfish, but was not seen elsewhere. Smaller lesions, chiefly on the claws of adult males, were probably the scars of agonistic contact. No evidence was found for fungal plague (Aphanomyces) in Ireland.

INTRODUCTION

In recent years there has been interest in Ireland in the commercial feasibility of harvesting the only native species of crayfish, Austropotamobius pallipes, or of the culture of this or an introduced species as a luxury food (Reynolds, 1979). To date, there is no published information on the occurrence in Ireland of diseases which might affect wild or cultured stocks, although these have been shown to be of economic importance elsewhere (Schaperclaus, 1954). This paper reports on the identification of two crayfish diseases in Ireland and on the apparent absence of another.

Two crayfish diseases are important in Europe: the crayfish plague, caused by the fungus Aphanomyces astaci and porcelain or whitetail disease, caused by the genus Thelohania (Sporozoa: Microsporidia). Introduced to Italy in the 1860s, crayfish spread within fifty years to places as far away as Sweden and Russia, where it had catastrophic effects on indigenous crayfish populations (Unestam, 1973). However, there is no evidence that it ever reached Britain (Holdich et al., 1978).

The genus Thelohania is known from the U.S.A., mainland Europe and Britain (Unestam, 1973), Russia (Voronin, 1971) and from Australia (Carstairs, 1979). One species, T. contejeani, infects European crayfish, invading the muscle fibers and replacing them with masses of spores. These cause an opaque whiteness of the muscles which gives the disease its popular name. Infection is believed to result from consumption of infected crayfish, and leads to progressive loss of muscle function, paralysis and finally death (Cossins, 1973). The time taken for death to occur has variously been estimated as several months (Vey et al., 1971), or one or two years (Bowler and Brown, 1977).

A number of other fungal and bacterial diseases of crayfish have been reported (Vey and Vago, 1973; Vey et al., 1975). One of these is "burn-spot" disease, which may be caused by a number of host-specific fungi; infected specimens of Astacus astacus, Orconectes limosus (Mann and Pieplow, 1938) and A. leptodactylus (Mann, 1940) have been recorded. Infection is manifested as a brown or black spot, with a red margin, on the exoskeleton. The disease is not believed to have a seriously harmful effect on the crayfish, although it may affect their marketability (Unestam, 1973).

METHODS

In a general survey of Irish crayfish, records were kept of any diseased or abnormal specimens encountered. Crayfish were intensively sampled in 1979 and 1980 in a small lake and a hill stream, in a study of two contrasting crayfish populations (O'Keeffe, in prep.). White Lake, Co. Westmeath, (National Grid ref. N 52 74) is a 31 ha. hardwater lake enclosed by glacially-formed hills; it has been described previously by Moriarty (1973). Brittas River, Co. Wicklow, (National Grid ref. O 32 20) a tributary of the River Liffey, is a hill stream in an area of granitic deposits, whose waters are nevertheless of medium hardness. Lake specimens were sampled using unbaited eel traps, hand-nets or were caught by hand while SCUBA-diving or snorkelling. River crayfish were taken by hand-net or were hand-caught in torchlight at night. Most specimens were examined live, than marked and returned to the place of capture. Others were frozen in liquid nitrogen prior to laboratory examination, and apparently healthy individuals among these were examined as muscle smears under phase-contrast microscope in an attempt to quantify the instance of developing infections of Thelohania. Dead crayfish were rarely seen; however any dead crayfish found in the lake were also examined microscopically, since putrefaction will lead to a whitish colouration of the flesh.

RESULTS

Crayfish infected by Thelohania were found in small numbers in both Brittas River (1.2% of sample) and White Lake (0.7%). Microscopic examination revealed two infected animals in a sample of 158 individuals which had not been diagnosed by mere visual examination. Of six crayfish found dead and examined microscopically, two had advanced Thelohania infection (Table 1).

The depth at which each crayfish sample was taken in the lake was recorded. Of the 12 diseased individuals trapped, seven were taken at a depth greater than 10m, while five of the six SCUBA-caught diseased animals were found below 7m. Thus 78% of the crayfish found to be infected came from the deepest part of the lake sampled.

In the general survey, one of three males captured in the River Trough, a small tributary of the River Shannon near Limerick, showed an

300

Table 1. Diseases found in Irish Crayfish, 1979-1980.

Mode of Examination	Site	Number Examined 0+	all others	Σ	No. Diseased (% of catch) Thelohania	Burn-spot
Macroscopic	White Lake	100	2470	2570	19 (0.74%)	78 (3.21%)*
	Brittas River	130	733	863	9 (1.04%)	0
Microscopic	White Lake	12	94	106	0	**
	Brittas River	0	52	52	2 (3.85%)	**
Dead	White Lake	0	6	6	2 (33.3%)	**
TOTAL	White Lake	112	2562	2676	19 (0.71%)	--
	Brittas River	130	784	914	11 (1.20%)	--

*only 2427 individuals examined

** not recorded

Table 2. Occurrence of Thelohania in samples collectd by various methods at White Lake.

Sampling Method	0+	Number Examined All others	Σ	No. Thelohania Infections	% of Sample
TRAP	0	2081	2081	12	0.57%
SCUBA	3	304	307	6	0.98%
SNORKEL	9	162	171	1	0.58%
HANDNET	94	23	117	0	--

advanced infection of Thelohania, as did one of about twenty in the River Suir, near Thurles, Co. Tipperary. A single infected specimen was among 117 crayfish taken at Lough Lene, in the upper reaches of the River Boyne system.

"Burn-spot" disease was recorded in about 3% of White Lake cray-fish, but not in the Brittas River population (Table 1). Lesions were observed on the dorsal surface of both the abdomen and the carapace. A relatively high incidence of what we termed "small burn-spot" was also noted on crayfish from White Lake. This takes the form of small pits, less than 1 mm in diameter, in the shell, surrounded by an area of melanization. Neither cuticle nor the muscle beneath are affected, however, and since approximately 90% of these lesions occurred on the claws of adult males, it seems likely that these are simply the scars of agonistic contact between crayfish.

No evidence was discovered of the occurrence of crayfish plague. Indeed, there were no obvious heavy infestations of any parasites other than those mentioned above.

DISCUSSION

A. THELOHANIA in Ireland appears to be widespread but occurring in low levels. In other European countries, the occurrence of porcelain disease has been variously estimated as less than 2% in Finland (Sumari and Westman, 1969), 10% in Lozère, France (Vey and Vago, 1973), 18% in the Oxford area in England (Pixell-Goodrich, 1956) and as high as 30% in Germany (Schaperclaus, 1954). A level of about 30% infection was noted in streams near Limoges, France, by one of us in September, 1979 (J.D.R., unpublished).

These results are not directly comparable for a number of reasons. First, there is evidence that the disease is density-dependent (Brown and Bowler, 1977), which would suggest that infection may persist at a low, stable level for a period of years but could, under certain conditions, reach relatively high levels and thereby bring about a population crash. If this be the case, reports of instantaneous levels of infection are of limited value.

Secondly, little if any information has been published on the size/age frequencies of infected animals. Of all the infected animals found in this study, the youngest was estimated to be in its fourth year of life and the majority of diseased animals were at least five years old. The influence of age upon the probability of infection may be explained in part by the period taken after host colonization begins for the characteristic whiteness to become visible in the muscles, but it seems likely that differences in feeding behaviour may also be important. Adult crayfish are known to remain quiescent during the day and to emerge from hiding to forage at night (Ingle, 1977). We have found, however, that yearlings held in the laboratory display less rigid patterns of activity than do older individuals, and spend much of each 24-hour period under cover. Under natural conditions, these crayfish rarely leave cover day or night, preferring to hide while feeding intermittently on small cladocerans, ostracods and chironomids. From this it may be inferred that young crayfish are relatively unlikely to encounter or consume the flesh of an infected crayfish, and for this reason, the number of yearling crayfish examined in each sample is included in Table 1.

The selective nature of most methods for sampling populations is well known (e.g Seber, 1973). In decapod crustaceans, the size and sex of animals in a sample from a natural population is often biased, and selectivity in trapping crayfish populations has been documented (Brown and Brewis, 1977). It has further been suggested that disease may influence the likelihood of crayfish entering traps (K. Bowler, quoted in Unestam, 1973). Analysis of the occurrence of diseased crayfish in samples taken by different methods in White Lake supports this theory (Table 2), as significantly more diseased animals were taken by SCUBA than by any other method. This may be explained in terms of the reduced mobility of infected crayfish (Cossins, 1973) which might render them unable to search for and climb into the entrances of traps. It is also possible, however, that diseased animals are less likely to escape from a diver than are healthy individuals. There is also some evidence that in White Lake diseased animals retreat, or are ousted, to less favourable parts of the lake. Apart from the yearlings which frequent the stony littoral, crayfish are most numerous on the sublittoral shelf and upper parts of the slope (see Fig. 1) where they burrow into the dense Chara or into bare mud. However, 78% of diseased animals came from deeper parts of the lake, where there is no shelter and reduced availability of food.

The microscopic examination of 158 smears of crayfish muscle which were judged to be free of Thelohania when visually inspected, yielded two infected specimens (1.3%). It is clear that microscopic examinations are necessary to obtain an accurate estimate of Thelohania levels in a crayfish population.

B. OTHER DISEASES. The only other disease encountered was burn-spot disease, which was only encountered in White Lake. Its appearance was distinct from the small melanized punctures frequently found on the chelae of males. However, incidence of burn-spot was significantly higher in males than in females (17 cases to 6; p< 0.05), suggesting a link with agonistic behaviour. Fighting may lead to shell damage, which in turn may facilitate the entry of a fungal parasite. There is scope for further research in this interesting area.

No evidence was found in this study of the occurrence of the plague fungus, Aphanomyces astaci in Ireland. Unestam (1973) has pointed out that diseased specimens may be difficult to identify by visual inspection alone, but he also states that wherever the disease is present, it exterminates all European crayfish, and A. pallipes is not known to be any more resistant than Astacus astacus. From this fact, and the widespread distribution of crayfish in Ireland (Reynolds, 1978), a generally low-lying country whose river systems are often interconnected at their headwaters, it is inferred that crayfish plague has not reached this country. Crayfish are known to have disappeared from some Irish waters in recent times and Moriarty (1973) suggests that the abrupt loss of sizable populations from Pallas Lake, Co. Offaly in 1954 is evidence that "some form of crayfish disease is present in Ireland." Duffield (1933) has documented the apparently total extermination of crayfish populations in England, but which proved to be only temporary, with gradual recovery of the populations occurring over a period of years. Thelohania was postulated as one of

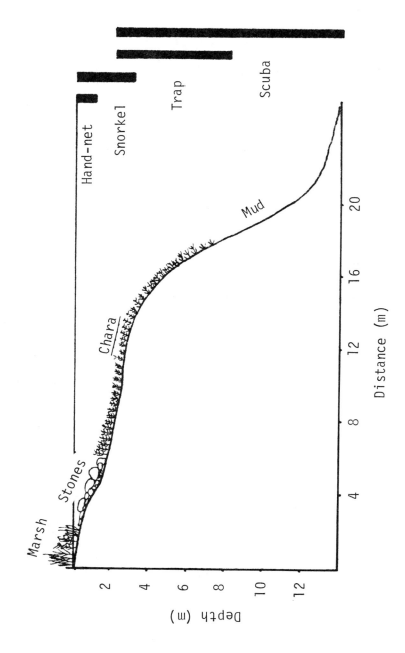

Fig. 1. Profile of White Lake showing benthic type and crayfish sampling methods.

several possible causes. Exhaustive searching of Pallas Lake has shown that no such recovery has taken place there. As crayfish occur in nearby water bodies, Aphanomyces is contraindicated and Thelohania, which might cause a population collapse in crowded conditions, is unlikely to have eradicated the Pallas Lake crayfish totally. We feel that other agents, such as organic pollution, application of rotenone or the inwash of biocides may have led to the population collapse, rather than disease.

Under present conditions in Ireland, with as yet unharvested wild crayfish stocks and an absence of crayfish culture, we see few signs of serious crayfish disease, and this situation should continue if Aphanomyces is not introduced. The governmental fisheries departments in both the Republic of Ireland and Northern Ireland are attempting to maintain this situation by jointly banning all live crayfish imports. At present, A. pallipes is widespread in Ireland and occurs in possibly a wider array of biotopes than elsewhere in its geographic range. This may be a reflection of the restricted island freshwater fauna and may have corresponding genetic importance for the genetic diversity of stocks.

ACKNOWLEDGMENTS

We are grateful to Martin Linnie and other members of the Dublin Unviersity Sub-Aqua Club for assistance with underwater collections. This work was carried out while C. O'Keeffe was in receipt of awards from the Departments of Fisheries and Education of the Irish government.

LITERATURE CITED

Brown, D.J. and K. Bowler. 1977. A population study of the British freshwater crayfish Austropotamobius pallipes (Lereboullet), p. 33-50. In O. Lindquist (ed.) Freshwater Crayfish III, Kuopio, Finland (1976).

Brown, D.J. and J.M. Brewis, 1979. A critical look at trapping as a method of sampling a population of Astropotamobius pallipes (Lereboullet) in a mark and recapture study, p. 159-164. In P.J. Laurent (ed.) Freshwater Crayfish IV, Thonon-les-Bains, France (1978).

Carstairs, I.L. 1979. Report of the microsporidial infestation of the freshwater crayfish Cherax destructor, p. 343-348. In P.J. Laurent (ed.) Freshwater Crayfish IV, Thonon-les-Bains, France (1978).

Cossins, A. 1973. Thelohania contejeani (Henneguy), microsporidian parasite of Austropotamobius pallipes (Lereboullet) – an histological and ultrastructural study, p. 151-164. In S. Abrahamsson (ed.) Freshwater Crayfish I, Austria (1972).

Duffield, J.E. 1933. Fluctuation in numbers among freshwater crayfish, Potamobius pallipes Lereboullet. J. Anim. Ecol. 2:184-196.

Holdich, D.M., Jay, D. and J.S. Goddard. 1978. Crayfish in the British Isles. Aquaculture 15:91-97.

Ingle, R.W. 1977. Laboratory and SCUBA studies on the behaviour of the freshwater crayfish Austropotamobius pallipes (Lereboullet). Rep. Underwater Ass. 2:1-15.

Mann, H. 1940. Die brandfleckenkrankheit beim sumpfkrebs (Potamobius leptodactylus Esch.). Zeitschr. Parasitenk. 11:430-432.

Mann H. and U. Pieplow. 1938. Die brandfleckenkrankheit bei krebsen und ihre erreger. Zeitschr. Fisherei 36:225-240.

Moriarty, C. 1973. A study of Austropotamobius pallipes in Ireland, p. 57-68. In S. Abrahamsson (ed.) Freshwater Crayfish I, Austria (1972).

O'Keeffe, C. (in prep). The ecology and aquaculture potential of freshwater crayfish in Ireland. Ph.D. thesis.

Pixell-Goodrich, H. 1956. Crayfish epidemics. Parasitology 46:480-483.

Reynolds, J.D. 1978. Crayfish ecology in Ireland, p. 215-220. In P.J. Laurent (ed.) Freshwater Crayfish IV, Thonon-les-Bains, France (1978).

Reynolds, J.D. 1979. The introduction of freshwater crayfish species for aquaculture in Ireland, p. 57-64. In Kernan, R.P. et al. (eds.), The Introduction of Exotic Species, Advantages and Problems. Royal Irish Academy, Dublin.

Schaperclaus, W. 1954. Fischkrankheiten. Akademie-Verlag, Berlin.

Seber, G.A.F. 1973. The estimation of animal abundance. Griffin, London.

Sumari, O. and K. Westman. 1969. The crayfish parasite Thelohania contejeani Henneguy (Sporozoa, Microsporidia) found in Finland, Ann. Zool. Fenn. 7:193-194.

Unestam, T. 1973. Significances of diseases on freshwater crayfish, p. 135-150. In S. Abrahamsson (ed.) Freshwater Crayfish I, Austria (1972).

Vey, A. and C. Vago. 1973. Protozoan and fungal diseases of Austropotamobius pallipes Lereboullet in France, p. 165-180. In S. Abrahamsson (ed.) Freshwater Crayfish I, Austria (1972).

Vey, A., Boemare, N. and C. Vago. 1975. Recherches sur les maladies bacteriennes de l'ecrevisse Austropotamobius pallipes Lereboullet, p. 287-298. In: J.W. Avault (ed.), Freshwater Crayfish II, Louisiana, U.S.A. (1974).

Voronin, V.N. 1971. New data on microsporidiosis of the crayfish Astacus astacus (L. 1758). Parazitologiya 5:186-191.

ULTRASTRUCTURE AND TAXONOMIC POSITION OF THE CRAYFISH PARASITE PSOROSPERMIUM HAECKELI HILGENDORF

Viljo Nylund and Kai Westman
Finnish Game and Fisheries Research Institute,
Fisheries Division,
P.O. Box 193, SF-00131, Helsinki 13, Finland
and
Kari Lounatmaa
Department of Electron Microscopy
University of Helsinki,
Mannerheimintie 172,
SF-00280, Helsinki 28, Finland

ABSTRACT

The parasite Psorospermium haeckeli was described as occurring in crayfish as far back as 1857, but was first observed in Finland in the native crayfish Astacus astacus in 1975. In spite of numerous studies, its taxonomic position is still uncertain. The groups to which it has been tentatively assigned include the Protozoa, Trematoda, and Nematoda. Our recent electron microscope studies failed to reveal any features characteristic of one of the above groups. On the other hand, the thick-shelled, spore-like stage occurring in the crayfish showed certain structural similarities to dimorphic pathogenic fungi. Little information is available on the life cycle of the parasite, its possible intermediate hosts, or its significance to the crayfish.

INTRODUCTION

Psorospermium haeckeli, a parasite on crayfish, was described as early as 1857 (Haeckel), and was named by Hilgendorf in 1883. The first observation of the parasite in Finland, on the native crayfish, Astacus astacus, was made in 1975 (Nylund and Westman 1977, 1978, 1979). Since then, research on P. haeckeli has been carried out in the Finnish Game and Fisheries Research Institute. As far as is known to the authors, Psorospermium has been found in four freshwater crayfish species, including the American Crayfish (Pacifastacus leniusculus), in four European countries besides Finland (Nylund and Westman 1979). Although a number of studies have been made on the parasite, uncertainty still exists regarding its taxonomic status, its life cycle, and its significance to its crayfish host. Haeckel (1857) and Hilgendorf (1883) did not specify the taxonomic position, but the name Psorospermium haeckeli may refer to Sporozoa. In his extensive review of Sporozoa, Labbé (1899) placed P. haeckeli in the group "species incertae." Unestam (1975) has suggested that it may belong to the Protozoa; it has also been proposed that it is a Sporozoon or the eggs of Trematoda (Schäperclaus 1954, 1979) or Nematoda (Ljungberg and Monné 1968). Grabda (1934) has published the most extensive investigations on P. haeckeli, including studies of its occurrence in the crayfish, and also of various amoeba-like and spore-like stages of its life cycle. Nylund et al. (1979) studied its occurrence in the crayfish, its deleterious effects on the host, and the ultrastructure of its spore-like stage. The aim of this paper is to discuss the

307

taxonomic status of P. haeckeli on the basis of electron microscopic investigations (Nylund et al. 1979).

MATERIALS AND METHODS

The crayfish (Astacus astacus) were caught with baited traps in Lake Pojanjärvi in southern Finland on 17-18 November 1977 and 4-7 September 1978. They were held in fiberglass basins at Evo Inland Fisheries and Aquaculture Research Station and at Porla Fish Culture Station. Samples were examined at different seasons of the year, but the ultrastructural study was made on samples from a crayfish fixed on 29 November 1977. Electron microscopic samples were prefixed with 3% glutaraldehyde (Leiras, Finland) in 0.1 M sodium phosphate buffer (pH 7.2) and postfixed for 2 h with 1% osmium tetroxide in the same buffer. The sections were obtained from Epon 812-embedded samples and stained with uranyl acetate and lead citrate. The transmission electron micrographs were taken with a JEM-100B electron microscope operating at 80 kV.

RESULTS AND DISCUSSION

According to Grabda (1934), the first stage of P. haeckeli is an amoeba-like organism with a homogeneous inner structure. He observed several stages in the life cycle: a mononuclear "amoeba," a binuclear "amoeba" and an oval thick-walled sporelike form. He also found parasites at different stages in the life cycle in the same crayfish individual. Grabda (1934) considered that he had fully described the life cycle of the parasite, but he did not consider its taxonomic position. The amoeba-like stages in the life cycle suggest that the parasite might be one of the Protozoa.

The electron microscopic and histological studies we have carried out have so far failed to reveal structural similarities to any of the Protozoa or the Sporozoa. Nor have we been able to observe any life stages of the parasite other than the spore-like thick-walled organism (Fig. 1), although crayfish infected by the parasite have been cultivated for several years and samples have been taken in different seasons.

Our electron micrographs of the structure of the parasite likewise failed to show any similarities to eggs of the Nematoda or Trematoda (personal communications, 22.5.1978, Veikko Huhta and Eeva Ikonen).

We found, however, that the structure of the wall and the internal structure of the spore-like form of the Psorospermium parasite showed a surprising resemblance to that of resting conidia of the dimorphic fungi. The wall of the parasite consists of thick, irregularly formed plates, thinning along the seams between these plates (Fig. 2). The wall is remarkably thick (ca 8 μm) in relation to the size of the parasite (on average 94 μm x 55 μm), and when viewed under the microscope appears to consist of five distinct layers.

In species of the dimorphic pathogenic fungi, e.g. Blastomyces dermatitidis, the wall of the conidium consists of at least four layers (Garrison and Boyd 1978). The ultrastructure of this multilayered wall

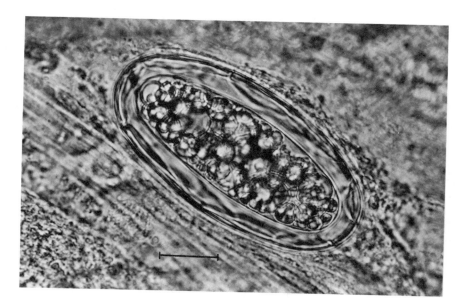

Figure 1: Fresh mount of connective tissue from inside of carapace of Astacus astacus, showing the spore-like stage of Psorospermium haeckeli. Bar equals 20 µm. (V. Nylund).

Figure 2: A broken specimen of P. haeckeli in fresh mount. The joints of the irregular plates are visible in the empty shell. Bar equals 20 µm. (V. Nylund).

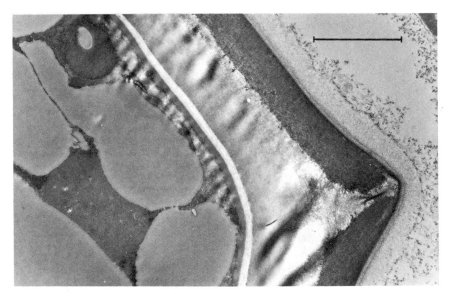

Figure 3: Electron micrograph of the thick, multilayered wall of P. haeckeli. Bar equals 5 μm. (V. Nylund).

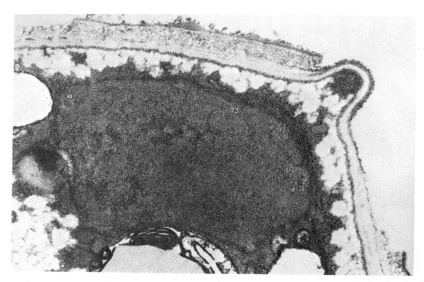

Figure 4: Portion of a conidial yeast mother cell complex showing the multilayered wall and the origin of the wall of a yeast-like cell initial arising from the innermost wall layer of the YMC. (Magnification x 52 900, shown at 80%.) (Courtesy of Robert Garrison.)

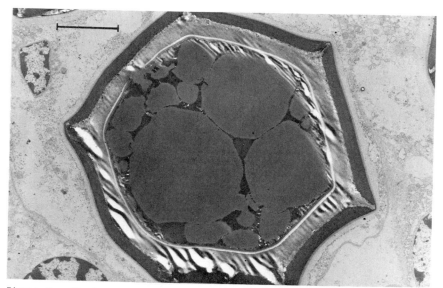

Figure 5: Electron micrograph of a transverse section of <u>P. haeckeli</u>. Bar equals 10 μm. (V. Nylund).

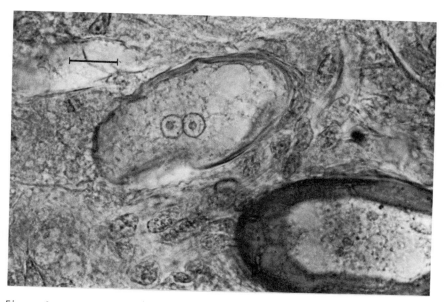

Figure 6: <u>Psorospermium haeckeli</u> with its double nucleus. Bar equals 20 μm. (V. Nylund).

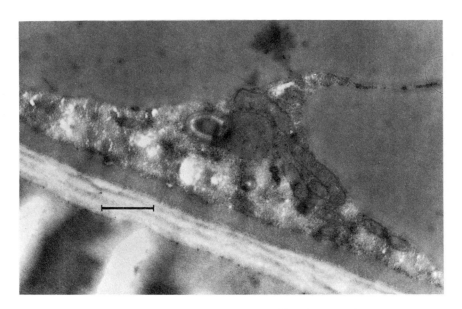

Figure 7: Several mitochondria between the cell wall and lipid bodies of _P. haeckeli_. Bar equals 1 μm. (V. Nylund).

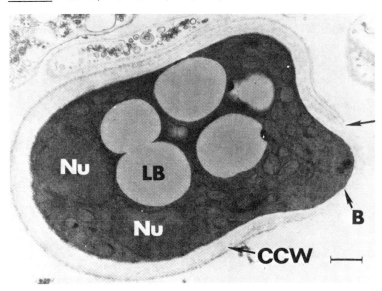

Figure 8: Conidium of _Blastomyces dermatitidis_. The conidial cell wall is rupturing and a bud-like process is emerging (B); nuclei (Nu) and lipid bodies (LB). Bar equals 0.5 μm. (From Garrison, R.G. and Boyd, K.S. 1978. With permission.)

is very similar to that of the wall of the spore-like stage of P. haeckeli (Figs. 3 and 4). The cytoplasm of the Psorospermium parasite contains large lipid droplets, mitochondria, double nucleus, and scattered electron opaque bodies (Nylund et al. 1979), just as in B. dermatitidis (Garrison and Boyd 1978) (Figs. 5, 6, 7 and 8). A further observation indicating the P. haeckeli is of vegatative origin is Wierzejski's (1888) report that the wall contains cellulose. No further information is available at present on the chemical composition of the wall. The dimorphic pathogenic fungi are unique in that their ability to invade animal tissue and cause disease is associated with a concomitant morphogenesis to a yeast-like or other "tissue" phase. These are the only fungi that are able, under natural circumstances, to induce disease in an uncompromised host. Thus morphogenesis and disease production are intimately related (Rippon 1980).

However, in spite of these similarities to the dimorphic pathogenic fungi, P. haeckeli cannot be classified with any certainty before the organism has been isolated in pure culture and studied in more detail.

LITERATURE CITED

Garrison, R.G. and K.S. Boyd. 1978. Role of the conidium in dimorphism of Blastomyces dermatitidis. Mycopathologia vol. 64, 1:29-33.

Grabda, E. 1934. Recherches sur un parasite de l'écrevisse (Potamobius fluviatilis L.), connu sous le nom de Psorospermium haeckeli Hlgd. - Mém. de l'acad. Pol. des sciences et des lettres. S.B., Math. u. Naturw. 6:123-142. Cracovie.

Haeckel, E. 1857. Ueber die Gewebe des Flusskrebses. Arch. Anat. Physiol. Med. p. 561-562.

Hilgendorf, F. 1883. Bemerkungen über die sogenannte Krebspest ins besondere über Psorospermium haeckeli. S.B. Ges. naturf. 9:179-183. Berlin.

Labbé, A. 1899. Sporozoa. In O. Butschli (ed.) Das Tierreich, 5:1-180. Berlin.

Ljungberg, O. and L. Monné. 1968. On the eggs of an enigmatic Nematode parasite encapsulated in the connective tissue of the European crayfish, Astacus astacus in Sweden. Bull. Off. int. Epiz. 69 (7-8):1231-1235.

Nylund, V. and K. Westman. 1977. Psorospermium haeckeli - ravun loistauti löydetty Suomesta. Suomen Kalastuslehti 7:162-165. (In Finnish).

Nylund, V. and K. Westman. 1978. Thelohania contejeani och Psorospermium haeckeli - tvä kräftparasiter. (Summary: Thelohania contejeani and Psorospermium haeckeli - two parasites on the European crayfish, Astacus astacus, in Finland). Information från Sötvattenslaboratoriet, Drottningholm. 14:72-81.

313

Nylund, V. and K. Westman. 1979. Psorospermium haeckeli, a parasite on the European crayfish, Astacus astacus, found in Finland, p. 385-390. In P.J. Laurent (ed.) Freshwater crayfish IV, Thonon-les-Bains, France (1978). 473 p.

Nylund, V., K. Westman, J. Wartiovaara and K. Lounatmaa. 1979. Ultrastructure and effects of Psorospermium haeckeli Hilgendorf, a crayfish (Astacus astacus L.) parasite found in Finland. The Second Scandinavian Symposium on Freshwater Crayfish, Lammi, Finland, 1979. (Manuscript).

Rippon, J.W. 1980. Dimorphism in pathogenic fungi. CRC Critical Reviews in Microbiology vol. 8 (1):49-97.

Schäperclaus, W. 1954. Fischkrankheiten. Akademie - Verlag. Berlin. 708 p.

_____ 1979. Fischkrankheiten. Akademie - Verlag. Berlin. 1089 p.

Unestam, T. 1973. Significance of diseases on freshwater crayfish, p. 135-150. In S. Abrahamsson (ed.) Freshwater crayfish I. Austria (1972). 252 p.

_____ 1975. Defense reactions in crayfish towards microbial parasites, a review, p. 327-336. In J.W. Avault, Jr. (ed.) Freshwater crayfish II. Baton Rouge, Louisiana, U.S.A. (1974). 676 p.

Wierzejski, A. 1888: Kleiner Beitrag zur Kenntnis des Psorospermium Haeckelii. Zool. Anz. Bd. 11:230-231.

EFFECTS OF RICE PESTICIDES ON Procambarus clarkii IN A RICE/CRAWFISH POND MODEL[1]

S. B. Ekanem, James W. Avault, Jr., Jerry B. Graves,[2] and H. Morris[3]
Fisheries Section
School of Forestry and Wildlife Management
Louisiana State University
Baton Rouge, Louisiana 70803 USA

ABSTRACT

In a field study, growth, survival, yield, and reproduction of crawfish (Procambarus clarkii) were compared in replicated tanks planted with untreated rice seed (control), in tanks planted with Difolatan®-treated rice seed (Treatment 2), and in tanks planted with Difolatan®-treated rice seed which also received Propanil®, Ordram®, and Furadan® applications (Treatment 1). Each treatment and the control were stocked with 236 adult crawfish (male and female in equal proportions), and the tanks were modeled to simulate natural conditions in a rice/crawfish pond. The sequence of rice planting, application and amount of chemicals, and harvesting were the same as farmers follow. Tail meat of crawfish harvested from Treatment 1 and the control were analyzed for Propanil, Ordram, and Furadan residues.

Six adult crawfish (all male) were harvested from Treatment 1, all tanks combined; no young were produced. Seven adult crawfish, consisting of five males and two females, were harvested from treatment 2; over 389 young crawfish with great variation in sizes were produced. Twenty-eight adult crawfish were harvested from the control, 18 being males and 10 females; over 275 young were produced. There were no significant differences (p > 0.05) in growth, survival, and yield among treatments. No pesticide residues were detected in the meat.

INTRODUCTION

Crawfish are currently being farmed in rice fields, wooded ponds, and open ponds; of these, rice fields offer the most readily adaptable area for expansion in hectarage. In rice fields, farmers rotate crawfish with rice. Chien's study (1978) revealed that the two crops complemented each other and enhanced productivity of both crops. Some farmers have already converted rice lands into rice/crawfish farms, and according to the Soil Conservation Service (1980) rice/crawfish double cropping now involves over 8,607 ha of Louisiana land. One cooperative near Eunice has over 1,200 ha in rice/crawfish.

Ridgway et al. (1978) reviewed the cost, benefit, and risk of pesticide use in agriculture. Pesticides are normally used to control pests which affect rice production. Crawfish grown in rice fields must meet the pesticide residue standards set by the Environmental Protection Agency. For example, crawfish cultured in ponds where mirex

---

[1] Study supported by the Louisiana Sea Grant Office, NOAA: and by the Louisiana Agricultural Experiment Station.
[2] Department of Entomology
[3] LSU Feed and Fertilizer Lab

was used to control fire ants accumulated mirex in excess of the tolerance level established by the EPA (Graves et al. 1977). Farmers normally use fungicides, herbicides, and insecticides in the presently accepted production system for rice, but use of pesticides and planting of treated rice seeds in a rice/crawfish rotation have been controversial issues among farmers. Some farmers claim that planting of treated rice seeds exterminated crawfish (LaCaze 1976) but a recent study (Cheah et al. 1979-80) indicated that fungicides as a group are generally not very toxic to crawfish. With this in mind, one becomes curious to know what effect the recommended application used in farming or excess of any given pesticide has on crawfish. A combination of pesticides is said to be more effective in rice culture than treatment with only one pesticide (Smith et al. 1977). Possible synergistic or additive effects of such combinations on crawfish have not been studied.

This study focuses on four pesticides commonly applied to rice: Difolatan®, a fungicide; Propanil® and Ordram®, both herbicides; and Furadan®, an insecticide. The objectives of this study were twofold; first, to determine the effects of these four pesticides on growth, survival, yield, and reproduction of crawfish when the recommended quantities for farmers are applied in a rice/crawfish pond model; and second, to detect and quantify any residues of Propanil, Ordram, and Furadan which might accumulate in the flesh of surviving crawfish.

MATERIALS AND METHODS

A field study was conducted at the Louisiana State University Ben Hur Farm to address both objectives. To prevent interaction between tanks and pesticides, metal tanks used in this experiment were painted with epoxy paint 1 month before the experiment commenced. The bottom of each tank was covered with 25 cm of top soil. Lime was applied to each tank at the rate of 368 kg/ha, and urea and 0-24-24 (N:P:K) at 52 and 74 kg/ha respectively as recommended by the LSU Feed and Fertilizer Laboratory for growing rice.

Between 15 May and 17 May, 1979, Starbonnet variety of the long grain rice (Oryza sativa) was planted at a rate of 153 kg/ha in three sets of tanks with six tanks per set. Four of the six tanks in each set were 7.3 m^2 and two were 2.6 m^2. Two sets of tanks, designated as Treatments 1 and 2, were planted with Difolatan®-treated rice seeds. The third set of tanks was planted with untreated seeds and served as a control. Seeding followed the time and methods recommended in the USDA Agricultural Handbook No. 289 (1966). The soil was kept wet to facilitate rice germination; the actual flooding was initiated after rice plants had reached at least 7.5 cm in height. Levels of water in tanks were raised periodically (about 2 cm below plant tips) to keep up with rice growth. The tanks were modeled to simulate conditions in a rice/crawfish pond.

Red swamp crawfish (Procambarus clarkii) were introduced into tanks when rice plants reached a height of 10 cm. From 8 June to 11 June, 1979 crawfish, approximately 8 cm in length and 10-25 g in weight, were stocked into each tank. Large tanks were stocked with 25 males and 25 females, and small tanks with nine males and nine females at an overall stocking rate of 68,587/ha. The stocking rate here exceeded the maximum (50,000/ha stocked by Smitherman et al. (1967) because the purpose and duration of the experiments were different.

Four days after tanks were stocked, the water level in all tanks was lowered to 2.5 cm above soil level, and liquid Propanil® was applied to those tanks in Treatment 1 on 16 June 1979, at a rate of 6.8 kg AI/ha. One day after application of Propanil®, the water level in all tanks was raised to 15 cm which was the level maintained throughout the experiment. On 20 June 1979, 10% granular (10G) formulation of Ordram® was applied to Treatment 1 at the rate of 3.4 kg AI/ha, and Furadan® (3G) was applied to Treatment 1 on 23 June at 0.7 kg AI/ha. The application rates of all pesticides used were those recommended for farmers use.

Water was pumped occasionally into tanks following evaporation to maintain the 15 cm water level farmers usually maintain in rice fields. Temperature, dissolved oxygen, total hardness, total alkalinity, and pH of water in each tank were checked at approximately 2-week intervals. Lime was applied when necessary to maintain the appropriate water hardness (100 ppm or above, de la Bretonne et al. 1969) for crawfish growth.

Rice was harvested from all tanks between 20 and 27 September 1979, with stubble left standing in tanks as crawfish feed. Between 26 February and 6 March 1980, crawfish from all treatments were harvested, counted, measured, and weighed. Lengths, weights, and numbers of adult crawfish harvested from the control and the two treatments were analysed with an analysis of variance to determine differences in growth, survival, and yield among treatments. A chi-square test was used to verify if survival was equal in the three treatments. Adult crawfish from Treatment 1 and the control were sacrificed to obtain tail meat which was frozen in preparation for pesticide residue analysis. Each tank was also examined for the presence of juveniles which indicated reproduction. An unpaired t-test was used to determine if the lengths of juvenile crawfish produced in Treatment 2 were greater than those produced in the control.

Pesticide Residue Analysis

Many methods of analyses have been established for determining residue levels of Propanil, Ordram, and Furadan (Gordon et al. 1964; Onley and Yip 1971; Patchett et al. 1972; Cook 1973), but the gas chromatographic technique is preferred because of its selective nature which enhances repetition and accuracy (Cook 1973). The methods established thus far were designed to detect residues of only one pesticide at a time. No method so far has been established which can detect these three pesticides at the same time from a small sample as the one used in this analysis. The method and procedure used in this analysis were modified from that of Lawrence and Leduc (1977).

Apparatus and Operations

The gas-liquid chromatograph used was a Perkin-Elmer model 3920 equipped with a Perkin-Elmer nitrogen detector. This chromatograph has a 0.91 m glass column (4 mm internal diameter) packed with 3% OV-101 on Gas-Chrom Q (80-100 mesh). The Gas-Chrom provided support for the OV-101 which did the separation. The column temperature was maintained at 170° C, but other operational parameters varied with time. Adjustments were made to optimize response to pesticides versus base line noise as follows: The injection port temperature was set between 180° and 200° C. Range and attenuation combination were selected from 1 x 8 to 10 x 4. Valve readings on gas cylinders were 78 PSI for carrier gas (helium), 45 PSI for air, and 16 PSI for hydrogen. The detector temperature ranged from 650° to 800° C.

Sample extraction

Seventeen grams of crawfish tail meat from Treatment 1 and from the control were blended separately with 100 ml of acetone for 5 minutes with a Sorvall® homogenizer. The blended mixture was suction-filtered through 11 cm (diameter) Buchner funnel fitted with a shark skin filter. The blending cup, the filter paper, and the filter cake were thoroughly rinsed with acetone before the total filtrate was transferred into a 1,000 ml separatory funnel containing 100 ml of dichloromethane and 100 ml of hexane. The funnel was shaken vigorously for 2 minutes and the phases were allowed to separate. The aqueous (lower) phase was drawn into a 250 ml separatory funnel containing 15 ml of saturated sodium chloride solution. This solution was extracted twice further with 70 ml portions of dichloromethane. The organic extract was combined with the upper extract left in the 1,000 ml separatory funnel. The content of this funnel was transferred into a 1,000 ml round-bottomed flask through a funnel containing 5 cm of anhydrous sodium sulfate. The funnel and the sodium sulfate were rinsed three times with dichloromethane.

Each sample extract was evaporated by rotary evaporation at 30° C to 0.5 ml and transferred into a graduated centrifuge tube with three rinsings using 30% methylene chloride in hexane. The volume was adjusted to 5 ml before Florisil® clean up.

Clean-up

Five grams of deactivated Florisil® were placed in a 1.5 cm (internal diameter) glass buret containing a wad of glass wool at the bottom. About 1 g of anhydrous sodium sulfate was added to the top of the Florisil®. The column was rinsed with 50 ml of hexane before the addition of 1 ml of sample extract. The column was then eluted with 100 ml of 15% acetone in hexane. The beaker in which the filtrate collected was placed in a water bath at low temperature, and the content was allowed to evaporate to 1 ml. The sample was then transferred through anhydrous sodium sulfate into a graduated centrifuge tube with several rinses of the beaker and the sodium sulfate. The volume was reduced to 5 ml under nitrogen before analysis by gas chromatography. Out of this final volume, 5 µl was injected into the chromatograph.

RESULTS AND DISCUSSIONS

Mean lengths of adult crawfish harvested from Treatment 1, 2, and the control were 10.5, 11.1, and 10.5 cm respectively (Table 1). Analysis of variance showed no significant difference (P> 0.05) in growth among treatments. Lengths were used here as a measure of growth.

318

Table 1. Lengths, weights and sex of adult crawfish harvested from Treatments 1, 2, and the control.

No.	Treatment 1[a]			Treatment 2[b]			Control[c]		
	Length (cm)	Weight (g)	Sex[d]	Length (cm)	Weight (g)	Sex	Length (cm)	Weight (g)	Sex
1	12.5	57.51	M	12.5	65.40	M	12.5	55.60	M
2	11.7	44.6	M	9.6	34.40	M	12.1	52.40	M
3	9.8	39.04	M	10.8	36.48	M	12.1	52.44	M
4	9.7	37.90	M	12.3	44.20	M	11.1	36.09	M
5	9.5	34.52	M	11.5	35.82	M	12.0	52.81	M
6	9.8	36.18	M	10.9	34.94	F	12.5	62.91	M
7				10.1	30.71	F	10.5	30.64	M
8							10.8	33.40	M
9							11.4	43.16	M
10							11.1	55.10	M
11							9.8	24.23	M
12							10.0	25.90	M
13							10.2	27.16	M
14							9.2	22.11	M
15							10.6	34.28	M
16							12.9	53.51	M
17							9.7	29.40	M
18							9.1	31.87	M
19							9.1	18.89	F
20							9.0	22.80	F
21							10.3	30.36	F
22							9.8	36.97	F
23							10.0	27.27	F
24							10.0	24.93	F
25							9.2	20.85	F
26							10.0	27.15	F
27							9.2	20.75	F
28							10.1	23.82	F
Mean	10.5	41.64		11.1	40.3		10.5	34.50	

[a]Treatment 1 was planted with Difolatan®-treated rice seed and also received Propanil®, Ordram® and Furadan® at rates of 6.8 kg AI/ha, respectively.

[b]Treatment 2 was planted with Difolatan®-treated rice seed only.

[c]The control was planted with untreated rice seed only.

[d]F indicates female, M indicates male.

Four times as many adult crawfish survived in control tanks (28) as did in Treatment 1 (six) or in Treatment 2 (seven). All crawfish which survived in Treatment 1 were male, five out of seven adult crawfish that survived in Treatment 2 were male, and 18 out of 28 that survived in the control were male (Table 1). A chi-square test indicated no significant difference (P > 0.05) in survival among the three treatments. Analysis of variance showed a significant difference (P < 0.05) between survival of the two sexes favoring males; the reason for this is unclear since Muncy and Oliver (1963) found no differences in mortality of sexes subjected to 10 insecticides.

Yield of adult crawfish from all control tanks combined (966 g) was about four times that from Treatment 1 (249.8 g) and over three times that of Treatment 2 (282 g), but the differences were not statistically significant (P > 0.05). Yield in this context is given in terms of total weight of adult crawfish only.

No juvenile crawfish were produced in Treatment 1. Over 389 juvenile crawfish were found in one tank in Treatment 2 and 275 juvenile crawfish were produced from two tanks in the control. The lengths of juvenile crawfish produced in Treatment 2 ranged from 2.2 to 7.1 cm, and those from the control ranged from 2.5 to 3.7 cm. An unpaired t-test indicated that lengths of juvenile crawfish from Treatment 2 were significantly greater (P < 0.05) than those from the control.

Pesticide Residue Analysis

There were no peaks on the chromatograms to indicate the presence of Propanil, Ordram, or Furadan residues in either the treated sample or the control sample. Samples analyzed here were obtained more than 8 months after exposure to the pesticides. The lowest residue levels this technique could detect were 0.01 ppm of Propanil, 0.15 ppm of Ordram and 4.0 ppm of Furadan. Tolerance levels of these pesticides in crawfish meat were not available for reference. The detectable residue level here is less than the 2 ppm of Propanil and more than the 0.1 ppm of Ordram allowed in rice grain by the EPA (Smith et al. 1977). The lowest detectable level of Furadan is more than the tolerance level of 0.02 ppm for fat, meat, and other cattle products (Code of Federal Regulation 1979). The limited sample of crawfish flesh used made it impossible to obtain more than one chromatogram, thus rendering detection level of Furadan relatively high.

In another study by the authors where a greater quantity of crawfish was exposed to 0.5 ppm of Furadan for 31.5 days and put through the same processes, Furadan could not be detected at 0.5 ppm. The extraction and clean up techniques used in this analysis were modified from that published by Lawrence and Leduc (1977). The meat was blended for 5 minutes instead of 4, and filtered through 11 cm (internal diameter) buchner funnel fitted with a shark skin filter. Since the lowest detectable level of Furadan is more than the tolerance level of 0.02 ppm for fat, meat, and other cattle products, the possibility of lower levels of pesticide accumulation should not be overlooked. Much lower levels can be detected by this method of analysis using an even greater quantity of meat. It is necessary to detect lower levels of these pesticides to determine if there is an accumulation, so consumers know what they eat. This is especially important when one considers that a consumer may be undergoing therapy with drugs and may show a poisoning response to accidental consumption of pesticides in food. Extremely small doses of pesticides ingested as residues in food may not affect drug metabolism, but those who are persistently exposed to high intakes of pesticides may respond adversely to drugs. Fishbein (1978) remarked that pesticide residues in human food may constitute health hazards. Even though Menn (1978) gave the consolation that animals (including humans) have efficient excretory systems that help to eliminate pesticides and their transformation products through urine and feces, great care needs to be taken not to overload the system.

SUMMARY AND CONCLUSIONS

1. No significant differences were detected in crawfish growth among the control (planted with untreated rice seed), Treatment 1 (planted with Difolatan®-treated rice seed and also received applications of Propanil®, Ordram®, and Furadan®), and Treatment 2 (planted with Difolatan®-treated rice seed only).

2. Only male crawfish survived the combination of the four pesticides, but both male and female crawfish survived the Difolatan® treatment and the control. There was a significant difference ($P < 0.05$) between the survival of the two sexes.

3. There was no significant difference ($P > 0.05$) in yield in total weight among treatments.

4. Failure to detect differences in growth, survival, and yield among treatments was probably due to high variability in test results. Further tests with less variability will be required to determine if significant differences actually exist.

5. The complete kill of female crawfish in most tanks prohibited reproduction. Any of the chemicals, or their combinations could have caused this complete kill of female crawfish.

6. Residues of Propanil, Ordram, and Furadan were not detected at the limit of sensitivity of the instrument 8 months after exposure to the chemicals.

7. Conclusions reached here hold true for a rice/crawfish pond model and in the laboratory under conditions in this study; some may and some may not apply in the actual rice/crawfish farm. Conclusions about the field experiment did not take into consideration the breeding condition and genetic factors of individual crawfish, number of rice plants, and the concentration of dissolved oxygen in each tank which might have had some effects on growth, survival, yield, and reproduction.

LITERATURE CITED

Cheah, M.L., J.W. Avault, Jr., and J.B. Graves. 1979-80. Some effects of rice pesticides on crawfish. La. Agr. 23(2):8-9 and 11.

Chien, Y.H. 1978. Double cropping rice Oryza sativa and crawfish Procambarus clarkii (Girard). Masters Thesis. LSU. 84 p.

Code of Federal Regulations. 1979. 40 CFR 180.254.

Cook, R.R. 1973. Carbofuran. Analytical method for pesticides and plant growth regulators. Academic Press, New York. 7:187-210.

de la Bretonne, L., J.W. Avault, Jr., and R.O. Smitherman. 1969. Effects of soil and water hardness on survival and growth of red swamp crawfish, Procambarus clarkii, in plastic pools. In Proc. Ann. Conf. Southeastern Assoc. of Game and Fish. Commissioners. 23:(629-633).

Fishbein, L. 1978. Overview of potential mutagenic problems posed by some pesticides and their trace impurities. Environ. Health Perspect. 27:125-131.

Gordon, C.F., A.L. Wolfe, and L.D. Haines. 1964. Stam. Analytical methods for pesticides and plant growth regulators. Academic Press, New. 4:235-241.

Graves, J.B., K.M. Hyde, F.L. Bonner, P.E. Shilling, and J.F. Fowler. 1977. Effect of mirex on crawfish production in rice fields. La. Agr. 20(2):8-9 and 11.

LaCaze, C. 1976. Crawfish farming, revised. Louisiana Wildl. and Fish Comm., Fish Bull. No. 7. 27 p.

Lawrence, J.F. and R. Leduc. 1977. Direct analysis of carbofuran and two nonconjugated metabolites in crops by high-pressure liquid chromatography with UV absorption detection. J. Agr. Food Chem. 25(6):1362-1365.

Menn, J.J. 1978. Comparative aspect of pesticide metabolism in plants and animals. Environ. Health Perspect. 27:113-124.

Muncy, R.J. and A.D. Oliver, J. 1963. Toxicity of 10 insecticides to the red swamp crawfish, Procambarus clarkii (Girard). Trans. Am. Fish. Soc. 92:428-431.

Onlye, J.H. and G. Yip. 1971. Herbicidal carbamates: extraction, clean up, and gas chromatographic determination by thermionic, electro capture, and flame photometric detectors. JAOAC 54(6):1366-1370.

Patchett, G.G., D.L. Shelman, and W.J. Smith. 1972. Ordram. Analytical methods for pesticides and plant growth regulators. Academic Press, New York. 6:668-670.

Ridgway, R.L., J.C. Tinnery, J.T. MacGregor, and N.J. Starler. 1978. Pesticide use in agriculture. Environ. Health Perspect. 27:103-112.

Smith, R.J., Jr., W.T. Flinchum, and D.E. Seaman. 1977. Weed control in U.S. rice production. U.S. Dept. Agr., Agr. Handb. 497, 78 p.

Smitherman, R.O., J.W. Avault, Jr., L. de la Bretonne, Jr., and H.A. Loyacano. 1967. Effects of supplemental feed and fertilizer on production of red swamp crawfish, <u>Procabarus clarkii</u>, in pools and ponds. In Proc. Ann. Conf. Southeastern Assoc. of Game and Fish. Commissioners. 21:452-458.

Soil Conservation Service. 1980. Billy Craft, compiler. Inventory of crawfish farmers.

USDA. 1966. Agr. Handb. 289:84-91.

TOXICITY OF SELECTED RICE HERBICIDES TO RED SWAMP CRAWFISH, PROCAMBARUS CLARKII (GIRARD)

C. G. Lutz and R. P. Romaire
Fisheries Section,
School of Forestry and Wildlife Management
Louisiana State University
Baton Rouge, Louisiana 70803, U.S.A.

ABSTRACT

Multiple cropping of crawfish (Procambarus clarkii) with other agricultural crops such as rice and soybeans is gaining wide spread acceptance in Louisiana, Mississippi, and Texas. However, a major problem in rotating crawfish with other agricultural crops or culturing crawfish on land used previously in other agronomic endeavors is the deleterious effects of agricultural pesticides on crawfish. Three herbicides - oxadiazon (Ronstar®), bentazon (Basagran®), and bifenox (Modown®) - presently labelled for weed control in rice were evaluated for their acute lethal toxicity to juvenile (25-40 mm total length) red swamp crawfish.

EFFECTS OF FORMULATED DIETS ON TWO SPECIES OF CRAYFISH:
AUSTROPOTAMOBIUS PALLIPES (L.) and PACIFASTACUS LENIUSCULUS (DANA) UNDER LABORATORY CONDITIONS

R. Fernández, C. López-Baissón, L. Ramos[1] and L. Cuellar[2]

ABSTRACT

Two different crayfish species were fed artificial diets with and without supplementation of fresh food, to determine the maintenance and survival of both species of crayfish under laboratory conditions in the absence of natural food.

The crayfish P. leniusculus survived and molted successfully with and without supplementation of the artificial diet. On the other hand, A. pallipes fed the diet without supplementation were unable to survive. Dead animals showed symptoms of Aphanomyces disease and hyphae of the crayfish plague fungus (Aphanomyces astaci) were found in these animals when examined at the microscope.

INTRODUCTION

Nutritional deficiencies observed in most crustaceans fed exclusively with artificial diets makes it necessary to supplement with some kind of fresh food if natural food sources are unavailable. These deficiencies seem to be related to active factors or growth- promoting substances (Foster and Beard, 1973; Fernández and Puchal, 1979). The necessity for identification and isolation of these substances is quite clear. Recently, Conklin et al. (1980) and Kanazawa et al. (1979) working with different diets and two different crustacean, lobster and shrimp respectively, identified lecithin as one of these factors (in the case of lobster, inclusion of lecithin was critical for survival).

On the other hand, dependence for these factors seems to vary according to different crustacean species, showing symptoms of nutritional deficiencies in different ways. While lobsters are unable to survive, shrimp growth is reduced drastically in relation to several fresh foods: mussel, clam, Artemia salina, etc. (Kittaka, 1975; Kanazawa et al., 1970).

The purpose of this work is to analyze the response of two less nutritionally studied crustaceans: crayfish P. leniusculus and A. pallipes, to a formulated diet with and without supplementation.

MATERIALS AND METHODS

One hundred juveniles P. leniusculus from Quiñon hatchery and 100 juveniles A. pallipes from Icona facilities at El Chaparrillo (Cuidad

[1]Quiñon S.A. - San Esteban de Gormaz. Soria. Spain

[2]Instituto Nacianal para la Conservación de la Naturaleza (ICONA). Gran Vía San Francisco, 35 Madrid-5. Spain.

Real) were divided in eight groups of 25 individuals averaging 14,5+0,6 mm. in cephalothorax length.

Two troughs of 5x1x0,15 m. were parcelled into four identical compartments with continuously passing-through water at the level of 2 litre/min.

The diet chosen for the experiment was a commercial one, elaborated with typical feed ingredients. Roughly it consisted of: fish meal 16%, soybean meal 20%, casein 5%, wheat flour 30%, corn starch 20%, lipid mix 4%, vit. mix 2%, mineral mix 3%. The diet was fed daily in excess to all groups. Supplement elected was fresh liver, fed once a week to only four groups of crayfish (two groups P. leniusculus and two groups A. pallipes).

The experiment lasted 11 weeks with the following experimental conditions: Temperature, 11°C. at the beginning 14,5°C at the end. Dissolved oxygen, 7 ppm. pH, 7,4. Photoperiod, 12 h.

Because Aphanomyces is an important consideration when working with crayfish, it is important to point out that water was taken from a spring in which a natural population of native crayfish A. pallipes graze.

RESULTS AND DISCUSSION

As can be seen from Table 1 and Figure 1, responses of both crayfish to the formulated diet vary in relation to the presence or absence of fresh liver in the ration. P. leniusculus fed exclusively the pelleted diet showed no statistical differences in survival relative to the ones receiving the supplement. All of them molted at least once during the experiment. Nevertheless, due to low temperatures, growth was generally poor and it was not possible to detect significant differences.

In contrast fresh liver was critical for A. pallipes survival. Absence of fresh food in the ration in conjunction with a slow increase in temperature resulted in mass mortality of crayfish in one week.

Close observation of dead animals showed symptoms of Aphanomyces infection. Posterior examination at the microscope of brown stained parts and soft membranes of the exoskeleton confirmed the presence of hyphae of the crayfish plague fungus Aphanomyces astaci.

A. pallipes is a known species for his lethal susceptibility to the crayfish plague fungus. Considering the origin of water, outbreak of the disease could be expected. What was surprising was the ability of A. pallipes to resist the fungus infection when receiving a supplement of fresh food once a week.

According to Unestam and Weiss (1970) and Unestam et al. (1977), the transfer from Astacus to Astacus is always a lethal infection, unless low-virulent laboratory strains of fungus are used. Nevertheless, infection was natural in this case.

326

Table 1. Survival of two crayfish species (P. leniusculus and A. pallipes) fed artificial diet with and without supplementation.

Species	Replicate	% Survival	
		with suppl.	without suppl.
A. pallipes	1	92%	16%
	2	96%	12%
P. leniusculus	1	92%	88%
	2	96%	96%

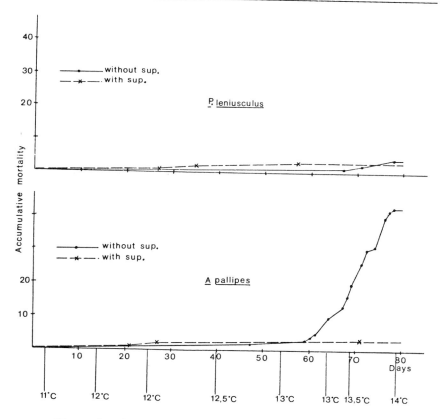

Figure 1. Accumulative mortality along the experiment.

Further experiments are being carried out trying to complement these results.

LITERATURE CITED

Conklin, D.E., L.R. D'abramo, C.E. Bordner and N.A. Baum. 1980. A successful purified diet for the culture of juvenile lobsters: the effect of lecithin. Aquaculture, 21.

Fernádez, R. and F. Puchal. 1979. Studies on compounded diets for Penaeus kerathurus shrimp. Proc. World Mericul. Soc. 10:781-787.

Foster, J.R.M. and T.W. Beard. 1973. Growth Experiments with the Prawn Palaemon serratus Fed with Fresh and Compounded Foods. Fish. Invest Ser. 11,22(7), 16 p.

Kanazawa, A., M. Shimaya, M. Kawasaki and K. Kashiwada. 1970. Nutritional requirements of prawn - 1: Feeding on artificial diet. Bull Jap. Soc. Sci. Fish, 37:711-715.

Kanazawa, A., S. Teshima, S. Tokiwa, M. Endo and F. Abdel Razel. 1979. Effects of short necked clam phospolipids on the growth of prawn. Bull. Sap. Soc. Sci. Fish, 45:961-965.

Kittaka, J. 1975. Food and growth of penaeid shrimp. Proc. First Int. Conf on Aquaculture Nutrition, p. 149-285.

Unestam, T., K. Söderhäll, L. Nyhlen, E. Svensson and R. Ajaxon. 1977. Specialization in crayfish defence and fungal aggressiveness upon crayfish plague infection p. 321-331. In O.V. Lindquist (ed.) Freshwater crayfish III, Kuopio, Finland (1976). 504 p.

Unestam, T. and D.W. Weiss. 1970. Studies on the host parasite relationship between freshwater crayfish and the crayfish disease fungus Aphanomyces astaci. Responses to infection by a susceptible and a resistant species. J. Gen. Microbiol. 60:77-90.

VI
PRODUCTION
PROBLEMS OF CRAYFISH

CRAWFISH, PROCAMBARUS CLARKII, GROWTH AND DISPERSAL
IN A SMALL SOUTH LOUISIANA POND PLANTED WITH RICE, ORYZA SATIVA

John F. Witzig,[1] and James W. Avault, Jr.
Fisheries Section, School of Forestry and Wildlife Management
Louisiana State University
Baton Rouge, Louisiana 70803

Jay V. Huner
Department of Biology
Southern University
Baton Rouge, LA 70913

ABSTRACT

Crawfish were cultured in a 0.8 ha earthen pond flooded on 29 September 1978. Unharvested rice served as the energy source and cover. Peak vegetative biomass was 6710 kg ± 178 kg/ha (dry wt.). Decomposition was linearly related to time after flooding. Crawfish production was 1266 kg/ha (wet wt.). Crawfish distribution was significantly affected by water depth, vegetation density, pond area, and time after flooding. Interactions between the four factors presented a complex picture of changing crawfish distribution over time in relation to resource availability within the pond. We found that shallow water (0.6-0.8 m), dense vegetation (greater than 646 g/m^2, dry wt.) and center areas had a higher mean yield of crawfish than did the deep water (0.8-1.2 m), light vegetation (less than 646 g/m^2, dry wt.) and edge areas. Roughly 200 days after flooding, distinct differences between preference for deep and shallow areas disappeared. A significant quadratic relationship between crawfish weight and time after flooding was found. The regression equation indicates that mean crawfish weight will begin to decline in March, half way through the trapping season. Feeding, to avoid stunting, should begin no later than mid-February.

INTRODUCTION

Commercial crawfish ponds are managed as self-sustaining systems with each pond producing an annual havestable crop and a breeding stock for the following year (Huner and Avault 1976). The ponds are controlled in such a way that the natural annual hydrologic regime is maintained in the pond system. In southern Louisiana the hydrologic cycle can be divided into a wet season when there is a water surplus and a dry season when there is no surplus. The dry season coincides with the summer months and the wet season with the rest of the year (Byrne et al. 1976). Commercial ponds are operated under a similar pattern of draining and reflooding. They are drained in late spring or early summer and reflooded in the autumn (Huner and Barr 1981). During the dry summer months a forage crop of annual grasses and sedges or cultivated rice is grown in the pond. Additionally, many ponds are dominated by perennial, semi-aquatic forbs, the principal species being

---

[1]Present Address: National Marine Fisheries Service, Beaufort, NC 28516 USA.

alligator weed, Alternanthera philoxeroides. The life history of the commercially important crawfish species, Procambarus clarkii and Procambarus acutus acutus, is intimately associated with the annual hydrologic cycle (Penn 1943, 1956; Huner, 1976).

Changes in the amount and type of vegetation present in the ponds results in an alteration in the basic arrangement of the food chain within the pond and the availability of cover and substrate for crawfish. During the autumn, the most obvious characteristic of a pond dominated by annual vegetation is the abundance of standing vegetation. This situation changes over the course of a few months to a situation in which there is little or no standing vegetation by mid-spring. The combined actions of microbial mineralization and detritivore, principally crawfish, grazing are responsible for this situation (Huner and Barr 1981, Miltner and Avault 1981). This elimination of vegetation has been associated with stunting in crawfish populations at a time when wild crops of crawfishes normally depress prices even for the largest of crawfishes (Avault et al. 1975). There are, however, few quantitative data on the absolute amount of vegetative biomass in ponds and the relationships between that vegetation and crawfish growth in ponds. In addition, crawfish prefer areas where vegetative cover is available and do move about in ponds (Huner and Barr 1981), but, again, few data are available on spatial and temporal distribution of crawfish in crawfish ponds.

The purposes of this study were to: 1) estimate the standing vegetative biomass in a small crawfish pond dominated by annual vegetation over time and predict its rate of removal; 2) describe spatial and temporal changes in crawfish distribution with respect to changes in vegetative biomass; and 3) address the relationship between vegetative biomass and crawfish growth.

MATERIALS AND METHODS

Pond B-3 is located at the Louisiana State University Ben Hur Farm. It is impounded by levees and is rectangular in shape (148 m by 54 m) with a surface area of 0.8 ha. The bottom is flat with a slight grade from the north and to the south end. Water ranges from a maximum of 1.2 m in the south end to 0.6 m in the north end. The sides slope sharply toward the bottom creating a 4 m wide transition zone of rapidly increasing water depth around the perimeter.

Since construction in 1974 pond B-3 has been managed using typical methods for commercial crawfish ponds. During the 1977-78 crawfish season 675 kg of crawfish were harvested. Over 99% were P. clarkii with the small remainder being P. a. acutus. The pond was drained at the end of the trapping season and replanted with 250 kg of weevil-spoiled rice seed during the last week of July. Three-cornered sedge, Cyperus spp., also contributed to the vegetative cover. Reflooding commenced on September 29, 1978 and was completed by mid-October.

The pond was divided into an interior and an edge area. Eighty percent (6383 m^2) of the bottom area was included in the interior zone. The interior was divided into 20 (28 m by 12 m) sampling areas each of which was further divided into quadrants. The edge zone was divided into 18 sampling 28 m by 4 m and 12 m by 4 m along the long and short

axes of the pond, respectively. The edge area was considered to be the transition zone; approximately 20% (1616 m^2) of the bottom area was included in this zone.

Vegetation

Sampling. Density of above ground vegetation was estimated periodically. Systematic vegetation cuttings (929 cm^2) were taken on September 25, and on an approximate biweekly basis thereafter through March 1979. Fifty-eight cuttings were collected on September 25. Single samples were taken from each edge section and from two different quadrants within each interior section. Between October 1, 1978 and January 1, 1979, 38 cuttings were taken, one sample from each edge and interior section. Location of sampling points within a section or quadrant was random. After January 1, 10 biweekly samples were taken. Sampling was discontinued after March because by then most of the vegetation had disintegrated.

All samples were dried in a forced air oven for 48 hours at 75°C and weighed within three hours after removal from the oven. Samples which could not be dried immediately after cutting were frozen until they could be processed. Results are expressed as grams of dry matter per m^2.

The two pond areas were divided into two vegetation density zones based on samples collected in October and November (Fig. 1). SYMAP (Dougenik and Sheehan 1975) was employed to delineate these zones. The low density zone had less than 646 g of vegetation (dry weight) per m^2; the high density zone had greater than 646 g of vegetation (dry weight) per m^2. The high and low density zones comprised 52% and 48%, respectively, of the entire pond area.

Total dry weight of vegetation in the pond was estimated by dividing the pond into 5 density zones, and multiplying the area in each zone by the average grass weight in each zone.

Statistical Analysis. Two-way analysis of variance was performed to test differences in dried vegetation weights of samples between sample periods. Simple linear regression techniques using least squares means of the vegetation for each sampling period were used to obtain equations for predicting average vegetation weight in two vegetation density categories over time. The first two sampling dates were excluded from regression analysis because the mean vegetation weights were significantly less than the mean vegetation weights for the third through fifth sample dates ($P < 0.01$).

Crawfish

Sampling. Crawfish distribution with respect to vegetation density, pond area, and water depth was studied during harvesting of the pond. The pond was divided into two areas (edge and interior), two vegetation zones (less than 646 g per m^2 and greater than 646 g per m^2, as discussed above) and two water depths. The southern half of the pond served as the deep water zone (depth 0.8 to 0.6 m). Thus, there were 8 strata or factor combinations.

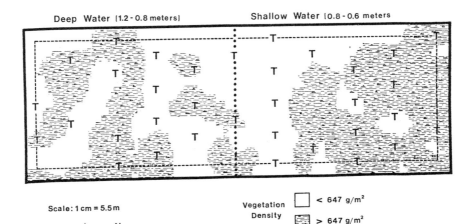

Deep Water (1.2 - 0.8 meters)   Shallow Water (0.8 - 0.6 meters)

Scale: 1 cm = 5.5 m

◄ ━ ◆ ━ ► N

Vegetation
Density   □ < 647 g/m²
               ▨ > 647 g/m²

T = Crawfish Trap Locations

Figure 1. Vegetation density map for pond B-3 with crawfish trap locations for 1979 trapping season. The two density categories were determined from vegetation samples taken during October and November 1978.

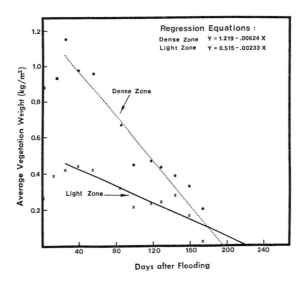

Regression Equations :
Dense Zone   Y = 1.219 - .00624 X
Light Zone   Y = 0.515 - .00233 X

Figure 2. Relationship of average vegetation weight to the number of days after the pond was flooded for two vegetation density categories (day 0 = September 29, 1978). Vegetative biomass (dry weight) during October and November was > 646 g/m² for the dense zone and < 646 g/m² for the light zone. X's and ·'s indicate observed average weight for the light and dense zones respectively.

Trapping began in late January and continued until June. Thirty-four stand up funnel traps constructed of 7.5 mm chicken wire were used throughout the study (Bean and Huner 1979). Trap locations are indicated in Figure 1. All strata had at least two traps. Traps were baited with 0.25 kg of fish scraps in the afternoon and run 16-24 hours later. The number of crawfish caught in each trap was recorded and total catch was weighed after each run.

Statistical Analysis. The effect of three environmental factors on crawfish distribution was tested with a two-way analysis of variance procedures for a 2 x 2 x 2 factorial arrangement of treatments. Least squares means were estimated for each factor and the significant factor interactions. Thirty trap runs were included in the analysis.

RESULTS AND DISCUSSION

Vegetation

Estimates of average vegetation density over time for the two vegetation strata are presented in Figure 2. The regression coefficients for the equations for both vegetation densities were highly significant ($P < 0.01$). A test of homogeneity of the regression coefficients showed that there was a highly significant difference between the two equations. The r^2 values for the two equations are 0.82 and 0.88 for the light and dense zones, respectively, indicating that the number of days after flooding account for much of the variation in the average vegetation weights in the two vegetation zones.

The average vegetation density in the two density zones increased for 4 weeks after flooding and then declined at a linear rate. The linear regression equations were used to predict the average vegetation weight remaining in an area of the pond using the number of days after the pond was flooded (DAF) as the independent variable. Approximately 180 days after flooding, the two areas contained the same density of vegetation, and 220 days after flooding all vegetation was virtually gone.

Peak vegetative biomass for the entire pond was estimated at 5368 ± 143.1 kg (dry weight). This represents roughly 676 g/m^2 which falls in the range of 500-1200 g/m^2 reported (Miltner and Avault 1981) from experimental ponds planted with rice or millet to serve as crawfish forage. It is well below the 1000-1200 g/m^2 cutoff reported to represent a significant BOD (biological oxygen demand) danger in crawfish ponds (Morrissy 1979). Therefore, it is not surprising that we encountered no significant dissolved oxygen depletions during the course of our study.

Vegetation density declined three times faster in the densely vegetated zones than in the lightly vegetated zone. Cameron (1972) showed that herbivore diversity in a salt marsh was greatest when plant biomass was greatest. Crawfish have also been found to prefer densely vegetated areas over lightly vegetated areas (Covich 1978). These findings indirectly suggest that the densely vegetated areas in a crawfish pond should contain proportionally more organisms than the sparsely vegetated zones. Thus, the nutritional resources are depleted

335

at a faster rate in the dense vegetation zone than in the light zone.

Crawfish Dynamics

Water depth, vegetation density, trap location, and trapping date had significant effects (P < 0.05) on crawfish distribution. An explanation of these results is complicated by significant two-factor interactions (water depth x pond area; pond area x vegetation density; trapping date x water depth). Tables 1, 2, and 3 list the mean numbers of crawfish caught in all two-factor strata. The marginal averages are the least squares estimates of the main effects. The reliability of these means as measured of the true main effects is decreased because of the presence of the significant interactions. The analysis does imply that the shallow water, dense vegetation, and center areas had a higher mean yield of crawfish than did the deep water, light vegetation, and edge areas.

The presence of significant interactions in any experiment indicates that the responses of the variable being measured is not consistent over all levels of the factor being tested. This is obvious in distribution of crawfish in response to water depth - pond area, and pond area - vegetation density combinations (Tables 1 and 3, respectively). The interaction of water depth x pond area was shown by the switch from a higher mean number of crawfish being caught in the deep water - edge strata (16.8 crawfish per trap) than in the deep water - center strata (16.0 crawfish per trap) to a higher mean number of crawfish being caught in the shallow water - center strata (22.9 crawfish per trap) than in the shallow water - edge strata (18.9 crawfish per trap) (Table 1). A similar situation was found in the interaction of vegetation density x pond area in which higher mean numbers of crawfish were caught in the light vegetation - center zone and the dense vegetation - edge zone. The apparent preference of the crawfish for the edge in deep water may not be due to an edge being present but due to the fact that the edge area is shallower than the center area and there is an actual preference for shallow water.

Figure 3 is illustrative of the interaction between date (DAF) and water depth. Throughout most of the season, the harvest from the deep water zone is less than that from the shallower zone. Iterative divisions of the season into an early and late period showed that a critical period occurs between 200 and 210 days after flooding (April 15-25). Prior to this critical period, the crawfish had a highly significant (P < 0.01) preference for the shallow zone; after this period they showed no preference for either depth. Water depth ceased to be an important factor in controlling crawfish distribution 200 DAF - the point at which most of the vegetation had disappeared. The more even distribution of the crawfish throughout the two water depths during the latter part of the season may have been caused by depletion of the food resources in the shallow water and subsequent movement of some crawfish into the deeper water where forage was available and where there were fewer competitors for a dwindling food supply. The effect of water depth during the early trapping period may also have been because of the planting history of the pond. During the 1977-78 season, only the shallower northern half of the pond was planted with rice and densely vegetated. Thus, the early crawfish distribution may have been skewed away from the deep water zone. Finally, temperature could also affect distribution as it rises in the spring.

Table 1. Average number of crawfish captured in four water depth - pond area strata in pond B-3, during the 1978-79 season.[1]

Water Depth Strata[2]	Pond Area Strata[3]		Mean
	Edge	Interior	
Deep	16.82 ± 0.91 (135)[4]	15.99 ± 0.46 (401)	6.40 ± 0.51 (536)
Shallow	18.94 ± 0.74 (163)	22.87 ± 0.44 (434)	20.91 ± 0.26 (835)
Mean	17.88 ± 0.59 (298)	19.43 ± 0.32 (835)	19.55 ± 0.26 (1133)

[1]Averages for each strata and the marginal averages are expressed as the least squares estimate of the mean ± 1 standard error of the least square mean.

[2]Water Depth: Deep = 1.2-0.8 m; Shallow = 0.8-0.6 m.

[3]Pond Area: Edge = peripheral 4 meter area from the water's edge toward the center (20% of the total pond area); Interior = remainder of the pond (80% of the total pond area).

[4]Number in ( ) is the number of samples taken in each stratum.

Table 2. Average number of crawfish captured in four vegetation density - pond area strata in pond B-3, during the 1978-79 season. [1]

| Vegetation Density Strata[2] | Pond Area Strata[3] | | Mean |
	Edge	Interior	
Light	15.07 ± 0.98 (97)[4]	19.30 ± 0.64 (437)	17.18 ± 0.54 (534)
Dense	20.69 ± 0.65 (210)	19.56 ± 0.46 (398)	20.12 ± 0.40 (599)
Mean	17.88 ± 0.59 (298)	19.43 ± 0.32 (835)	19.55 ± 0.26 (1133)

[1]Averages for each strata and the marginal averages are expressed as the least squares estimate of the mean ± 1 standard error of the least square mean.

[2]Vegetation Strata: Light < 646 g/m^2 dry weight during October and November 1978. Dense > 646 g/m^2 dry weight during October and November 1978.

[3]Pond Area: Edge = peripheral 4 meter area from the water's edge toward the center (20% of the total pond area); Interior = remainder of the pond (80% of the total pond area).

[4]Number in ( ) is the number of samples taken in each stratum.

Table 3. Average number of crawfish captured in four water depth - vegetation density strata in pond B-3, during the 1978-79 season.[1]

Water Depth Strata[2]	Vegetation Density Strata[3]		Mean
	Edge	Interior	
Deep	15.02 ± 0.86 (233)[4]	17.18 ± 0.56 (401)	16.40 ± 0.51 (597)
Shallow	19.35 ± 0.65 (301)	22.46 ± 0.57 (296)	20.91 ± 0.44 (536)
Mean	17.18 ± 0.54 (534)	20.12 ± 0.40 (599)	19.55 ± 0.26 (1133)

[1]Averages for each strata and the marginal averages are expressed as the least squares estimate of the mean ± 1 standard error of the least square mean.

[2]Water Depth: Deep = 1.2-0.8 m; Shallow = 0.8-0.6 m.

[3]Vegetation Strata: Light < 646 g/m^2 dry weight during October and November 1978. Dense > 646 g/m^2 dry weight during October and November 1978.

[4]Number in ( ) is the number of samples taken in each stratum.

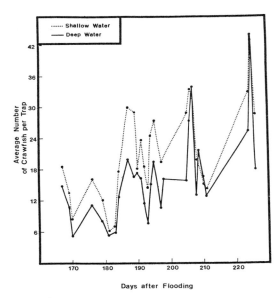

Figure 3. Interaction of water depth and number of days after flooding in pond B-3, January - May 1979. The two water depths used were 1.2-0.8 m and 0.8-0.6 m. (The average is expressed as the least squares estimate of the mean for each water depth on each trapping run).

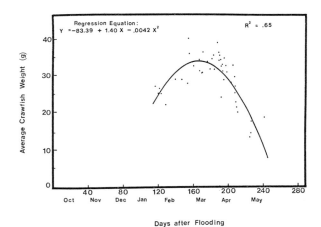

Figure 4. Change in the average crawfish weight harvested from pond B-3 as a function of the number of days after initial pond flooding. The regression curve indicates the predicted average weight on each day; the •'s indicate actual observations. The regression equation is given in the figure.

Crawfish are generally considered to be harvestable at lengths greater than 75 mm (12 g) (Huner and Barr 1981); however, larger crawfish are more marketable. A major problem for the commercial farmer is that crawfish show a great range of size during harvest and generally become smaller toward the end of trapping season (Huner 1978, Huner and Romaire 1979). This trend is apparent in pond B-3 (Figure 4). The quadratic relationship between mean weight and number of days after flooding was highly significant (P < 0.01; r^2 = 0.65). The regression equation given in the figure predicts that a peak in mean weight will occur approximately in mid-March (167 DAF) and decline thereafter. Crawfish harvested during the early part of the season were either young-of-the-year just reaching harvestable size or holdovers from the previous season (juveniles and adults).

The decline in size of crawfish has usually been attributed to reduction in the amount of food available. Supplemental feeding may be of use in reducing late season stunting (Goyert and Avault 1977; Huner and Barr 1981). The quadratic relationship between mean size of crawfish and DAF indicates that mean crawfish weight began declining midway through the trapping season (167 DAF) at which time the mean vegetation weight was approximately 150 g/m^2 (dry weight) (predicted from equations in Figure 2). This is, however, a delayed response. Potential growth of most organisms is determined early in their development; thus, what a culturist sees as stunting 167 DAF was probably the end result of a process that began several weeks prior to the first appearance of stunted animals in the catch. Therefore, should supplemental feeding be advisable and effective, then it should begin well before stunted crawfish are caught and probably no later than early February.

Total yields of crawfish from pond B-3 was 1032 kg (1266 kg/ha). A ratio of 0.1653 ± 0.0395 was calculated for conversion of dry grass biomass to crawfish yield over the entire trapping season. Approximately 6 grams of dried vegetation were required to produce 1 gram of crawfish (wet weight). This assumes that all crawfish were caught which they were not. Therefore, the actual conversion ratio was higher. Unharvested crawfish were removed by various predators, migrated from the pond, or remained in burrows following pond draining. We feel that this did not exceed 100 kg total.

Our "food conversion" ratio is higher than the usually acceptable conversion ratio of 0.1 between trophic levels (Odum 1971). This discrepancy arises because we used a ratio comparing dried vegetation to live crawfish weight rather than dry weight to dry weight as is usually used.

The conversion ratio of dried vegetation biomass to crawfish calculated in this study is approximately double that calculated for the large Australian crawfish, Cherax tenuimanus (Morrissy 1979). Morrissy used artificial grain-based feeds to promote detritus formation rather than natural vegetation. He normally applied the material once at the beginning of a study. This would suggest that crawfish may be more efficient in utilizing natural feeds than nutrient rich, artificial grain-based feeds applied to form detritus. However, a logical explanation for this phenomenon would involve the relatively rapid decomposition of uneaten, nutrient rich feeds with rapid loss to foraging crawfish. Terrestrial vegetation decomposes much more slowly

341

(Goyert and Avault 1977; Miltner and Avault 1981) and would be available for a longer period of time. This factor, then, should be considered when deciding the economic feasibility of using supplemental feeds to prevent stunting if one time application of detritus generating material is to be made. It should be noted that high protein, animal feeds are quickly consumed by crawfishes with rapid effect on growth but use of such feeds has not yet been shown to be economically feasible in extensive crawfish culture systems in the USA (Huner and Barr 1981).

ACKNOWLEDGMENTS

This study was funded, in part, by the Louisiana Agricultural Experiment Station and the Louisiana Sea Grant Program.

LITERATURE CITED

Avault, J. W., Jr., L. W. de la Bretonne, Jr. and J. V. Huner. 1975. Two major problems in the culture of crayfish in ponds, pp. 149-157. In: J. W. Avault, Jr. (ed.) Freshwater Crayfish II, Baton Rouge, LA (1978).

Bean, R. A. and J. V. Huner. 1979. An evaluation of selected crawfish traps and trapping methods, pp. 141-151. In: P. J. Laurent (ed.) Freshwater Crayfish IV, Thonon-les-Baines, France (1978). 473 pp.

Byrne, P., M. Boregasser, G. Drew, R. Mueller, B. L. Smith, Jr. and C. Wax. 1976. Barataria Basin: Hydrologic and climatologic processes. Louisiana State University, Center for Wetland Resources, Baton Rouge, LA, Sea Grant Pub. No. LSU-T-76-010.

Cameron, G. N. 1972. Analysis of insect trophic diversity in two salt marsh communities. Ecology 53:58-73.

Covich, A. P. 1978. Activity analysis of radio-monitored crayfish. Bull. Ecol. Soc. Am. 59:70 (abstract).

Dougenik, J. A. and D. E. Shechen. 1975. SYMAP user's reference manual. 5th ed. Laboratory for Computer Graphics and Spatial Analysis. Graduate School of Design, Harvard University, Cambridge, MA.

Goyert, J. C. and J. W. Avault, Jr. 1977. Agricultural by-products as supplemental feed for crayfish, Procambarus clarkii. Trans. Am. Fish. Soc. 106: 629-633.

Huner, J. V. 1976. Observations on the life histories of recreationally important crawfishes in temperate habitats. Proc. Louisiana Acad. Sci. 38:20-24.

Huner, J. V. 1978. Crawfish population dynamics as they affect production in several small, open crawfish ponds in Louisiana. Proc. World Mariculture Soc. 9:619-640.

Huner, J. V. and J. W. Avault, Jr. 1976. Sequential pond flooding: A prospective management technique for extended production of bait size crawfish. Trans. Am. Fish. Soc. 105:637-642.

Huner, J. V. and J. E. Barr. 1981. The red swamp crayfish. Biology and exploitation. Louisiana State University, Center for Wetland Resources, Baton Rouge, LA, Sea Grant Pub. No. LSU-T-80-001.

Huner, J. V. and R. P. Romaire. 1979. Size at maturity as a means of comparing populations of Procambarus clarkii (Girard) (Crustacea: Decapoda) from different habitats, pp. 53-64. In: P. J. Laurent (ed.) Freshwater Crayfish IV, Thonon-les-Bains, France (1978). 473 pp.

Miltner, M. and J. W. Avault, Jr. 1981. Rice and millet as forages for crawfish. Louisiana Agriculture 24(3):8-10.

Morrissy, N. M. 1979. Experimental pond production of marron, Cherax tenuimanus (Smith) (Decapoda: Parastacidae). Aquaculture 16:319-344.

Odum, E. P. 1971. Fundamentals of Ecology. 3rd ed. W. E. Saunder and Co., Philadelphia, PA.

Penn, G. H. 1943. A study of the life history of the Louisiana red crawfish, Cambarus clarkii Girard. Ecology 24:1-18.

Penn, G. H. 1956. The genus Procambarus in Louisiana (Decapoda: Astacidae). Amer. Midl. Nat. 56:406-422.

EFFECTS OF FLOODING DATES AND TYPE OF DISPOSAL OF RICE STRAW

ON THE INITIAL SURVIVAL AND GROWTH OF CAGED JUVENILE CRAYFISH,

Procambarus clarkii, IN PONDS

Yew-Hu Chien and James W. Avault, Jr.
Fisheries Section
School of Forestry and Wildlife Management
Louisiana State University
Baton Rouge, Louisiana 70803, USA

ABSTRACT

A study was conducted to determine how flooding dates and type of rice straw disposal affect survival and growth rate of caged juvenile (< 20 mm) red swamp crayfish (Procambarus clarkii). Eighteen ponds were planted with rice (Oryza sativa), which was later cut to a height of 30 cm. Ponds were assigned to six treatments involving two flooding dates (20 September and 15 October) and three types of rice straw disposals: (1) Baled -- the cut part of the rice straw was collected from ponds and piled into heaps on the banks. (2) Standing -- the rice straw was left as it was after cutting. (3) Disked -- rice straw was incorporated into the soil about 5-7.5 cm deep by disking. Juvenile crayfish were stocked into aluminum screen cages (10 x 10 x 20 cm) at 10 per cage, with two cages per pond. Crayfish survival was significantly (P < 0.05) higher in late-flooded ponds (84%) than in early-flooded ponds (70%). No significant differences in survival rates were found among baled (74%), standing (80%), and disked (76%) rice straw treatments at the end of a four-week period. Growth rate was significantly (P < 0.05) correlated with average diurnal dissolved oxygen (DO) concentration and average water temperature. High DO and temperature favored growth, though high temperatue may also have stimulated cannibalistic predation and reduced survival. A highly significant (P < 0.01) inverse correlation between survival rate and growth rate demonstrated density-dependent growth.

INTRODUCTION

Oxygen depletion is a major problem in commercial crayfish farming (Avault et al. 1975). When rice field ponds are flooded in early September, DO depletion may result from rapid decomposition of rice straw under warm water conditions. Consequently, juvenile crayfish (< 20 mm in total length) flushed out of their burrows at flooding may be under stress from low concentrations of DO. Conversely, when rice ponds are flooded in late October, colder water temperatures increase oxygen solubility and reduce the biochemical oxygen demand (BOD), thus minimizing DO depletion. However, reduced water temperature may result in slow crayfish growth. Acharya (1935a,b) demonstrated that decomposition of rice straw was least pronounced under complete anaerobiosis and more pronounced under aerobic and waterlogged conditions. Avault et al. (1975) suggested that terrestrial vegetation be mowed and if possible disked underground to minimize DO depletion.

MATERIALS AND METHODS

The study was conducted during 1978 at Ben Hur Farm, Louisiana State University (LSU), Baton Rouge, Louisiana. Eighteen 0.05 ha ponds with an average depth of 0.8 m were planted with rice (_Oryza sativa_), variety LaBelle, in mid-April. The rice was cut 1 September to a height of 30 cm. The grain was not harvested. Ponds were randomly assigned to six treatments with three replications each, in a 2 x 3 factorial arrangement with two flooding dates and three types of rice straw disposals. The treatments were as follows:

Rice Straw Disposal	Flooding Dates	
	Early Flooding (09/20/78)	Late Flooding (10/15/78)
Baled	3	3
Standing	3	3
Disked	3	3

Ponds were flooded either 20 September or October 1978 and were referred to as early- and late-flooded ponds, respectively. The rice straw was disposed of in the following manners: (1) Baled -- the cut part of the rice straw was collected from ponds and piled into heaps on the banks. The rice straw was not returned to the ponds until February 1979. (2) Standing -- the rice straw was left as it was after cutting. (3) Disked -- rice straw was incorporated into the soil about 5-7.5 cm deep by disking.

Crayfish, approximately 14.5 mm long, were confined in cages in order to observe initial growth and survival. They were first acclimated to saturated DO and field water temperature in an outdoor 60-liter drum for 24 hours and then stocked in aluminum screen cages (10 x 10 x 20 cm) at 10 crayfish per cage. Two cages were placed in each pond in late afternoon when DO was highest. The top 5 cm of each cage projected above the water surface in order to give the crayfish access to the air-water interface in the event of DO depletion. Water temperature and DO were measured twice weekly, at dusk and at dawn, at a depth of 10 cm. Crayfish were fed a pelleted ration (supplied by Aquatic Diet Technology, Inc.) every other afternoon at a rate of 0.5 g per crayfish to minimize cannibalism from food deficiency. Survival and growth (total length) were determined at the second and fourth week after stocking.

RESULTS AND DISCUSSION

Dissolved Oxygen and Water Temperature

No significant difference ($P > 0.05$) of average dawn DO during the four-week period was found either between early- (1.1 mg/l) and late-flooded ponds (1.4 mg/l) or among baled (1.5 mg/l), standing (1.2

mg/l), and disked ponds (1.0 mg/l) (Table 1). During the four-week period, average diurnal DO (average of dawn and dusk DO) was higher in early-flooded ponds (4.4 mg/l) than in late-flooded ponds (3.6 mg/l) even though the average water temperature in early-flooded ponds was 2.9°C higher than late-flooded ponds. Rice straw disposal had a significant ($P < 0.05$) effect on diurnal DO concentration. The average diurnal DO levels were 4.9, 3.8 and 3.4 mg/l for ponds with baled, standing, and disked rice straw, respectively. Dissolved oxygen was highest in ponds with baled rice straw, since less decomposition took place in the water column than in the standing and disked treatments. Lowest diurnal DO was observed in ponds where rice straw was disked into the soil. Disking the soil increased the soil-water contact area and water turbidity. The enlarged oxidizing surface of the soil and high concentration of suspended particulate matter and colloidal organic matter in the water probably resulted in greatest oxygen consumption.

The average water temperature in early-flooded ponds (23.4°C) was significantly higher ($P < 0.05$) than in late-flooded ponds (20.5°C). Shading by standing rice straw resulted in the lowest average temperature (21.4°C). The highest average water temperature of disked ponds (22.6°C) was mainly because of greater absorption of solar radiation by suspended particles in turbid water (Boyd 1979). The average temperature in ponds with baled forage was 21.8°C).

Survival and Growth

Regardless of treatment, no significant correlation was found between crayfish survival and average diurnal DO or average dawn DO during the four-week study. The insignificant correlation was because of several factors. First, transfer of juvenile crayfish was carried out in the afternoon when DO was at its highest level, and the gradual decline in DO could have given the crayfish time to adjust. Secondly, conditions of low DO were not persistent. Jaspers and Avault (1969) noted that DO in water of crayfish burrows ranged from 1.4 to 0.2 mg/l without notable mortality in young crayfish. They indicated that the water may have acted as a humidity source to keep gills wet. Melancon and Avault (1977) observed that, when subjected to an abrupt DO change from 6.0 to 0.4 mg/l and from 6.0 to 1.1 mg/l, crayfish of 9-12 mm in total length had survival rates after 60 hours of 0 and 58%, respectively. Death may be from the inability of crayfish to regulate oxygen consumption when suddenly placed under low-oxygen conditions (Maloeuf 1937).

Survival rate was inversely related to average water temperature -- that is, high water temperatures resulted in low survival and vice versa. Mason (1978) found an adverse effect of high temperature on the survival of crayfish Pacifastacus leniusculus. Frequent molting during fast growth under high temperatures could make crayfish more vulnerable to cannibalism. Hoffman et al. (1975) found that as temperature increased from 5 to 10°C, the aggressive behavior of the lobster (Homarus americanus) also increased. High temperature might not only increase activity, and thus encounters, but also make the hard-shelled individuals more aggressive and cause them to prey on the soft-shelled ones.

The survival rate of juvenile crayfish was significantly ($P <

Table 1. The effects of flooding dates and disposals of rice straw on average diurnal temperature, dissolved oxygen (DO), survival, and growth rate of juvenile (total length less than 20 mm) crayfish, Procambarus clarkii, during a four-week period. Flooding dates: E = early flooding (9/20/78); L = late flooding (10/15/78). Disposals of rice straw: B = baled; S = standing; D = disked.

Duration of test	Treatments	Average Diurnal Temp (°C)	Average Diurnal DO (mg/l)[1]	Average Dawn DO (mg/l)	Survival Rate (%)	Growth Rate (mm)
0-14 days	EB	26.1	4.9	1.2	78	6.9
	ES	25.1	2.5	0.4	95	4.8
	ED	26.6	2.6	0.4	85	5.7
	LB	21.4	4.5	1.0	95	3.8
	LS	21.1	3.9	1.2	95	4.7
	LD	21.9	2.2	0.6	96	4.3
15-28 days	EB	20.6	6.5	2.4	83[2]	2.9
	ES	20.1	4.3	1.2	81	3.2
	ED	21.6	5.7	0.8	79	4.6
	LB	19.0	3.7	1.4	87	4.6
	LS	19.2	4.3	2.0	87	4.5
	LD	20.1	3.0	2.1	89	4.4
0-28 days	EB	23.4	5.7	1.8	65	9.7
	ES	22.6	3.4	0.8	77	8.0
	ED	24.1	4.2	0.6	67	10.3
	LB	20.2	4.1	1.2	83	8.3
	LS	20.2	4.1	1.6	83	9.2
	LD	21.0	2.6	1.4	85	8.7

[1]Average of dawn DO and dusk DO.

[2]The last two weeks survival rate =

$$\frac{\text{The number of crayfish that survived at the end of the fourth week}}{\text{The number of crayfish that survived at the end of the second week}} \times 100$$

0.05) higher in late-flooded ponds (84%) than in early-flooded ponds (70%) at the end of the four-week period. Survival rates during the first two weeks were 95% to 86% in late-flooded and early-flooded ponds, respectively, and 88% and 81% the last two weeks.* High survival in late-flooded ponds was primarily because of a water temperature significantly lower than in early-flooded ponds. No significant ($P > 0.05$) differences in survival rates were found among baled (74%), standing (80%), and disked (76%) rice straw treatments at the end of the four-week period.

Crayfish growth rate was inversely correlated with survival, i.e., high survival resulted in low growth and vice versa. Goyert (1978) reported that crayfish averaging 33 mm in total length, raised at a low density (10 crayfish/ m^2), grew 14.2 mm larger than crayfish grown at a high density (40 crayfish/m^2) over 65 days. Abrahamsson and Goldman (1970) observed that body length was inversely related to population density. The food supply in our study was adequate; thus, the inverse relationship between growth and survival may have been because of a "space factor" (Hile 1936), whereby crowding impedes growth independent of food abundance.

Crayfish generally reduce feeding and growing if under persistent low DO (LaCaze 1976). Growth rate during the four-week period was positively correlated with average diurnal DO and average water temperature. Increasing DO concentration and water temperature resulted in faster growth.

No significant differences in growth rates were found between early-flooded ponds (9.3 mm/4 weeks) and late-flooded ponds (8.7 mm/4 weeks) at the end of the four-week period. However, during the first two weeks, growth rate was significantly higher in early-flooded ponds (5.8 mm/2 weeks) than in late-flooded ponds (4.3 mm/2 weeks). This resulted from 4.4°C higher water temperature and/or 9% lower survival in early-flooded ponds than in late-flooded ponds. During the last two weeks, growth rate was higher in late-flooded ponds (4.5 mm/2 weeks) than in early-flooded ponds (3.6 mm/2 weeks). The growth rates at the end of the four-week period were 9.0, 8.6, and 9.5 mm/4 weeks for baled, standing, and disked ponds, respectively. Although there were no significant differences in crayfish growth among treatments of rice straw disposals, there was significant ($P < 0.05$) interaction between flooding dates and rice straw disposals. The highest average growth rate (10.3 mm/4 weeks) was observed in early-flooded ponds in which rice straw was disked. Mason (1978) found that the combination of high temperature and short photoperiod gave the best growth of juvenile crayfish P. leniusculus, with temperature being the predominant influence. The highest average initial growth, observed in early-flooded ponds in which rice straw was disked, was attributed to the higher water temperature and lower survival rate. Like the effects of the photoperiod, the weaker photointensity caused by turbidity in disked ponds may have resulted in a higher growth rate.

---

*The last two weeks' survival rate =

$$\frac{\text{The number of crayfish survived at the end of the fourth week}}{\text{The number of crayfish survived at the end of the second week}} \times 100$$

CONCLUSIONS

Temperature had dominant effects on both the growth and survival of crayfish. Although high water temperature favored growth, it may also have stimulated cannibalism and reduced survival. In this study, the average dawn DO was as low as 0.4 mg/l but, since crayfish had access to the water surface, their survival was not detrimentally affected. In real situations, however, crayfish that climbed to the water surface in response to DO shortage would be exposed to predation.

ACNKOWLEDGMENTS

This research was supported by the Louisiana Sea Grant College Program, maintained by the National Oceanic and Atmospheric Administration, U.S. Department of Commerce, and the Louisiana Agricultural Experiment Station.

LITERATURE CITED

Abrahamsson, S. and C. R. Goldman. 1970. Distribution, density, and production of the crayfish, Pacifastacus leniusculus, in Lake Tahoe, California-Nevada. Oikos 21:83-91.

Acharya, C. 1935a. Studies on the anaerobic decomposition of plant materials. I. Anaerobic decomposition of rice straw. Biochem. J. 29:528-541.

Acharya, C. 1935b. Studies on the anaerobic decomposition of plant materials. II. Some factors influencing the anaerobic decomposition of the rice straw. Biochem. J. 29:953-960.

Avault, J. W., Jr., L. de la Bretonne, Jr. and J. V. Huner. 1975. Two major problems in culturing crayfish in ponds: Oxygen depletion and overcrowding. In: Papers from the Second Int. Symp. on Freshwater Crayfish 2:139-143.

Boyd, C. E. 1979. Water quality in warmwater fish ponds. Auburn University Ag. Exp. Sta. Auburn, Ala. 359 pp.

Goyert, J. C. 1978. The intensive culture of crayfish, Procambarus clarkii (Girard), in a recirculating water system. Ph.D. dissertation. Louisiana State University, Baton Rouge, LA.

Hile, R. 1936. Age and growth of the Cisco, Leucichthys artedi (Le Sueur), in the lakes of the North-eastern Highlands, Wisconsin. Bull. U.S. Bur. Fish. 48:211-317.

Hoffman, R. S., P. J. Dunham and P. V. Kelly. 1975. Effects of water temperature and housing conditions upon the aggressive behavior of the lobster (Homarus americanus). J. Fish. Res. Board Canada 32(5):713-718.

Jaspers, E. and J. W. Avault, Jr. 1969. Environmental conditions in burrows and ponds of the red swamp crayfish, Procambarus clarkii, near Baton Rouge, Louisiana. In: Proc. 23rd Ann. Conf. Southeast Assoc. Game Fish Comm. 23:634-647.

349

LaCaze, L. 1976. Crawfish farming. Louisiana Wildlife and Fisheries Commission. Fisheries Bulletin No. 7, 27 pp.

Maloeuf, N. S. F. 1937. Studies on the respiration (and osmoregulation) of animals. I. Aquatic animals without an oxygen transporter in their internal medium. A. Verleich. Physiol. 25:1-28.

Mason, J. C. 1978. Effects of temperature, photoperiod, substrate, and shelter on survival, growth, and biomass accumulation of juvenile Pacifastacus leniusculus in culture. In: Papers from the Fourth Int. Symp. on Freshwater Crayfish 4:73-82.

Melancon, E. J. and J. W. Avault, Jr. 1977. Oxygen tolerance of juvenile red swamp crayfish, Procambarus clarkii. In: Papers from the Third Int. Symp. on Freshwater Crayfish 3:371-380.

A COMPARISON OF DELTA DUCKPOTATO
(SAGITTARIA GRAMINEA PLATYPHYLLA) WITH RICE (ORYZA SATIVA)
AS CULTURED RED SWAMP CRAYFISH (PROCAMBARUS CLARKII) FORAGE

W.B. Johnson, Jr.[1], L.L. Glasgow,
and J.W. Avault, Jr.
Fisheries Section
School of Forestry and Wildlife Management
Louisiana State University
Baton Rouge, Louisiana

ABSTRACT

This study compares delta duckpotato (*Sagittaria graminea platphylla*) with rice (*Oryza sativa*) as forage for use in crayfish (*Procambarus clarkii*) culture. Eight 0.014 ha ponds, four planted in rice and four in delta duckpotato, were used. Duckpotato required mid-summer standing water whereas rice was moistened only in the summer. All ponds were flooded in October. Water quality, primary production, and crayfish population dynamics and yield were evaluated. Dissolved oxygen, pH, and carbonate alkalinity were higher in duckpotato ponds than in rice ponds. Primary production was much higher in duckpotato ponds in the fall, but differences disappeared after that. Crayfish population trends reflected the different hydrologic regimes. Crayfish from the duckpotato treatment were larger and more abundant in the fall; form I crayfish appeared in abundance in November and crayfish growth slowed soon after. Crayfish from the rice treatment matured later. Total yield did not differ significantly between treatments because of the aforementioned treatment-time interaction in crayfish abundance.

INTRODUCTION

Major problems in crayfish (*Procambarus clarkii*) husbandry are poor water quality, especially low dissolved oxygen (Avault et al. 1974; LaCaze 1976) and a deficient food supply (Avault et al. 1974; LaCaze 1976; Goyert 1978). It was believed that *Sagittaria graminea platyphylla*, delta duckpotato possessed several promising attributes justifying its examination as a crayfish forage: (1) *S. graminea* is a prolific freshwater aquatic plant that survives inundation (Wooten 1970). (2) *Sagittaria falcata*, a similar species, has an extremely high turnover rate (Hopkinson et al. 1978). (3) *Sagittaria* spp. supports a dense aufwuch community (Odum 1957). (4) *S. platyphylla* has a succulent tissue which appears to attract the red swamp crayfish *Procambarus clarkii* (Viosca 1931). (5) *Sagittaria* sp. also has a rapid decay process (Wunder 1949). (6) red swamp crayfish grow faster eating duckpotato than smart weed, *Polygonum punctatum* (Johnson 1980). (7) duckpotato tubers and rhizomes are relished by puddling waterfowl (Martin and Uhler 1939) making it potentially attractive for use in combined waterfowl and crayfish impoundments.

---

[1]Present address: Center for Wetland Resources, Louisiana State University, Baton Rouge, Louisiana.

The objective of this study was to compare delta duckpotato, a wild fresh water marsh plant, with rice, *Oryza* *sativa*, an established crayfish forage and aquacultural crop. Their impact on water quality, primary production, and crayfish population dynamics and yield in small earthen impoundments was examined.

MATERIALS AND METHODS

The investigation was conducted at Ben Hur Farm, Louisiana State University, Baton Rouge, Louisiana, in eight impoundments, each 0.014 ha in surface area, with an average depth of 0.75 m. The water was drained in early June and no crayfish were stocked, as resident populations were found to be homogeneous (Johnson 1980).

Four randomly selected ponds were planted in rice and four ponds in duckpotato. Both species were planted (rice with seeds, duckpotato with tubers and transplants) in mid-June; replanting continued through the summer to ensure an adequate forage density. Duckpotato required deeper water than rice could stand. All ponds were fully inundated at 1 October, 1978.

Diel dissolved oxygen (DO) and temperature were monitored on 14 alternate weeks beginning 19 October, 1978, and ending 26 May, 1979. Measurements were made *in* *situ* with a polarographic oxygen meter (Yellow Springs model 57) in each pond at the surface, at 35 cm, and at 75 cm. A mean of the three readings was assumed to be representative of the entire pond. Three sets of oxygen-temperature readings were made at each diel sampling. The first set was taken not more than two hours before sunset; the second was taken the following day within one hour of sunrise; and the third was taken that afternoon again within two hours of sunset. These data were then used to estimate primary production, following the techniques described in Harris (1978).

In addition, the mid-afternoon, biweekly sampling of surface water DO, pH, and alkalinity began on 6 January, 1979. Ammonia-nitrogen and turbidity determinations began on 29 April, 1979. DO was measured *in* *situ* with the aforementioned meter. Other water quality parameters were assessed in the laboratory from surface samples. Alkalinity titrations and pH determinations were made with a Beckman Zeromatic II pH meter model 96 A, as outlined in APHA et al. (1975). Alkalinity titrations required 0.05N HCl in a 200 ml water sample. Turbidity was measured in Jackson Turbidity Units (JTU) with a Bausch and Lomb Spectronic 20, according to methods outlined by the Hach Chemical Company (1976).

Dip net, small-mesh cylindrical traps (6 mm mesh), and large-mesh cyclindrical traps (19 mm mesh) were fished throughout the study so that the entire crayfish population could be sampled. [Hunner and Barr (1980) describe cylindrical trap construction.] Small-mesh traps were baited with carp (*Cyprinus* *carpio*) through 22 January, 1979. Thereafter the heads of channel catfish (*Ictalurus* *punctatus*) were utilized. Beginning in January, large-mesh traps were also used, baited with channel catfish heads. A trap set consisted of one baited trap per pond always placed at the same location. All eight traps were set in the afternoon and retrieved at the same time 20 to 26 hours later. Upon retrieval, all crayfish were enumerated, measured from rostrum to telson to the nearest 4 mm and identified. Males were evaluated for form I.

All crayfish landed in large-mesh traps were harvested. Those crayfish less than 76 mm in total length, not form I, and caught by small-mesh traps were returned to the appropriate pond within three hours of removal; all others were harvested. Harvest was wet weighed to the nearest gram. White river crayfish (P. acutus acutus) composed less than 5% of the total landings and are not discussed.

Trapping periodicity followed no schedule; however, large-mesh traps were often retrieved on consecutive days to monitor the decrease in catch per unit effort (CPUE). The mean decline in CPUE over a trapping sequence was used to estimate crayfish abundance (Carle and Strub 1978). All other statistical comparisons were made using a completely random design, factorial arrangement of treatments, analyzed by an F test with treatment, month, and treatment by month as effects.

RESULTS AND DISCUSSION

Water quality and primary production

Mean mid-afternoon (p <0.02), morning (p <0.05), and evening (p <0.01) DO levels in the duckpotato ponds were significantly greater than in the rice ponds (Table 1). These differences were dramatic; for example, the morning DO in rice ponds during October and November was a third of the mean DO level in duckpotato ponds. This can be attributed to several factors. Duckpotato ponds were maintained with deeper water levels throughout the previous summer so that much decay had already occured. As predicted duckpotato did not die upon flooding; moreover, duckpotato and its aufwuch algae were an oxygen source. This trend did not continue, as evidenced by a highly significant (p < 0.01) treatment by month interaction. While the morning DO levels in the duckpotato treatment were much greater in the fall, there was no mean treatment difference (p > 0.05) in DO for the remaining months.

Coincident with the DO differences between treatments were other related water quality differences. The pH and carbonate alkalinity were significantly greater (p > 0.01) in duckpotato ponds than in the rice ponds. Correspondingly, mean bicarbonate and total alkalinity levels were significantly less (p <0.01) in the duckpotato ponds than in the rice ponds (Table 1). These findings are a reflection of the elevated primary production rates in the duckpotato ponds both early and late in the study (Fig. 1).

Changes in DO concentration in surface waters are acceptable for utilization in the calculation of primary productivity, provided atmospheric diffusion is monitored (Odum 1960; Hornberger and Kelly 1974; Harris 1978). Accordingly, readings were made on calm days, and the ponds had a small fetch and large freeboard (50 cm).

Analysis of variance indicated a significant (p < 0.01) treatment by month interaction of mean primary production levels. Duckpotato ponds in the fall and late spring had higher mean primary production rates than rice ponds but were not different during the interim (Fig. 1).

Duckpotato plants that remained alive after flooding and the associated aufwuch community provided additional oxygen. In contrast,

353

Table 1. Mean surface water characteristics.

		Dissolved oxygen				Alkalinity			
		Morning	mg l^{-1} mid-afternoon	evening	pH	mg l^{-1} HCO$_3$	CO$_3$	CaCO$_3$ Total	JTU
	Oct	3.6		7.9					
	Nov	4.1		7.5					
	Jan	11.5	10.0	7.8	8.4	223	10	223	
Duck-	Feb	6.9	8.1	8.3	8.0	183	9	192	
potato	Mar	7.1	10.3	9.1	8.3	157	15	172	
	Apr	5.2	11.2	10.2	8.3	108	20	128	194
	May	5.1	8.9	10.8	8.1	280	8	288	141
	Oct	1.4		5.3					
	Nov	1.2		4.2					
	Jan	9.7	7.9	12.4	7.7	227	2	229	
Rice	Feb	7.2	8.1	8.4	7.9	194	2	196	
	Mar	3.8	6.7	5.3	7.9	188	0	188	
	Apr	2.6	7.8	4.8	8.1	205	2	207	119
	May	6.7	7.4	11.2	7.9	412	3	415	85

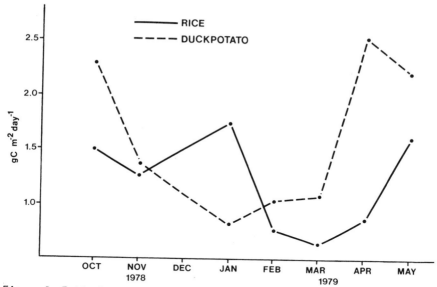

Figure 1. Estimates of mean monthly primary production using diel oxygen changes according to Harris (1978).

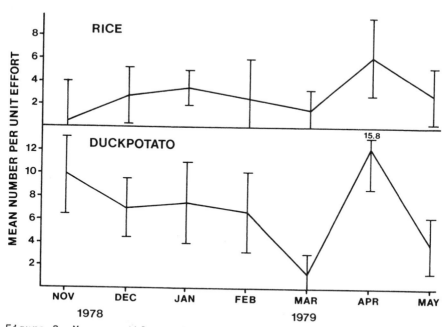

Figure 2. Mean monthly number per unit effort of form I red swamp crayfish caught in small mesh (6 mm) cylindrical traps.

the rice began senescing after seed production in early September; moreover, decay was more intensive in the rice ponds as evidenced by the low morning DO levels. During the spring duckpotato plants sprouted and, although of a low density, existed with submersed phyllodial leaves. This contributed to higher late spring DO levels in the duckpotato ponds.

This suggests that the crayfish in duckpotato ponds were foraging near the sediment surface. Further evidence is the significantly higher ($p < 0.01$) JTU levels in the duckpotato ponds than in the rice ponds (Table 1). Low densities of food in duckpotato ponds forced crayfish to explore sediments for food.

Crayfish dynamics

Form I represents an adult stage of sexual maturity in male crayfish of the subfamily Cambarinae (Penn 1959). Only form I crayfish are capable of successful fertilization (Crocker and Barr 1968). Form I males have cornified apical elements on the first pair of pleopods and distinct hooks on the ischiopodites of at least one pair of periopods (Penn 1959). Other workers have used the form I catch as an index of population status in feral (Penn 1943; Huner 1975; O'Brien 1977) and cultured (Huner and Avault 1976; Huner 1978) red swamp crayfish populations. Although referring to female sexual maturity, Steele and Steele (1976) and Wenner et al. (1974) regarded mean size at the attainment of sexual maturity as a useful index of population growth in crustacean populations.

Because of earlier flooding, duckpotato ponds received young-of-the-year cohorts earlier, resulting in November landings of red swamp crayfish that were significantly larger ($p < 0.05$). Nearly five times more individuals were landed in duckpotato ponds with small mesh traps than in rice ponds (Fig. 2). These findings are similar to those of Huner and Avault (1976) who observed a December pulse of form I crayfish in ponds flooded in September. Fall flooding delays crayfish development to the degree that immature crayfish growth is temporarily slowed by cooler temperatures. As a result, cohorts dominated by form I crayfish do not develop until spring (Huner 1975; Huner and Avault 1976). Form I red swamp crayfish in small- and large-mesh trap landings in the duckpotato ponds were smaller than those from rice ponds (Table 2) which suggests slower growth (Wenner et al. 1974; Momot et al. 1978). Insufficient food in duckpotato ponds undoubtedly contriubted to the shorter mean crayfish length. A food source that will be sustained throughout the season is essential; Huner and Romaire (1978) emphasized the importance of adequate food in growing large crayfish. In addition, Flint and Goldman (1977) showed that growth rate in *Pacifasticus leniusculus* can be depressed by increased competition for food.

Trends in the catch of all crayfish (*Procambarus spp.*) parallel the catch of form I crayfish. Initially, because of the maintenance of high summer water levels, almost five times more crayfish (individuals and weight) in November, four times more in December, and two times more in January were trapped from duckpotato ponds than from rice ponds. Beginning in February and continuing through the remainder of the study, no meaningful differences in the numbers or weight of harvested crayfish were witnessed (Fig. 3).

Table 2. Mean total lengths (mm) of form I red swamp crayfish caught in small (6 mm) and large (19 mm) mesh traps.

		Nov	Dec	Jan	Feb	Mar	Apr	May
Duck-potato	Small mesh	78.8	79.4	80.4	80.1	85.0	77.6	75.4
	Large mesh	--	--	79.1	85.5	82.3	78.6	76.1
Rice	Small mesh	84.0	84.5	89.4	85.5	87.9	81.4	81.7
	Large mesh	--	--	84.1	87.7	86.5	84.5	77.7

Table 3. Estimates of the mean number of harvestable red swamp crayfish from 0.014 ha culture ponds using the weighted maximum likelihood removal method. Normally estimated 95% confidence limits are in parentheses.

	Duckpotato	Rice
January 15-17	84.5 (76.5 - 94.3)	48.8 (43.8 - 56.5)
March 2-6	53.0 (33.0 - 84.0)	156.8 (47.8 - 321.5)
March 15-17	49.7 (35.7 - 71.0)	152.0 (86.0 - 230.5)
April 8-12	42.8 (35.8 - 53.5)	333.8 (79.8 - 698.5)
April 24-29	55.8 (48.8 - 65.6)	41.5 (34.5 - 52.1)
May 13-19	47.3 (40.3 - 57.6)	39.0 (34.5 - 52.1)
May 24-28	65.0 (62.0 - 69.8)	39.0 (37.1 - 40.9)

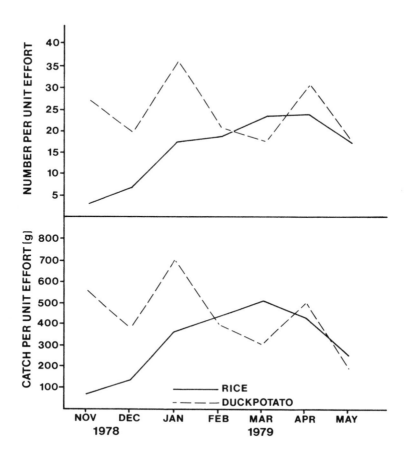

Figure 3. Mean monthly numbers per unit effort and catch per unit effort of crayfish (_Procambarus_ spp.) caught in small mesh (6 mm) traps.

Changes in the mean monthly CPUE are not entirely revealing. Patterns of catch depletion over a trapping sequence can provide insights into the dynamics of population abundance. A weighted maximum likelihood removal method (Carle and Strub 1978) was used to estimate abundance; removal sequences were done on consecutive days to satisfy the assumption of a stable population. Only large-mesh traps were used, and estimates, therefore, reflect the abundance of harvestable crayfish (75 mm in total length).

January estimates revealed that duckpotato ponds had significantly greater (p < 0.05) numbers of red swamp crayfish than rice ponds (Table 3). This is a reflection of the aforementioned early flooding of the duckpotato treatment. By 24-28 March, crayfish became significantly more abundant (p < 0.05) in the rice ponds. This indicates that the late fall young-of-the-year crayfish were entering the harvestable population after the crayfish in the duckpotato ponds had matured and been removed. This difference continued until 13-19 May, when again there were no treatment differences. Estimates made from 24-28 May removals showed that the situation was again reversed; the abundance of crayfish in duckpotato ponds significantly (p < 0.05) exceeded the abundance of crayfish in rice ponds. This is because the second duckpotato cohort was growing into the harvestable size range. This cohort is the product of the large form I populations that were found in duckpotato ponds in late autumn.

Finally, although total yield was 320 kg ha^{-1} greater in rice ponds than in duckpotato ponds, the difference was not statistically significant (p > 0.05). This can be explained by the time/treatment interaction that was found in many of the aforementioned analyses.

In conclusion, there were two important findings regarding the use of duckpotato relative to rice in crayfish ponds. First, oxygen levels were higher immediately after fall inundation in the duckpotato ponds than in rice ponds, because of reduced amounts of decay and oxygen producing plants and aufwuch algae. Second, populations developed sooner in the duckpotato ponds because of the deeper water required for summer development of duckpotato stands. This allowed development of two significant cohorts, in late fall and late spring, in contrast to the single mid-spring cohort in rice ponds.

ACKNOWLEDGMENTS

Contribution No. LSU-CEL-81-19 of the Coastal Ecology Laboratory, LSU Center for Wetland Resources, Baton Rouge, La. This research was supported by the Louisiana Sea Grant College Program and the Louisiana Agricultural Experiment Station.

LITERATURE CITED

American Public Health Association, American Water Works Association and Water Pollution Conrol Federation. 1975. Standard methods for examination of water and wastewater, 14th ed. APHA, Washington, D.C. 1193 p.

Avault, J.W., Jr., L. de la Bretonne, Jr., and J.V. Huner. 1975. Two major problems in culturing crayfish in ponds: oxygen depletion and overcrowding, pp. 139-144 In J.W. Avault (ed.). Freshwater Crayfish II, Baton Rouge, Louisiana (1974). 676 p.

Carle, F.L. and M.R. Strub. 1978. A new method for estimating population size from removal data. Biometrics, 34:621-630.

Crocker, D.W. and D.W. Barr. 1968. Handbook of crayfishes of Ontario. Univ. Wisconsin Press, Madison, Wis. 200 p.

Flint, R.W. and C.R. Goldman. 1977. Crayfish growth in Lake Tahoe, California-Nevada. USA. Effects of habitat variation. J. Fish. Res. Board Can. 34:155-159.

Goyert, J.C. 1978. The intensive culture of crayfish _Procambarus clarkii_ (Girard) in a recirculating water system. Ph.D. Dissertation, Louisiana State University. 92 p.

Hach Chemical Company. 1976. Colorimetric procedures and chemical list for water and wastewater analysis, 6th ed. Hach Chemical Company, Ames, Iowa. 78 p.

Harris, G.P. 1978. Photosynthesis, productivity, and growth: the physiological ecology of phytoplankton. Arch. Hydrobiol. Beih. 10:1-171.

Hopkinson, C.S., J.G. Gosselink, and R.T. Parrondo. 1978. Above-ground production of seven marsh plant species in coastal Louisiana. Ecology 59:760-769.

Hornberger, G.M. and M.G. Kelly. 1974. A new method for estimating productivity in standing waters using free oxygen measurements. Water Res. Bull. 10:265-271.

Huner, J.F. 1975. Observations on the life histories of recreationally important crawfishes in temporary habitats. La. Acad. Sci. 37:20-24.

Huner, J.V. 1978. Crawfish population dynamics as they affect production in several small, open commercial crawfish ponds in Louisiana. Proc. World Mric. Soc. 9:619-640.

Huner, J.V. and J.W. Avault, Jr. 1976. Sequential pond flooding; A prospective management technique for extended production of bait sized crawfish. Trans. Am. Fish. Soc. 105:637-642.

Huner, J.V. and J.E. Barr. 1980. Red swamp crawfish: Biology and exploitation. Center for Wetland Resources, Louisiana State University, Baton Rouge, LSU-T-80-001. 148 p.

Huner, J.V. and R.P Romaire. 1979. Size at maturity as a means of comparing populations of _Procambarus clarkii_ (Girard) (Crustacea, Decapoda) from different habitats, p. 53-65 In P.J. Laurent (ed.) Freshwater Crawfish IV, Thonon-les-Bains, France (1978). 493 p.

Johnson, W.B. 1980. Evaluation of poultry waste, rice (Oryza sativa), and delta duckpotatoe (Sagittaria platyphylla) as crayfish (Procambarus spp.) forages. M.S.,Thesis Louisiana State University. 94 p.

LaCaze, C.G. 1976. Crawfish farming. La. Wildl. and Fish. Comm. Fish. Bull. 7: 27 p.

Martin, A.C. and F.M. Uhler. 1939. Food of game ducks in the United States and Canada. USDA Bull. 634. 157 p.

Momot, W.T., H. Gowing, and P.D. Jones. 1978. The dynamics of crayfish and their role in ecosystems. Am. Midl. Nat. 99:10-35.

O'Brien, T.P. 1977. Crawfishes of the Atchafalaya Basin, Louisiana with emphasis on those species of commercial importance. M.S. Thesis, Louisiana State University. 79 p.

Odum,H.T. 1960. Analysis of diurnal oxygen curves for the assay of reaeration rates and metabolism in polluted marine bays. Proc. First Int. Conf. on Waste Disposal in the Marine Environment. 1:547-555.

Odum, H.T. 1957. Trophic structure and productivity of Silver Springs, Florida. Ecol. Monogr. 27:55-112.

Penn, G.H. 1943. A study of the life history of the Louisiana red crawfish, Cambarus clarkii (Girard). Ecology 24:1-18.

Penn, G.H. 1959. An illustrative key to the crawfishes of Louisiana with a summary of their distribution within the state. Tulane Stud. Zool. 7:3-20.

Steele, D.H. and V.J. Steele. 1969. The biology of Grammarus (Crustacea, Amphipoda) in the northwestern Atlantic. I. Gammarus duebeni Lilly. Can. J. Zool. 47:235-444.

Viosca, P. 1931. The bullhead, Ameriurus melas catus, as a dominant in small ponds. Copeia 1:17-19.

Wenner, A.M., C. Fusara, and A. Oaten. 1974. Size at onset of sexual maturity and growth rate in crustacean populations. Can. J. Zool. 52:1095-1106.

Wooten, J.W. 1970. Experimental investigations of the Sagittaria graminea complex: Transplant studies and genecology. J. Ecol. 58:233-238.

Wunder, W. 1949. Fortschrittliche Karpfenteichwirtschaft, E. Schweit Zerbart'sche. Verlagsbuchandlugh, Erwin Nagele, Stuttgart.

FEEDS AND FORAGES FOR
RED SWAMP CRAWFISH, Procambarus Clarkii,:
15 YEARS RESEARCH AT LOUISIANA STATE UNIVERSITY REVIEWED

James W. Avault, Jr., Robert P. Romaire,
and Michael R. Miltner
Fisheries Section
School of Forestry and Wildlife Management
Louisiana State University
Baton Rouge, Louisiana 70803 USA

ABSTRACT

The red swamp crawfish (Procambarus clarkii) feeds well on pelleted rations, green plants, periphyton, detritus, and agriculture by-products. The ideal feed/forage regime must produce high yields of crawfish per hectare at a cost to allow for maximum profit. Many pelleted rations, though producing high yields, appear uneconomical. Planted forages such as rice give both high yields and high monetary returns. A major problem with rice forage is that it decomposes completely before the crawfish growing season is completed. A combination of two or more forages with differing decomposition rates should prove to be the best feeding regime. In selecting the best forage(s) the following parameters should be considered: carbon:nitrogen ratio, decomposition rate, biomass production, lodging rate, and water quality.

INTRODUCTION

When the crawfish farming industry in Louisiana began measured development circa 1966, less than 500 ha were in production. Crawfish were not fed formulated rations but subsisted on natural plants and associated organisms. Production per ha then was low, seldom exceeding 1000 kg/ha, and the price the farmer received per kg of live crawfish was also low, usually less than $0.88.

Research at Louisiana State University (LSU) began in 1966 to focus on improving feeds and feeding regimes for increasing crawfish production at a favorable cost. This paper summarizes 15 years research.

NATURAL VEGETATION

Early crawfish farmers relied on volunteer natural vegetation - such as alligator weed (Althernanthera) (philoxeroides), water primrose (Jussiaea spp.) and smart weed (Polygonum spp.) to provide food for crawfish. Clark (1973) and Clark et al. (1974) conducted experiments in metal pools with two species of plants, with feeds, and with use of fertilization. Replicated pools were stocked at a rate of 31,000 and 62,000 young crawfish/ha in winter and with 62,000 per ha in summer. Pools receiving vegetation were planted with alligator weed or smart weed. Fertilization pools received several applications of 8-8-8 fertilizer at 130 kg/ha/application. Pools designated for feed received either catfish pellets (35% protein) or an extruded ration (32% protein). Crawfish were fed daily at 3% of bodyweight. Overall, crawfish production was higher in summer than in winter. Optimal production and growth occurred for crawfish given pelleted feed (1,186 kg/ha and 22 g/crawfish) and for those given extruded feed (1,135 kg/ha and 21 g/crawfish). Average production and growth were similar for crawfish maintained on smart weed (777 kg/ha and 13 g/crawfish), those maintained on alligator weed (676 kg/ha and 12 g/crawfish), and those with fertilization (694 kg/ha and 13 g/crawfish). From a practical standpoint, these results did not provide a good method of growing crawfish. Production of crawfish on feed did exceed 1000 kg/ha but the cost was prohibitive. Production of crawfish with plants, though much lower, was still the better method for growing crawfish at a profit. However, use of natural plants on commercial farms is unpredictable. One year a given pond may grow an abundance of vegetation, the next year virtually none.

In other studies with natural vegetation, delta duckpotato (Sagittari platyphylla) was planted in replicated ponds (Johnson 1980). Mean production of crawfish was 945 kg/ha, still less than the desired minimum production of 1000 kg/ha.

HAYS, MANURES, AGRICULTURAL BY-PRODUCTS AND FERTILIZATION

A laboratory study was conducted to determine growth of crawfish fed six agricultural by-products (Goyert 1975; Goyert et al. 1975-76; Goyert and Avault 1977). Crawfish fed dried sweet potatoe (Ipomoea batatas) vines attained the highest final average weight followed by sweet potatoe trimmings, rice (Oryza sativa) stubble, and rye (Secale cereale) hay. Crawfish fed soybean (Glycine max) stubble and sugar cane (Saccharum officinarum) bagasse grew poorly. In further studies, ten different materials were analyzed to determine nutrient content and decomposition rate over time. The materials were dried at 60°C for 24 hours, ground, then analyzed for protein and other parameters. For the decomposition study, the dried materials were placed in flasks containing pond water. Periodic samples were analyzed for carbon and nitrogen.

Results are summarized as follows:

Item	Initial Protein%	Carbon:Nitrogen Ratio Initial	2 weeks	6 weeks	12 weeks
Soybean Stubble	18.8	13	11	10	7
Sweet Potato Vines	11.8	20	17	11	10
Rice Stubble	5.5	45	38	22	13
Rye Hay	9.3	18	18	10	7
Sugar Cane Bagasse	3.6	57	89	36	38
Sweet Potato Trimmings	7.1	25	14	6	6
Tree Leves, Mixed	6.4	61	35	27	28
Soybean	44.8	6	7	6	7
Annual Grasses	10.4	19	18	11	10
Corn Leaves	17.1	15	12	8	8

When considering agricultural by-products as feed substrate for crawfish, one must consider not only their initial nutritional value but also their nutritional value after entrance into the detrital system. When organic material enters the detrital system, micro-organisms use the carbon as a source of energy, with large amounts liberated from the system as carbon dioxide. Through mineralization, nitrogen is released by decomposition but instead of being liberated (lost) from the system, it is immobilized into microbial tissue. Microbes ultimately die and nitrogen is released (mineralization) back into the system which in turn is immobilized by other microbes. In essence, over time there is a net loss of carbon with a "relative" increase of nitrogen. When the C:N ratio drops to 17:1, the detrital material is considered nutritious for crawfish.

Romaire (1976) reported crawfish production in replicated ponds receiving rice hay (736 kg/ha), bahiagrass hay (733 kg/ha), fertiliztion (192 kg/ha), and control ponds receiving neither hays nor fertilization (202 kg/.ha). He concluded that it was not practical to feed crawfish hay from the outset, but hay may provide feed substrate when vegetation is depleted.

Rivas (1978), Rivas et al. (1978) and Avault et al. (1978) reported on sugar cane wastes and chicken manure as feed substrate for crawfish. In one study four treatments in replicated ponds gave the following results: (1) sugar cane bagasses (wastes after processing) was applied twice at a rate of 2000 kg/ha each time, yielding an average of 280 kg/ha of crawfish, (2) chicken manure was applied at a rate of 20 kg/ha/day for 147 days, yielding an average of 438 kg/ha, (3) sugar cane bagasse was applied twice at a rate of 2000 kg/ha each time plus 110 kg/ha of 8-8-8 fertilizer was applied 10 times. An average yield of 459 kg/ha of crawfish was obtained, (4) the last treatment consisted of two applications of 2000 kg/ha of sugar cane bagasse each and 20 kg/ha/day for 147 days of chicken manure. The average yield of crayfish was 395 kg/ha.

In a second study conducted from November to May, there were four treatments: (1) sugar cane bagasse at a rate of 4000 kg/ha in one application gave the lowest average yield of 359 kg/ha of crawfish, (2) chicken manure applied at a rate of 20 kg/ha/day gave the highest average yield of 600 kg/ha, (3) the third treatment consisting of sugar cane filter cake at 40 kg/ha/day gave the second lowest average production with 370 kg/ha, (4) sugar cane filtered cake at 40 kg/ha/day plus chicken manure at 20 kg/ha/day gave the second highest average yield with 581 kg/ha.

To summarize these two studies, significant differences were found among treatments in both studies. Nevertheless, in no instance did crawfish production reach the minimal acceptable level of 1000 kg/ha.

Johnson (1980) and Johnson and Avault (1980) examined use of chicken manure in combination with rice forage. All ponds were planted with rice. Ten ponds received manure biweekly, five at a rate of 10/kg/day and five at a rate of 102 kg/ha/day. Four rice ponds received no manure and served as controls. There was no significant difference in mean crawfish yields among rice-low manure (1349 kg/ha), rice-high manure (1631 kg/ha) and rice-control (1264 kg/ha) ponds. Though no significant differences existed between treatments, one outstanding fact was noted - all ponds containing rice as forage yeilded well in excess of 1000 kg/ha.

PLANTED FORAGES

The first in-depth studies on use of planted forages as crawfish feed were reported by Miltner (1980) and Miltner and Avault (1980, 1981). Chien (1978) and Chien and Avault (1978, 1980) reported on the value of rice straw as feed substrate for crawfish. Japanese millet and a domestic rice, variety La Belle, were evaluated as planted fodder for crawfish. In addition, the effects of two alternate planting dates, June 15 and August 1, and mowing versus forage left standing were examined. Crawfish production for the various treatments is given below in kg/ha:

	Rice		Millet	
	Standing	Mowed	Standing	Mowed
Early Planted	1609	1201	treatment lost	943
Late Planted	1208	871	948	1015

Use of Supplemental Feeds

Though rice planted as a fodder gave satisfactory yields of crawfish, most fodder was completely depleted in March, with 2 to 3 months growing season left. Cange et al. (1982) evaluated use of a low-cost ($100/ton) cattle range pellet (crude protein 9%) as supplemntal feed for crawfish. Sixteen ponds were used; 12 were planted with rice and four were not. In treatment one, four ponds without planted vegetation received a pelleted ration from 25 September to 3 May. In treatment two, four ponds planted with rice received the pelleted ration from 25 September to 3 May and, in treatment three, four rice-planted ponds received the pelleted ration only from 1 March through 3 May. Four rice-planted ponds received no supplemental feed and served as controls. Production of crawfish in kg/ha was:

Pond Treatment	Average yield (kg/ha)
Pellets only	881
Planted-rice only	1274
Planted-rice and pellets (March-May)	2016
Planted-rice and pellets (September-May)	2130

When pellets were supplemented, production increased significantly. From this study, use of feed pellets may be practical on a commercial basis, but cost of pellets plus daily cost of labor needs documentation.

DISCUSSION OF PLANTED FORAGES

Food Sources

In ponds planted with rice as forage, the crawfish derives its food from eating green plants, periphyton, benthos, plankton, animal detritus and plant detritus. All food sources are interrelated. When a pond containing planted rice is first flooded, crawfish begin feeding on tender, green rice shoots and on periphyton. Newly hatched young may feed on plankton. As time passes, periphyton increases in biomass. Chien (1980) for example, noted that periphyton in grams/m^2 ws 229, 267, and 333 at 6, 12, and 18 weeks after flooding, respectively. Meanwhile, rice plants begin dying, decomposing, and enter the detrital system. When detrital material reaches a C:N ration of 17:1, the material is deemed nutritious for crawfish. Ultimately, rice-forage detritus becomes depleted. In Louisiana this may occur in February or early March. The detrital food web now gives way to a plankton food web (Witzig 1980; Witzig et al. 1980), and crawfish growth plummets (Avault et al. 1974). The detrital food web, then, is the foundation for growth and yield of crawfish, yet rice forage is inadequate to provide crawfish enough food to last through the crawfish growing season (October to May).

Dynamics of Rice Forage

When evaluating rice plants or any other plant as crawfish forage, one must consider: biomass production, lodging, decomposition rate, and C:N ration (Miltner 1980; Milner and Avault 1980). In one study, the rice variety, La Belle, produced a peak standing crop or biomass production of 0.78 kg/m^2 prior to flooding; whereas Japanese millet produced only 0.57% kg/m^2. It would seem logical that in this case rice was more desirable than millet because it produced more detrital substrate, and this was true based on crawfish yield. Use of rice varieties, such as Saturn, which produce more biomass than La Belle would seem desirable; and use of fertilizers to grow rice also seems practical.

The timing of availability of whole plant matter to the detrital system is important in crawfish culture. Rice has evolved a vascular system that permits it to remain viable in submerged soils, whereas Japanese millet dies when flooded. Thus, rice lodges over an extended period of time, raining substrate into the water. Millet, on the other hand, lodges rapidly, decomposes quickly, and is quickly depleted.

The decomposition rate is governed by the C:N ratio of the plant material. Rice, for example, which has a much lower C:N ratio than sugar cane leaves decomposes faster. Rice-plant detritus may be in excess of crawfish needs early in the season but becomes depleted later on. Ideally, two or more plant species with differing decomposition rates (C:N ratios) should be used to stagger food (detritus) substrate over time.

Other Considerations

Water quality, treatment of rice straw, and crawfish harvesting are important considerations. When flooding rice plants in the fall, the plants should not be submerged completely. Otherwise they may all begin dying at once and lead to an oxygen depletion. Moreover, one must carefully consider the time of fall flooding. Flood to early when it is still very hot, and decomposition is accelerated. Flood to late when it is much cooler, then the crawfish growing season is shortened. Generally, ponds are flooded in Louisiana from 15 September to 15

October. In one study where rice was harvested (Chien 1980; Chien and Avault 1980) 18 ponds were used to determine the effect of fall flooding and the effect of rice straw disposal on crawfish yield. Half the ponds were flooded on 20 September and half on 15 October. Rice straw was either disked under, left standing, or baled and added back periodically. Results in kg/ha of crawfish were:

Flooding Date	Baled	Disked	As Is	Average
20 September	1554	821	1173	1183
15 October	1457	784	1141	1127
Average	1506	803	1157	1155

Harvesting plays an integral part in total yield of crawfish and average size of crawfish. Generally, most crawfish farmers underharvest by not trapping intensively enough (number of traps per ha) or often enough. When rice substrate begins to become depleted and/or when crawfish growth lessens, it is virtually impossible to overharvest using current methods - traps and bait. If crawfish are not removed quickly, they stunt off and may create an overcrowding problem which will carry over the following year.

CURRENT RECOMMENDATIONS AND FUTURE RESEARCH

(1) In the South, the best practice is to seed approximately 100 kg of high yielding (biomass) rice variety per ha on a well prepared seed bed. Rice should be fertilized to enhance growth. In Louisiana, rice used as forage may be planted from April through July.

(2) The second year, the soil need not be disked. After crawfish harvest ends, late May in Louisiana, the water should be removed slowly over 1 to 2 weeks to enhance crawfish burrowing. Rice seed can be broadcasted when the water level is down to 10 cm. Following planting, the remaining water should be removed. This no-till practice saves labor and energy.

(3) Future research should concentrate on use of high biomass producing rice plants and other domesticated plants and on use of two or more plants species with differing decomposition rates. Research on water quality is concomitant with research on forages.

LITERATURE CITED

Avault, James W., Jr., Larry W. de la Bretonne, and Jay Huner. 1974. Two major problems in culturing crayfish in ponds: oxygen depletion and overcrowding. In Proceedings 2nd International Crayfish Symposium 2:139-148.

Avault, James W., Jr., Ricardo Hernandez, and Mike Giamalva. 1978. Sugar cane waste products and chicken manure as supplemental feed for crawfish. Sugar y Azucar 73(6):42.

Cange, Stephen, Michael Miltner, and James W. Avault, Jr. In Press. Use of range pellets as supplemental feed for crawfish. Progressive Fish-Culturist.

Chien, Yew-Hu. 1978. Double cropping rice Oryza sativa and crawfish Procambarus clarkii (Girard). M.S. Thesis. LSU. 94 p.

Chien, Yew-Hu. 1980. Effects of flooding dates and disposals of rice straw on crayfish. Procambarus clarkii (Girard) culture in rice field. Ph.D. Dissertation. LSU. 120 p.

Chien, Yew-Hu, and James W. Avault, Jr. 1978. Double cropping rice, Oryza sativa, and red swamp crawfish Procambarus clarkii. In Proceeding 4th International Crayfish Symposium 4:263-276.

Chien, Yew-Hu, and James W. Avault, Jr. 1980. Effects of flooding dates and type disposal of rice, Oryza sativa, straw on the crawfish, Procambarus clarkii (Girard), culture in rice fields (Abstract only). In Abstracts of Fish Culture Section of the American Fisheries Society. 1980:14.

Chien, Yew-Hu, and James W. Avault, Jr. 1980. Production of crayfish in rice fields. The Progressive Fish-Culturist. 42(2):67-70.

Clark, Donald Francis. 1973. Effects of feeding, fertilization, and vegetation on production of red swamp crawfish, Procambarus clarkii. M.S. Thesis. LSU. 86 p.

Clark, Donald F., James W. Avault, Jr., and Samuel P. Meyers. 1974. Effects of feeding, fertilization, and vegetation on production of red swamp crayfish, Procambarus clarkii. In Proceedings 2nd International Crayfish Symposium 2:125-138.

Goyert, Jonathan Craig. 1975. Agricultural by-products as supplemental feed for crawfish, Procambarus clarkii. M.S. Thesis. LSU. 38 p.

Goyert, Jonathan C., James W. Avualt, James E. Rutledge, and Teme E. Hernandez. 1975-76. Agricultural by-products as supplemental feed for crawfish. Louisiana Agriculture 19(2):10-11.

Goyert, Jonathan C., and James W. Avault, Jr. 1977. Agricultural by-products as supplemental feed for crayfish, Procambarus clarkii. Transactions of the American Fisheries Society. 106(6):629-633.

Johnson, Woodruff Barnes, Jr. 1980. Evaluation of poultry waste, rice (Oryza sativa), and delta duckpotato (Sagittaria platyphylla) as crayfish (Procambarus spp.) forages. M.S. Thesis. LSU. 94 p.

Johnson, W. Barnes, and James W. Avault, Jr. 1980. Some effects of poultry manure supplementation to rice/crawfish experimental earthen ponds (Abstract only). In Abstracts of Fish Culture Section of the American Fisheries Society. 1980:15.

Miltner, Michael Ronald. 1980. Evaluation of domestic rice, <u>Oryza</u> <u>sativa</u>, and Japanese millet, <u>Echinocloa</u> frumentacea, as fodder for red swamp crawfish, <u>Procambarus clarkii</u>. M.S. Thesis, LSU. 106 p.

Miltner, Michael R., and James W. Avault, Jr. 1980. An evaluation of rice (<u>Oryza sativa</u>) and Japanese millet (<u>Echinocloa frumentacea</u>) as forage for red swamp crawfish (<u>Procambarus clarkii</u>) (Abstract only). In Abstracts Fish Culture Section of the American Fisheries Society,. 1980:15.

Miltner, Michael R., and James W. Avault, Jr. 1981. Rice and millet as forages for crawfish. Louisiana Agriculture 24(3):8-10.

Rivas, Ricardo. 1978. Poultry manure, and sugar cane by-products as supplemental feed and fertilizer for crawfish <u>Procambarus clarkii</u> (Girard). M.S. Thesis. LSU. 38 p.

Rivas, Ricardo, Robert Romaire, James W. Avault, Jr., and Mike Giamalva. 1978. Agricultural forages and by-products as feed for crawfish, <u>Procambarus clarkii</u>. In Proceedings 4th International Crayfish Symposium 4:337-342.

Romaire, Robert Paul. 1976. Population dynamics of red swamp crawfish, <u>Procambarus clarkii</u> (Girard), in ponds receiving fertilization, and two agricultural forages as supplemental feed. M.S. Thesis. LSU. 103 p.

Witzig, John. 1980. Spatial and temporal patterns of macroinvertebrate communities in a small crawfish pond. M.S. Thesis. LSU. 113 p.

Witzig, John, James W. Avault, Jr., and Jay V. Huner. 1980. Insect dynamics in a crawfish pond with emphasis on predaceous insects (Abstract only). In Abstracts of Fish Culture Section of the American Fisheries Society. 1980:14.

AN APPROPRIATE FOOD DELIVERY SYSTEM FOR LOW-LEVEE POND CULTURE OF Procambarus clarkii, THE RED SWAMP CRAYFISH

Michael R. Miltner and James W. Avault, Jr.
Fisheries Section
School of Forestry and Wildlife Management
Louisiana State University
Baton Rouge, Louisiana 70803 USA

ABSTRACT

Conventional approaches to feeding aquatic species in mainstream aquaculture often require purchase and delivery of a high-cost processed rations, feed storage facilities, and labor and equipment inputs to dispense feeds. Energy costs and other resource constraints are forcing both agriculturalists and aquaculturalists to move toward more conservative practices such as minimum or no tillage, double cropping, polyculture, crop rotation, heat recovery, and on-farm recycling of waste products. Research at LSU has recently focused on use of pond-planted forage crops as an appropriate feed delivery system in crawfish farming. Production of harvestable ($\geq$ 75 mm) crawfish has ranged as high as 2000 kg/ha in forage-planted ponds. A partial list of plant species currently being evaluated includes rice (Oryza sativa), sorghum (Sorghum spp.), American jointed vetch (Aeschynonema americana), Japanese millet (Echinocloa frumentacea), and alligator weed (Alternanthera philoxeroides). Development of a Detritus-based forage/feeding system appears to be an appropriate approach based on present biological knowledge regarding crawfish food, water quality, and behavioral requirements, and also in the larger socio-agricultural context.

INTRODUCTION

Farming of the crayfish, Procambarus clarkii, is seeing a continued expansion in the south central states, foremost Louisiana, Texas and Mississippi. There are an estimated 60,000 acres in crayfish production in these states currently, with the bulk (55,000 acres) occurring in Louisiana (Census, 1980).

New and existing crustacean farmers are confronted with many effects of energy and resource constraints as agriculturalists. Rapidly mounting production costs and diminished returns are forcing the farmer to move toward more resourceful, conservative on-farm practices such as minimum or no-tillage, double cropping or polyculture, crop rotation, range pasturing, heat re-use, and recycling of farm waste products, etc. The crayfish farmer is not immune to these pressures, and key concerns include increasing efficiency of harvest technology, cost effective improvement of water quality, and development of an appropriate food delivery system for culture of the red swamp crayfish. It is the latter concern that this paper addresses.

Feed Delivery Systems

Conventional approaches to feeding aquatic species in U.S.

370

aquaculture often require purchase and delivery of a high-cost processed ration, feed storage facilities, and labor and equipment inputs to dispense feeds. For example, the annual operating cost of feed for commercial catfish production in Mississippi in 1980 was $1072/acre or from 51% to 58% of the total annual production costs (Waldrop and Smith 1980). At the present time feed costs account for the second largest expense in producing freshwater prawns, M. rosenbergii, in Hawaii. Prawn culture is currently the only other viable crustacean farming industry in the U.S. Farmers feed an average of 40-60 lb/ acre/day which, at an average cost of $0.13/lb, results in a yearly operating cost of almost $50,000 for a 20 acre prawn farm (Stahl 1979).

In the mid-sixties, LSU Agricultural Experimental Station researchers tested the use of formulated fish rations and special crustacean diets for increasing crayfish production in simulated culture environments (Clark et al. 1974). These experiments demonstrated that crayfish will readily accept such feeds and grow rapidly, but cost of nutritionally complete rations and associated equipment/laboratory costs does not justify their use in pond culture.

More recently, attention has shifted toward use of locally available agricultural by-products, livestock manures, and hays as feed and substrate for crayfish. Goyert et al. (1975) evaluated a number of byproducts from Louisiana agricultural crops as supplemental feeds, including rice and rye hays, soybean cuttings, sweet potato trimmings, and sugar cane wastes, in a soilless laboratory study. Romaire (1976) compared rice hay and bahiagrass hay treatments to those ponds receiving inorganic fertilization and to those where no food or substrate were added. Avault and Hernandez (1978) tested two by-products of the sugar cane industry, bagasse and filter cake, alone and combined with dried poultry manure as supplemental feeds in non-vegetated crayfish ponds.

Some of these materials have shown promise, but in many cases there are logistical problems associated with their labor-intensive collection, transportation, and application. For the small-scale backyard or subsistence culturalist such a waste feeding scheme would be appropriate, and also for those commercial producers who have integrated operations where both agricultural crops and/or confined livestock are raised. A number of researchers have investigated this integration of livestock husbandry with low-trophic level fish culture (Buck 1978, Stickney et al. 1979). Still, the costs associated with the handling logistics may be prohibitive. The farmer would need to determine suitable loading times and rates in order to insure adequate feedstock and maintain pond water quality. Schroeder (1974) has investigated the problem of oxygen depletion in manured ponds and has derived a means of estimating BOD based on percent dry matter in manure. Romaire and Boyd (1978) have also derived a practical technique for predicting oxygen depletion, which with some adaptation, might be employed to estimate optimum loading rates for crayfish ponds.

An additional problem also surfaces regarding the adoption of feeding innovations among crayfish farmers in south Louisiana. In many cases the crayfish pond represents a minor diversification of a total farming operation and the major crops, i.e. rice, sugar cane, soybean, have priority (de la Bretonne and Fowler 1976). Under these circumstances, crayfish are not always given the status of livestock, requiring care, attention, and feeding. The prevalence of the Cajun concept of crayfish as "lagnaippe", i.e. "a little something extra," is something that continues to affect the farmer's willingness to adopt innovation. In this socio-cultural context, caretaking of the pond crayfish, which includes feed delivery, is subsequently given less attention than in other commercial finfish culture scenarios. It becomes apparent that in considering any new feed delivery approach, what comes to be accepted as appropriate will be defined by socio-cultural forces as well as dollar cost and return logic.

Pond Planted Forage Crops as Appropriate Feed System

A number of studies have been undertaken at LSU to evaluate the use of pond-planted forage crops as an appropriate feed delivery system in crayfish farming. Miltner et al. (1981) compared domestic rice (Oryza sativa) and Japanese millet (Echinocloa frumentacea) as planted fodder for the red swamp crayfish. This study also examined the effects of alternative management practices such as planting date and mowed fodder versus standing fodder as they affected water quality, food availability, and crayfish yields.

MATERIALS AND METHODS

Forage Propagation, Management

The 24 ponds used in the study, each measuring 0.125 acre (0.05 hectare), were drained slowly between June 4 and June 14 to allow resident crayfish to burrow for completion of their reproductive cycle. On June 15, 12 ponds were designated "early planted," six of which were planted to rice and six to millet at 100 pounds per acre and 45 pounds per acre, respectively. Presprouted seed was broadcast by hand on mucky pond bottoms as they were drained. Planting during and immediately after pond drawdown allows the farmers to take advantage of the moisture and mucky tilth that aid in establishing a good stand of forage.

The remaining 12 ponds, designated as "late planted," were planted in a like manner on August 1, 45 days after pond draining. These ponds were managed in much the same way as early planted ponds with two exceptions: 1) weed control was necessary, using both manual and chemical (Paraquat) means, since volunteer weeds invaded plots; and 2) additional water pumping was necessary to insure that pond bottoms were mucky at planting time. Cultural inputs were minimized in all ponds; no tillage was done and no fertilizer was applied.

Forages in all ponds were allowed to grow and remain standing until September 28 when ponds assogned to the "mowed" treatment were cut with a rotary mower, leaving loose hay and a 5-inch stubble within ponds. All 25 ponds were reflooded on October 6, at which time adult crayfish and their young were flushed from burrows.

Key parameters were measured to determine whether there were differences in forage crop productivity and forage behavior under flooded conditions. The peak forage standing crop (biomass in kg/m^2) was measured using a clip method at the time of grain panicle formation or "heading out." Ten (10) 0.25 m^2 samples were taken from each pond with all forage above ground being clipped, oven dried, and weighed.

Forage biomass is important since it affects water quality and reflects the amount of substrate available for periphyton attachment and for conversion to detritus. Crayfish are known to subsist on periphyton associated with macrophytes and, most significantly, on the particulate detritus produced by the fragmentation and decomposition of the plant material (Goldman 1974, Avault et al. 1975, Goyert et al. 1975, Chien 1980).

It was also hypothesized that differences in forage lodging rate and decomposition among treatments might have an effect on availability of detrital feedstock for crayfish and on water quality. Lodging is defined by a decrease in standing vegetative cover as senescence and weathering cause the forage to fragment and fall into the water.

Lodging rate was measured on a monthly basis using a modified point quadrant technique (Goodall 1951). This method allowed recording of the monthly decrease in emergent stem/leaf material in ponds where forage was left standing. Forage decomposition rates were determined using a litterbag loss technique (Odum and de la Cruz 1967), in which pre-weighed samples were enclosed in small mesh bags, immersed in ponds, and sampled monthly to determine loss of whole plant material as forage decomposed to particulate detritus.

Dissolved oxygen (DO) and temperature measurements were made at dawn twice a week in each pond using YSI a meter with a thermistor.

Crayfish were harvested from January 10 until May 19. Four double-throated, 3/4-inch standup traps were used in each pond, baited with catfish heads, and run on 18- 24-hour sets. Total weight of harvestable crayfish was recorded for each pond sample on each trapping date, and the cumulative yield of all fishing days was taken as total yield per pond. The total harvest effort amounted to 120 trap days per pond.

RESULTS AND DISCUSSION

The superior forage/management combination tested in this study was domestic rice planted June 15 (early planted) and left unmowed. The average crayfish yield from this treatment was 180 pounds per pond (1436 pounds per acre), significantly greater ($p < 0.01$) than from all other treatments. Early planted rice produced the highest forage biomass, significantly greater ($p < 0.05$) than other treatments (Table 1).

Average crayfish yields from ponds planted with millet (842 lb/acre) were generally inferior to those from ponds planted with rice (1165 lb/acre). Key forage parameters showed that millet produced from 20 percent to 50 percent less forage biomass, lodged at a more rapid rate, and that decomposition was slow and incomplete when compared with rice.

Millet that was left standing was 90 percent lodged by mid-November, while 41 percent of the emergent early planted rice and 64

Table 1. Crawfish production, forage parameters, and water quality data from 0.125 acre ponds planted with rice or millet.[1]

Treatment[2]	Mean crawfish yield (lbs/A)	Mean forage biomass (kg/m²)	Mean forage lodging rate[4]	Forage decomposition rate[5]	Mean dawn DO[6] (ppm)
Early Rice					
Mowed	1,073	0.86	Complete[7]	Intermediate	1.8
Standing	1,436	1.03	Intermediate	Intermediate	4.1
Late Rice					
Mowed	778	0.46	Complete	Rapid	2.7
Standing	1,079	0.76	Slow	Intermediate	3.3
Early Millet					
Mowed	842	0.70	Complete	Slow, incomplete	3.2
Standing[8]	--	--	--	--	--
Late Millet					
Mowed	907	0.52	Complete	Intermediate	2.2
Standing	847	0.50	Rapid	Slow, incomplete	4.0

[1] Rice planted at 100 lbs/A, millet at 45 lbs/A.
[2] Each treatment had three replications. "Early" forages were planted June 15, "late" forages were planted August 1.
[3] Biomass measured initially October 2 before flooding.
[4] Rate of decrease of emergent measured monthly using point quadrant method.
[5] Rate of decay of whole plant material to less than 2 mm particles measured monthly.
[6] Dissolved oxygen measured biweekly at 6 a.m., 3 inches from pond bottom. Average taken for first 66 days after flooding.
[7] Mowed treatments were 100% lodged at flooding by definition.
[8] No data available due to crop loss.

percent of the late planted rice remained standing at the time. The rate of lodging was significantly slower (p < 0.01) in unmowed rice ponds than it was in unmowed millet ponds. Rice has evolved a vascular system that allows it to remain viable in submerged soils. Japanese millet has no such vascular system and therefore died in response to flooding. It is desirable that lodging take place over an extended period to provide a continuous "raining of food for crayfish during the entire growing season.

LSU Experiment Station researchers have found that the carbon:nitrogen (C:N) ratio of forage material is a good indicator of its nutritional value to crayfish. The initial C:N ratio affects the rate of decomposition of plant material to microbially enriched detritus, P. clarkii's main source of nourishment in culture ponds.

Past research has shown that forage becomes nutritionally complete after decomposition to a C:N ratio of 17:1 or less. Monthly proximate analysis of the forage material in this study showed that millet remained too high in fibrous carbonaceous materials (cellulose and lignin) from February through May, when crayfish depend on detritus most. Rice, on the other hand, had a more optimal low C:N ratio and continued to decompose into particulate detritus throughout the study.

Dissolved oxygen was adequate to sustain juvenile crayfish growth and survival in ponds where forage was left standing, but dawn DO levels dipped below the critical threshold of 1 part per million on numerous occasions in mowed ponds during October and November. Standing forage provides, in addition to superior water quality, more surface area for periphyton growth in the fall, extends food/substrate later into the season, and allows for spatial segregation to alleviate crowding of crayfish.

A propagated detrital feedstock such as the forage planting system outlined here seems to have certain advantages over other food delivery scenarios in crayfish culture. Replicated studies and field tests have shown that if forage planting is done during and immediately following crayfish pond draining in June, tillage, pesticide applications, and inorganic fertilizer inputs are not necessary for production of a good stand of forage. Soil moisture, tilth, and fertility are normally adequate in the drained crayfish pond to sustain germination and growth of forage crop for the following year's crayfish population.

Rice forage, or the combination of rice and other suitable crops require only an initial labor input for planting whereas conventional approaches to feeding aquatic species often require routine labor and equipment inputs to dispense feed at pond side. Any cultural practice that would increase labor requirements and costs on the crayfish farm is problematic (de la Bretonne and Fowler 1976). A no-labor technology for planting seed is already available in the form of small agricultural aircraft often used to sow rice seed for farmers in S.W. Louisiana (Hill, personal communication). Total costs for rice seed and aerial planting are approximately $35 per acre (Musick et al. 1980).

An emergent semi-aquatic forage crop such as the domestic rice varieties used at LSU also eliminates the need for a standard feeding schecule. Rice forage planted and left standing has its own built in "feeding schedule" due to the gradual staggered lodging and fragmentation of the material into the detrital pool.

The timing of decomposition and delivery of forage material to the benthic region appears to be important as it affects both the availability of plant matter for conversion to detritus and also the availability and configuration of substrate in the pond (Chien 1980, Miltner 1980). This latter factor, the amount and configuration of vegetative substrate, determines the extent of the surface area for periphyton attachment and also the opportunity for spatial segregation among individuals in the expanding population. Goyert et al. (1975), Chien (1980) and Miltner et al. (1981) all concluded that substrate (standing forage) and its configuration affect crayfish population density and survival. Similar affects have been noted with _Macrobrachium rosenbergii_ (George 1977).

Though the standing rice crop appears to meet most criteria for a "good" crayfish forage in this bioregion, it has been observed to "play out" or become depleted later in the season when crayfish grazing pressure and microbial activity are highest. Studies by Johnson and Avault (1980) and Cange et al. (1981) have investigated the use of poultry manure and range pellet supplements, finding both to increase crayfish production over ponds planted to rice forage alone. An alternative approach might be to plant a combination of forage crops with different C:N ratios and decay rates to sustain adequate nutritional food, substrate, and periphyton throughout the season.

CONCLUSION

To conclude, we believe that the forage propagation method described is an appropriate feed delivery system for crayfish farmers. It is "appropriate" (Congdon 1977) in the sense that it appears to satisfy food/substrate requirements of the crayfish population and sustain consistent yields of over 1000 kg/ ha; it is replicable, having been employed on both an experimental and commercial scale by model farmers in Louisiana and Texas (Davis, personal communication); it is simple, low cost, and adaptable with existing technology and resources available to the local agricultural community; and it can be implemented over a wide range of farm sizes from homestead subsistence to large commercial operations. Also, research suggests that cultural inputs such as tillage, fertilizer and pesticide applications are often unnecessary if pond drainage and forage propagation are properly managed.

Propagation of a forage crop is an effective alternative for crayfish farmers to conventional feed delivery approaches for other aquatic species which require the purchase and delivery of an expensive processed ration, construction of storage facilities, and labor and equipment to dispense feed at pond site.

LITERATURE CITED

Avault, J. W., Jr., L. de la Bretonne, Jr. and J. V. Huner. 1975. Two major problems in culturing crayfish in ponds: oxygen depletion and overcrowding. In Papers from the 2nd Int. Symp. on Freshwater Crayfish 2: 139-144.

Avault, J. W., Jr., R. Hernandez and M. Giamalva. 1978. Sugar cane waste products and chicken manure as supplemental feed for crawfish. Sugar Y Azucar 73(6):42.

Buck, H., R. J. Baur and C. R. Rose. 1978. Polyculture of Chinese carps in ponds with swine wastes. In Papers from Symposium on Culture of Exotic Species. Amer. Fish. Coc., Atlanta. 24 pp.

Chien, Yew-Hu. 1978. Double cropping rice Oryza sativa and crawfish Procambarus clarkii (Girard). M.S. Thesis. LSU. 94 pp.

Chien, Yew-Hu and J. W. Avault, Jr. 1978. Double cropping rice, Oryza sativa, and red swamp crawfish, Procambarus clarkii. In Proceedings 4th International Crayfish Symposium 4:263-276.

Clark, Donald F., James W. Avault, Jr. and Samuel P. Meyers. 1974. Effects of feeding, fertilization, and vegetation on production of red swamp crayfish, Procambarus clarkii. In Proceedings 2nd International Crayfish Symposium 2:125-138.

Congdon, R. J. 1977. Introduction to Appropriate Technology. Rodale Press. Emmaus, PA. 205 pp.

de la Bretonne, L., Jr. and J. F. Fowler. 1976. The Louisiana crawfish industry -- its problems and solutions. Presented 30th Ann. Conf. Southeastern Assoc. Game and Fish Comm. 23 pp.

Davis, J. Personal communication with Texas A&M Extension Service. March 1981.

George, M. J., K. H. Mohammad and N. K. Pillay. 1967. Observations on the paddy-field prawn cultivation of Kerala, India. FAO World Scientific Conf. Biology and Culture of Shrimps and Prawns 67/E/18.

Goldman, C.R., J.C. Rundquist and R.W. Flint. 1974. Ecological Studies of the California crayfish, Pacifastacus leniusculus, with emphasis on their growth from recycling waste products. In Papers for the 2nd Int. Symp. on Freshwater Crayfish 2:481-487.

Goodall, S. W. 1951. Some considerations in the use of point quadrants for the analysis of vegetation. In Aust. J. Sci. Res. 5(1):1-41.

Goyert, J. C. and J. W. Avault, Jr. 1977. Agricultural by-products as supplemental feed for crayfish, Procambarus clarkii. Transactions of the American Fisheries Society 106(6):629-633.

Johnson, W. V., Jr. 1980. Evaluation of Poultry waste, rice (Oryza sativa), and delta duckpotato (Sagittaria platyphylla), as crayfish (Procambarus spp.) forages. M.S. Thesis. LSU. 94 pp.

Miltner, M. R. 1980. Evaluation of domestic rice, Oryza sativa and Japanese millet, Echinocloa frumentacea, as fodder for red swamp crawfish, Procambarus clarkii. M.S. Thesis. LSU. 106 pp.

377

Miltner, M. R. and J. W. Avault, Jr. 1981. Rice and millet as forages for carwfish. Louisiana Agriculture 24(3):8-10.

Musick, J. et al. 1980. Cost and returns in rice production in Louisiana. Pub. No. 231 from Dept. of Ag. Econ. Louisiana State University Extension Service. Baton Rouge, LA.

Odum, E. P. and A. A. de la Cruz. 1967. Particulate organic detritus in a Georgia salt march ecosystem, pp. 383-388. In George H. Lauff (ed.) Estuaries. AAAS, Washington, DC.

Romaire, R. P. 1976. Population dynamics of red swamp crawfish, Procambarus clarkii (Girard), in ponds receiving fertilization, and two agricultural forages as supplemental feed. M.S. Thesis. LSU. 103 pp.

Schroeder, G. 1974. Use of fluid cowshed manure in fishponds. Bamidgeh. The Bulletin of Fish Culture in Israel. 26(3):84.

Stahl, Margo. 1979. The role of natural productivity and applied feeds in the growth of Macrobarchium rosenbergii. Unpublished manuscript.

Stickney, R. R. 1979. Principles of Warmwater Aquaculture. John Wiley and Sons. New York. 375 pp.

Waldrop, J. and R. D. Smith. 1980. An economic analysis of producint pond-raised catfish for food in Mississippi. Dept. of Agric. Econ. Miss. State Univ. 36 pp.

CRAWFISH, PROCAMBARUS SPP., PRODUCTION FROM SUMMER FLOODED EXPERIMENTAL PONDS USED TO CULTURE PRAWNS, MACROBRACHIUM ROSENBERGII, AND/OR CHANNEL CATFISH, ICTALURUS PUNCTATUS, IN SOUTH LOUISIANA

Jay V. Huner
Department of Biological Sciences
Southern University
Baton Rouge, Louisiana 70813

Michael Miltner and James W. Avault, Jr.
School of Forestry and Wildlife Management
Louisiana State University
Baton Rouge, Louisiana 70803 USA

ABSTRACT

Resident crawfish were harvested in ponds used to culture prawns and/or catfish fingerlings in the summers of 1979 and 1980. Stocking rates were: prawns, 25,000/ha, 1979; 25,000 or 50,000/ha, 1980; catfish fry, 50,000/ha, 1979; 150,000/ha, 1980. P. clarkii was dominant but P. a. acutus was abundant in some ponds. Crawfish yeilds were 136-443 kg/ha, 1979, and 338-636 kg/ha, 1980. Differences were attributed to pond management practices, presence of prawns, and harvesting methods. In another, related study, resident crawfish were harvested from ponds stocked in September 1980 with 10-15 cm catfish fingerlings at rates of 1785, 3573, 8073, or 14,285/ha. P. clarkii dominated most ponds but many P. a. acutus were present. Yields were 192-1207 kg/ha with minimal differences attributable to catfish stocking densities. Green sunfish, Lepomis cyanellus, contamination was the primary factor affecting crawfish production.

INTRODUCTION

Open pond culture of crawfish, Procambarus spp., in the southern USA involves duplication of the region's natural hydrological cycle in earthern ponds (Huner and Barr, 1981). That is, ponds are dried in the summer when rainfall is low and evaporation is high and they are refilled in the fall when heavy rains resume and evaporation rates decline. This process eliminates aquatic predators, permits growth of terrestrial vegetation that will serve as food and substrate for the crawfish, and phases reproductive activity of the self-perpetuating crawfish populations.

Much information is available on crawfish population biology in southern culture ponds (Huner and Romaire, 1979; Huner, 1978; Huner and Avault, 1976); however, very little is known about their responses to systems that are flooded throughout the summer period when ponds are normally drained. While summer pond draining has positive attributes, it is not mandatory with respect to completion of the crawfish life cycle (Huner and Barr, 1981).

While crawfish ponds are dry in the summer, regional finfish culture ponds are operating at maximum capacity. Crawfish do occur in these ponds but are normally treated as pests (Huner, 1976). Growing demand for crawfish and lowered profit margins are causing farmers to reconsider the status of crawfish present in their fish culture ponds. Management practices must be developed as fishes present in the ponds often prey on crawfish sometime in their lives. Green et al. (1979) studies the simultaneous culture of channel catfish (Ictalurus punctatus), buffalofish (Ictiobus spp.), paddlefish (Polyodon spathula), and golden shiner minnows (Notemigonus crysoleucas) with crawfish, Procambarus spp. They found that experimental fish culture ponds could be managed to produce large crawfish crops in addition to valuable finfish crops. Briefly, such ponds were treated like crawfish ponds during the summer of the first year. They were refilled in the fall, and crawfish were harvested into June of the following year. Potentially predaceous catfish fingerlings were stocked in cages in March or April and released into the pond in June. Buffalofish and paddlefish were stocked loose into the pond as the time that catfish were stocked in cages. Fish were harvested the following fall when the ponds were drained. The researchers found that young crawfish recruited to such a pond if it was refilled immediately but they did not reach harvestable size in the second year, probably because of the absence of adequate food.

The dominant crawfish in most southern culture ponds is Procambarus clarkii although Procambarus acutus acutus, an equally desirable species, can be found in significant numbers in some ponds (Huner, 1980). Very little is known about the relationships between the two species although circumstantial evidence suggests that P. clarkii is more prolific and tolerant of poor water quality. Growth rates for the two species are virtually identical but P. a. acutus has a more northerly distribution implying that it tolerates lower environmental temperatures (Huner and Barr, 1981).

In the summer of 1979 and 1980, we cultured prawns, Macrobrachium rosenbergii, with and without channel catfish fry in earthen ponds (Huner et al., 1980; Huner et al., 1981). Resident crawfish, P. clarkii and P. a. acutus, were harvested from these ponds that had been previously used in crawfish experiments. During the fall-winter-spring period of 1980-1981, we cultured 10-15 cm channel catfish fingerlings in nearby ponds in an overwinter grow-out study. These ponds had been filled with water continuously for a year before being stocked with catfish. Crawfish, P. clarkii and P. a. acutus, were harvested during the winter and spring of 1981.

The purpose of this paper is to discuss our observations with respect to crawfish yields in fish culture ponds treated unlike conventional crawfish ponds. We will address crawfish yields, apparent fish-prawn-crawfish and fish-crawfish interactions, and crawfish species composition.

MATERIALS AND METHODS

Prawn-Catfish Studies

Twelve 0.04 ha earthen ponds designated "T-ponds" located at the LSU Ben Hur Biological Research Center were used to culture prawns with or without channel catfish fry during the spring-summer-fall periods of 1979 and 1980. These ponds had well established populations of P. clarkii and P. a. acutus and have been used to culture crawfish in past years. In 1979 stocking rates were: newly metamorphosed prawn post larvae, 25,000/ha with or without 50,000 swim up channel catfish fry/ha and swim up channel catfish fry alone, 50,000/ha. In 1980, stocking rates were: newly metamorphosed prawn post larvae, 25,000 and 50,000/ha, singularly, or 25,000/ha with 150,000/ha swim up channel catfish fry/ha and swim up channel catfish fry alone, 150,000/ha. There were three replications per treatment. Prawns were stocked in mid-May and catfish were stocked in mid-June of each year. All were harvested in late September/early October of each year.

In both years ponds were dry during most of the winter but were refilled to about one half normal level by rainfall by early April. Very little vegetation was present. In 1979, ponds were drained in mid-April and left fallow until being refilled in mid-May to accommodate prawns. In 1980, ponds were drained 90% dry and refilled in the same week in early May just before prawns were stocked.

Crawfish were harvested with traps in both years; however, in 1979, two 18.8 mm mesh traps were used in each pond while two 9.3 mm mesh traps and two 3.1 mm mesh traps were used in each pond in 1980. Traps are normally baited with fish offal one afternoon and run the following morning. Ponds were not sampled before draining in the spring of 1979 but were trapped five times before draining in late April/early May of 1980. Five to ten trappings were made each month, June-September, in both years. Weight and number of crawfish caught were recorded in 1979. The same data plus species identification were recorded in 1980. Crawfish caught in cast nets or seines used to sample prawns and catfish were retained but did not exceed 10% of the total harvest. All crawfish caught in 1979 were considered to be large enough to eat being 12 g in size or larger. Crawfish larger than 7 g in size were considered to be edible in 1980 because such crawfish were marketed commercially in Louisiana. Those smaller were considered to fish bait quality.

Overwinter Catfish Grow Out Study

Fifteen, 0.014 ha earthen ponds designated "R-ponds" and located at the Ben Hur facility were used in this study. Catfish stocking rates were 1785, 3573, 8073, and 14,825/ha. The pupose of the study was to determine if these fish could be grown to a harvestable size in a period of one year from the time that they hatched. The lowest denisty was replicated three times. The others were replicated four times. Fish at the lowest density received no feed. Those at the higher denisities were fed a floating ration (35% protein) to satiation each day when water temperatures exceeded 21° C. Fish were fed about 2% of estimated body weight every other day when temperatures were lower with the floating feed (10-20° C) and sinking feed (35% protein) (below 10° C).

One 8.8. mm mesh trap was used in each pond to trap crawfish. Each was normally baited on one day and emptied one or two days later. Traps were run 2-6 times in January, February, March and June and 10-15 times in April and May. Ponds were seined with a 13 m, 3.1 mm mesh bag seine, 2-4 times during the winter and spring to check catfish growth. All crawfish caught in the seine were harvested. Crawfish larger than 7 g were classed as edible. Those smaller were classed as fish bait size. Ponds were drained in mid-June to harvest the catfish.

Observations ended then.

In December, it became obvious that several ponds were heavily contaminated with 50-125 mm green sunfish, Lepomis cyanellus, known crawfish predators (Huner and Barr, 1981). All sunfish caught in traps and in the seine were recorded and destroyed.

The ponds had been used in a crawfish feeding study from September 1979-May 1980. In May 1980, they were used in an insecticide toxicity bioassay study (Permethrin and Fenvalerate) and in June were treated with the same insecticides to kill all fish and crawfish present in the water column. These are non-residual pyrethroids. We stocked young-of-the-year channel catfish fingerlings, mean size 10-15 cm (18-29 g), in late September. At that time the ponds had been filled with water for approximately one year.

RESULTS

Prawn-Catfish Studies

Prawn and catfish yields may be summarized as follows. In 1979, there were no significant differences in prawn and catfish yields whether stocked alone or together. Mean prawn yields were 14,364/ha (SD=359) and 442 kg/ha (SD=54.7). Mean catfish yields were 17,282/ha (SD=3722) and 529 kg/ha (SD=92.1). Details of this study may be obtained by consulting Huner et al. (1980). In 1980, prawn yields were well below those of 1979 in all stocking combinations. Catfish yields were slightly higher than in 1979 because more catfish were stocked but survival was so low that absolute yields were similar. Mean prawn yields were: 25,000/ha, alone - 4483/ha (SD=949) and 154 kg/ha (SD=22); 50,000/ha, alone - 11,092/ha (SD=3270) and 269 kg/ha (SD=84.3); 25,000/ha+150,000 catfish fry/ha - 2626/ha (SD=4243) and 92 kg/ha (SD=85.2); mean catfish yields were: 150,000/ha, alone - 21,683/ha (SD=6717) and 641 kg/ha (SD=127); 150,000/ha+25,000 prawns/ha - 27,992/ha (SD=21,911) and 768 kg/ha (SD=275). Details of this study will be forthcoming (Huner et al., 1981).

Crawfish yields were much lower in 1979 (125-433 kg/ha, mean=297 kg/ha, SD=77) than in 1980 (337-661 kg/ha, mean=531 kg/ha, (SD=90)) (Table 1). Analysis of variance showed no significant differences in crawfish yields in either year regardless of treatment; however, there was a substantial difference in catch per unit effort with respect to treatment through time (Table 1). That is, the greatest catch per unit effort (CPUE) was realized in July of each year with drastic declines into September in all treatments but the presence of prawns in a treatment dramatically accelerated the decline.

Fifty to eighty percent of the crawfish caught in 1980 were edible size (Table 1). All caught in 1979 were edible size. Ten of the twelve ponds were used in each year although stocking rates varied as ponds were assigned to treatments randomly. We could find no meaningful relationships between crawfish yields between years. That is, there was just as much chance for a pond having a high relative yield in 1979 having a low or a high relative yield in 1980. Additionally, in 1980 there was no discernible relationship between catch of bait-sized or edible-sized crawfish before draining and

Table 1. Crawfish Yield and Catfish Per Unit Effort Data, 1979-1980 Prawn-Catfish Fingerling Studies.[1]

Treatment	July 1979	July 1980	August 1979	August 1980	September 1979	September 1980
Catfish Alone[2]						
CPUE (no.)	50.0 (7.1)	39.9 (17.6)	24.4 (0.07)	38.6 (23.3)	12.9 (0.21)	18.8 (12.0)
Crawfish Yield kg/ha	432 (16.3)	545 (43)				
%Edible-Sized	100	67 (27)				
Prawns-25,000/ha						
CPUE (no.)	37.3 (14.9)	57.9 (16.5)	19.4 (13.1)	32.1 (22.4)	5.1 (4.5)	3.6 (1.0)
Crawfish Yield kg/ha	318 (90.0)	558 (70.0)				
%Edible-Sized	100	52 (18)				
Prawns-50,000/ha						
CPUE (no.)	----	30.3 (13.8)	----	20.2 (7.2)	----	1.6 (0.3)
Crawfish Yield kg/ha	----	547 (91)				
%Edible-Sized	----	88 (1.5)				
Catfish+Prawns[3]						
CPUE (no.)	22.8 (13.2)	36.4 (25.2)	18.1 (10.9)	29.8 (7.9)	6.0 (2.1)	3.6 (1.3)
Crawfish Yield kg/ha	235 (92.0)	486 (164)				
%Edible-Sized	100	67 (27)				

[1]Values in parentheses are standard deviations.
[2]Catfish fry stocked at 50,000/ha in 1979 and 150,000/ha in 1980.
[3]Prawn post larvae stocked at 25,000/ha with 50,000 catfish fry/ha in 1979, and 150,000 catfish fry/ha in 1980.

refilling of ponds. Crawfish caught in both years had distinct but bearable off-flavor. It was best described as an earthy flavor. All were suitable for use as fish bait, scientific specimens or stocking ponds.

P. a. acutus accounted for less than 3% of the crawfish harvested in 1980 but they accounted for roughly 23% of the crawfish in T-2 and T-8. There were fewer than 1% P. a. acutus in the other ponds. We observed the same trend, qualitatively, in 1979.

Most of the P. a. acutus harvested in both years were mature adults. No recently released young-of-the-year P. a. acutus were present. In contrast, multi-structured populations of P. clarkii were present in all ponds. All ponds had adults, juveniles, and newly released young-of-the-year as of early June. Ponds with the greatest yields, number and biomass, had proportionally more young-of-the-year and juveniles present.

Overwinter Catfish Grow Out Study

Table 2 details the association between catfish densities and sizes (initial/final), feed added to ponds, relative green sunfish abundance, and crawfish yields. Several points are readily apparent. When relatively large numbers of green sunfish were present in the December-February period, crawfish yields were invariably low. The presence of larger catfish and green sunfish correlates well with lower relative crawfish yields. Only one pond (R-14) had high numbers of green sunfish early in the season and generated a substantial crop of crawfish; however, that pond was unique. A dense growth of filamentous algae was present throughout the winter. Very few green sunfish escaped when the pond was seined to sample crawfish. In addition, we observed a substantial number of young-of-the-year P. clarkii recruit to the pond in late February after many green sunfish had been removed. Many green sunfish were recovered from ponds R-17 and R-5 late in the season but they gained entrance through flooded drains after the bulk of the young-of-the-year, P. clarkii and P. a. acutus had recruited to the ponds. Green sunfish in pond R-10 were less than 30 mm long during the winter and early spring, too small to affect young crawfish.

Catfish received more feed than necessary to account for observed growth. Food conversion values exceeded 2.5:1 in all ponds and reached levels greater than 10:1 in several of the high density ponds where survival was poor. Early morning dissolved oxygen concentrations often reached critical levels of 1-2 ppm. Major fish kills occurred in ponds T-12 and T-13. These resulted from a combination of low oxygen levels and ectoparasitic protozoans on gills of affected fish. Ectoparasitic protozoans were a problem in all ponds but R-3. With the exception of pond R-3, all fed ponds were treated with potassium permanganate at least three times in the spring to control the parasites. Catfish fed poorly for several days before and after treatments so additional feed was wasted. Catfish growth was apparently retarded when compared with that which would be expected after waters warmed in April (Boyd, 1979). An average of 16.5% of all crawfish were bait-sized (Table 3). Most, average 82%, were caught with the seine. Some 29% of the edible-sized crawfish were harvested with the seine. No discernable off-flavors were detected in crawfish harvested in this study.

Table 2. Crawfish Yields from Ponds Stocked with Channel Catfish Fingerlings in September 1980 and Harvested in June 1981.

Rank-Pond	Crawfish Yield kg/ha	No. Green Sunfish, Dec.-Mar.	Catfish Stocked-Harvested/ ha	Catfish Size Stocked-Harvested, g.	Feed Applied kg/pond
1-2	1207	0	8073-2286	29-95	17.4
2-13	1138	2	14,285-7500	27-95	37.2
3-17	1135	1	14,285-6571	21-88	36.7
4-10	1058	32	1785-1357	28-80	---
5-18	1044	3	1785-1000	28-86	---
6-5	1024	10	1785-1714	28-118	---
7-9	1014	0	8073-5000	29-153	24.2
8-8	964	0	8073-6786	29-165	28.4
9-14	951	761	14,285-8643	18-66	25.8
10-11	816	119	8073-5571	19-192	28.5
11-2	700	541	14,285-9000	20-123	33.7
12-16	644	300	3573-2429	28-180	26.1
13-15	437	439	3573-2643	28-251	18.5
14-1	328	630	3573-2929	28-224	17.2
15-3	192	79[1]	3573-2857	28-214	18.2

[1]Pond could not be seined because of dense plant growths. Large numbers of green sunfish were present throughout the study.

385

Table 3. Composition of Crawfish Harvest from Ponds Stocked with Channel Catfish Fingerlings in September 1980 and Harvested in June 1981.

Pond[1]	Mean Size Edible Crawfish	% Bait Crawfish	%Bait Crawfish Caught with Seine	%Edible Crawfish Caught with Seine	% P.a. acutus
18	14.4 g	30	80	21	39
17	16.0 g	24	88	31	22
16	18.3 g	25	91	40	7
15	25.8 g	1	83	27	22
14	11.5 g	11	87	28	5
13	14.0 g	20	76	21	11
12	17.4 g	15	80	19	42
11	16.4 g	13	77	38	25
10	16.7 g	19	87	45	70
9	19.4 g	16	84	35	25
8	20.8 g	12	85	34	75
5	12.3 g	33	57	17	25
3[6]	30.0 g	--	--	--	50
2	22.9 g	11	88	37	16
1	23.1 g	1	0	11	3

[1] Catfish stocking densities: 1785/ha, ponds 18, 10, 5; 3573/ha, ponds 16, 15, 3, 1; 8073/ha, ponds 12, 11, 9, 8; 14,285/ha, ponds 17, 14, 13, 2;

[2] Mean = 16.5; SD = 9.17;

[3] Mean = 75.9; SD = 23.4;

[4] Mean = 28.9; SD = 9.95;

[5] Mean = 29.7; SD = 9.95;

[6] Dense growths of vegetation prevented seining this pond.

Only two age classes of P. a. acutus were present, adults and one distinct group of young-of-the-year. P. clarkii populations were multi-structured including adults, juveniles, and one or more distinct groups of young-of-the-year. All P. acutus had matured by the time ponds were drained. Young-of-the-year and juvenile P. clarkii were still present in all ponds when they were drained, being most abundant in ponds with the greatest yields.

P. a. acutus was very abundant in some ponds, accounting for up to 70% of the trap harvest (Table 3). The R-ponds are laid out in two rows. Most of the ponds with high numbers of P. a. acutus, R-5, 8, 9, 10, 11 and 12, are located adjacent to each other at the northern end of the two rows.

DISCUSSION

Differences in crawfish yields between 1979 and 1980 in prawn-catfish ponds were undoubtedly due to differences in pond management practices. Harvesting effort was doubled and trap mesh was reduced in 1980. Ponds were not fully drained in 1980. Each activity should have had a positive influence on crawfish yields but we were unable to assess the impact of each factor. Certainly, thin-shelled, small-clawed young-of-the-year and juvenile crawfish are less able to respond to pond draining by burrowing. Thus, they should be expected to respond better to partial pond draining than to total pond draining. Increasing harvest effort and reducing mesh size will not have a positive effect on yield, however, unless crawfish are available in the first place.

Although there were no significant differences in crawfish yields in ponds with prawns and/or catfish in the summer of 1979 and 1980, average crawfish yields 297 and 531 kg/ha were low compared to the 500-1000 kg/ha yields in conventionally managed ponds (Huner and Barr, 1981). It is impossible to assess the degree to which either species depressed crawfish yields as control ponds were not available.

The period of precipitous decline in CPUE in the prawn ponds corresponded to the time when the first large, 25-40 g, male prawns appeared in the ponds. These animals have very large chelae and are very aggressive (Hanson and Goodwin, 1976). There is no way to determine if crawfish were driven from the water column into burrows or destroyed by these prawns but circumstantial evidence strongly suggests that prawns were responsible for their absence from trap catches.

Previously, Huner (1978) had noted a general decline in crawfish numbers through the summer in a pond that was left filled. Therefore, the general decline in numbers observed in the prawn and/or catfish ponds is probably a natural event.

Mean prawn numbers at harvest varied greatly over the two summers. This indicates that the absolute number of prawns necessary to impact on crawfish does not have to be especially great.

Young catfish produced in both summers were certainly large enough at 10-30 g to consume newly recruited young-of-the-year crawfish in late August and September when they achieved those sizes. That is the time that major recruitment of young crawfish begins in most crawfish ponds (Huner, 1978). If left alone at those densities, 17,000-27,000/ha, they may well have done that. However, none of the densities of 1738-14,285 catfish/ha seemed to have affected relative crawfish production in the winter catfish grow-out study. Our observed maximum crawfish yields to 1000 kg/ha were indeed comparable to yields obtained in well managed crawfish ponds treated in the conventional manner (Huner and Barr, 1981). This observation must be tempered with the observation that 40-50% of all crawfish were harvested with a seine and about 20% of the total was bait-sized. Very few bait-sized crawfish are harvested commercially in Louisiana. Then too, while most harvestable crawfish were generally larger than 12 g, the commonly accepted minimum edible size, they were hardly choice animals.

Green sunfish clearly decimated crawfish populations but the presence of larger 175-250 g catfish even at low densities seemed to accentuate the effect. When green sunfish are actually present in a pond is a critical factor. If large numbers of 50-125 mm fish are present during the fall-winter crawfish recruitment period, crawfish production will be greatly affected. Physical removal, however, can permit successful crawfish production when late winter-early spring recruitment takes place.

One might speculate that crawfish yields in the overwinter catfish grow-out study could be attributed to wasted feed. Food conversion values ranged from 2.5 to 16 to 1 depending on catfish survival. These are well above the 1.5-2.0 to 1 values considered normal for most catfish operations (Meyers et al., 1973). Yet no feed was added to the ponds with the lowest catfish stocking rates and these had autochthonous organic reserves large enough to generate crawfish yields as great as those in any fed pond. There is no doubt that the crawfish did, in fact, eat catfish food as they were frequently observed doing just that. Why then, was crawfish production not higher in fed ponds? Perhaps this is indirect evidence of catfish predation, especially on recently released young-of-the-year crawfish.

Poor water quality and parasite problems may have also affected catfish predation on crawfish in the spring in two ways. Lethargic fish would be less likely to feed on any food source. Smaller fish even when feeding actively would have been less able to consume the larger crawfish present at that time. Thus, better growth performance of catfish may have adversely impacted crawfish yields.

Only two significant age classes of P. a. acutus were present in the R-ponds, adults and young-of-the-year. All of the young-of-the-year matured by June. Thus, it is not surprising that only adults were found in the two summer prawn-catfish studies. Interestingly, the greatest number of P. a. acutus found in the T-ponds used in the prawn-catfish studies were found in ponds T-7 and T-8. These are adjacent to the R-ponds that had the greatest number of that species. Our data, however, do little to clarify the relationship between P. clarkii and P. a. acutus. They do suggest that P. a. acutus with its limited recruitment period is more vulnerable to predator pressure than the more prolific P. clarkii.

A certain degree of caution must be exercised in reviewing our data on the percentages of P. clarkii and P. a. acutus in ponds. That is, both species are very fecund producing 50-600 or more offspring per female, depending on size and environmental conditions (Penn, 1956). Thus, it would not take many females of either species to skew species distributions in ponds as small as those that we employed in our studies.

Must ponds be drained in the summer in order to produce crawfish crops? The data from our overwinter catfish grow-out study suggests that the answer is no as long as predaceous fishes are absent, but crawfish will tend to be small.

Practical Applications

Application of our data to fish culture situations must be done with caution. It appears, however, that south Louisiana catfish farmers can stock as many as 12,500, 20-30 g fingerlings per ha. in the fall in a previously fish free pond with resident crawfish and realize a significant crawfish crop. The farmer would do well to consider use of a small mesh seine to harvest the crawfish as we found a seine to be the most efficient means of harvesting crawfish from our ponds.

Persons contemplating summer prawn and/or catfish fingerling culture in ponds with resident crawfish in south Louisiana can expect to obtain crawfish from those ponds. Yields should be significantly lower than those obtained from conventionally managed crawfish ponds. Such crawfish may have an off-flavor. Our prawn-fish stocking rates were, it should be noted, very low so it is impossible to project crawfish yields if higher stocking densities are used.

ACKNOWLEDGMENTS

This study was funded, in part, by the Louisiana Sea Grant Program and the Louisiana Agricultural Experiment Station. The senior author wishes to acknowledge Dr. Clyde E. Johnson, Chairman Department of Biological Sciences, Southern University, Baton Rouge, LA, for granting him release time to conduct the research reported in this paper.

LITERATURE CITED

Boyd, C.E. 1979. Water quality in warmwater fish ponds. Auburn University, Auburn, Alabama.

Green, L.M., J.S. Tuten, and J.W. Avault, Jr. 1979. Polyculture of red swamp crawfish (Procambarus clarkii) and several North American fish species, p. 287-298. In P. J. Laurent (ed.) Freshwater Crayfish IV, Thonon-les-Bains, France (1978). 473 p.

Hanson, J.A. and H.L. Goodwin. 1977. Shrimp and prawn farming in the Western Hemisphere, State-of-the-art reviews and status assessment. Dowden, Hutchinson, & Ross, Stroudsburg, PA

Huner, J.V. 1976. Raising crawfish for fish bait and food: A new polyculture crop with fish. The Fisheries Bulletin 1(2):7-9.

Huner, J.V. 1978. Crawfish population dynamics as they affect production in serveral small, open crawfish ponds in Louisiana. Proceedings of the World Mariculture Society 9:619-640.

Huner, J.V. 1980. Red crawfish population dynamics in culture ponds: A summary. Proceedings of the First National Crawfish Culture Workshop. University of Southwestern Louisiana, Lafayette, LA.

Huner, J.V. and J.W. Avault, Jr. 1976. Sequential pond flooding: A prospective management technique for extended production of bait size crawfish. Transactions of the American Fisheries Society 105:637-642.

Huner, J.V. and J.E. Barr. 1981. Red swamp crawfish. Biology and exploitation. Center for Wetland Resources, Louisiana State University, Baton Rouge, LA. Sea Grant Publication No. LSU-T-80-001.

Huner, J.V. and R.P. Romaire. 1979. Size at maturity as a means of comparing populations of Procambarus clarkii (Girard) (Crustacea:Decapods) from different habitats, p. 53-64. In P. J. Laurent (ed.) Freshwater Crayfish IV, Thonon-les-Bains, France (1978). 473 p.

Huner, J.V., W.G. Perry, Jr., R.A. Bean, M. Miltner, and J.W. Avault, Jr. 1980. Polyculture of prawns, Macrobrachium rosenbergii, and channel catfish fingerlings, Ictalurus punctatus, in Louisiana. Proceedings of the Louisiana Academy of Sciences 43:95-103.

Huner, J.V., M. Miltner, J.W. Avault, Jr., and R.A. Bean. 1981. Freshwater prawn culture studies in ponds near Baton Rouge, Louisiana, 1980 (In Preparation).

Meyers, F.P., K.E. Sneed, and P.T. Eschmeyer. 1973. Second report to the fish farmers. The status of warmwater fish farming and progress in fish farming research. Bureau of Sport Fisheries and Wildlife, Washington, D.C. Resource Publication 113.

Penn, G.H. 1956. The genus Procambarus in Louisiana (Decapoda:Astacidae). American Midland Naturalist 56:406-422.

AN EVALUATION OF SELECTED CRAWFISH TRAPS AND TRAPPING TECHNIQUES IN LOUISIANA

V. Pfister and R. P. Romaire
Fisheries Section
School of Forestry and Wildlife Management
Louisiana State University
Baton Rouge, Louisiana   70803, U.S.A.

ABSTRACT

Harvesting crawfish (_Procambarus_ _clarkii_) with traps is the largest variable expense to commercial producers in Louisiana, generally consuming 40 to 60% of the gross revenues. Most producers harvest crawfish with baited wire traps constructed from 19 mm hexagonal mesh poultry netting. The traps are usually set for 24 hours at a rate of 25 traps/hectare. Seven traps presently used in the commercial industry and three new trap designs are being evaluated for harvest efficiency in three commercial ponds during the 1980-1981 crawfish season. Concomitantly, trap placement, trap density, frequency with which traps are emptied, retentative ability of various trap designs and manipulation of environmental parameters, such as water circulation, are being investigated to ascertain those combinations of harvesting strategies that minimize cost and maximize catch of crawfish.

A COMPARATIVE STUDY ON THE PRODUCTION OF CRAYFISH (ASTACUS ASTACUS L.) JUVENILES IN NATURAL FOOD PONDS AND BY FEEDING IN PLASTIC BASINS

M. Pursiainen
Evo Inland Fisheries and Aquaculture Research Station
SF-16970 EVO
Finland

T. Järvenpää and K. Westman
Finnish Game and Fisheries Research Institute
Fisheries Division
P.O. Box 193, SF-00131, Helsinki 13, Finland

ABSTRACT

As the mortality of newly hatched 1-cm-long crayfish juveniles has been found to be very high in natural waters, methods for hatchery production of stage 2 juveniles and further rearing methods have been investigated. Fertile females were stocked in 100 m^2 ponds together with males in 1979, and 95% of surviving females were egg-bearing in the summer of 1980. Production of newly hatched stage 2 juveniles was found to be about 68 individuals per egg-bearing female. In 15 m^2 earthen ponds the juveniles feed only on natural food. With initial density 100 ind./m^2 the survival rate at the end of a 3-month growing period was 67%, while at an initial density 300 ind./m^2 it was only 32%. The mean size and weight at the end of the growing period were 21.4 mm and 221 mg at the lower density, which means an increase in biomass of 11.1 g/m^2. At the higher density, the mean size was 19.5 mm and weight 166 mg, or there was a biomass increase of only 4.7 g/m^2. In 4 m^2 plastic basins the juveniles were fed zooplankton and a mixture of fish, vegetables, and shrimp shell waste. The survival rate at an initial density of 100 ind./m^2 was 58% and at 300 ind./m^2 50%. The mean weight at end of growing period in basins was only half that of pond-reared juveniles and the increase in biomass only 2.0 and 1.3 g/m^2 for respective densities.

INTRODUCTION

Since 1893, the disastrous crayfish plague, Aphanomyces astaci, has caused great losses to the populations of Astacus astacus L., the only endemic crayfish species in Finland (Westman, 1973a, 1975). Pollution of inland waters, together with various engineering works undertaken on them has futher aggravated the situation, making it more and more difficult for crayfish to thrive (see e.g., Niemi, 1977).

In order to reduce the effects of crayfish plague, plague-resistant American signal crayfish Pacifastacus leniusculus Dana have been stocked in various parts of Europe, including Finland (e.g. Abrahamsson, 1973, Westman, 1973a, Brinck, 1977). In many waters where the plague has disappeared after destroying all the crayfish or the environmental conditions have improved after pollution or construction operations the original crayfish population can be restored. Restoration, however, requires efficient stocking measures and abundant stocking material.

As far as indigenous crayfish are concerned, stocking in Finland has been effected as transfer stocking, in which most frequently crayfish with a total length of less than 10 cm have been transferred from one lake or river to another; this has required special permission as the law prohibits catching crayfish smaller than 10 cm in length. The procedure has often produced very good results; this is clearly demonstrated by the expansion during this century of the area in which crayfish occur in Finland (Westman, 1973a). Such stocking material is, however, constantly in short supply, and is also now very expensive.

Newly hatched juveniles have been primarily used in the stocking of American signal crayfish, both in Finland and elsewhere in Europe. Fürst (1977) observed a very high mortality rate following such stocking; he recommends that this method be used only in those lakes in which all the fish have been deliberately destroyed. The mortality rate of newly hatched juveniles has been observed as being very high in indigenous crayfish populations (see e.g. Tcherkasina, 1977). Apparently the high mortality of juveniles is caused by intraspecific competition, by invertebrate predation (Dye & Jones, 1975), or by fish predation (see Svärdson, 1972).

Due to the high morality of juveniles, several countries have investigated possibilities for the rearing of crayfish juveniles to a larger size for stocking purposes (Westman, 1973a; Tcherkasina, 1977; Anwand, 1978; Cuellar & Coll, 1978; Müller, 1978; Cukerzis et al., 1978). During the second half of the 1970s the Evo Inland Fisheries and Aquaculture Research Station of the Finnish Game and Fisheries Research Institute has developed rearing methods, which are based either on natural food or artificial feeding, to produce one-summer old crayfish juveniles for stocking purposes. The present report, covering the period 1979-1980, discusses the production methods currently in use.

MATERIALS AND METHODS

CULTIVATION OF CRAYFISH BROOD STOCK

Adult male and female crayfish (carapace lengths greater than 35 mm) were caught from a 50-hectare lake, and stocked in two earthen ponds of 100 m^2 each in early August, 1979. The maximum depth of the ponds was about 140 cm. To provide shelters for the crayfish, either brick tubes or stones were used on the bottom of the ponds. Only those females whose ovaries were fully developed, and those males whose sperm tubes were visible under their carapaces, were selected for the ponds. Stocking density was 4 individuals per square metre, at a ratio of 1 male for every 3 females.

Crayfish in the earthen ponds were dependent mainly on natural food. Fish and vegetable material were fed to them on average once a week during the open water season water season from August to the end of October 1979, and from May to June 1980. Temperatures in the ponds varied from a maximum of 22° C in the summer to a low of 0.5^o C in the winter. The oxygen balance was good, according to measurements made, it remained at over 80% saturation throughout the entire period. The pH varied between 5.8 and 6.4.

The ponds were emptied at the end of June (17-20) 1980. All live crayfish were collected, and females with pleopod eggs were transferred to a crayfish hatchery to await the hatching of juveniles.

HATCHING

Hatching was carried out using two methods. In the first method, a unit accommodating 12 egg-bearing female crayfish was built, in which each female had its own net-bottomed, 12 x 10 cm compartment. The unit was set on salmonid hatching trays, four of which were then immersed in plastic hatching troughs in running water. When juvenile crayfish dropped from beneath the female's tail, as stage 2 juveniles, they fell through the net-bottomed compartment onto the hatching trays, where they had their own living space. Altogether 240 egg-bearing females were used in this compartment method.

In the second method, compartments for egg-bearing females were constructed as above, but the juveniles did not fall into a separate juvenile "nursery" but stayed with the mother in the same compartment. It was then possible to count juvenile production for each female. In this method 223 individual egg-bearing females were used.

On the basis of previous experience the separation of the females from each other has been found to be necessary. If there are several females in the same space, a considerable portion of the eggs and stage 1 juveniles are lost as a result of quarreling among the females.

The egg-bearing females were fed every second day with fish and vegetable matter. The hatching spaces were also cleaned and the progress of egg development was checked. Feeding was stopped entirely when the eggs began to hatch, to prevent excrement and decompositing food from polluting the water.

Collection of juveniles in the first method was carried out by removing each unit housing females from one hatching tray to another, and the juveniles which had fallen into the space underneath were collected and counted from the original holder. In the second method (i.e. individual hatching compartment), the female crayfish was moved to the next empty compartment, and the juveniles were collected from the bottom of the original compartment. Collection was carried out almost daily, beginning from the first day of hatching, until all stage 2 juveniles had been released from underneath the abdomen of the mother.

JUVENILE REARING

Newly-hatched, stage 2 juvenile crayfish were placed in ponds and basins for further rearing immediately after being collected from the hatching units on July 18. Rearing took place in four 15 m^2 earthen ponds, and in two rectangular, 4 m^2 plastic basins. The initial density was 100 individuals per m^2 in two of the earthen ponds and one of the plastic basins, and 300 individuals per m^2 in the other ponds and basin. The bottom of the earthen ponds was composed of gravel and sand. In addition, bricks were laid down to provide the juveniles with hiding places. The ponds were bordered by a belt of aquatic vegetation, which provided food as well as shelter for the juveniles. The bottoms of the plastic basin were covered with crushed shells and the same type of bricks as in the earthen ponds.

In the earthen ponds, the juveniles were entirely dependent on natural food. To increase the primary production of the ponds, they were fertilized during the growing season every second week with a product specially prepared for gardens from compressed chicken manure. The exchange of water in the ponds was kept to the absolute minimum; in practice, only evaporated water was replaced with new water. Therefore, daily temperatures were fairly high, up to 26° C. The oxygen content of the water varied from 70-100% saturation, and the pH from 5.7 to 6.7. Due to the fertilization , the dissolved phosphorus and nitrogen concentrations were two to three times higher than in the incoming water, or 80-100 μg P/liter and 900-1000 ug N/liter.

Juveniles in the plastic basins were fed with zooplankton and frozen artificial feed made especially for the purpose (1/3 fish, 1/3 vegetable matter, and 1/3 shrimp shell waste). Food was given almost every other day. The basins had a water flow rate of about 0.3 1/sec. The maximum temperatures were slightly over 22° C, and the oxygen and pH values were the same as in the earthen ponds.

The rearing experiment was concluded on October 7 by draining the water from the ponds and basins and collecting the juveniles. The juveniles were counted, and for each group, a sample of 50 individuals were collectively weighed and the length of the carapace determined in order to compute average size.

RESULTS AND DISCUSSION

BROOD STOCK CULTIVATION

Of the total of 767 sexually mature crayfish placed in the two ponds (572 females and 195 males), 92%, or 706 crayfish, were recovered. Of these, 534 were females, 95% of which had pleopod eggs. The most apparent cause of mortality during the winter period was cannibalism, although some crayfish almost certainly remained hidden in the bottom of the ponds during collection as a result of the unavoidable collapse of holes dug by the crayfish. The stocking density of four individuals per square metre was not too high, as no significant losses occurred. The method in which only reproductive females were stocked in the ponds also gave good results, since almost all the females spawned. It should be mentioned that in natural populations not all of the sexually mature females spawn every year. Abrahamsson (1972) found that the proportion of sexually mature females spawning varied from 53% to 97% in different localitites (cf. Westman & Pursiainen 1982).

JUVENILE HATCHING

At Evo, crayfish normally mate at the end of October, and lay eggs at the end of October - beginning of November. Because the females in these experiments were placed in earthen ponds, it was not possible to discover the exact dates of egg laying. If we assume that the eggs appeared in early November, the number of day-degrees required to hatch

the juveniles (July 7) was 1150. From the beginning of hatching until
the last stage 2 juveniles detached themselves (July 18), a total of 180
day-degrees were required. The number of day-degrees from egg-laying to
hatching corresponds well with observations made by Cukerzis et al.
(1978).

Altogether 31,660 crayfish juveniles hatched; of these, 87.2%, or
27,783 individuals, survived to become stage 2 juveniles. The number of
juveniles per mother is presented in Table 1. The average weight of
stage 2 juveniles was 38 mg immediately after detachment from the
mother, based on the weighing of two lots of 200 juveniles.

Abrahamsson (1972) has calculated the number of pleopod eggs just
before hatching to be 90 to 170 eggs per female in various populations.
Thus, the number of hatched juveniles, 71 per female in compart-
mentalized incubation, and 61 per mother'in individual incubation,
(average of all mothers = 68 individuals), is clearly lower than one
would expect from the number of eggs found in natural populations. It
should, however, be noted that the average size of the female crayfish
used for cultivation was small, all the females being less than 10 cm
long. The number of juveniles produced with these methods may be
considered rather good when compared, for instance, to those of Cukerzis
et al. (1978) who had 37-42 juveniles per female.

JUVENILE REARING

From the two earthen ponds with an initial density of 100
individuals per square metre, 2025 one-summer old crayfish were
recovered from the 3000 newly hatched juveniles. The survival rate was
67.5%. In the plastic basin with a similar initial density, 234 out of
an initial 400 individuals survived, or 58.5% (see Table 2).

The survival rate was only 32.4% in the two earthen ponds with an
initial density of 300 individuals per m^2, i.e., of the 9000 juveniles
that were stocked, only 2912 crayfish were alive at the end of the
rearing period. The mortality in the plastic basin with the same
initial density was not so high; out of an initial 1200 individuals, 604
crayfish were alive at the end of the experiment, or 50.3%.

Mortality in the earthen ponds with the lower denisty may be
primarily due to predatory invertebrates and other external factors.
With the higher density, the reasons for mortality may include, in
addition to predation, cannibalism and occasional exhaustion of the food
supply (see Dye & Jones, 1975). Survival rates in the plastic basins
were about the same with both initial densities. As the mortality
caused by the predation in the fiberglass basins is practically nil, due
to the absence of predators, the remaining reasons for mortality are
thus cannibalism and the inadequacy or unsuitability of the artifical
feed provided.

The growth of crayfish juveniles in the earthen ponds of both
initial densities was clearly better than in the fiberglass basins. The
final average weight of crayfish in the earthen ponds with the lower
initial density was 221 mg, and in those with the higher density, 166

Table 1. The mean juvenile production and the percentage survival of the juveniles per egg bearing female using different hatching methods.

Method	Number of females	Hatched juveniles no/female	Living stage 2 juveniles no/female	Percentage survival
Compartments	240	75.5	63.4	84.0
Individual compartments	223	60.7	56.4	92.9

Table 2. The survival and growth of newly hatched juveniles during the first summer in different rearing conditions and the increase of crayfish biomass per area of the rearing basin.

Method	Stocking density no/m²	mean weight mg	biomass g/m²	Density no/m²	survival percent	mean weight mg	Increase of mean weight mg	Yield biomass g/m²	Increase of biomass percent
Pond with natural food	100	38	3.84	67.5	67.5	221	481.6	14.92	288.5
Pond with natural food	300	38	11.40	97.1	32.4	166	336.8	16.12	41.4
Plastic basin	100	38	3.84	58.5	58.5	100	163.2	5,85	52.3
Plastic basin	300	38	11.40	151.0	50.3	84	121.1	12.68	11.2

mg. The average weight of crayfish in the fiberglass basins were 100 mg and 84 mg respectively (Table 2). This means that when the initial average weight was 38 mg, at best the crayfish juveniles increased their weight by nearly 6 times in less than three months.

The weight of crayfish juveniles at the end of the rearing period in the earthen ponds with an initial density of 100 individuals per m^2 also corresponds well to observations made of natural populations (Cukerzis, 1975).

The lengths of the carapace varied, so that juveniles from ponds with an initial density of 100 ind. per m^2 had on average a carapace length of 10.7 mm long (total length 21.4 mm), and juveniles raised in the plastic basin with a corresponding density 8.5 mm long (total length 17.0 mm). The respective figures for the higher initial density were: in earthen ponds 9.8 mm (19.6 mm), and in the plastic basin 7.9 mm (15.8 mm). The maximum growth result corresponds almost exactly to the observations of carapace lengths obtained from one-summer old juveniles in natural populations in Finland (Westman & Pursianen, 1982).

The crayfish biomass at the commencement of rearing was 3.84 g/m^2 at an initial density of 100 ind. per m^2, and 11.40 g/m^2 at an initial density of 300 ind. per m^2. The maximum production per square metre was obtained in earthern ponds with a density of 100 ind. per m^2, i.e. the biomass increased four times by the end of the experiment as compared to the initial situation. Production in plastic basins was clearly inferior to that in earthen ponds (see Table 2).

The individual growth of crayfish juveniles and the production of the growing space provided are significantly influenced by the number of day-degrees during the rearing period. The fairly high temperatures (26° C) measured during this period did not cause discernible mortality. The sum of day-degrees in the earthen ponds was, depending on location, 1264-1278, and that in plastic basins 1134, or about 10% lower. This, however, may not be the explanation for the lower production and smaller individual growth in the plastic basins, at least not entirely.

CONCLUSIONS

Not all of the sexually mature female crayfish reproduce every year in natural populations (Abrahamsson, 1972; Westman and Pursiainen, 1982). Therefore, crayfish to be used for cultivation purposes should be caught after the female crayfish have molted, i.e. in Finland normally in August. By inspecting the developmental stage of the crayfish it is then easily possible to select only those females that will spawn. Using this procedure, 95% of the females kept in the earthen ponds over the winter had pleopod eggs in the next June-July.

In those cases where the female crayfish will constantly be reared in hatchery conditions, they should be checked in August-September, and reproductive females moved into separate ponds.

Cultivation densities for female crayfish are primarily dependent on the amounts of food and hiding places available. In the Evo trials, fairly good survival rates were obtained by stocking 4 individuals per m^2 (92% wintered successfully), although feeding was not intensive. Increasing the number of shelters and increasing feeding should make it possible to increase the density considerably above that used in the trials.

Although the ratio between females and males (3:1) seemed fairly low, it is evident that there were enough males, since almost one hundred percent fertilization of eggs was observed.

For the collection of newly hatched juveniles to be successful, a separate hatching space should be used for the egg-bearing females. Trials carried out previously at Evo suggested that egg-bearing females should be kept separate from each other, otherwise eggs and stage 1 juveniles fall off and are destroyed during the latter part of the hatching period as a result of quarreling among the females.

It has proved to be most practical to place each female in her own 10 x 12 cm netted compartment. Stage 2 juveniles are able to pass through the 6 mm holes in the net into their own space, from which they are easy to collect. It is possible to combine individual female compartments to conform to various sizes and shapes, according to the space available.

The number of hatched juveniles in these trials, which averaged 68 per female, is smaller than the number of eggs observed in natural conditions just before hatching (Abrahamsson, 1972). The result can, however, be improved if larger females are available. In the Evo trials, the crayfish were only 7.5-10 cm long.

It has been concluded to be necessary to raise newly hatched juveniles until they are at least one-summer old before using them for stocking, as juvenile mortality in the natural state during the first summer is very high, sometimes even 90% (Tcherkasina, 1977).

Most of the crayfish were collected without difficulty from the 15 m^2 ponds by slowly draining the water overnight, so that the juveniles walked into the collection box which was positioned at the mouth of the drainage pipe.

The best results in the rearing and recovery of juveniles were obtained from earthen ponds when the stocking density was 100 individuals per square metre. The survival rate in this trial group was 67.5%, while in the higher density (300 ind./m^2) it was only about half this. In the plastic basins, depending on the density, the survival rate varied between 50.3-58.5%. Growth of crayfish juveniles in the group with the lower initial density in earthen ponds corresponded fairly well to observations made in natural populations (Cukerzis, 1975; Westman and Pursiainen, 1982). It seems obvious that it would not make sense to rear crayfish juveniles in earthen ponds at a density of over 100 individuals, because at higher densities their mortality increases sharply. Rearing in fiberglass basins requires further refinements in feeding methods and basin structure before it can be used successfully.

400

LITERATURE CITED

Abrahamsson, S. 1972. Fecundity and growth of some populations of *Astacus astacus* L. in Sweden. Rep. Inst. Freshw. Res. Drottningholm, 52:23-37.

Abrahamsson, S. 1973. Methods for restoration of crayfish waters in Europe. The development of on industry for production of young of *Pacifastacus leniusculus*. Freshwater Crayfish 1:203-210.

Anwand, K. 1978. Der Signalkreps (*Pacifastacus leniusculus* Dana) Untersuchungen über seine Einburgerungswürdigkeit. Z. Binnenfishei DDR 25 (1978):373-381.

Brinck, P. 1977. Developing crayfish populations. Freshwater Crayfish 3:211-228.

Cuellar, L. and Coll, M. 1978. First essays of controlled breeding of *Astacus pallipes* (Ler.). Freshwater Crayfish 4:273-276.

Cukerzis, J. 1975. Die Zahl, Struktur and Produktivitat der Isolierten Population von *Astacus astacus* L. Freshwater Crayfish 2:513-527.

Dye, L. and Jones, P. 1975. The influence of density and invertebrate predation on the survival of young-of-the-year *Orconectes virilis*. Freshwater Crayfish 2:529-538.

Fürst, M. 1977. Flodkraftan och signalkraftan i Sverige 1976. Inf. fran Sotv. lab. Drottningholm 10(1977):1-32.

Müller, G.J. 1978. Studie zur Neueinburgerung des gegen die Krebspest resistanten Signalkrebses (*Pacifastacus leniusculus* Dana). 300 pp. Gottingen 1978.

Niemi, A. 1977. Population studies on the crayfish *Astacus astacus* L. in the River Pyhajoki, Finland. Freshwater Crayfish 3:81-94.

Svärdson, G. 1972. The predatory impact of eel (*Anguilla anguilla* L.) on populations of crayfish (*Astacus astacus*). Rep. Inst. Freshw. Res. Drottningholm 52:149-191.

Tcherkasina, N. Ya. 1977. Survival, growth and feeding dynamics of juvenile crayfish (*Astacus leptodactylus* cubanicus) in ponds and River Don. Freshwater Crayfish 3:95-100.

Westman, K. 1973a. The population of the crayfish *Astacus astacus* L. in Finland and the introduction of the American crayfish *Pacifastacus leniusculus* Dana. Freshwater Crayfish 1:41-55.

Westman, K. 1973b. Cultivation of the American crayfish *Pacifastacus leniusculus*. Freshwater Crayfish 1:211-220.

Westman, K. 1975. On crayfish research in Finland. Freshwater Crayfish 2:65-75.

Westman, K. and Pursiainen, M. 1981. Size and structure of crayfish (_Astacus astacus_ L.) populations on different habitats in Finland. In: Ilmavirta, V., Jones, R.J., and Persson, P.-E. (ed.), Lakes and Water Management. Proc. 30 Years Jub. Symp. Finn. Limnol. Soc. Helsinki 1980, Developments in Hydrobiology 7, Hydrobiologia 86:67-72.

THE COMMERCIAL FISHERY FOR PACIFASTACUS LENIUSCULUS DANA IN THE SACRAMENTO-SAN JOAQUIN DELTA

Darlene McGriff
California Department of Fish and Game
1701 Nimbus Road, Rancho Cordova, California 95670

ABSTRACT

The signal crayfish (Pacifastacus leniusculus Dana) has been commercially fished in the Delta region since 1970. Catch statistics have been maintained by the California Department of Fish and Game since that time. The number of boats fishing at least part of each season ranged from 9 to 59 ($\bar{X}$ = 32). Between 1970 and 1974, 50% of the catch was taken by 24.4% of the boats; between 1975 and 1980, 15.5% of the boats took 50% of the total catch. Seven age classes (0+ to 6+) are postulated for the Delta population. The commercial catch is composed primarily of ages 2+ and 3+ crayfish. Ages 4+ to 6+ represent about 15% of the catch. Regulations adopted by the California Fish and Game Commission after the 1975 season include a minimum legal size of 92 mm TL, a requirement that fishermen sort their catch on their boat and return short crayfish to the water immediately, and standards for holding and transporting crayfish to minimize loss. The mean size of crayfish over 92 mm TL has remained at about 105 mm between 1976 and 1980. Optimum effort appears to be about 300,000 trap-sets per year. Effort in excess of this appears only to reduce catch-per-unit-effort.

INTRODUCTION

The Delta population of Pacifastacus leniusculus, with annual landings well in excess of 200,000 kg, may well be the most heavily exploited wild population anywhere. The commercial fishery began in 1970 in response to a new market in Sweden (Nicola 1971). In 1970, information was collected on the size range and sex ratio of the crayfish caught. No further data were collected and the fishery remained unregulated until 1975 when the California Department of Fish and Game initiated a study of the impact of the commercial fishery on the crayfish stocks with the goal of developing a plan to insure the protection and proper utilization of the crayfish resource. This paper describes the Delta fishery from 1970 through 1980 and the impact of regulations adopted to manage the crayfish stocks.

THE DELTA REGION

The Sacramento-San Joaquin Delta is formed by the confluence of the south-flowing Sacramento River and the north-flowing San Joaquin River in the Central Valley of California. This area was once a vast tidal marsh, but levee construction in the 19th century transformed it into a network of navigable channels enclosing below-sea-level agricultural islands (Figure 1).

Most of the commercial crayfish catch is taken in the waterways from Rio Vista northward to Colusa. In the sloughs south of Rio Vista, fishing success varies considerably from season to season. There is virtually no fishing south of Frank's Tract. For purposes of this

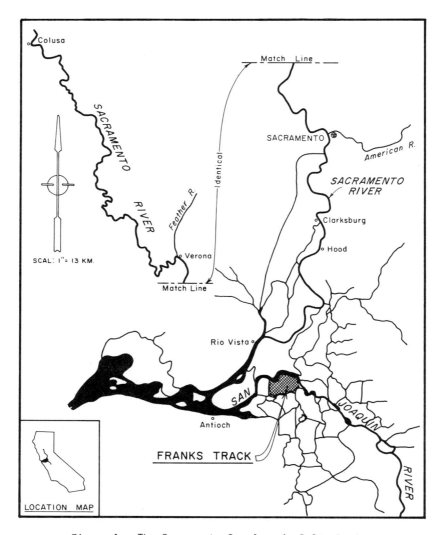

Figure 1. The Sacramento-San Joaquin Delta Region.

404

study, the Sacramento River winding southward through a wide alluvial plain from Colusa to Sacramento is considered part of the Delta region.

The Delta region is a complex and diverse habitat. It is influenced by semidiurnal tides that reverse flows as far upriver as Clarksburg (69 river km) and alter water levels and velocities as far north as Verona (129 river km). The rising tides also carry sea water into the Delta region. The extent of this saltwater intrusion depends on the flow of fresh water released from upstream dams and varies annually and seasonally. The Delta waterways vary in width from nearly 1 km in the lower Sacramento River at Rio Vista, to less than 0.1 km in some interior sloughs. Depths vary from about 16 m in parts of the San Joaquin River to less than 3 m in some interior sloughs. The waters in the Delta region are well mixed at all times of the year by wind and tidal currents and show little or no vertical stratification (Kelley 1966). Water temperatures throughout the region are fairly uniform and range from a mean summer temperature of 21.5°C to a mean winter temperature of 9.2°C. Dissolved oxygen levels have an annual mean of 9.2 mg/l (range = 5.0 mg/l to 12.4 mg/l), but vary seasonally with water temperature (Calif. Dept. Water Resources 1962, 1977, 1978, 1979; Bureau of Reclamation 1972, 1973, 1974). The bottom composition of the rivers and sloughs is influenced by the currents but is generally some combination of sand, silt, and clay with everchanging configurations of submerged snags and other river borne debris.

MATERIALS AND METHODS

Commercial Crayfish Operations

Gear. Crayfish are captured with baited traps set in series on a line. The trap commonly used is a cylinder 30.5 cm in diameter and from 61 to 76 cm long, constructed of 2.5 x 1.25 cm welded wire mesh. An inverted cone with a 7.5 cm opening is set into each end of the trap body and a door is cut into the side for baiting and crayfish removal. Traps are baited with whole, frozen herring and/or perforated cans of dog food. Traps are attached about 15 m apart on a polypropylene rope about 60 to 350 m in length and set parallel to the bank in 1.2 to 9 m of water. The traps are pulled and emptied at about 24-hour intervals during the height of the season, and every 48 to 72 hours when fishing is slow. Most boats used in the fishery are equipped with a motorized winch for hauling trap lines and a sorting device for culling crayfish less than 92 mm TL.

Season. Although legally open all year, the fishing season is determined primarily by the behavior of P. leniusculus. Fishing operations start when the water has warmed and the crayfish begin actively foraging and thus become susceptible to trapping. The onset of activity can vary from mid-April to late June. Fishing success remains high through the first half of August. From mid-August to mid-September there is a sharp decline in catch as the crayfish undergo their late summer molt. By late September fishing revives, but mating, egg extrusion, and declining water temperatures reduce crayfish vulnerability and the fishing season normally ceases by late October.

Records. The Department of Fish and Game requires the initial purchaser of all commercially-landed species to fill out a fish ticket

405

showing the name of the fisherman, the Fish and Game boat registration number, the total weight of fish landed, price paid, gear used, and area of capture of the fish. Yearly summaries of these tickets yield total yearly catch and total boats fishing each season. In addition, the crayfish fishermen are required to submit a log of their daily fishing activities showing the location fished, the number of traps worked, and the weight landed.

Age Composition

Age class boundaries for the commercial crayfish catch were set using Cassie's (1954) probability paper method of separating polymodal length-frequency data into its component groups or ages. The analysis was run on the commercial catch samples from July 1975 and the entire 1976 season.

Sampling the Commercial Catch

Between 1975 and 1980, from 1,000 to 7,000 crayfish were sampled each season from randomly selected landings taken from all crayfish processors. At the processing plant, the total weight of crayfish in selected containers was obtained to the nearest 0.1 kg. Then, approximately 1/4 to 1/3 of the crayfish were scooped by hand from the middle of the container, from top to bottom to help insure an unbiased sample. This subsample was weighed to the nearest 0.1 kg and each individual crayfish was sexed and measured to the nearest 1 mm TL (tip of acumen to end of telson). Yearly summaries were made from the data collected.

Effort

Estimates of total fishing effort in trap-sets were obtained from the fishing logs submitted monthly from the fishermen, and the annual summary of the fish tickets according to the equations:

$$CPUE_{total} = \frac{\sum_{i=1}^{n} C_1}{\sum_{i=1}^{n} F_1} \quad \text{and} \quad F_{total} = \frac{C_t}{CPUE_{total}}$$

where: C_1 = catch reported in the logs

F_1 = effort reported in the logs

n = no. of months of fishing

C_t = annual catch from the fish ticket summary

F_{total} = annual effort in trap-sets.

A trap-set is defined as the period of time a trap is in the water, from the time it is first baited and set, until it is pulled and emptied. During the height of the season, one trap-set of effort equals one trap in the water over one night. During periods of slow fishing, such as during the late summer molting period, one trap-set of effort equals one trap in the water over two or three nights. Effort measured in trap-nights would be a more sensitive measure of true effort and would more accurately reflect changes in crayfish vulnerability throughout the season. Unfortunately, all effort data were obtained from the fishing logs, which did not provide information on trap-nights of effort.

Estimates of the overall annual mean catch-per-unit-effort (CPUE) were obtained from the fishing logs according to the equation:

$$CPUE_{mean} = \frac{\Sigma \ C/F_m}{N}$$

where: C/F = individual fishermans monthly CPUE
         N = number of months of log data for the season

The variance, standard deviation, standard error of the mean, and 95% confidence interval were computed for the overall annual mean CPUE. This figure was ued to follow year-to-year fluctuations in CPUE.

RESULTS

Commercial Crayfish Operations

Gear. Since 1970 the number of boats fishing at least part of the season ranged from 9 to 59 ($\overline{X}$ = 32). The mean number of traps fished per boat increased steadily from 131 traps per boat in 1975 to 194 traps per boat in 1980.

The catch is not divided equally among all boats. Between 1970 and 1974, 50% of the catch was taken by 24.4% of the boats (range = 14.9 to 34.4%), between 1975 and 1980 the proportion of the boats taking 50% of the catch decreased to 15.5% (range = 12.7 to 18.6%).

Season. More than 98% of the total yearly catch is taken between the months of May and October. However, the seasonal pattern of the harvest has varied since 1975 (Table 1). Between 1975 and 1977, the landings rose sharply in the spring, peaked in June or July, then steadily declined (Figure 2). In 1978, marketing problems postponed most fishing until August and September. As a result, the peak of fishing effort occurred at the low point of crayfish catchability. Thus, the harvest was less than 25% of the total annual catch in any year since 1975. In 1979 and 1980, the catch was relatively constant throughout the season, although a noticeable slump occurred in August during the late summer molting period. Also, in recent years a greater proportion of the total catch was harvested in September and October.

Records. The catch reported in the fishing logs is considered representative of the total commercial catch. An average of 80.6% (range = 47.2 to 99.9%) of the total catch derived from the fish tickets was reported by the fishermen in their logs between 1975 and 1980.

Age Composition

Determining crayfish age is difficult. Analysis of size frequency distributions is helpful, but the wide variation in lengths within any

Table 1. Percent of total yearly crayfish catch taken each month from May through October.

	1975	1976	1977	1978	1979	1980
May	0.4	0.2	0.7	2.7	4.8	10.4
June	24.4	2.0	42.4	6.6	21.5	20.7
July	39.8	44.1	38.0	8.0	26.9	23.2
August	0.7	31.0	16.5	35.2	10.4	11.6
September	5.4	19.2	0.2	40.7	19.3	12.6
October	8.9	3.4	1.9	5.4	16.4	19.5
Total	99.6	99.9	99.7	98.6	99.3	98.0

Table 2. Commercial landings of crayfish (kg) in the Sacramento-San Joaquin Delta.

Year	Gross	Net	Rejected	% Rejected
1970	48,000	48,000	Not Avail.	--
1971	63,000	56,000	7,000	11.1
1972	33,000	24,000	9,000	27.3
1973	47,000	34,000	12,000	25.5
1974	112,000	81,000	31,000	27.7
1975	242,000	169,000	73,000	30.2
1976	249,000	226,000	23,000	9.2
1977	243,000	215,000	28,000	11.5
1978	47,000	45,000	2,000	4.3
1979	206,000	197,000	9,000	4.4
1980	224,000	Not Avail.	Not Avail.	--

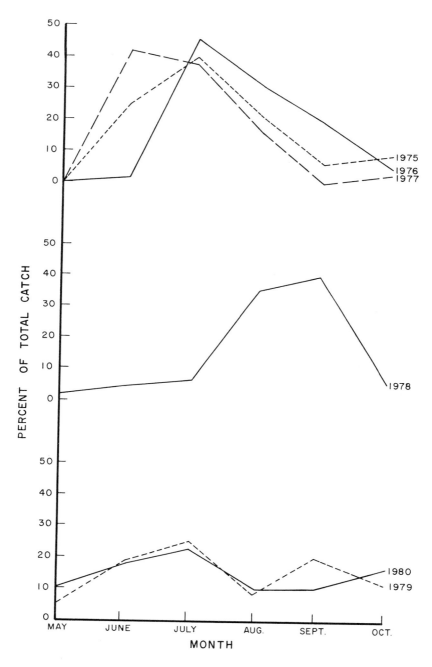

Figure 2. May to October Crayfish Harvest Pattern, 1975 to 1980.

age class, and the overlap of sizes between age classes are long-recognized problems (Andrews 1907; Miller 1960). Assigning discreet boundaries between age classes is arbitrary. Boundaries are presented here only as a tool for analyzing the commercial catch. Trends in age composition are the most important consideration and undue significance should not be given to any particular age class boundary. Growth is assumed to be relatively constant over the period and area of study.

Seven age classes are postulated for the Delta (ages 0+ to 6+) and boundaries have been set at:

Age 0+ = $\leq$ 64 mm

Age 1+ = 65 - 78 mm

Age 2+ = 79 - 98 mm

Age 3+ = 99 - 112 mm

Ages 4+ - 6+ = $\geq$ 113 mm

The differences in age class cutoff points between the 1975 and 1976 samples in the Cassie analyses ranged from 2 mm for the younger age classes to a maximum of 6 mm for the older age classes. Because of this increasing overlap in the boundaries for ages 5+ and 6+, ages 4+ to 6+ were grouped together as "older age classes".

Regulations and Harvest

Commercial fishing began in 1970 with a harvest of 48,000 kg of crayfish (Table 2). Catches remained below 100,000 kg until 1974 when 112,000 kg were landed. This more than doubled the following year with a harvest of 242,000 kg. Except for 1978, the catch has remained above 200,000 kg since 1975. In 1978, financial problems between Swedish brokers and California processors left the fishermen without a market until the middle of August. This abbreviated season was responsible for the extremely low catch of 47,000 kg. It was not reflective of a decrease in crayfish population levels. Because 1978 was atypical with regard to catch and effort, it has been deleted from most subsequent analyses.

Nicola (1971) found a mean length of 104.2 mm TL in the commercial catch samples in 1970 (Table 3). He did not give the age composition of the catch but visual inspection of his graphs indicate, assuming the age class boundaries presented above, that the catch was composed primarily of 2- and 3-year-olds with a considerable percentage of ages 4 and older crayfish.

By 1975 the catch had increased to 242,000 kg, but samples taken from the commercial catch showed that the mean length decreased to 89.8 mm TL and the age composition shifted to 60.4% 2-year-olds, 19.2% 3-year-olds, 16.6% 0- and 1-year-olds, and 3.8% ages 4 and older (Table 4). This decline in mean length and the shift to younger crayfish may have been due to the fishing methods used in the early years of the fishery.

Between 1970 and 1975, the fishermen delivered their entire catch to the processing plants even though the processors would not buy crayfish less than 92 mm TL. The crayfish were held in large, unvented

410

Table 3. Mean length in millimeters of crayfish in the commercial catch samples.

	1970*	1975	1976	1977	1978	1979	1980
No. sampled	1,019	4,752	5,160	7,029	6,142	5,802	1,089
Mean length all crayfish (± SE)	104.2	89.8 (±0.35)	102.2 (±0.33)	99.8 (±0.27)	102.8 (±0.22)	104.3 (±0.25)	101.4 (±0.65)
Mean length legal- + sized crayfish (±SE)		101.2 (±0.33)	105.4 (±0.31)	103.6 (±0.24)	103.8 (±0.20)	105.4 (±0.24)	104.5 (±0.59)
Percent crayfish + < 92 mm		57.0	16.6	20.6	5.9	6.2	16.4

*From Nicola (1971).
+No minimum legal length in 1970.

Table 4. Percent age composition of the commercial crayfish catch.

	1975	1976	1977	1978	1979	1980
Ages 0 & 1	16.6	1.4	2.9	0.4	0.4	1.4
Age 2	60.4	40.4	46.1	32.4	29.0	40.2
Age 3	19.2	40.4	38.7	54.2	52.1	42.9
Ages 4 and older	3.8	17.8	12.3	13.0	18.5	15.5
Total	100	100	100	100	100	100

Table 5. Fishing effort in Delta commercial crayfish fishery from 1975 to 1980, except for 1978.

Year	Trap-sets (to nearest 1,000)	Number of boats	Mean number traps set/boat
1975	275,000	21	131
1976	302,000	45	137
1977	331,000	50	155
1979	309,000	39	170
1980	466,000	47	194

containers and many suffocated before reaching the processing plants. Dead and short (< 92 mm TL) crayfish were sorted out and returned to the river. However, many dead crayfish were observed at the release sites and live car tests showed that less than 50% of the live rejected crayfish survived longer than 1 week. These losses plus other concerns about the unregulated fishery prompted the Fish and Game Commission to adopt several regulations in 1975 designed to remedy these problems.

Beginning with the 1976 season, the minimum size for crayfish was set at 92 mm TL. Fishermen were required to sort their catch on the boat. They were allowed to possess a maximum of 20% by weight of undersized crayfish. Additionally, the new regulations required that the containers used to hold or transport crayfish be vented on the bottom and the amount of crayfish in any container was limited to a maximum depth of 33 cm.

In 1976, the first year the new regulations were in effect, the mean length of the catch rose to 102.2 mm (Table 3). The age composition shifted upward so that 2- and 3-year-olds each accounted for 40.4% of the catch, the percentages of age 4 and older increased to 17.8%, and age 0 and 1 crayfish decreased to 1.4% (Table 4, Figures 3a and 3b).

The number of undersized (< 92 mm TL) crayfish ultimately reaching the processing plants depends on a variety of factors such as the experience of the fisherman, how carefully the catch is sorted, the type of sorter used, and the relative number of short crayfish to legal crayfish in the area being fished. The percentage of short crayfish in the catch samples between 1976 and 1980 varied from 5.9 to 20.6% (Table 3). This fraction in turn affects the overall mean length of the catch. A more consistent measure of mean length for the commercial catch is the mean length of the legal-sized ($\geq$ 92 mm TL) crayfish since it is less affected by variations in the numbers of short crayfish in the sample.

The mean length of the legal-sized crayfish for 1976 was 105.4 mm. This is significantly larger than the mean of 101.2 mm for 1975 (t-test; 1-tail $p < 0.0005$). The mean length of legal crayfish in the catch varied little since 1976 (range = 103.6 to 105.4 mm), and it has remained significantly greater than the 1975 mean of 101.2 mm (t-test; 1 - tail $p < 0.01$).

The age profile of the catch has remained relatively constant since 1976 (Table 4). Two- and 3-year-olds account for an average of 83.3% of the catch (range = 80.0 to 86.6%), and ages 4 and older contribute an average of 15.4% (range = 12.3 to 18.5%).

Since the new regulations on sorting and holding crayfish were the only major changes in the fishery between 1975 and 1976, they were likely responsible for the changes noted in the catch statistics.

In 1977 and 1979, further refinements were made in the crayfish regulations. In 1977, the maximum allowable percentage of short crayfish was changed from 20% by weight to 20% by number for greater ease of enforcement. In 1979, dead crayfish were included with shorts as "unacceptables" and the maximum allowable percentage was lowered to 10% by number.

412

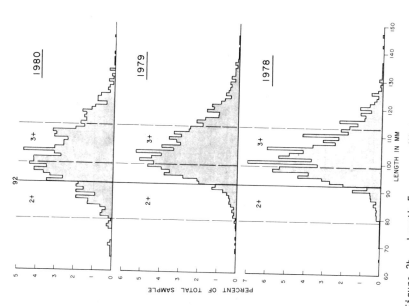

Figure 3a. Length Frequency Histograms of Commercial Crayfish Catch Samples for 1975 to 1977, Showing Age Composition and Minimum Legal Length.

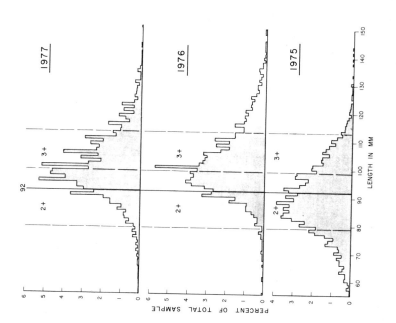

Figure 3b. Length Frequency Histograms of Commercial Crayfish Catch Samples for 1978 to 1980, Showing Age Composition and Minimum Legal Length.

These regulations are acceptable to both the fishermen and the processors; they are enforceable, and they appear to be adequately protecting the crayfish stocks.

Effort

The catch-per-unit-effort (CPUE) in trap-sets declined gradually between 1975 and 1980 (Figure 4). The drop in CPUE from 0.87 kg/trap-set in 1975 to 0.78 kg/trap-set in 1976 is a result of the on-board sorting requirement. In 1975, 57% by number of the catch sampled at the processing plants was below 92 mm TL as opposed to only 16.6% in 1976 (Table 3). The gradual decline in overall annual mean CPUE between 1976 and 1979 probably reflects the increase in efficiency of the sorters retaining fewer short crayfish, year to year population fluctuations, and the gradual increase in total fishing effort. The drop in CPUE between 1979 and 1980 is primarily the result of a large increase in total fishing effort in 1980 (Table 5). For the 1980 season, total effort increased 50.8% over 1979, the number of boats fishing increased by 20.5%, and the mean number of traps set per boat increased by 14.1%. The total fishing area did not increase, however, and the increased number of traps most probably became physical competitors as described by Ricker (1975).

CPUE is negatively correlated with increasing fishing effort ($r = -0.901$) (Figure 5). It appears that in this static fishing area CPUE is maximized at about 300,000 trap-sets of effort, and increased effort serves only to reduce CPUE.

Another factor affecting the CPUE is the frequency of gear rotation, Miller and Van Hyning (1962) found that the crayfish catch declined rapidly in an area fished repeatedly even though adults were still abundant in the immediate vicinity. During experimental 5-consecutive-day sets for population estimation (McGriff, unpubl. data), the catch declined at 14 of 15 sampling stations from day 1 to day 5. This decline averaged 50.2% (range = -27.7 to -67.3%). If the areas were not fished again for about a week, the catches again revived. Both the frequency of gear rotation and the distance moved, as noted in the fishing logs, decreased in the commercial fishery over the years with increases in fuel costs, the number of fishermen, and the number of traps worked per boat.

DISCUSSION

CPUE can often be used as an index of stock abundance, in which case declining CPUE could indicate declining stock density (Ricker 1975). However, for the Delta crayfish fishery, past a certain point additional units of gear in a static fishing area become physical competitors for the same fish and CPUE drops independently of total stock density. The relatively constant age composition of the catch, especially with regard to the older age classes, and the relatively consistent mean lengths of crayfish in the catch, indicate a stable population despite declining CPUE.

The Delta crayfish population appears able to sustain an annual harvest of about 225,000 kg.

414

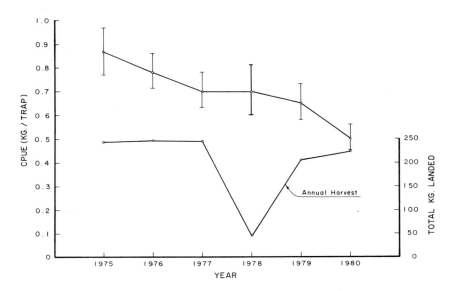

Figure 4. Overall Annual Mean CPUE with 95% Confidence Limits and Annual Harvest for the Years 1975 to 1980.

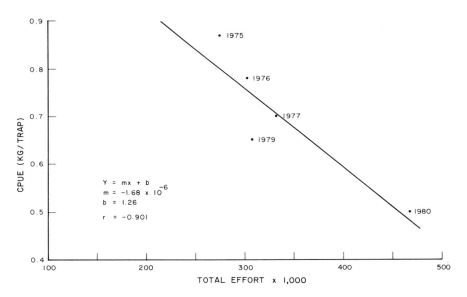

Figure 5. Linear Regression of CPUE versus Total Fishing Effort, 1975 to 1980, Except for 1978.

415

The fishery responded well to regulations requiring on-board sorting of the catch for minimum length and setting standards for holding and transporting crayfish. The Department of Fish and Game has tried to keep the crayfish regulations simple while still protecting the stocks. The present regulations appear adequate for this purpose.

LITERATURE CITED

Andrews, E. A. 1907. The young of the crayfishes Astacus and Cambarus. Smithsonian Inst., Contrib. to Knowledge 359(1718):1-79.

Bureau of Reclamation. 1972. Delta-Suisun Bay surveillance program. A water quality progress report on the Central Valley operations sampling program. Methods and data appendix.

Bureau of Reclamation. 1973. Delta-Suisun Bay surveillance program. A water quality progress report on the Central Valley operations sampling program. Methods and data appendix.

Bureau of Reclamation. 1974. Delta-Suisun Bay surveillance program. A water quality progress report on the Central Valley operations sampling program. Methods and data appendix.

California Department of Water Resources. 1962. Sacramento River water pollution survey. Bull. 111 plus appendices A-D.

California Department of Water Resources. 1977. Sacramento-San Joaquin Delta water quality surveillance program 1976. 304 pp.

California Department of Water Resources. 1978. Sacramento-San Joaquin Delta water quality surveillance program 1977. 300 pp.

California Department of Water Resources. 1979. Sacramento-San Joaquin Delta water quality surveillance program 1978. 276 pp.

Cassie, R. M. 1954. Some uses of probability paper in the analysis of size frequency distributions. Aust. J. Mar. Freshwater Res. 3:513-522.

Kelley, D. W. 1966. Description of the Sacramento-San Joaquin Estuary, pp. 8-17. In: C. R. Hazel and K.W. Kelley (comps.) Ecological studies of the Sacramento-San Joaquin Estuary. Calif. Dept. Fish and Game, Fish Bull. (133):1-133.

Miller, G.C. 1960. The taxonomy and certain biological aspects of the crayfish of Oregon and Washington. Thesis, Oregon State Coll. Corvallis. 216 pp.

Miller, G.C. and J. M. Van Hyning. 1962. The commercial fishery for freshwater crawfish, Pacifastacus leniusculus (Astacidae), in Oregon 1893-1956. Res. Repts. of the Fish Comm. of Oregon 2:77-89.

Nicola, S. J. 1971. Report on a new crayfish fishery in the Sacramento River Delta. Calif. Dept. Fish and Game, Inland Fish. Admin. Rept. 71-7. 22 pp.

Ricker, W. E. 1975. Computation and interpretation of biological statistics of fish populations. Bull. Fish. Res. Bd. Can. 191:382.

THE CRAYFISH PROCAMBARUS CLARKII FROM CALIFORNIA RICEFIELDS: ECOLOGY, PROBLEMS, AND POTENTIAL FOR HARVEST

Ted R. Sommer and Charles R. Goldman
Institute of Ecology and
Division of Environmental Studies
University of California
Davis, California 95616

ABSTRACT

An investigation of the feasibility of harvesting Procambarus clarkii from California rice fields has been made. The life history of the animal and its negative and positive effects on the rice industry were examined in field studies and grower surveys.

The ecology of this crayfish species in California appears to be at least qualitatively similar to descriptions from other states, and its distribution has expanded greatly in the last 20 years. Its range now extends from San Diego to Tehama County. Population sizes are quite variable. Crayfish biomass ranged from 1.4 kg/ha to 204 kg/ha in the two fields surveyed.

Four general types of problems with crayfish were reported by farmers in ten rice-growing counties. The growers experienced displacement of rice irrigation boxes, seedling damage, as well as diffiuclties with herbicide management and rice harvest. Sixty percent of the growers had taken specific measures to deal with crayfish. These methods include structural changes in the field, pesticide use, and sometimes small-scale harvest. The problems have clearly become concerns to growers, and annual statewide costs to growers may exceed a quarter-million dollars annually. Further investigations of harvest as a control method are warranted, for pesticide use may not be wholly effective and harvest offers the possibility of profit rather than expenditure.

INTRODUCTION

The California crayfish industry is currently centered around a fishery for the Signal Crayfish, Pacifastacus leniusculus in the Sacramento River. The present annual harvest remains around 500,000 pounds and supports a sizeable export market and a small local distribution. However, recent data indicates that the fishery may be near optimum yields (McGriff 1982). Thus, new fishing methods will have to be developed and/or new sources of crayfish found in order for the industry to expand. In this research study the feasibility of utilizing a new resource, Procambarus clarkii, the Red Swamp Crayfish, will be investigated. This species is harvested primarily in the southern United States, and accounts for a major portion of the U.S. market (Huner and Barr 1981).

In California, our particular area of interest is ricefields, which farmers report contain large natural populations that are proliferating throughout the state (F. Dubois, pers. comm.). There are presently over 580,000 acres of rice in production in California, which

418

may contain large populations of harvestable crayfish.

Two important factors must be examined in order to determine whether a ricefield crayfish fishery is desirable: 1) the life history of the crayfish and 2) the effects of crayfish on rice production. This project has been divided into two studies in order to deal with each of these questions.

Very little is known about the natural history of this species in western states, even though it has been here at least 55 years (Baker 1980). Some brief observations in California have been made by Riegel (1959), Lange and Chang (1967) and Baker (1980), but the animal has never been subject to intensive study. A primary objective of this investigation is to collect information on the life cycle of the ricefield populations.

The second study will focus on some of the possible positive and negative impacts of crayfish on California rice production. The compatability of crayfish and crayfish harvest with rice-growing techniques is a major consideration, for owners of ricefields seem to have the most to gain or lose by their presence. Crayfish are considered by many farmers to be pest species, primarily because of the burrowing activity of large populations around irrigation boxes (Burton et al. 1980). A few other problems have also been mentioned and in some cases their control has been costly to the rice producer (J. Hill, pers. comm.). However, with recent interest in crayfish harvest, this liability may be turned into an asset. It is possible that crayfish harvesting may serve to reduce negative effects of crayfish and produce a valuable food.

The effects of crayfish on rice production will be studied by first pinpointing the problems caused by crayfish, then describing the different methods available to deal with the problems. Included in this section will be crayfish harvest, for it offers the possibility of profit, rather than expenditure. Finally, an estimate of the costs of crayfish to growers will be given.

METHODS AND MATERIALS

Study 1: Ecology

Life history - qualitative. The life history of P. clarkii was studied in 24.7 hectare (61 acre; Field 56) and 80.5 hectare (199 acre; Field 9) commercial ricefields, approximately 20 miles north of the U.C. Davis campus. Crayfish populations were monitored through monthly trappings, burrow excavations and field observations. Animals were also held for behavioral observations in aquaculture ponds located at the U.C. Davis Student Experimental Farm.

Distribution. An updated distribution of the species in California was obtained by supplementing past and recent trap data with phone interviews with 30 rice growers in different parts of the state (see Methods, Study 2). Crayfish were counted as being "present" in a given county if confirmed by at least two growers.

Population sizes. The population size of crayfish in commercial ricefields was estimated at the beginning and the end of the rice-growing season. The population in Field 9 was computed during April-May, 1981, immediately after flooding and before establishment of vegetative cover. Water clarity was quite high, and occupied burrows could be counted directly.

The late season count was done in Field 56 during field draining. As the fields were drained the crayfish were concentrated into "borrow pits," which were 0.3 - 1.0 meter deep ditches that run alongside the levees of a field. Quantitative samples were taken by randomly placing a 29 cm plastic ring into the water and collecting all crayfish caught in it. Crayfish were counted, sexed and measured to the nearest 0.05 mm carapace length with vernier calipers. Weights were computed using known length-weight relationships for this species (Huner and Barr 1981).

Topographical maps adapted from aerial photographs and a Hewlett-Packard 9810A Digitizer were used to find paddy perimeters and areas in Fields 9 and 56. The figures for crayfish/meter of levee and crayfish/ meter^2 from these samples were used to compute the total population size. Early and late season counts were not done in the same field, but the physical areas were somewhat comparable. The fields were geographically adjacent and similar rice culture was used.

Study 2: Problems, Methods and Costs

The compatability of crayfish and crayfish harvest with rice growing techniques was studied by interviewing growers in seven of the ten major California rice-growing counties and in three minor ones. Selection of growers was non-random, as no complete list of rice growers was available. Instead, a listing of 25 members of the Rice Research Board was obtained. These members were contacted, interviewed, and asked to give the names of other growers in their areas, thereby increasing the sample size. A 23-question format was developed and growers were interviewed by telephone during the summer of 1981. Twenty-one of the questions were multiple-choice and two were open-ended. The survey contained questions on farm size, the presence or absence of crayfish in the grower's ricefields, effects of crayfish on the rice crop and field, an estimate of the economic impact of crayfish on production, pesticide (including herbicide use) and crayfish control measures if any, as well as the harvest, consumption and market potential of crayfish.

RESULTS AND DISCUSSION

Ecology

Qualitative life history. The observed life history of the animal appears to be qualitatively similar to descriptions from the southern United States (Huner and Barr 1981, Penn 1943). P. clarkii's "active" season began in April, when dry fields and ditches were reflooded. The crayfish emerged from burrows, moved to all parts of the field and began feeding on decomposing plant matter and other unidentified forages. Egg-carrying females were present almost immediately after reflooding, and were found throughout the rice-growing season.

420

Reproduction appeared to be greater during certain times of the year. The reproductive peak of the animal may be detected indirectly by noting the frequency of females in trap catches. A lower percentage of females generally means that many are ovigerous and are therefore less susceptable to capture (Penn 1943). The frequency of females caught in traps randomly placed in fields and ditches decreased from 52.6% (n = 97) April 20 - May 21, to 44.2% (n = 215) July 22 - August 6, indicating that there was a peak in reproduction during the summer. Distinct size classes were very difficult to distinguish because of the somewhat continuous reproduction.

One of the major causes of mortality in P. clarkii appeared to be bird predation. Large groups of crows (Corvus brachyrhynchos), great blue herons (Ardea herodias), egrets (Casmerodius albus), and gulls (Larus sp.) were frequently seen feeding on crayfish during the rice-growing season. Other important sources of mortality included rats (Ratus), whose populations inhabited the levees and ditches, and crayfish cannibalism. In recent years, human predation and pesticides have also become significant factors.

Field draining occurred during late August and early September, and many crayfish moved out of the fields into adjacent ditches either overland or with the drainage water. During this time of the year large numbers of redswamp crayfish was observed migrating across roadways to find new sources of water. Many crayfish died in these migrations, for they were easy prey for birds, rodents, humans and dessication.

Most fields are not reflooded after rice harvest in the fall, so the surviving animals must retreat to burrows as the fields and ditches dry up. Burrow analyses conducted by both Louisiana researchers (Jaspers and Avault 1969) and ourselves have demonstrated that the animal has a strong capacity for survival. The crayfish were able to live in burrows throughout late fall, winter, early spring and probably much longer. Furthermore, burrow excavations revealed that the tunnels contained no free water, or water that was close to being anoxic. The crayfish were also able to live through agricultural burnings, which were done in the fields to dispose of rice straw and to kill a parasitic fungus of rice. The life cycle of the crayfish resumed in spring when the fields were reflooded.

Distribution. The past and updated distributions for P. clarkii are shown in Figure 1. The range of the animal has clearly expanded northward and eastward since 1959. Figure 1 is a combination of data from 4 different sources: the northern limit of the species in California (Tehama Co.) was recently established by Baker in 1980. The central California distribution is a mixture of personal trap data and interviews with rice growers. The southern distribution is mostly data from Riegel (1959), but is presently unconfirmed. The range of the red swamp crayfish has clearly expanded northward and westward since 1959, and will probably continue to do so, for many inhabitable counties remain open.

Population sizes. In 63% of the field perimeter in Field 9 the early-season population size was very low. The average number of crayfish per meter of levee was 0.0144 and the total perimeter was 38748 meters. The estimate of total population size was 5810 animals.

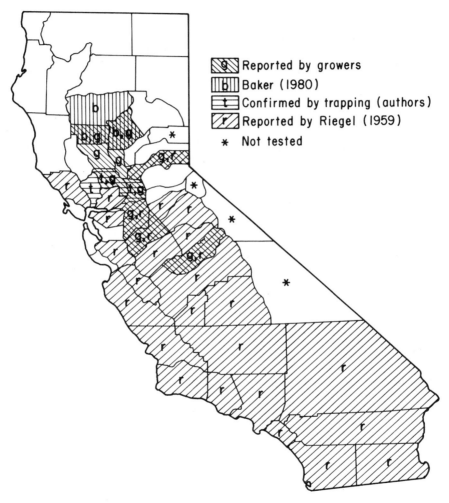

Figure 1. Counties known to contain P. clarkii, 1981

The mean size of crayfish trapped was 43.0 carapace length, or 19.3 grams (n = 78). Estimating the total biomass, a figure of 112 kg or 1.4 kg/ha was obtained.

On the other hand, field 56 had much higher densities of crayfish at the end of the season. The average density of crayfish under water in the borrow pits was 30.7 crayfish/m^2 during draining. Burrows above the water line were also counted, adding an additional 0.65 crayfish/meter of levee. The total area and levee perimeter containing crayfish (above and below water) were 11,970 m^2 and 19,100 m, respectively. A total of 380,000 crayfish were estimated to be present in the entire field. The average carapace length was 38.16 mm, or 13.25 g (n = 288), so the field density was estimated to be 5030 kg or 204 kg/ha.

Study 2

A wide range of areas and growers were interviewed in this study. The 30 growers interviewed included individuals from Yolo, Colusa, Sacramento, Glenn, San Joaquin, Sutter, Butte, Madera, Placer and Stanislaus Counties. Sixty percent of those contacted had "1000 or more acres," 37 percent had "500-1000 acres" and 3% had "less than 500 acres" of land devoted to rice.

Types of problems. All of the growers reported crayfish in at least some of their fields and 66 percent had found crayfish in all of them. When asked "what is the economic impact of crayfish on your levees and crop," the vast majority of these growers felt that the impact was "negative" (89%), rather than "positive" or "no effect" (11%). The descriptions of these effects by growers can be classified into four general groups (Table 1).

The first effect occurred around rice irrigation boxes, wooden structures placed in levees for management of irrigation water. Rice boxes are essential for optimal rice stand establishment and ultimately rice yields. Crayfish however, are apparently attracted to them and animals frequently burrow along the edge of the boxes, causing the boxes to float away and irrigation water to be lost. The box must be immediately replaced when this occurs. There may be over 20 of these structures in a given field, and growers have to check them frequently. In extreme cases extra labor may be hired specifically for this task at a considerable cost.

Another problem related to water management occurs when the herbicide Ordram® (molinate) is applied. During this period, the flow of water into the field must be held back in order to maximize the time that the herbicide will be exposed to the plants. Rice boxes are closed, and inflows and outflows are shut off to prevent herbicide loss. If crayfish burrow through the levees at this time, herbicide-containing water may seep out. All of the growers questioned use this chemical in at least some capacity (Table 2), and 27% had problems with crayfish.

A relatively minor complaint was that crayfish interfered with the establishment of rice seedlings. This problem is supported by a study on crayfish confined with seedlings (Grigarick 1980). A significant reduction in seedling establishment was noted, but it was not clear

423

Table 1.

Type of Problem	% Growers Interviewed with Problem
Rice Irrigation Boxes	69%
Herbicide Management	27%
Seedling Damage	3%
Harvest Difficulties	10%

Table 2. Percentages of fields containing crayfish for users and non-users of chemical treatment methods. All figures are percentages.

Fields Containing Crayfish	Parathion		Ordram		Sevin		CuSO$_4$	
	Growers Using	Non Users	Growers Using	Non Users	Growers Using	Non Users	Growers Using	Non Users
25	13.3	3.3	13.1	0	0	13.3	3.3	10
50	0	6.6	3.3	3.3	0	6.6	3.3	3.3
75	0	0	0	0	0	0	0	0
100	53.3	20	70	0	13.2	60	40	33.3
Mean % fields with crayfish	83	87.5	86.5	50	100	83	91	73

Table 3: Assessment of crayfish effects by users and non-users of chemical treatment methods. All figures are percentages.

Effects on Levees	Parathion		Ordram		Sevin		CuSO$_4$	
	Growers Using	Non Users	Growers Using	Non Users	Growers Using	Non Users	Growers Using	Non Users
0-none		0	3.3	3.3	0	0	3.3	3.3 0
n-negligible	26.6	10	33.3	0	6.6	30	16.5	20
m-moderate	30	10	40	0	3.3	36.6	20	20
e-extensive	10	3.3	13.2	0	3.3	10	6.6	6.6
Mean effect of crayfish	n-m	n-m	n-m	0	n-m	n-m	n-m	n-m

Table 4. Estimated costs of crayfish activities to levees and crops.

	$0-1/acre	$1-5/acre	over $5/acre
% Growers Responding	48	44	8

424

whether burrowing, consumption of seedlings, or a combination of the two was primarily responsible.

A final complaint was that crayfish burrowing activity sometimes made rice harvest difficult. Crayfish may construct "mud-plugs" or "chimneys" above their burrows which are taken up by the harvest machinery. Burrow-mud harvested with the rice plants may cause mechanical problems and/or may reduce rice quality so that cleaning is required. Nevertheless, it appears that recent advances in rice-breeding may eliminate this difficulty. New short stature varieties of rice "lodge" less in the soil and are harvestable without lowering the machine's blades to burrow level. Growers report that other work in rice technology may also reduce some of the problems with rice boxes and herbicide management. New laser-planing technology allows the production of straight rather than contoured rice paddies (Rutger and Brandon 1981). Straight levees require fewer rice boxes than contoured levees and have less surface area, thereby reducing the number of structures exposed to damage. In addition laser leveled fields may be maintained with more uniformily low water thus reducing pressure against rice boxes.

Methods to deal with problems. To deal with crayfish problems, growers resorted to a number of tactics. Sixty percent of those interviewed said that they had "taken specific measures to deal with crayfish" in their ricefields at one time or another. Rice box damage was often dealt with (15% of growers) by using either gravel, boards or plastic liners alongside the structures to inhibit burrowing activity. A new fiberglass box with a "lip" on the bottom has also been tried by one grower. The lip lodges in the soil better, reducing the effect of burrowing activity.

Other growers rely on chemical methods to combat crayfish. No compounds are presently registered for use against crayfish (Burton et al. 1980), although some growers treat fields for other pests and kill crayfish in the process. However, it appears that these methods may not be altogether successful. Tables 2 and 3 show some of the substances commonly used by rice growers and the percentages of interviewed growers using each type. All of the compounds are known to be toxic to crayfish (Huner and Barr 1981).

The farmers were also asked "what percentage of your fields contain crayfish? -- 0, 25, 50, 75, or 100%?" and whether the effect of crayfish on their levees was "none," "negligible," "moderate," or "extensive." Although these estimations of damage and population frequency are highly subjective (Table 3), they illustrate that there were no marked differences between groups using and not using specific substances. Crayfish were shown to inhabit a very large percentage of the fields of both "users and "non-users" of Parathion, Ordram, Sevin and CuSO$_4$, indicating that none of the compounds were causing large-scale extinctions (Table 2). Furthermore, the mean effect of crayfish on levees was very similar between users and non-users. Table 3 reveals that although the treatment methods used by the growers were fairly diverse, the mean effect of crayfish was between "negligible" and "moderate" in all cases. The lack of market differences in P. clarkii abundance (Table 2) and effects (Table 3) when chemical treatment methods were used was also supported by observations by growers. A number of those interviewed said that pesticides initially killed off large numbers of crayfish, but in subsequent years the populations were as abundant, or often more numerous than before.

Harvest. Of those interviewed, 6.6% had tried harvest specifically as a control measure. Two methods were used: 1) trapping around levees and rice-boxes during the rice-growing season, and 2) harvest of crayfish as fields were drained at the end of the season to prevent population buildup the following year.

The effectiveness of these methods would probably depend largely upon population densities. For example, low densities such as those found in Field 9 would probably be very inefficient to harvest, while Field 56 would be much easier. But it is probably also true that sparse populations would not cause enough damage to provide incentive to harvest. Preliminary results were obtained from Field 56 on the potential for harvest from borrow pits at the end of the season. Eight test harvests were done by 2 to 3 unskilled collectors using only crude hand and net methods. The average collection rate was 219 individuals or 4.2 kg/person/hour. Because these harvests were done using only crude hand methods, they probably represent a minimum figure for the harvestability of P. clarkii. It is likely that with increased interest in harvest, more efficient hand or mechanized methods will be developed. The fact that population densities and sizes in the borrow pits were so high indicates that this method warrants further investigation.

It is interesting to note that although 6.6% of the farmers questioned had tried harvest as a means of control, 63.3% had harvested crayfish on a small scale for personal consumption. Also, growers were asked "Do you think that there is a general market for crayfish in California?" More than half (56.6%) of the growers felt that there was. These data illustrated the fact that most growers were not opposed to the idea of crayfish harvest, and that many frequently conducted it themselves on a small scale. Whether harvest can make the transition from small-scale to large-scale still remains to be seen.

Cost of crayfish to growers. In this section growers were asked to give an "estimate of the economic impact of crayfish" on their levees and crop. Costs could include crop damage and levee damage as well as labor and materials in control measures. Table 4 presents the choices given to the farmers and the percentages in each category.

Most responded that damages were $0-5/acre, but some said that costs were as high as $5 or more per acre. Although this method was subjective, it provided a general range of costs. Taking the growers' lowest cost/acre from each category (i.e., $0/acre, $1/acre and $5/acre) the minimum costs in California's 580,000 acres were well over a quarter of a million dollars, and possibly a half-million dollars annually.

CONCLUSIONS

The life history of Procambarus clarkii in California ricefields appeared to be at least quantitatively similar to descriptions from the southern United States (Penn 1943). Since its introduction, the animal has undergone an amazing proliferation in both population size and distribution. Much of its expansion has occurred in the past 20 years, and the known distribution now extends from the Mexican border to Tehama county in the north. Population sizes appeared to be quite variable, for early season estimates in Field 9 were far below late season estimates in Field 56.

In order for a fishery to be established in the ricefields, the negative effects of crayfish will have to be dealt with. They have clearly become a very expensive problem. Problems such as rice box damage may be easily solved using modifications in box design and both rice box and levee damage may be reduced by the new laser-planing techniques, which allow construction of straight paddies and lower water pressures. Advances in rice-breeding may also eliminate harvest of crayfish chimneys by machinery. Pesticides may be used if pesticide registrations are forthcoming, but there is some evidence that this may not be altogether effective.

Harvest is a desirable control measure, especially since it is a potential source, rather than sink, of farm income. Late-season population sizes in at least some paddies have been shown to contain large, harvestable populations, and it is likely that many others are similar. Many rice-growers already harvest crayfish on a small scale and are interested in a commercial market for them. Rice-crayfish double-cropping is another possibility that warrants further investigation. It is estimated that at least 85,000 acres of California rice fields are reflooded annually (Dave Zyner, pers. comm.) for duck hunting clubs. Much of this acreage could be used to extend the growing season of crayfish, thereby increasing total production.

Finally, there is some concern that public health may be affected by toxic accumulations of pesticides and herbicides in the animal. A successful fishery and market depends on a safe, minimum level of these materials. Many growers do not use these compounds and a clean harvest is certainly feasible in their fields. Preliminary analyses of crayfish from ricefields (Burk 1979) lab studies (Ekanem 1982), and rivers receiving ricefield water have all shown that accumulation is negligible. The results are promising, and future data on this and other aspects of ricefield crayfish ecology will help to initiate a new industry.

ACKNOWLEDGMENTS

This research was supported in part by the Jastro-Shields Foundation, California State Critical Applied Research Funds (SCAR), the University of California Appropriate Technology Program, and the U.C. Davis Student Experimental Farm. We also gratefully acknowledge the field assistance of A. Forcella, M. Snyder, H. Morales, E. Hallen and M. Parris.

LITERATURE CITED

Baker, R.A. 1980. The distribution and abundance of three introduced crayfish species in the northern Sacramento Valley. Masters Thesis. Chico State University.

Burk, D. 1979. A survey of pesticides in Procambarus clarkii. Unpublished manuscript.

Burton, V., A. Grigarick, D. Hall and R. Webster. 1980. Insect and disease control recommendations for rice. University of California. Leaflet 1748.

Ekanem, S.B., J.W. Avault, J.S. Graves and H. Morris. 1982. Effects of rice pesticides on Procambarus clarkii in a rice/crawfish pond model. In: C.R. Goldman (ed.), Papers from the Fifth International Symposium of Freshwater Crayfish. In Press.

Grigarick, A.A. 1979. Comprehensive Research on Rice. Annual Report. Project No. RP-3.

Huner, J. and J. Barr. 1981. The red swamp crawfish: biology and exploitation. Sea Grant Publ. No. LSU-T-80-001. 148 pp.

Jaspers, E. and J.W. Avault. 1969. Environmental conditions in burrows and ponds of the red swamp crawfish, Procambarus clarkii. Proc. 23rd Conf. S.E. Assoc. Game and Fish Comm. 28 pp.

Lange, W.H. and V. Chang. 1967. Laboratory and field evaluation of selected pesticides in control of the red crayfish in California ricefiels. J. Econ. Ent. V.60(2):473-477.

McGriff, D.O. 1983. The commercial fishery for Pacifastacus leniusculus in the Sacramento-San Joaquin delta and its management. In: C.R. Goldman (ed.), Papers from the Fifth International Symposium of Freshwater Crayfish. In press.

Penn, G.H. 1943. A study of the life history of the Louisiana red crawfish, Cambarus clarkii. Ecology 24:1-18.

Riegel, J. 1959. The distribution of crayfish in California. Masters Thesis. Univeristy of California at Davis.

Rutger, N.J. and D.M. Brandon. 1981. California rice culture. Sci. Am. Feb:42-51.

COMMERCIALIAZATION OF VERMONT CRAYFISH SPECIES (ORCONECTES IMMUNIS AND ORCONECTES VIRILIS): PUTTING THEORY INTO PRACTICE

James R. Nolfi
Associates in Rural Development, Inc.
Burlington, Vermont 05401, U.S.A.

ABSTRACT

Research and speculation over the last half dozen years has been concerned with the potential for commercial aquaculture and fishing of crayfish species in the northeastern part of the United States, with particular reference to indigenous Orconectes species. This report will test the theory that indigenous populations in Vermont can be successfully exploited for the human food market and for bait for angling. A comparison will be made between expected catch and observed catch on a small-scale sustained commercial basis, development of market for restaurant and retail sales, and comparison between the predicted economics and observed cash flow, cost and profit. The ability of wild stocks to sustain trapping pressure will be measured and the results summarized.

INTRODUCTION

Since 1975, research has been conducted on various aspects of biology and culture of native Vermont crayfish relative to the development of an indigenous crayfish industry (Nolfi, 1977, 1979a, 1979b, 1980; Nolfi and Miltner, 1979). These investigations which examined the full spectrum of potential aquaculture systems from closed-system culture to wild stock fishery have led to a hypothesis for the development of this potential industry (Nolfi, 1980). According to this scheme, two indigenous species have potential for further investigation: Orconectes immunis, which can be cultured in ponds in the northern United States (see Forney, 1968; Calala, 1976) and is desirable as bait for angling; and Orconectes virilis, a locally abundant species with potential for food and possibly fish bait. Because no industry exists at present and the potential market was nonexistent or unknown, the following process was proposed. For Orconectes virilis, a wild stock fishery would be developed first, requiring minimum capital expenditure, and the supply of crayfish made available would be utilized to develop the market for food and fish bait. Should sufficient interest in the local crayfish be generated, hatchery production of juveniles with pond grow-out would be developed to supply larger numbers with controlled production. For Orconectes immunis, wild stock fishery might supply initial market development phase, with sustained pond culture should sufficient market success obtain.

At the same time, it was clear that certain biological aspects of O. virilis in particular in Vermont lakes were as yet unclear. In particular, the food habits of this species in lakes with very limited macrophyte resources (State of Vermont, 1974) had not been delimited and the utilization of O. virilis as forage for important salmonid sport fisheries in Vermont lakes had not been investigated. While it seems likely that no ill effects would be visited on these lakes by an intensive fishery, since stock collapse from excess capacity has not been observed in a crayfish fishery, except under conditions of deterioration of water quality from pollution, lowering of lake levels, or introduction of disease organism (Nolfi, 1979a), there was concern that these biological issues be investigated in the course of the development of the potential industry. Consequently, a research project was proposed and accepted for the allocation of Vermont's modest portion of the federal commercial fisheries research funds (less than 0.5 percent). Unfortunately, all such funds were removed in the recent budget cuts.

Consequently, in the spring of 1981, a decision was made by Associates in Rural Development, Inc., a private research and consulting firm, located in Burlington, Vermont, to initiate investigation of the potential Vermont crayfish industry by attempting to put theory into practice.

MATERIALS AND METHODS

Orconectes virilis for this study were obtained for Caspian Lake, Caledonia County, Vermont. This lake was studied in detail by the state of Vermont, Agency of Environmental Conservation in 1974 (State of Vermont, 1974). The lake is 800 acres (324 hectares) with a maximum depth of 142 feet (43 meters) and an average depth of 71 feet (22 meters). The lake is presently oligotrophic and drains a region of glacial till low in limestone. Consequently, the lake has high secchi disc transparency (maximum of 9.5 meters), high dissolved oxygen levels (5.6 to 2.8 mg/L), and low total hardness (averaging 67 mg/L). The lake is characerized biologically by very low phytoplankton levels, rather insignificant macrophyte populations, significant visible numbers of O. virilis, and one of the finest lake trout (Salvelinus namaycush) fisheries in Vermont. Rainbow trout (Salmo gairdneri) and brown trout (Salmo trutta) are second in importance, along with small numbers of landlocked salmon (Salmo solar). Cold water forage species are present as well.

While no regulation presently limits the sport or commercial catch of crayfish in Vermont, trap size limits are imposed in designated "trout waters," such as Caspian Lake. This means, for the purposes of the study, that only standard minnow traps with a maximum length of 18 inches (45 cm) and maximum funnel opening of one inch (2.5 cm) were permitted. The commercially avaiable traps which were utilized in this study were constructed from galvanized metal with a mesh size of 1/4 inch (0.5 cm). Traps were arranged on lines of five with a float, baited with tins of pet food which had been perforated. The bait was attractive for approximately four to seven days.

Daily catch was sorted into "bait size," under 75 mm total body length, or over 75 mm total body length, "food size." Each size class was weighed daily. Since the market outlets were located one and a half hours' drive from the lake, the catch was held in cages in the Lamoille River and distributed to outlets once or twice each week.

Before initiating this study, pro forma income and expense sheets were developed for the potential fishery endeavor and cash flow

430

analyses were made for a number of scenarios of catch variation, and actual market price for bait and food crayfish. These calculations are presented in Table 1 and were based on best estimates of catch and market price from earlier investigations. Since this project was privately funded, our financial goal was complete payback of initial capital expenditures and operating costs, and a variety of potential profit levels were predicted.

Traps were set for Orconectes immunis in a number of locations in western Vermont where the species is easily captured by hand or with a minnow seine. After approximately 300 trap-night effort with no O. immunis captured in traps, this line of research was terminated.

RESULTS AND DISCUSSION

Expenses of Operation

After 10 weeks of operation, expenses are approximately 10 percent below predicted level (see Table 1). Differences in category distribution between expected and actual with somewhat higher labor costs (more hours) and lower captial and direct costs than predicted.

Catch of O. virilis

After 10 weeks of operation, total catch and yield per effort are below expectation. Total catch is low because only 26 days of trapping were accomplished in 10 weeks rather than 60 days anticipated. A total of 105 kg (230 lbs.) of crayfish over 75 mm, and 135 kg (296 lbs.) under 75 mm were obtained. Learning and start-up were more time-consuming than anticipated, and there were some equipment (boat) and weather problems. In addition, traps originally thought to be available were either slow in delivery or unavailable as anticipated, so that 100 traps were never set and only in the last three weeks were more than 50 traps utilized.

Catch per trap was undoubtedly lower than predicted because of the smaller trap size. Nonetheless, a reasonably accurate figure for yield to be used for future calculations was determined to be a mean of 0.09 kg (0.20 lbs.)/trap-night with a standard deviation of 0.01 kg (0.03 lbs.) for crayfish over 75 mm, and mean of 0.15+0.03 kg (0.32+0.07 lbs.)/trap-night under 75 mm, during the main portion of the trapping season. Using these values for income calculation, revised profit/loss and cash flow can be developed (see Tables 2-5). Under conditions developed in this model, there is always a positive cash flow during the season, all operating and capital expenses are paid with the 16-week season, and a net profit of approximately $1,200 is predicted per 100 traps.

Sales of Bait and Food Crayfish

Sales of the limited volumes of crayfish caught were below the supply available. The major breakdown in the systems fell on the bait sales which were nonexistent. Two factors were involved. A sizable volume of bait crayfish were allocated to a market for preserved crayfish to be packaged and sold by a dealer in Massachusetts. Preservation methods broke down, and the dealer refused to accept the

Table 1. Expenses--Vermont Crayfish

	PRE-START UP	WEEKLY 1	2	3	4	5	6	7	8	9	10	11	12	13	14	SHUT DOWN COSTS	TOTAL OPEN & CLOSE COSTS	TOTAL OPERATIONS COSTS	TOTAL ALL COSTS
Equipment Purchase*				704				704				704				704	704	2112	2816
Maintenance and Repair		10	20	20	20	20	20	20	20	20	20	20	20	20	10		0	260	260
Insurance 300,000/ Liability	35				35						30				35		35	65	100
Bait		30	60	60	60	60	60	60	60	60	60	60	60	60	30		0	780	780
Harvest Labor	56	42	42	42	42	42	42	42	42	42	42	42	42	42	56	56	112	588	700
Harvest Mileage	40	80	80	80	80	80	80	80	80	80	80	80	80	80	80	40	80	1120	1200
Delivery Labor			35	35	35	35	35	35	35	35	35	35	35		35		0	420	420
Delivery Mileage			44	44	44	44	44	44	44	44	44	44	44		44		0	528	528
Phone	50	10	7	5	5	5	5	5	5	5	5	5	5	5	5	15	65	77	142
Miscellaneous	100	50	50	50	50	50	50	50	50	50	50	50	50	50	50	40	140	700	840
Fringe?																			
TOTAL	281	222	338	1040	371	336	336	1040	336	336	366	1040	336	257	296	855	1136	6650	7786

*2700 @ 20% for four months -- total amortization on investment + $1000 start-up capital

Table 2. Production/Income--Vermont Crayfish

	1	2	3	4	5	6	7	8	9	10	11	12	13	14	TOTAL	
#s Catch	75	100	150	200	200	200	200	200	200	200	200	150	100	75	2250	200 lbs./wk. max. catch
#s Food	25	33	50	67	67	67	67	67	67	67	67	50	33	25	752	
#s Bait	50	67	100	133	133	133	133	133	133	133	133	100	67	50	1498	
Food @ 2.50/lb $s; Bait @ 1.75/lb	150	200	300	400	400	400	400	400	400	400	400	300	200	150	$4500	1A avg. 2.00 — gross is 13% higher with higher price
Food @ 3.50/lb $s; Bait @ 1.75/lb	175	233	350	467	467	467	467	467	467	467	467	350	233	175	$5252	1B avg. 2.33 — 25% higher with higher yield
#s Catch	100	175	250	300	300	300	300	300	300	300	300	250	175	100	3450	300 lbs./wk. max. catch
#s Food	33	58	83	100	100	100	100	100	100	100	100	83	58	33	1148	
#s Bait	67	117	167	200	200	200	200	200	200	200	200	167	117	67	2302	
F @ 2.50; B, 1.75	200	350	500	600	600	600	600	600	600	600	600	500	350	200	$6900	2A 2.00
F @ 3.50; B, 1.75	233	408	583	700	700	700	700	700	700	700	700	583	408	233	$8048	2B 2.33
#s Catch	134	234	334	400	400	400	400	400	400	400	400	334	234	134	4604	400 lbs./wk. max. catch
#s Food	67	117	167	200	200	200	200	200	200	200	200	167	117	67	2302	
#s Bait	67	117	167	200	200	200	200	200	200	200	200	167	117	67	2302	
F @ 2.50; B, 1.75	285	497	710	850	850	850	850	850	850	850	850	710	497	285	$9784	3A 2.13
F @ 3.50; B, 1.75	352	614	877	1050	1050	1050	1050	1050	1050	1050	1050	877	614	352	$12086	3B 2.63
#s Catch	167	292	417	500	500	500	500	500	500	500	500	417	292	167	5752	500 lbs./wk. max. catch
#s Food	100	175	250	300	300	300	300	300	300	300	300	250	175	100	3450	
#s Bait	67	117	167	200	200	200	200	200	200	200	200	167	117	67	2302	
F @ 2.50; B, 1.75	367	642	917	1100	1100	1100	1100	1100	1100	1100	1100	917	642	367	$12652	4A 2.20
F @ 3.50; B, 1.75	467	817	1167	1400	1400	1400	1400	1400	1400	1400	1400	1167	817	467	$16102	4B 2.80

433

Table 3. Cash Flow—Vermont Crayfish

4/20/81
Doesn't include tax impact.

Scenario		START-UP	1	2	3	4	5	6	7	8	9	10	11	12	13	14	15	TOTAL
1A	Cash to Start	$1000	719	647	509	(231)	(202)	(138)	(74)	(714)	(650)	(586)	(552)	(1192)	(1228)	(1285)	(1431)	
	Expenses	281	222	338	1040	371	336	336	1040	336	336	366	1040	336	257	296	855	7786
	Income	0	150	200	300	400	400	400	400	400	400	400	400	300	200	150	0	4500
	End Week	719	647	509	(231)	(202)	(138)	(74)	(714)	(650)	(586)	(552)	(1192)	(1228)	(1285)	(1431)	(2286)	(3286)
1B	Cash to Start	1000	719	672	567	(123)	(27)	104	235	(338)	(207)	(76)	25	(548)	(534)	(558)	(679)	
	Expenses	281	222	338	1040	371	336	336	1040	336	336	366	1040	336	257	296	855	7786
	Income	0	175	233	350	467	467	467	467	467	467	467	467	350	233	175	0	5252
	End Week	719	672	567	(123)	(27)	104	235	(338)	(207)	(76)	25	(548)	(534)	(558)	(679)	(1534)	(2534)
2A	Cash to Start	1000	719	697	709	169	398	662	926	486	750	1014	1248	808	972	1065	969	
	Expenses	281	222	338	1040	371	336	336	1040	336	336	366	1040	336	257	296	855	7786
	Income	0	200	350	500	600	600	600	600	600	600	600	600	500	350	200	0	6900
	End Week	719	697	709	169	398	662	926	486	750	1014	1248	808	972	1065	969	114	(886)
2B	Cash to Start	1000	719	730	800	343	672	1036	1400	1060	1424	1788	2122	1782	2029	2180	2117	
	Expenses	281	222	338	1040	371	336	336	1040	336	336	366	1040	336	257	296	855	7786
	Income	0	233	408	583	700	700	700	700	700	700	700	700	583	408	233	0	8048
	End Week	719	730	800	343	672	1036	1400	1060	1424	1788	2122	1782	2029	2180	2117	1262	262

		START-UP	1	2	3	4	5	6	7	8	9	10	11	12	13	14	15	TOTAL
3A	Cash to Start	1000	719	782	941	611	1090	1604	2118	1928	2442	2956	3440	3250	3624	3864	3853	
3A	Expenses	281	222	338	1040	371	336	336	1040	336	336	366	1040	336	257	296	855	7786
3A	Income	0	285	497	710	850	850	850	850	850	850	850	850	710	497	285	0	9784
3A	End Week	719	782	941	611	1090	1604	2118	1928	2442	2956	3440	3250	3624	3864	3853	2998	1998
3B	Cash to Start	1000	719	849	1125	962	1641	2355	3069	3079	3793	4507	5191	5201	5742	6099	6155	
3B	Expenses	281	222	338	1040	371	336	336	1040	336	336	366	1040	336	257	296	855	7786
3B	Income	0	352	614	877	1050	1050	1050	1050	1050	1050	1050	1050	877	614	352	0	12086
3B	End Week	719	849	1125	962	1641	2355	3069	3079	3793	4507	5191	5201	5742	6099	6155	5300	4300
4A	Cash to Start	1000	666	758	1009	833	1509	2220	2931	2938	3649	4360	5041	5048	5576	5908	5926	
4A	* Expenses	334	275	391	1093	424	389	389	1093	389	389	419	1093	389	310	349	938	8664
4A	Income	0	367	642	917	1100	1100	1100	1100	1100	1100	1100	1100	917	642	367	0	12652
4A	End Week	666	758	1009	833	1509	2220	2931	2938	3649	4360	5041	5048	5576	5908	5926	4988	3988
4B	Cash to Start	1000	666	858	1284	1358	2334	3345	4356	4663	5674	6685	7666	7973	8751	9258	9376	
4B	* Expenses	334	275	391	1093	424	389	389	1093	389	389	419	1093	389	310	349	938	8664
4B	Income	0	467	816	1167	1400	1400	1400	1400	1400	1400	1400	1400	1167	817	467	0	16102
4B	End Week	666	858	1284	1358	2334	3345	4356	4663	5674	6685	7666	7973	8751	9258	9376	8438	7438

*Include a 15% increase in all now fixed expenses (on S. L. computation)
($745 = 53/mo)

Table 4.   Expected Catch -- "1982" Season (100 traps, six days/week).

Week	Lbs. Bait	$ Bait*	Lbs. Food	$ Food**
1	110	$ 193	60	$ 210
2	110	193	60	210
3	110	193	60	210
4	130	228	80	280
5	130	228	80	280
6	150	263	100	350
7	150	263	100	350
8	190	333	120	420
9	190	333	120	420
10	190	333	120	420
11	190	333	120	420
12	190	333	120	420
13	190	333	120	420
14	150	263	100	350
15	120	210	100	350
16	100	175	80	280
Total	2,590	$4,533	1,540	$5,390

*   -- @ $1.75/lb.
** -- @ $3.50/lb.

Table 5. Cash Flow—Vermont Crayfish (Based on 1981 Catch Figures)

	Start Up								Week									Total
		1	2	3	4	5	6	7	8	9	10	11	12	13	14	15	16	
Cash to Start	1000	719	900	965	328	465	637	914	487	904	1321	1738	1451	1868	2285	2562	2865	3024
Expenses	281	222	338	1040	371	336	336	1040	336	336	336	1040	336	336	336	257	296	855 8428
Income	0	403	403	403	508	508	613	613	753	753	753	753	753	753	613	560	455	0 9597
End Week	719	900	965	328	465	637	914	487	904	1321	1738	1451	1868	2285	2562	2865	3024	2169 1169 1169

437

product. While we believe the problems have been resolved, the dealer has lost interest in our supply. Second, a live bait dealer who handles northern New England and New York state, who promised to buy "all that we could supply," failed to carry through his commitment. Partially, this was exacerbated by supply shortages on our part at precisely the time most sensitive to this dealer -- the initiation of the bass season in early June. This loss was critically important because the dealer was a major bait wholesaler.

For some time, it has been apparent that sales of small O. virilis captured in traps were necessary to make the economics of a capture fishery worth the effort. Consequently, the failure of bait market development for 1981 was critical. For future years, it would be essential to develop a stronger relationship with buyers of live and preserved bait and to have live bait-size crayfish available in early June. Clearly, there is a preference for the "paper crab" (O. immunis) and should culture of crayfish in Vermont progress, O. immunis is a natural since young-of-the-year are ideal market size, and the culture population continues without restocking. Seining wild stock O. immunis is possible, but not economical, and trapping seems futile.

It is likely that small O. virilis which possess very delicate carapaces during the period of rapid growth (July to August) are suitable for bait and that proper "education" could result in consideration of this species as equivalent to O. immunis.

Sales of food crayfish lagged for several reasons. There was some difficulty in obtaining adequate stocks initially and reluctance to obtain orders without a reasonable certainty of a regular supply. Probably, the learning curve could have been shortened by more supervision of the trapper in the first few weeks, and adequate stocks for initial sales obtained.

The basic marketing approach followed the work of Griffin (1975) and Carroll et al. (1976), and focused on the gourmet restaurants and markets, which catered to predominantly middle and upper middle class, relatively affluent, well educated, families. Initially, our efforts focused on "raw bars," which serve fresh oysters and shrimp, with an emphasis on serving whole crayfish as an hors d'oeuvre or light meal. Later, we found restaurants interested in crayfish for garnishes, and as components in soups, etc. In a number of cases, we found interest in peeled tails for sauces, crepes, etc., but difficulty in incorporating new items into the menu with less than a three-month notice. Clearly the approach to marketing is the correct one. Restauranters, chefs, managers of seafood sales were interested, open, even excited about fresh local crayfish. The price was considered acceptable $7.70/kg ($3.50/lb.).

Two major problems developed, however. First, public acceptance lagged behind the "decisionmakers." Much greater effort needs to be mustered in this area. Some work was done, but it was clearly too little and probably too late, as well. Second, we needed a full, or at least half, time effort in marketing and sales, and that was not forthcoming. Three staff members split less than quarter-time on this aspect of the project. This was very definitely too little and too scattered an effort. It appears that in Vermont, at this point in history, one needs to plan on sales of 2.5-5 kg (5 to 10 lbs)/site/week. Thus, for the sales figures presented in Tables 2-5, 27 to 55 kg(60 to 120 lbs)/week, one would need 12 to 24 sites.

At present, several local newspapers, the magazine published by the state Tourist Agency, and other publications have indicated interest in articles on this gourmet delicacy. A concerted public relations campaign needs to be mustered if a local food market is to be developed.

There are possiblities which have become apparent to develop mid-range metropolitan markets (Nolfi, 1980) in New York City and Montreal. Supply and sales management difficulties this year have made such an approach impossible, but, nonetheless, the market is available should other difficulties mentioned above be overcome.

Effect of Harvest on Stock

While it is too early to see any significant effect of harvest on stock of crayfish in Caspian Lake, it seems very likely that the level of effort for this year was far below anything necessary to affect the population significantly. Figure 1 shows yield/effort figures, and the impression given for this data for the early and middle season is that the population is behaving perfectly normally. That is, relatively low catches early in the season, rising until mid to late August, then falling off in September to low catches in early October. Moreover, relative to 75 to 100 traps, one gets the distinct impression that this is still a very large lake.

CONCLUSIONS

The results of the first attempts to commercially exploit Orconectes virilis in Vermont are somewhat disappointing, but not without clear lessons. Predictions on the business/financial aspects of the problem were as good as could be expected given the information available at the time. The decision to go forward with the project based on those predictions was probably sound. Inadequate management effort was expended in initial training and supervision of harvesting, and inadequate effort was expended in sales. Insufficient consideration was given to the necessity to develop public awareness of crayfish as a food.

Crayfish did sell in some locations, and there is interest in the media in developing public awareness. Sales to restaurants of peeled tails might increase use. Price did not seem to be the limiting factor in most sales, either at the supplier level or consumer level. There was some public reluctance that "they weren't worth the effort, too small." However, these folks wanted two-inch steaks and two pound lobster, so 100 to 125 mm crayfish would have been too small as well.

Whether ARD, Inc. will continue the effort is a management decision which has still to be made. Probably, for our small firm, the critical question will be the extent to which we can commit management time to the project, without which it would not bear continuation.

439

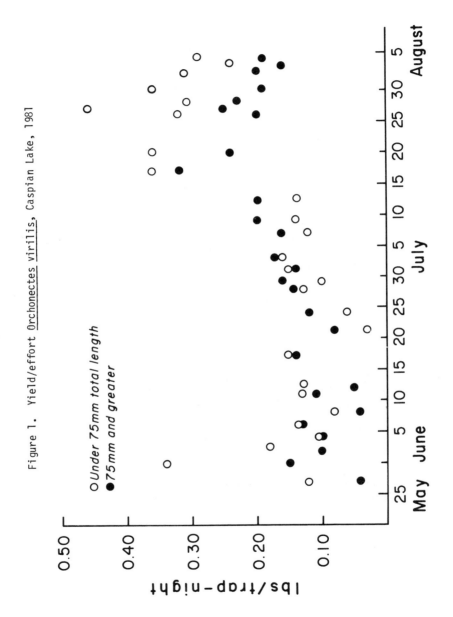

Figure 1. Yield/effort Orchonectes virilis, Caspian Lake, 1981

○ Under 75mm total length
● 75mm and greater

ACKNOWLEDGMENTS

I wish to thank the Associates in Rural Development, Inc. staff for support on this project, particularly Rick Davis, Alan Newman, Richard Donavan, and George Burrill.

LITERATURE CITED

Calala, L. 1976. Crawfish: keeping crawfish for bait, making soft shell crawfish and their care. New London, OH. 24 p.

Carroll, J.D., R. Franz, H. Blades, and T. Griffin. 1976. Crayfish test market study. Houston, Texas. Research Series No. 42, Marketing. University of Southwestern Louisiana, Lafayette, LA. 73 p.

Forney, J.L. 1968. Raising bait fish and crayfish in New York ponds. Cornell University Extension Bulletin 986. Cornell University, Ithaca, NY. 31 p.

Griffin, T. 1975. An identification of early adopters and heavy consumers of crawfish among non-natives of south Louisiana, p. 631-634. In J. Avault (ed.) Freshwater Crayfish II, Baton Rouge, LA (1974). 676 p.

Nolfi, J.R. 1977. Preliminary studies in closed-systems crayfish culture, p. 181-186. In Ossi V. Lindquist (ed.) Freshwater Crayfish III, Kuopio, Finland (1976). 504 p.

Nolfi, J.R. 1979. The social ecology of crayfish fisheries, p. 207-214. In P.J. Laurent (ed.) Freshwater Crayfish IV, Thonon-les-Bains, France (1978). 473 p.

Nolfi, J.R. 1979. Prospects for commercial aquaculture of crayfish in Vermont. Presented at the 35th Annual Northeast Fish and Wildlife Conference, Providence, RI, April, 1979. 15 p.

Nolfi, J.R. 1980. Commercial aquaculture systems for crawfish in the northeastern United States. Proceedings of the World Mariculture Society 11:151-162.

Nolfi, J.R. and M. Miltner. 1979. Preliminary studies on a potential crayfish fishery in Vermont, p. 313-322. In P.J. Laurent (ed.) Freshwater Crayfish IV, Thono-les-Bains, France (1978). 473 p.

State of Vermont. 1974. Caspian Lake water quality report. Lake Eutrophication Series, Report No. 10. Montpelier, VT. 94p.

VII
SOVIET CONTRIBUTIONS

MICROFLORA OF THE DIGESTIVE TRACT OF THE CRAYFISH ASTACUS ASTACUS L.

L. Mickeniene
Institute of Zoology and Parasitology of the
Academy of Sciences of the Lithuanian SSR,
Vilnius 232699, USSR

INTRODUCTION

Microbiological processes which occur in the digestive tract of animals are of great significance to their feeding (Ciurlys, 1959; Leonovich, 1964; Sosnovskaya, 1953). Development of microflora of the digestive tract depends on many factors, including the season, environment, age of the animal; however, the greatest influence is exercised by the peculiarities of feeding (Lubianskiene & Jankevicius, 1975; Lubianskiene, Jankevicius, Trepsiene, & Zableckis, 1977; Syvokiene, 1977). The literature available on the microflora of the digestive tract of crustaceans is limited and in most cases the studies are carried out for sanitary purposes (Sousa & Caland, 1968; Caland & Sousa, 1968; Saha & Raychandhuri, 1973; Kergosien, 1969). The feeding habits of Astacus astacus L. have not been studied fully and there is a lack of information concerning the microflora of the digestive tract of A. astacus. Because of the decrease of A. astacus resources in natural water bodies in Lithuania it was necessary to breed juveniles artificially. Thus, we had to study thoroughly the feeding habits of crayfish under natural and laboratory conditions. In order to control the processes of metabolism in the organism of A. astacus it is necessary to investigate not only feeding, food sources and the habitat but the microflora of the digestive tract and its activity as well, as all these factors are apparently interconnected. The purpose of this study was to determine the total number of bacteria, the amount of microorganisms of some physiological groups, species composition and physiological-biochemical activity of the microflora of the digestive tract of A. astacus, depending on feeding peculiarities.

MATERIALS AND METHODS

The experiments were carried out in Laboratory of Carcinology in Institute of Zoology and Parasitology of the Academy of Sciences of the Lithuanian SSR. 25 mature males from Lake Juodis (Trakai district, Lithuanian SSR) and 10 crayfish maintained in aquariums with running water at 17-18° C, were studied. The crayfish were fed a mixture of fish and Chara rudis. The digestive tract of crayfish was divided into two parts: the stomach and the mid-intestine and hind-intestine. The contents of the digestive tract from 3-4 crayfish were mixed for one sample. The total amount of bacteria was determined by the method of direct calculations on membrane ultrafilters (Razumov, 1932). Physiological groups of microorganisms were determined by the Rodina (1965) method, Odintsova (1956) method, Kuznetsov and Romanenko (1963) method, Krasilnikov's (1949) and Bergey's (1957) manuals were used for determination of species composition of microorganisms.

445

RESULTS AND DISCUSSION

A large number of microorganisms was found in the contents of the digestive tract of the crayfish studied. It was noted that the total amount of bacteria depended more on the feeding intensity than on food composition. The largest total amount of bacteria was found in cells per/gm of the contents of the digestive tract. The study on crayfish after starvation revealed that the amount of bacteria declined 20-50 fold when compared to the amount of bacteria during intensive feeding. The results obtained revealed the presence of definite and constant microflora in the digestive tract of A. astacus and dependence of the amount of bacteria on the feeding intensity of crayfish (Table 1).

The activity of microorganisms of every physiological group in decomposition of definite organic combinations is specific. The following physiological groups of microorganisms were studied: protein-mineralysing, starch-hydrolysing, proteolytic and lactic bacteria, cellulose-splitting microflora, actinomycetes and some micromycetes (yeasts and moulds). The following physiological groups of microorganisms were detected in the contents of the digestive tract of A. astacus under natural and laboratory feeding conditions: protein-mineralysing, starch-hydrolysing, proteolytic, lactic bacteria and some micromycetes. The amount of microorganisms of the physiological groups studied directly depend on the peculiarities of feeding. Protein-mineralysing and proteolytic bacteria constituted 92.2% of all the physiological groups under laboratory feeding conditions, while under natural feeding conditions bacteria were found to be more varied (Fig. 1). In quantitative respect protein-mineralysing bacteria predominated and split various organic combinations. There were more bacteria of all the physiological groups indicated in 1 g of the contents of the mid-intestine and hind-intestine than in 1 g of the stomach contents. No actinomycetes were found in the digestive tract of crayfish, while micromycetes (yeasts and moulds) were found in small amounts in a few tests. Apparently, micromycetes get into the digestive tract with the food and cannot be considered constant microflora in the given species of crayfish. Cellulose-splitting microflora hydrolyses cellular tissue forming sugar and organic acids which are of great significance in feeding of other bacteria. The hydrolysis of cellular tissue in the contents of the digestive tract was rather weak. This suggests that the crayfish consume food of an animal origin containing fats which slow down the process of cellulose-splitting (Brooks et al., 1954).

Morphological variety of microorganisms found in the digestive tract of crayfish wasn't great. Gram-negative rods predominated, whereas cocci and gram-positive rods were found in considerably smaller amounts.

The majority of strains of microorganisms studied, grown on beef-extract agar, exhibited fluorescence characteristic of the genus Pseudomonas.

Food variety determines species composition of bacteria. The bacteria present in the digestive tract of crayfish were those of the genera Pseudomonas, Bacillus, Bacterium, Micrococcus, Sarcina. The composition of bacteria decreased in laboraotry fed animals or during starvation; however, in all cases the genus Pseudomonas predominated, including Ps. denitrificans and Ps. fluorescens. It is evident from the

446

Table 1. Physiological-biochemical activity of microorganisms in the digestive tract of Astacus astacus L. under natural and laboratory feeding conditions.

| Strains studied | Gelatine liquified | | Protein peptonized | | Starch hydrolyzed | | Carbohydrate assimilated | | | | | | | | | | Mineral N assimilated | | Molec-ular N fixed | | Nitrates reduced to nitrites | |
| | | | | | | | glucose | | lactose | | saccharose | | maltose | | mannitol | | | | | | | |
	1	2	1	2	1	2	1	2	1	2	1	2	1	2	1	2	1	2	1	2	1	2
							a) natural feeding															
42	10	23,8	26	61,9	10	23,8	41	97,6	2	4,7	9	21,4	40	95,2	38	90,5	40	95,2	26	61,9	35	83,4
							b) laboratory feeding															
53	31	58,5	36	67,9	14	26,4	50	94,3	7	13,2	28	52,8	51	96,2	50	94,3	52	98,1	42	79,2	47	88,7

1 - the amount of strains in fermentative processes

2 - % of the total amount of strains

447

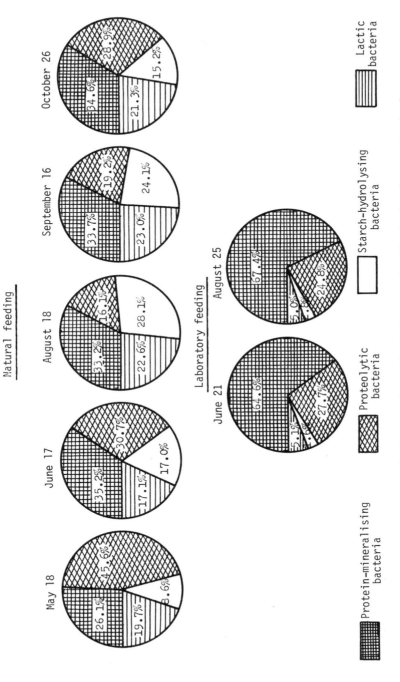

Figure 1. Effect of feeding on the ratio of the amount of bacteria of separate physiological groups in the digestive tract of *Astacus astacus* L. of the total amount of bacteria of the studied physiological groups (expressed in per cent).

results obtained that feeding is the main criterion which determines the formation of microbic population and physiological-biochemical activity of microorganisms.

The analysis of physiological-biochemical activity suggests that considerably high activity in splitting the nutrient substance is characteristic of the studied microorganisms. Proteolytic activity was typical of the majority of the microorganisms studied (Table 1). Many strains appeared to be able to assimilate carbohydrates. They actively assimilated glucose, maltose, mannitol. Many microorganisms developed well in nitrogen source, many strains were nitrogen-fixing. The highest physiological-biochemical activity was revealed in microorganisms from the digestive tract of laboratory fed crayfish. This is further evidence of the increased microbiological activity which is necessary for compensation of the lack of nutrient substances.

CONCLUSIONS

1. As a result of microbiological studies on the digestive tract of A. astacus carried out in 1979 it was revealed that feeding intensity was the main factor determining the amount of microorganisms.

2. The amount of microorganisms of some physiological groups are directly dependent on feeding peculiarities. In laboratory fed animals protein-mineralysing and proteolytic bacteria amounted to 92% of the studied microorganisms, while in freshly collected crayfish physiological groups of microorganisms proved to be more varied.

3. The peculiarities of crayfish feeding affected species composition of microorganisms.

4. The genera Pseudomonas, Bacillus, and Bacterium predominated in the contents of the digestive tract of the studied crayfish.

5. High physiological-biochemical activity is characteristic of the studied microorganisms due to which they are of great significance in splitting the nutrient substances in the digestive tract of crayfish.

LITERATURE CITED

Bergey's Manual of Determinative Bacteriology. 1957. London.

Brooks, C.C. et al. 1954. J. Animal Sci., N 13.

Caland, C.M. and T.T Sousa. 1968. Arg. Estac. biol. mar. Univ. fed. Ceara, vol. 8, N. 1.

Ciurlys, T.K. 1959. Kormleniye selskokhoziaystvennykh zhyvotnykh, M.-L.

Kergosien, E.E. 1969. A Thesis presented to the Graduate Faculty of the University of Southwestern Louisiana.

Krasilnikov, N.A. 1949. Opredelitel bacteriy i actinomitsetov, M.-L.

Kuznetsov, S.I. and V.I. Romanenko. 1963. Microbiologicheskoye izucheniye vnutrennikh vodoyomov, M.-L.

Leonovich, V.V. and T.M. Poperekova. 1964. Tr. VNII selskokhoz. microbiol., vyp. 18.

Lubianskiene, V.N. and K.K. Jankevicius. 1975. Tr. AN LitSSR. Ser. V., t. 4 (72).

Lubianskiene, V.N., K.K. Jankevicius, O.P. Trepsiene, and J.I. Zableckis. 1977. Tr. AN LitSSR. Ser. V., t. 1 (77).

Lubianskiene, V.N., K.K. Jankevicius, O.P. Trepsiene, and J.I. Zableckis. 1980. Tr. AN LitSSR. Ser. V., t. 3. (91).

Odintsova, R.N. 1956. Microbiologicheskiye metody opredeleniya vitaminiov, M.

Razumov, A.A. 1932. Microbiologiya, t. 1, vyp. 2.

Rodina, A.G. 1965. Metody vodnoy microbiologiyi, M.-L.

Saha, N. and D.N. Raychandhuri. 1973. Sci. and Cult., vol. 39, N 8.

Sosnovskaya, E.A. 1953. Issledovaniye microflory rubtsa ovets v sviazi s razlichnymi ratsionami kormleniya. Avtoref. kand. dis., M.

Sousa, T.T. and C.M. Caland. 1968. Arg. Estac. biol. mar. Univ. fed. Ceara, vol. 8, N 1.

Syvokiene, J.S. 1977. Tr. AN LitSSR. Ser. V., t. 2 (78).

CHEMICAL REGULATION IN CRAYFISH BEHAVIOUR
DURING POSTEMBRYONIC DEVELOPMENT

A. Burba
Institute of Zoology and Parasitology of the
Academy of Sciences of the Lithuanian SSR
Vilnius 232600, USSR

INTRODUCTION

Reproduction is of exceptional significance in any animal's life, including crayfish. Many adaptational mechanisms contribute to the successful solution of this important problem. Behavioural mechanisms are of great importance among them. They are labile and easy to carry out, thus they help animals to adapt themselves to changing environmental conditions.

The period of postembryonic development is very important in ontogeny of crayfish. It is connected with a number of dangers, thus labile behavioural mechanisms are of great significance at this period. Hatching Astacus astacus L. and Astacus leptodactylus Esch. larvae takes place in June. According to Cukerzis (1970) during first 5-8 days (until the first moulting) A. astacus larvae take hold of the abdominal pleopods of the female, feed on the remains of yolk and perform only occasional, spontaneous movements with the abdomen or instinctively grasp with chelae. After the first moulting larvae gradually leave the female, yet return to her when threatened. After the second moulting (18-20 days after the hatching) the young crayfish become completely similar to the adult animals and start independent life. Mason (1970) notes that egg laying, hatching, the first and the second moulting, the beginning of independent life are the crucial phases in the crayfish Pacifastacus trowbridgi life. Females become more aggressive before the time of hatching. According to Mason's data larvae hold on the female for 21-25 days.

What are the young crayfish guided by when they cluster on the abdomen of the female? Little (1975) indicates that females of Orconectes sanborni, Cambarus virilis, Procambarus clarkii release chemical products which attract larvae to their mother. These attractants are species specific and larvae can be attracted to a "strange" brooding female of the same species. Crayfish females begin to produce an attractant when they deposit eggs and it becomes maximally effective when the eggs hatch (Little, 1976).

Females' behaviour changes at this period. Pacifastacus trowbridgi females become less active and more aggressive (Mason, 1970). A female carrying eggs exhibits the maximum aggressiveness before the moment of hatching. Such aggressiveness of females can have a bad effect when they are placed together under laboratory conditions or in artificial breeding of crayfish in high denisty. Dunhan (1972) reports that in decapods aggressiveness decreases when they are placed together. However, it can't be attributed to the period of reproduction, as other literary sources prove.

In intraspecific contacts between Orconectes virilis individuals the intensity of encounters depends on the position of antennae, antennules and chelae fingers of the contacting animals (Heckenlively, 1970). Territorial behaviour which is especially expressed in repro-duction period is connected with aggressive actions. Some authors assume that males of the crab <u>Ocypode ceratophthalmus</u> don't allow another crab to make a burrow nearer than 71 cm from his if the density is 0.6 specimens/m^2. If the density of 0.6 specimens/m^2 territorial behaviour is absent (Lighter, 1976). According to Zucker (1977) males of the crab <u>Uca musica</u> even fill in the burrows of other crabs if they are 25 cm away for their own.

All these adaptational mechanisms help to solve problems arising to crustaceans at the period of reproduction. A lack of the data on behavioural mechanisms in the crayfish A. astacus and A. leptodactylus during postembryonic development encouraged us to begin experimental investigations.

MATERIALS AND METHODS

Females of <u>A. astacus</u> and <u>A. leptodactylus</u> and their larvae hatched in indoor tanks with artesian water at 14°C were used for the experiments. The females were fed on chilled fish and <u>Chara</u> sp. The experiments were carried out in June-July, 1977-1979.

To examine the territorial phenomenon, 63 females of <u>A. leptodactylus</u> with hatched larvae were kept in a tank with bottom area of 3.5m^2. The layer of water was 12 cm. In the course of seven days the distance between the females was measured daily, the nature of their aggressive encounters and the principle of their distribution on the bottom were observed.

The presence of attractants attracting larvae to the female was determined by giving the larvae a choice between the chemical stimuli. In the first series of experiments the larvae after the first moulting were washed by water taken from under the female's abdomen and released into the tank with bottom area of 900 cm^2 and layer of water of 5 cm. The mother and a control crayfish, an exuvia or algae were placed in the same tank. The number of larvae that had clustered under the mother and under the control crayfish or an object was counted.

In the second series of experiments the larvae could use only chemical cues, as in the two-choice maze (Little, 1975) visual or tactile contacts with the mother and control crayfish were absent. During the experiments water flowed continually and the larvae were counted in test chambers when not more than 10% of larvae remained in the main chamber.

RESULTS AND DISCUSSION

Our observations have revealed that the crayfish A. astacus and A. leptodactylus when held in groups exhibit no aggressiveness. They share the same burrows gathering as many as 7 adults in a burrow (12 cm x 10.5 cm x 3.5 cm). However, the situation in the tank with 63 females changes as soon as the larvae had hatched. The females spread all over the bottom, though usually the majority of the females preferred the place of water influx, corners, walls. The females were motionless but distinguished themselves in maximum aggressiveness - the slightest movement of one individual stimulated furious response of the others. The females raised their chelae, extended them outwards, opened the fingers, stretched out the antennae, stamped with walking legs, turned. Only a few individuals (usually the smallest females) gathered at the walls of the tank and even touched one another. Apparently, they lacked free area, as the bottom was divided among larger and more aggressive females. The measurements demonstrated that "tolerance" distance between A. leptodactylus females was 15.3 cm on the average. Thus, the territory defended by one female was about 0.6 m^2. The females kept distance defending the territory they had occupied. After a slight movement of a female, a process resembling a chain reaction took place. The neighbouring females caught the movement of one of the females regardless of their position. They instantly assumed threatening poses, turned around, moved a few centimetres aside or back. In doing this they broke the distance between the neighbouring females and aroused their aggressive response. The females remained almost in their places, yet demonstrated aggressiveness very intensively.

The borders of the territory defended by a female pass close to the other females and the major priority to keep distance. Direct encounters are rare, they are rather peaceful - the females try to push back one another with raised and extended chelae. Here biological importance of demonstrative movements is expecially revealed - they help to cause the least damage. In our experiment territorial behaviour, based on observation of aggressive poses of neighbouring individuals, is the result of especially increased aggressiveness which is observed from the moment of hatching. Thus hatching is a signal to the female and we make a suggestion that this signal is of chemical origin. Apparently, as an egg breaks chemical products are released into the water. They are perceived as a signal by females.

The results of the experiments of larvae's preference for the mother and for other control crayfish or objects (Chara sp., an exuvia) are given in Table 1. The results of the studies indicate the presence of an attractant which helps larvae to discriminate females of their species. It appears that on the average 72.6% of the larvae cluster under the mother.

A. astacus and A. leptodactylus are closely related species. Evidently, chemical products being attractants in these closely related species have similar or identical composition, for larvae were often attracted to either species.

The same table indicates that the sex of crayfish in view of the fact that females release attractants is significant in larval choice.

The relative body length of a model crayfish is also very important in choice. Relatively more larvae cluster under larger individuals (females and males). Attention should be drawn to increased aggressiveness of females at this period. When a mother and a "strange" crayfish are placed in the same aquarium the mother attacks "the strange" one. The relative body length is of great importance in their fight. The mother chases the smaller individual, while individuals of the same body length or larger can resist her attacks. Larvae have more chances to cluster under a quiet, non-chased

453

individual, especially if her abdomen is larger (morphological sign of females).

For larvae the female is primarily a shelter and protection from unfavourable environmental conditions. This is proved by the data given in Table 1 which reveal that larvae gather under large exuviae, algae, large nonbrooding females, and adult males. Eventually larvae abandon the female's abdomen. Control calculations have revealed that on the 21st day after hatching 87.0% of all the larvae cluster on the female. The rest 13% gather in algae or spread on the bottom. On the 24th day after hatching 77.4% of the larvae clustered on the female, and only 15.9% on the 28th day.

In their first days, larvae gather under the female's abdomen, probably stimulated by an attractant. Later they begin to climb the mother's body. If the female is not disturbed she remains motionless at this time. Investigation of the mother gradually grows into investigation of the surrounding space, searching for other shelters and food. Apparently, the attractant gradually becomes less effective. The female releases the attractant from under her abdomen. The beginning of its activity is connected with the moment of hatching. Thus, the appearance of chemical products which change female behaviour and attract larvae are closely connected with hatching. The products which are released when the egg breaks might be the source of attractants. Their further activity is closely related to their localization on the remains of the outer covering of the egg. Larvae break these coverings while gathering under the abdomen. The source attractants is gradually destroyed.

The presence of attractants is proved also by the tests in the two-choice maze. In this case visual and tactile contacts between larvae and the mother or a control individual are absent. Only one-sided chemical contact is possible. The results of the experiemnts are given in Table 1. Larvae of both A. astacus and A. leptodactylus respond to the attractants of the females. 64.2% of A. astacus larvae and 68.3% of A. leptodactylus larvae moved into the test chamber with water in which brooding females had been held. Less than 10% of larvae remained in the main chamber. Table 2 shows how larvae responded to water in which separate model crayfish had been held. The percent of the larvae gathering in the chambers with water in which the mother had been held is always higher than in the chambers containing water in which males or nonbrooding females had been held. When the test chamber contained water with brooding females of both species no significant difference was observed in the choice of larvae. The data of the experiment has proved that attractants of A. astacus and A. leptodactylus are similar or identical - larvae choose the chamber with water in which the mother had been held as well as the chambers containing the chemical stimulus of brooding females of both species.

The presence of the attractant is proved by the difference between the number of larvae gathered in the chamber containing water in which the mother had been held and that gathered in the chamber with water in which a control crayfish had been held. The fact that not all the larvae move into the mother's chamber indicate that investigatory-searching activity develops in larvae of this age and the mother is first of all a convenient shelter which is found due to the attractants. The attractants are especially important under natural

Table 1. Larval Choice of A. astacus mother or the control crayfish (an object).

Control crayfish (an object)				Number of clustered larvae in %		
Species	Sex	Relative Body length	The state of females	Number of larvae	Under the mother	Under the control crayfish (object)
A. astacus	female	equal of the mother	non-brooding	890	72.4	27.6
A. astacus	female	equal of the mother	brooding	276	83.7	16.3
A. astacus	male	larger than the mother	---	818	84.9	15.1
A. astacus	male	smaller than the mother	---	186	99.7	0.3
A. leptodactylus	female	larger than the mother	non-brooding	778	41.6	58.4
A. leptodactylus	female	larger than the mother	brooding	393	67.6	32.4
A. leptodactylus	female	equal of the mother	non-brooding	291	95.4	4.6
A. leptodactylus	male	smaller than the mother	---	358	89.5	10.5
Fresh exuvia of A. astacus	female	much larger than the mother	non-brooding	224	22.1	77.9
Dry exuvia of A. leptodactylus	female	much larger than the mother	---	181	31.4	68.6
Algae Chara sp	---	----	---	208	50.4	49.6

455

Table 2. The choice of A. astacus and A. leptodactylus larvae in the two-choice maze.

Species of larvae	Control crayfish (an object)				Number of clustered larvae in %	
	Number of larvae	Control crayfish			Number of larvae which chose the chambers, in % of the mother	Number of larvae which chose the chambers, in % of the control crayfish
		Species	Sex	The state of females		
A. astacus	91	A. astacus	male	---	72.5	27.5
A. astacus	86	A. astacus	female	brooding	55.8	44.2
A. astacus	92	A. astacus	female	non-brooding	72.0	28.0
A. astacus	139	A. leptodactylus	male	---	68.9	31.1
A. astacus	103	A. leptodactylus	female	brooding	54.3	45.7
A. leptodactylus	205	A. leptodactylus	male	---	68.8	31.2
A. astacus	76	A. astacus	male	---	67.2	32.8

conditions where the surrounding environment is much more complicated. It is very important to a larva to have a reliable guide which would help it to find the shelter, i.e. its mother.

CONCLUSIONS

The results obtained revealed that during postembryonic development of A. astacus and A. leptodactylus there is a number of important behavioural mechanisms which are brought into action and controlled by chemical products.

Increased aggressiveness is characteristic of females during reproduction. They are especially aggressive from hatching of eggs to the second moult of larvae after which they start independent life.

During aggressive encounters females assume aggressive demonstrative poses and movements. The cases of fight are rare, usually females are satisfied with pushing one another.

Females perceive the signal of hatching of eggs by chemical means. This signal starts working at the moment of hatching. It especially affects behaviour of females up to the first moulting of larvae.

After the first moulting the larvae which are able to move perceive the mother as a shelter which is found by means of attractants.

Products giving a signal to females and attracting larvae start working at the moment of hatching. The hypothetical place of their location is an egg. When the number of the remains of the outer covering of an egg (with a chemical signalizator) decreases on the pleopods of a female (larvae break them themselves clinging to the mother's abdomen), females become less aggressive and larvae wander further and further from their mother to feed and look for burrows. Such explanation of the origin of chemical signals seems to be quite trustworthy, as the only connection between the mother and larvae is the connection through environment and the most probable signal in water is a chemical one.

Chemical composition of attractants of the closely related species A. astacus and A. leptodactylus is similar or identical.

LITERATURE CITED

Cukerzis, J.M. 1970. Biologiya shirokopalogo raka, "Mintis," Vilnius.

Dunhan, Philip J. 1972. Some effects of group housing upon the aggressive behaviour of the lobster Homarus americanus. J. Fish. Res. Board Can., 29, N 5, 598-601.

Heckenlively, Donald B. 1970. Intensivity of aggression in the crayfish, Orconectes virilis (Hagen). Nature (Engl.), 225, N 55228, 180-181.

Lighter, Frederick. 1976. The social use of space in the Hawaiian ghost crab, <u>Ocypode</u> <u>ceratophthalmus</u>. Pacif. Sci., 30, N 3, 211-212.

Little, Edward E. 1975. Chemical communication in maternal behaviour of crayfish. Nature, 225, N 5507, 400-401.

Little, Edward E. 1976. Ontogeny of maternal behavior and brood pheromone in crayfish. J. Comp. Physiol., 112, N 2, 133-142.

Mason, John C. 1970. Maternal-off-spring behavior of the crayfish, <u>Pacifastacus</u> <u>trowbridgi</u> (Stimpson). Amer. Midland Natur., 84, N 2, 463-473.

Zucker, N. 1977. Neighbor disilodgement and burrow-filling activities by male <u>Uca</u> <u>musica</u> <u>terpsichores</u>. A spacing mechanism. mar. Biol., 41, N 3, 281-286.

CIRCADIAN VARIATIONS OF PROTEOLYTIC ACTIVITY IN THE DIGESTIVE SYSTEM OF THE CRAYFISH ASTACUS ASTACUS L.

G. Mackeviciene
Institute of Zoology and Parasitology of the
Academy of Sciences of the Lithuanian SSR,
Vilnius 232600, USSR

INTRODUCTION

The studies on circadian rhythms of behaviour, locomotor activity and physiological-biochemical processes in aquatic invertebrates are closely connected with the problem of rearing and feeding of industrial crustaceans under artificial conditions. Determination of circadian rhythms of metabolic processes in crayfish is directed to elucidation of mechanisms of adaptation and acclimation of animals to changing environmental conditions.

The majority of reports on biorhythms of decapod crustaceans deal with circadian rhythms of locomotor activity (Hammond and Fingerman 1975, Quilter and Williams 1977), the rate of oxygen consumption (Aldrich 1977), the daily rhythms of feeding (Tarvardiyeva 1978, Cherkashina 1972).

The data on circadian rhythms of metabolic processes in the organism of decapod crustaceans are scarce. The circadian rhythm of the colour changes in integument of the decapod crustacean Hippolyti various Leech is known (Chassard-Bouchaud and Hubert 1970). The circadian rhythm of the concentration of free fatty acids were found in the muscle tissue of the shrimp Palaemon serratus (Pennant 1977) (Martin and Ceccaldi 1977). The circadian rhythm of enzymatic activity of lactate dehydrogenase, α-glucosaminidase, glutamate dehydrogenase and alkaline phosphatase was revealed in the abdominal muscles of Palaemon squilta Linne (Trellu, Ceccaldi 1977). The report (Fingerman et al. 1978) gives the data about the circadian rhythm of monoamino-oxydase enzymes and 5-hydroxytryptophan decarboxylase in the eyestalks of the crab Uca pugilator.

It is known that the activity of the digestive enzymes in decapod crustaceans varies depending on the intensity of food intake, the stage and the period of the intermolt cycle (Van Wormhoudt 1975 and Mackeviciene 1979). The rhythmicity of the secretion of the gastric juice in the digestive gland connected with food intake is characteristic of the digestive system (Hirsh and Jacobs 1929, Van Weel 1960, Mackeviciene and Mazylis 1975).

The circadian rhythm of the activity of the digestive enzymes (amylase, protease, trypsine, etc.) with two maximum values was determined in the digestive tract of the shrimp Penaeus kerathurus (Van Wormhoudt et al. 1972) and in the digestive gland of Palaemon serratus (Crustacea, Natantia) (Rodrigues et al. 1976). Such data of the circadian rhythm of the activity of digestive enzymes in Astacus astacus L. were not found.

MATERIALS AND METHODS

The males of the crayfish A. astacus were kept in running water under a constant 13 hr dark/11 hr light regime. The crayfish were fed on fish and Chara rudis (L.). Feeding was stopped 2 weeks before the experiment. For the determination of the cyclic variations of the proteolytic activity the crayfish (31 to 42 g) were sacrificed every three hours during 24 hr period.

The gastric juice was collected by the method of Jordan (1914). For the analysis the gastric juice was diluted with 1:200 physiological solution. Samples of the digestive gland were homogenized by crushing and grinding in the cold.

Proteolytic activity (PA) was determined in the diluted gastric juice and in the supernatants of homogenates of the digestive gland according to a modification of the method of Anson (Kunitz 1947) at 25 C using casein of the Hammersten quality in the phosphate-citrate buffer at pH 6 (optimum pH). Optic density was determined by the spectrophotometer at 280 nm. The proteolytic activity (PA) was calculated according to Kaverzneva (1971) from 1 μ/g wet weight for 1 ml non-diluted gastric juice.

Protein content in the gastric juice and in the supernatants of homogenates in the digestive gland was determined according to the method of Lowry et al. (1951).

RESULTS AND DISCUSSION

The results obtained revealed that under constant environmental conditions, the proteolytic activity in the gastric juice of A. astacus crayfish varies within 1.09 to 3.2 μ/ml of the gastric juice (Table 1). The curve of the proteolytic activity in the gastric juice is of monophase nature with clearly expressed maximums at the beginning of the lighting period (09.00 and 12.00 hr). The minimum level of the proteolytic activity was determined at 15.00 and 21.00 hr. In comparison with the maximum at 09.00 hr the proteolytic activity in the gastric juice decreased to the minimum at 21.00 hr. Relatively high level of the proteolytic activity occurred at the beginning of the dark phase (18.00 hr) and at night (24.00, 03.00 hr).

The circadian rhythm of the proteolytic activity for the digestive gland is expressed more clearly. The maximum of the enzymatic activity, like in the gastric juice, occurred at 09.00 hr, the second peak was determined at midnight (Table 1). The PA in the digestive gland of A. astacus males gradually decreased from the maximum activity at the beginning of the light period to the minimum level at 21.00 hr. After half a 24 hr period the enzymatic activity in the digestive gland decreased 1.5 times on the average. The second minimum of the PA in the digestive gland was observed in the middle of the dark period (03.00 hr), contrary to the relatively high activity level in the gastric juice. It is evidently caused by rhythmic secretion of the digestive juice from the gastric gland to the stomach.

The results of our analysis agree with the data of other authors obtained when studying the circadian rhythm of the proteolytic activity

Table 1. Circadian rhythm of proteolytic activity (PA) in the digestive gland and gastric juice of the crayfish <u>Astacus astacus</u> L. males

Time	Proteolytic activity		Protein content	
hr	Gastric juice μ/ml	Digestive gland μ/g dry wt	Gastric juice mg/ml	Digestive gland mg/g dry wt
9	3.21± 0.77	0.393 ± 0.04	128.4 ± 24.64	81.43 ± 13.34
12	3.09 ± 0.46	0.318 ± 0.06	122.3 ± 15.42	65.53 ± 9.86
15	1.65 ± 0.30	0.321 ± 0.04	123.7 ± 14.58	81.20 ± 14.90
18	2.40 ± 0.35	0.289 ± 0.03	120.8 ± 13.60	76.57 ± 6.42
21	1.09 ± 0.19	0.237 ± 0.03	122.13 ± 25.17	91.77 ± 1.44
24	2.26 ± 0.26	0.295 ± 0.03	138.07 ± 3.70	89.63 ± 5.18
3	2.36 ± 0.29	0.240 ± 0.08	123.6 ± 15.64	69.35 ± 3.89
6	2.32 ± 0.33	0.299 ± 0.03	96.87 ± 27.17	82.60 ± 16.55

in other species of decapod crustaceans under laboratory conditions. Thus, the principle maximum of the activity of digestive enzymes in the digestive tract of the shrimp _Penaeus kerathurus_ is also noted at 09.00 hr (Van Wormhoudt et al. 1972). It is possible that the peak of the activity of digestive enzymes at the beginning of the light period (09.00 hr) results from the rhythmicity of the digestive juice secretion in the digestive gland of crayfish, as in the case of the crab _Thalamita crenata_ (Van Weel 1960). Analogously to the circadian rhythm of the activity of digestive enzymes, the enzymatic activity of alkaline phosphatase, glutamate dehydrogenase and the concentration of free fatty acids in the muscles of decapod crustaceans is highest in the morning (09.00 hr) (Martin and Ceccaldi 1977, Trellu and Ceccaldi 1977).

The increased level of metabolic processes in the organism of decapod crustaceans at the beginning of the light phase (09.00 hr) is evidently caused by the light stimulus at the beginning of lighting period.

LITERATURE CITED

Aldrich, J.C. 1977. Physiol. and Behav. Mar. Org., Proc. 12th Eur. Symp. Mar. Biol. Stirling. Oxford e.a., 1978, 3-10.

Chassard-Bouchaud, C. and M. Hubert. 1970. Experentia, 26, N 5, 542-543.

Cherkashina, N.Y. 1972. Tr. VNIRO 90, 55-71.

Fingerman, M., R.E. Schultz, B.P. Bordlee, and D.P. Dalton. 1978. Comp. Biochem. and Physiol., C 61, N 1, 171-175.

Hammond, R.D. and M. Fingerman. 1975. Chronobiologia 2, N 2, 119-132.

Hirsch, G.C. and W. Jacobs. 1929. Zeit. vergl. Physiol. 8, 102-144.

Kaverzneva, E.D. 1971. Prikladnaya biokhimiya i microbiologiya, t. 6, N 2.

Mackeviciene, G.J. and A.A. Mazylis. 1975. V sb: Osnovy bioproduktivnosti vnutrennikh vodoyomov Pribaltiki, 322-325.

Mackeviciene, G.J. 1979. V sb: Biologiya rechnykh rakov vodoyomov Litvy, 85-121.

Martin, B.J. and H.J. Ceccaldi. 1977. Comptes rendus des seances de la SOCIETE DE BIOLOGIE et de ses filiales, t. 171, N 3, 608-612.

Quilter, C.G. and B.G. Williams. 1977. J. Zool. 182, N 4, 559-571.

Rodrigues, D., A. Van Wormhoudt, and Y. Le Gal. 1976. Comp. Biochem. and Physiol., B 54, N 1, 181-191.

Tarverdiyeva, M.I. 1978. Biologiya moria, N 3, 91-95.

Trellu, J. and H.J. Ceccaldi. 1977. J. Interdiscipl. Cycle Res., 8, N 3-4, 357-359.

Van Weel, P.B. 1960. Zeit. Vergl. Physiol. 43, 567-577.

Van Wormhoudt, A., H.J. Ceccaldi, and Y. de Tal. 1972. C.r. Acad. Sci., D274, N8, 1208-1211.

Van Wormhoudt, A. 1975. Comp. Biochem. and Physiol., vol. 49 A, N 4, 707-715.

ON THE SYSTEMATICS OF PALAEARCTIC CRAYFISHES
(CRUSTACEA, ASTACIDAE)

S. Ya. Brodski
Ukrainian Research Institute of Fisheries
Kiev, UkrSSR

The classification of Palaearctic crayfishes of the family Astacidae is still imperfect. First of all it concerns the pontic crayfishes (Pontastacus), which among the Palaearctic Astacidae are the more widely and variedly represented. R. Bott (1950) considers the pontic crayfishes as a subgenus Pontastacus, while Mordukhay-Boltovskoy and Starobogatov (1977) considers them as an independent genus.

At the present time new information appeared on karyology (Silver and Cukerzi, 1970; Niiyama, 1962), biochemical genetics (Balakhnin et al., 1979; Romanov et al., 1976; Badio et al., 1979; Nement and Tracey, 1979), systematics (Mordukhai-Boltovskoy and Starobogatov, 1977), zoogeography (Starobogatov, 1975) and ecology of the crayfishes, which confirms a number of positions of the existing classification of the family Astacidae. At the same time, this provides a material for defining the taxonomic range of Pontastacus crayfish and the systematics of the family Astacidae.

In this report we shall try to show the necessity of the elevation of the pontic crayfishes to the independent genus.

Pontastacus

Crayfishes of the genus Pontastacus contain characters which are quite different from the representatives of genus Astacus. They include the following morphological characters: pontic crayfishes have a long grooved rostrum with an interrupted front slot in front of the eyes as compared with flat, narrowing towards the front rostrum, continuous front slot in the Astacus, two pairs of equally developed postorbital tubercules as compared with one well developed pair in the Astacus, inequilateral pointed pleura of the abdomen with a spinule as compared with the equilateral or rounded unarmed pleurae in the Astacus. These groups differ by the shape of the chela and carapace. The first one has a chela with narrow digits without a hollow and conoid tubercles on the inner blade of the immovable digit while the carapace is thin, armed on the sides and at the back of the cervical groove with large spinules. The second one has a wide chela with a hollow and conoid tubercles while the carapace is smooth and hard. P. pachypus which has the characters of both groups is an exception. The crayfishes of these groups differ by their coloration and other characters (Birstein, Vinogradov, 1934; Brodski, 1973-1977; Kessler, 1874; Skorikov, 1908; Schimkewitsch, 1884, 1886; Bott, 1950; Karaman M, 1963; Nordmann, 1842; Rathke, 1837).

They also differ by the genetic characters. An analysis showed that in Pontastacus crayfishes there are twice as many chromosomes -- 184 (Silver and Cukerzis, 1964) than in Astacus (Cukerzis, 1970) and probably 4 times more than in the Austropotamobius. The serum of their hemolymph contains genetic markers (albumens, isoferments, etc.) which make it possible to reliably distinguish separate taxa by the frequencies of the alleles. Thus, the high frequency of alleles S -0.86 of the hemolymph alkali phosphatase is a typical character of P. e. bessarabicus while for other taxa it does not exceed 0.11 (Romanov et al., 1976). The comparison of the data on the analysis of the hemolymph serum with the aid of special genetic methods provides every crayfish taxon with typical, polygonal figure which we shall discuss below.

Peculiar to Pontastacus crayfish are the zoogeographical and ecological characters connected with polyploidy which made them more resistant and plastic for living in the varying conditions of the medium and interspecific competition.

Connected with the polyploidy is a much later origin of Pontastacus crayfishes as compared with that of the Astacus. The crayfishes with different sets of chromosomes should have lived in different geological epochs: polyploidic genus Austropotamobus -- in the basins of palaeogenic or, possibly, cretacious Europe, diploid genera: Astacus, Cambaroides and Cambaurs -- in the early and middle Pliocene, tetraploidal genus Pontastacus -- probably, existed in the late pliocene-Pleistocene as a type adapted to the life in the myxohalinic medium of the Caspian Basin.

The large area of Pontastacus crayfishes which is inherent to young taxons (the entire Palaearctic territory) and their constant expansion and presence in various types of waterbodies in different geographical regions (from seas to ponds) also confirms their later origin as compared with the relict genus Astacus, which has a small, constantly decreasing area and lives in eutrophic lakes and small rivulets. This information at the same time shows that Pontastacus crayfishes probably originated after the formation of the river drainage in Europe, when the river Volga began to flow into the Caspian and the rivers Don and Dnieper into the Azov and Black Sea basins (Starabogatov, 1975).

There are also differences in the character of the spread of representatives of these groups in their generic areas in the water reservoirs. Astacids are not uniformly spread in their generic areas, but every taxon occupies a number of permanent, often remote localities with typical living conditions. Later origin of Pontastacus crayfishes is supported firstly by the territorial location of their areas nearer to the Caspian Sea (which is considered the birthplace of the Astacidae family) than the areas of the Astacus and the most remote genus Astropotamobius and, secondly, by the location of the areas in the remote reservoirs of mutual living of the Pontastacus and Astacus. The former occupies the central mainline and the outflow districts while the latter occupies the top districts and corners of the water systems and basins (small tributaries, flood lakes, ets.). It is interesting to note the character and "behavior" of the concentrations, formed under such conditions, of various species of crayfishes. The representatives of genus Astacus usually form small concentrations which are characterized by more or less spatial constancy during the vegetation period as compared with the large moving concentrations of Pontastacus crayfishes -- from "live spots" larger than the Astacus concentrations to "crayfish fields" of several sq. kilometers. For the transient types of P. pachypus the concentrations have a somewhat transitional

character: by their size they are nearer to the representaives of the
Astacus and by their mobility to the Pontastacus.

And finally it is also the ecological properties connected with
polyploidy which distinguish Pontastacus crayfishes. Pontastacus
crayfishes differ from Astacus crayfishes (mainly represented by the
steno-biotic types with a narrow ecological valency which are not as
flexible as the former) in their eurybionticy, eurychornocy, high
vagility and wide ecological valency. This is connected to their
ability to adapt themselves to different conditions and factors of
their medium. Pontastacus crayfishes, in contrast with Astacus
crayfishes are eurythemic, euryoxybiontic, euryhotic, hydrobionts,
possessing a higher activity during the day. They are vagile polytopic
species differing from the Astacus (which probably change their
ecotopes for winter time only) in that they eat, molt, mate and breed
on different but definite ecotopes. The Pontastacus crayfishes live
for a certain time leading, just as Astacus crayfishes, a single way of
life. However, they differ from the latter in that during their
migration to other districts, or depending on the conduct of the feed
organisms, they form shoals which behavior is not observed in the
representatives of the Astacus. If several species or subspecies of
crayfishes live in the water reservoirs, then a strict distribution
exists between them of ecotopes, food stations and foods, as well as
the feeding of the individuals of different sexes and ages.

An important property of Pontastacus crayfishes, further distinguishing them from the Astacus, is their ability to form ecological
types (ecotypes) which, despite morphological similarity, they differ
by their steno- or rheophility; found among Pontastacus are limno-
bionts, potabobionts, myxophilic and other types with ecological, as
well as physiological and other peculiarities such as euryedophilic
types (pelo-, psammo- and agrilophils). This variability is manifested
differently in crayfishes of different species: the Pontastacus lives
on different grounds, Astacus -- mostly on hard grounds, P. pachypus --
on stony grounds. The first one, as was already mentioned, changes
them several times a year, the second one -- once a year, the third
one, as it should be from the transient type, changes similarly to the
Pontastacus several times a year not only its main ecotopes but the
grounds as well; in autumn it lives for a short time on hard grounds
with overgrowth of red algae.

The data presented provide a basis for the elevation of the pontic
crayfishes to the separate genus Pontastacus. Such a solution of this
problem is important also for convenient definition of the interspecies
systematics of the given genus and the entire family. At the present
time the system of the analysed genus comprises such taxa as the
subgenus and race which, in our opinion, only encumber it since the
first taxon is introduced with the aim of observing the hierarchy of
this system; the second taxon is not protected by the International
Code of the Zoological Nomenclature. Therefore, the separation of the
pontic crayfish into an independent genus is expedient and justified.
However, in order to solve the problem of the independence of the
species or subspecies of the earlier described by us (Brodski,
1973-1977) races of genus P. leptodactylus, it is also necessary to
examine the characters which reliably characterize smaller taxa than
the genus, in particular, the species and subspecies as well as to
solve the problem of the species independence of former subspecies P.

leptodactylus cubanicus and P. l. eichwaldi. The species and subspecies of the analysed groups of crayfishes and particularily genus Pontastacus, differ as do the genera, by the morphological and genetic characters. It is significant that the morphological characters themselves are characterized by different coefficients of their variation (CV) and, as was already mentioned, by the frequencies of the alleles of the polymorphic systems of the albumens and alkali phosphatase of the hemolymph serum. An analysis of the widely spread types in the North-West Black Sea regions showed that the most variable of them is species P. leptodactylus (CV of the chela width 24.9%, width of the carapace 29%) and the least variable -- P. pachypus (respectively 11.4% and 7.9%).

P. e. bessarabicus, in addition to a high CV (20.5% and 16.5%), differ also by a high frequency of the alleles S-0.86 of alkali phosphatase, which in the other types is not more than 0.11, and by the shape of the polygon. However, the polygons of this shape in crayfishes P. cubanicus daucinus and P. leptodactylus differ from one another so much that all of them are probably truly referred to different species. The polygons of P. leptodactylus leptodactylus and P. l. salinus testify to the necessity of referring them only to different subspecies of the same species.

The analyzed taxa, in addition to the high vagility and other characters, listed above, of genus Pontastacus differ also by the shape of the concentrations and by the characteristics of the latter. In species P. kessleri and P. pylzowi which lead a similar mode of life as Astacus astacus, the shape of the concentrations approaches the latter, they form small immobile spots. In the other taxa of genus Pontastacus the concentrations differ from one another by the shape, density and other characters from small, but longer than in the mentioned two species, "live spots" (P. leptodactylus - 64 sp/ha in the Kiev water reservior) to "crayfish fields" with high densities, occupying areas from several hundreds of sq. m. (P. c. daucinus - 672 sp/ha in Lake Katlabukh and P. pachypus - up to 500 sp/ha in the Caspian Sea) to several sq. kilometers (P. e. bessarabicus - on the average 272 sp/ha in the Dniester estuary). These concentrations are not constant by their sizes and age structure. They change according to the seasons depending on the biotic (living cycle, food base, quantity of taxa of crayfishes in one water reservoir) and abiotic factors (landscape structure, ground quality, etc.), as well as the properties of the taxa. For example, P. leptodactylus populates different grounds. Astacus lives on hard, mainly clay grounds (agrilophis) while P. pachypus - mainly on stoney grounds (lithophils). But, as was already mentioned, the first changes its ecotopes and the grounds with them several times a year and the second once a year (for winter).

Finally, the analysed taxa are also distinguished from one another by their qualitative indicies and by their permissible optimum levels of removal of the producers from the population. The analysis showed that in crayfishes P. c. daucinus from Lake Katlabukh (Danube Basin), P. e. bessarabicus from the Dniester estuary and P. leptodactylus from the Kakhovka and Kiev water reservoirs of the Dnieper cascade the average linear size is 11.3; 10.7; 11.0 and 12.7 cm respectively, average weight - 48.6; 50.8; 46.7 and 82 g; fecundity - 287; 350; 242 and 423 eggs, survival of the offsprings from the eggs - 59.1; 68.2; 69.5 and 55.4% (long-term data). When more than 40% of the parent stock is removed from the population of P. e. bessarabicus, overfishing of the stock takes place. However, the population of P. c. daucinus from lakes Katlabukh and Kitai, located near the River Danube, continue to reproduce normally when more than 60% of the parent crayfish are removed from each of them.

These data testify to the independence of P. cubanicus and P. eichwaldi. The decisive argument in favor of this is their living mutually while maintaining definite morphological differences in the Danube basin, in lakes Yalpukh and Kugurlui, where each of these species is represented by its own subspeicies P. c. daucinus and P. e. danubialis. At the same time these data provide a basis for updating the earlier proposed classification of the Palaearctic family Astacidae which comprise 3 subfamiles with 6 genera, 18 species and smaller taxa.[1]

Subfamily ASTACINAE, 4 genera

Genus Austropotamobius, 2 subgenera

- Subgenus Austropotamobius (Austropotamobius) with 1 species - Austropotamobius (Austropotamobius) torrentium and Austropotamobius (Austropotamobius) Torrentium torrentium natio torrentiu, Ap. (Ap.) t. t. n. danubialias an Ap. (Ap). t. t. n. macefdonicus - 3 races

- Subgenus Austropotamobius (Atlantastacus) with 2 species: Austropotamobius (Atlantastacus) pallipes with 2 subspecies: Austropotamobius (Atlantastacus) palliupes pallipes and Ap. (At.) p. bispinosus and Austropotamobius (Atlantastacus) italicus with 3 subspecies: Austropotamobius (Atlantastacus) italicus italicus, Ap. (At.) I. lusitanicus and Ap. (At.) Corcisus.

Genus Astacus, 2 species: Astacus astacus with 2 subspecies - Astacus astacus astacus with 3 races: As. as. as. nation Astacus, As. as. as. n. pretzmani and As. as. as. n. canadsie, and Astacus astacus balcanicus with 2 races: As. as. b. nation balcanicus and As. as. b. n. graeca and Astacus colchicus.

Genus Pontastacus, 6 species: Pontastacus leptodactylus with 4 subspecies: Pontastacus leptodactylus leptodactylus, Pt. l. salinus, Pt. l. boreoorientalis, Pt. l. intermedius and morpha Pt. l. l. morpha angulosus (angular crayfish); Pt. c. daucinus; Pontastacus eichwaldi with 3 subspecies: Pontastacus eichwaldi eichwaldi, Pt. e. danubilais and Pt. e. bessarabicus; Pontastacus kessleri; Pontastacus pylzowi and Pontastacus pachypus with 2 subspecies: Pontastacus pachypus pachypus and Pt. p. notabilis.

Genus Pacifastacus, 1 species - Pacifastacus leniusculus and 2 subspecies: Pacifastacus leniusculus leniusculus and Pf. l. klamanthensis.

---

[1]The International Code of the zoological nomenclature and the idiomatics of crayfishes (Spritzy, 1974) have been used throughout the description.

CAMBAROIDINAE, 1 genus

Genus <u>Cambaroides</u>, 4 species - <u>Cambaroides dauricus</u> with 3 supspecies: <u>Cambaroides</u> dauricus dauricus, Cd. d. koshewnicowi and Cd. d. wladiwostokiensis and <u>Cambaroides</u> schrenckii with 2 subspecies: <u>Cambaroides</u> schrenckii schrenckii and Cd. sch. sachalinesis, <u>Cambaroides</u> <u>japonicus</u> and <u>Cambaroides</u> similis.

CAMBARINAE, 1 genus

Genus <u>Orconectes</u>, 2 species - <u>Orconectes</u> <u>limosus</u> and <u>O. virilis</u>.

The first version of this Classification was reported in 1978 at the International Limnology Congress on the Danube (Brodski, 1980).

LITERATURE CITED

Badio, C., Robotti, C., and Orsi, M. 1979. Probable effect of hatchery conditions on the genetic variability of <u>Astacus leptodactylus</u> Eschsch. In "Freshwater Crayfish." <u>4</u>:257-262.

Balakhnin, I.A., and Brodski, S.YA. 1979. Comparative study of the narrow-clawed crayfish (<u>Astacus Leptodactylus</u>) of the Dnieper, Dniester and Danube Rivers by means of genetic analysis of the albumin and alcali phosphatase systems of the hemolymph serum. In Biokhimicheskaja i populatsionnaja gentika ryb," Lenigrad, pp. 139-141 (in Russian).

Birstein, Ya.A., and Vinogradov, L.G. 1934. Freshwater Decapoda of the USSR and their geographic distribution. Zool. Zh., <u>13</u>(1):39-70 (in Russian, Engl. Summary).

Bott, R. 1950. Die Flusskrebse Europas (Decapoda, Astacidae). Abh. Senckenberg Naturf. Gesellsch., <u>483</u>, 36 s.

Brodski, S.Ya. 1973. Freshwater crayfishes (Crustacea, Astacidae) of the Soviet Union Vestnik Zoologii, 4:49-53: -1974, <u>4</u>:59-59; -1974, <u>6</u>:48-54; 1976, <u>4</u>:14-19; 1977, <u>6</u>:48-53 (in Russian).

Brodski, S.YA. 1980. Mixohaline decapod crayfishes as an object of the cultivation. Biol. morya (Vladivostok), <u>4</u>:61-67 (in Russian, Engl. summary).

Cukerzis, J.M. 1970. The biology of Crayfish <u>Astacus</u> <u>astacus</u>. Vilnius, Publishing House "Mintis," 204 p. (in Russian, Engl. Summary).

Kessler, K.F. 1874. Russian crayfishes. Tr. Russk. entomol. obschestva, <u>8</u> (3-4):228-320 (in Russian).

Mordukhai-Boltovskoy, F.D., and Starbogatov, Ya.I. 1977. Class Crustacea. In "Opredelitel' presnovodnyh bespozvonochnyh evropeiskoe chasti SSSR," Leningrad, "Gidrometeoizdat" Publishers, p. 211-212 (in Russian).

Nement, S., and Tracey, L. 1979. Allozyme variability and relatedness in six crayfish species. Heredity, 70:37-43.

Niiyama, H. 1962. On the unprecedentedly large number of chromosomes of the crayfish Astacus trowbridgii Stimpson. Annotat. zool. Japan, 35(4):219-233.

Nordmann, D. 1842. Observations sur la Fauua Ponticue. Paris, 1842.

Rathke, H. 1837. Beitrag zur Fauna Krim. Mem. Akad. Imp. Sci. Petersburg, 3:370 s.

Romanov, L.M., Balakhnin, I.A., and Brodski, S.Ya. 1976. Genetic analysis of population structure of crayfishes from the waterbodies of the Ukraman SSR. Genetika, 12(12):81-85 (in Russian).

Schimkewitsch, W.M. 1886. On the genus Astacus. Izv. Moscow obschestva l'ubiteley estestvoznanija i etnografii, 50(1):10-24 (in Russian).

Silver, D., and Cukerzis, M. 1964. Number of chromosomes of the narrow-clawed crayfish. Tsitologiya, 5:631-632 (in Russian, Engl. Summary).

Skorikov, A.S. 1908. On the systematics European and Asian Potamobiidae. Ezhegodnik zool. Mus. Saint Petersburg, Published by the Akad. Nauk, p. 115-118 (in Russian).

Spritzy, R. 1974. The use of idioms in Astacology. Intercraysymp II, Baton Rouge, 2 p.

Starobogatov, Ya.I. 1975. Freshwater malacofaunas of the East Europa, its dispersal and some questions of the late cainozoic history of the drainage systems. In "Aktual'nye voprosy zoogeografi," Kishinev, Publishing House "Stintsa," p. 213-214 (in Russian).

ETHOGENESIS OF CRAYFISH

J.M. Cukerzis
Institute of Zoology and Parasitology of the
Academy of Sciences of the Lithuanian SSR
Vilnius 232600, USSR

INTRODUCTION

Fluvial crayfish (Astacidae), including Astacus astacus L., are aquatic invertebrates, possessing rather perfect and specialized external organs of motion. Ten pairs of walking legs and pleopods ensure comparatively rapid moving on the bottom and swimming in all directions, while the tail fan is used when throwing the body back and escaping from danger or an obstacle.

Crayfish behavior may be regarded as a total combination of functions of external "working" organs of an animal definitely orientated in time and space (Fabry, 1976). Proceeding from A. Promptov's (1956) supposition that behaviour is founded on motor reactions and that motion is the most important objective criterion in investigation of behavior, we have carried out the studies on daily activity of crayfish juveniles. This was accomplished in investigations of movements of the body and first pair of appendages -- the chelae which serve for moving, grasping, defense and attacking, and play effectory as well as receptory roles.

In A. astacus hatching of larvae occurs at the end of June -- the beginning of July under conditions of Lithuania. The young crayfish, called larvae at this period, take hold with their claws (which look like hooks) of pleopods of the female and remain almost motionless for 5-9 days.

The first molting takes place after 6-10 days of life and larvae of the 2nd stage start independent life. The number of moltings during the first summer varies depending on environmental conditions, and A. astacus larvae undergo 5-6 stages of development.

The analysis of daily activity of crayfish juveniles is, as a matter of fact, investigation of ethogenesis of A. astacus during the first summer of life.

Ethogenesis is one of the main problems of modern ethology. However, studies on ethogenesis of animals are lacking, particularly in lower ones. The great variety of displays make it difficult not only to ascertain general characteristics of this process, but also to work out its division into periods. Difficulties increase to a great extent due to the fact that "establishment of adequate contacts of an organism with environment start from the very first days of life of an animal. They inflict themselves on processes of maturing which are independent, genetically stipulated and are under no infleunce of environment" (Panov, 1975).

During the prenatal (embryonic) period of ontogenesis the most important morpho-functional systems are formed and they prepare to perform normal vital functions. Functional correlations between parts

471

and organs of an embryo appear and have decisive significance in forming motive functions of the organism. By the end of embryogenesis, during an early postnatal period of ontogenesis innate, instinctive, stereotyped movements are formed in invertebrates, including crayfish. These movements are of paramount importance for survival.

Behavior of most invertebrates in the course of the whole life is determined mainly by innate movements, which is "the most important objective criterion in investigations of behavior. Various forms of this motion in quantitative and qualitative combinations show changes which occur both, in the organism itself and in its relations with external environment" (Promptov, 1956).

MATERIALS AND METHODS

With the purpose of the complete analysis of the activity of A. astacus juveniles, observations were carried out during prenatal period of their development. Eggs and larvae of A. astacus females were used in experiments and observations. The females were caught in Lake Sventas and Lake Juodis and placed in Laboratory of Carcinology in Institute of Zoology and Parasitology of the Academy of Sciences of the Lithuanian SSR. They were kept in aquariums and basins joined in automatic closed running water system at 14-21°C. It ensured optimal conditions for females with eggs, hatching of larvae, development and growth of juveniles.

Hatching of larvae occurred on June 15-18, 1979; June 28-July 2, 1980, the 1st stage lasted until June 24, 1979; July 10, 1980. After the first molting larvae of the 2nd stage, which at that time already begin moving and after leaving the female start independent life, were placed in separated aquariums and kept isolated or in groups of 20-60 specimens. The aim was to observe ethogenesis and daily activity of a single, artificially isolated specimen, as well as of those kept in groups, to carry out analysis of individual behaviour of an animal beyond communicative context and compare it with behavior of a crayfish juvenile in a group. Regarding instinct as "innate motor act of behavior" (Krushinsky, 1977), juveniles kept isolated and in groups should reveal innate and acquired movements of the body and chelae of A. astacus juveniles.

Besides "single" and "group" division the juveniles were divided into two groups according to feeding. For observations of investigatory and comfort behavior the juveniles were fed on plant (Chara rudis A. Br.) and animal (chironomids) food. Observations of feeding behavior were carried on hungry juveniles. The received food placed at a certain distance during the experiment.

One of the main tasks in the studies on behavior and its ontogenetic changes is the division of the united composite stream of behavior into elements easy to work on. Thus, we have distinguised the following movements of the body: "forward" (1), "backward" (2), "to the side" (3), "turn" (4), "turn over" (5), "reflex of freedom" (i.e. response of an animal to space restriction) (6), and "a stop" (7). The following movements of the chalae have been distinguised: "Straightening out" (forward) (1), "joining" (2), "expanding" (3), "pushing off" (from an obstacle) (4), "crossing" (5), and "towards the mouth": (6).

Observations of the body and chelae movements were carried out at different times of the day for 6 minutes and were repeated five times. The duration of longer movements was recorded with a stop watch. The stereotype of daily activity, i.e. the succession of movements and quantitative evaluation of the stereotype were determined by the method of matrices, where all types of the body and chelae movements were inserted horizontally and vertically. The data made it possible to find out the succession of movements, evaluate them quantitatively and make up the stereotype of daily activity of an animal. They also served as a basis for making schemes of the succession, the distribution of probabilities of the succession for drawing schemes of movement stereotype.

Besides, direct observations were made on the activity during intervals of six minutes and the speed of some movements of larvae in different stages of their development was fixed.

The observations were carried out from the beginning of June to the end of September and thus, covered the whole cycle of the development of A. astacus juveniles. In all 1,050 observations were made, 940 matrices were filled in. Six people took part in the observations to reduce chances of subjective evaluation and decrease the possibility of errors. For comparison analogous observations were carried out on closely related species of crayfish, i.e. Astacus Leptodactylus Esch. juveniles.

RESULTS AND DISCUSSION

Daily activity of A. astacus juveniles consists on the whole of locomotor activity and chelae movements (chelae activity). Locomotor and chelae activity are the main components of investigatory, feeding and comfort behaviour of A. astacus juveniles.

Locomotor and chelae movements enumerated above are characteristic of all the forms of behavior of A. astacus juveniles. However, such locomotor and chelae movements as "forward," "reflex of freedom," "straightening out" characterize investigatory behavior; "to the side," "joining," "towards the mouth" define feeding behaviour; expanding and crossing of the chelae are characteristic of comfort behavior.

All indicated movements are innate and form the repertoire of movements of larvae of all stages, except the 1st (Figures 1 and 2).

Larvae of the 1st stage (i.e. juveniles of four days of age on the average), motionlessly holding onto pleopods of females display no daily activity. Rare and irregular, wavy body and chelae movements and a stroke of the tail fan are the first nonspontaneous movements of endogenous or exogenous character.

Daily activity of larvae of the 2nd stage (i.e. juveniles of 12 days of age on the average) is expressed in locomotor and chelae movements characteristic of investigatory behavior: the contribution of the movement "forward" average 44.2% in the locomotor stereotype, the contribution of the movement "straightening out" amounts to 52.3% in the chelae sterotype. The movements "to the side" characteristic of feeding behavior are seldom used, while the movement "towards the

473

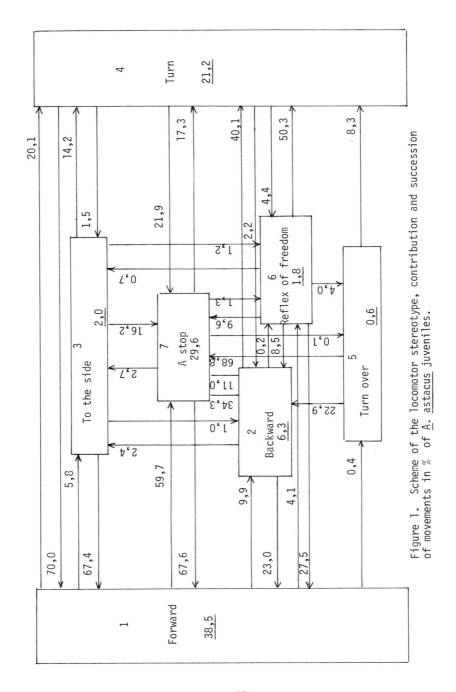

Figure 1. Scheme of the locomotor stereotype, contribution and succession of movements in % of A. astacus juveniles.

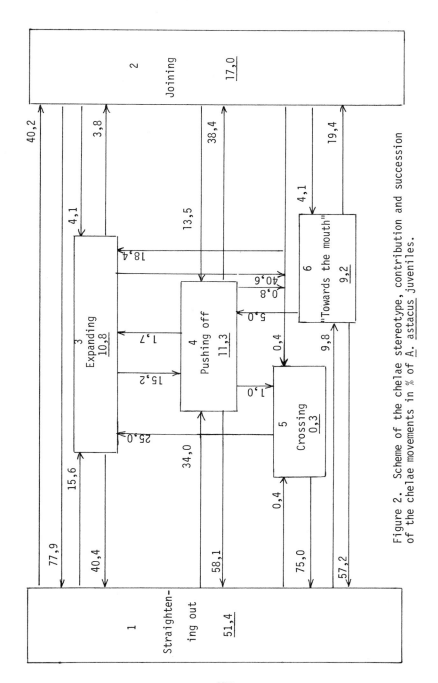

Figure 2. Scheme of the chelae stereotype, contribution and succession of the chelae movements in % of A. astacus juveniles.

mouth" is completely absent. Consequently, the whole daily activity of
larvae of the 2nd stage consists of investigatory behavior. Food
obtaining activity in A. astacus larvae of the 2nd stage is absent.
Daily activity of larvae of the 2nd stage gradually increases: in
larvae of one day of age it takes only 22.5% of the whole time, while
in larvae of 5 days of age it reaches 73.4%. The speed of movements
also increases: from 25.32 cm/min. in larvae of one day of age to 58.2
cm/min. in larvae of 5 days of age.

Daily activity of larvae of the 3rd stage (i.e. juveniles of 26
days of age on the average) consists of investigatory and food
obtaining activities and investigatory, feeding and comfort behavior.
Investigatory activity predominates in sated single and group larvae of
the 3rd stage, food obtaining activity prevails in hungry larvae.
Investigatory activity of sated larvae is characterized by a
considerably greater contribution (by 37% on the average) of the
movements "forward," "backward," straightening of the chelae than in
hungry larvae. Food obtaining activity in hungry larvae is
characterized by the contribution of chelae movements "expanding" and
"towards the mouth," which is almost two times greater than in sated
larvae. It should be noted that food obtaining activity is better
expressed in group larvae than in isolated; the movement "toward the
mouth" occurs almost twice as frequent as in isolated larvae. In
locomotor activity of group larvae of the 3rd stage the movement
"backward" becomes more frequent. It is apparently connected with the
retreat from danger when meeting other specimens. Movements of larvae
of the 3rd stage are more coordinated: the movement "turn over"
disappeared almost completely. Comfort behavior is expressed in more
frequent and longer stops attended by the chelae movements "joining"
and "crossing."

Daily activity of larvae of the 4th stage (i.e. juveniles of 45
days of age on the average) is characterized by further reduction of
locomotor activity and increase of stops after moving forward. The
contribution of the main movements ("forward," "backward," "to the
side," "turn") in locomotor activity continue to increase and reach
80.9%. Daily activity in sated larvae takes only 10%, in hungry larvae
it takes up to 30% of the experimental time. Investigatory activity
dominates in daily activity of sated larvae and food obtaining activity
prevails in hungry larvae.

Daily activity of larvae of the 5th stage (i.e. juveniles of 66
days of age on the average) continues to decrease. Rigidity of
fixation decreases, lability and variability of locomotor and chelae
movements increase: after almost every movement and a stop the whole
repertoire of movements is used in the following order. Such
distribution of successions in locomotor and chelae activity of group
hungry larvae is observed distinctly, in particular when the number of
succession variants reaches 26.

Daily locomotor activity of larvae of the 6th stage (one-summer
juveniles), i.e. juveniles of 83 days of age on the average, increases
from 60.1% to 78.3% in comparison with larvae of the 5th stage. The
increase of daily locomotor activity is the result of intensification
of food obtaining activity which is caused by preparation for
wintering. Therefore, the contribution of the chelae movement" towards
the mouth" increases from 18.4% in larvae of the 5th stage to 24.0% in

one-summer juveniles. Lability and variability of locomotor movements increase, rigidity of fixation of the chelae food obtaining movements intensify. It is, apparently, also connected with the general increase of food obtaining activity in one-summer juveniles of A. astacus.

ACKNOWLEDGMENTS

The author wants to express his sincere thanks to J. Sestokas for collection of the materials and to R. Luksaite for help in preparing the report.

LITERATURE CITED

Fabry, K.E. 1976. Principles of zoopsychology. (In Russian). Moscow, p. 286.

Hinde, Robert A. 1970. Animal behaviour. New York. p. 856.

Krushinsky, L.V. 1977. Biological principles of rational activity. (In Russian). Moscow. p. 272.

Panov, E.N. 1975. Ethology - its sources, formation and place in investigations of behaviour. (In Russian). Moscow. p. 64.

Promptov, A.N. 1956. Essays on the problems of biological adaptation of behaviour of passerine birds. (In Russian). Moscow, p. 312.

SELECTED LITERATURE ON THE LITHUANIAN CRAYFISH

Jakov Cukerzis
Institute of Zoology and Parasitology
of the Academy of Sciences of the Lithunanian SSR,
Vilnius 232600, USSR

Arnold I.N., M. Mulen. 1902. On export of Crayfish. (In Russian). -Vestnik ryboprom., No 6. p. 373-380.

Belianin, O.L., J.V. Doroshenko, T.A. Stepushkina. 1980. Usage of the principle of purposiveness for classification and analysis of behaviour of crustacean. (In Russian). Tr. Leningradskogo obshch. yestestvoispyt., vol. 80, issue 4, p. 8-17.

Biology of the crayfish of the Lithuanian inner waters. 1979. (In Russian, summ. in English). Vilnius, 146 p.

Bowkiewicz, J. 1926. Phenomenon of Heterochely in Potamobius astacus (L.). (In Polish). Prace Tow. Przyj. Nauk w Wilnie, vol. 3, p. 1-20.

Bowkiewicz, J. 1928. Crayfish. (In Polish). Wilno.

Cukerzis, J.M. 1956. Biology and production of Astacus astacus L. in the Lithuanian SSR. (In Russian). Akad, Nauk LitSSR, Inst. Biol., 15 p.

Cukerzis, J.M. 1958. On the problem of ousting of Astacus astacus L. by Astacus leptodactylus Esch. in the lakes of Eastern Lithuania. (In Russian). Tr. AN LitSSR. Ser. B, vol. 4, p. 249-260.

Cukerzis, J.M. 1959. A method for quantitative registration, reporduction and protection of Astacus astacus L. in water bodies of the Lithuanian SSR. (In Russian). Uchen. Zapiski Viln. Uchit. inst., issue 1, p. 143-163.

Cukerzis, J. 1959. Life, reproduction and protection of crayfish. (In Lithuanian). Vilnius, 52 p.

Cukerzis, J.M. 1961. Investigations into biology and production of crayfish in the Baltic republics and Byelorussian SSR. (In Russian). IX nauch. conf. po izuch. vod. Pribaltiki: Tez. dokl., Riga, p. 13-14.

Cukerzis, J. 1962. Astacus astacus L. and its industrial significance. (In Lithuanian). Zuvininkyste vidaus vandenyse, Vilnius, No 1(2), p. 3-7.

Cukerzis, J.M. 1962. Experiments on incubation of eggs of Astacus astacus L. (In Russian). Tr. AN LitSSR. Ser. V, vol. 2, p. 163-169.

Cukerzis, J.M. 1962. On interspecific relations between Astacus astacus L. and Astacus leptodactylus Esch. in the water bodies of Lithuania. (In Russian). II zool. conf. LitSSR, Vilnius, p. 140-141.

Cukerzis, J.M. 1963. Conference on restoration of crayfish resources. (work review). (In Russian). Rybovodstvo i rybolovstvo, No 2, p. 57.

Cukerzis, J.M. 1963. Present conditions and future prospects of investigations into biology and production of crayfish in the Baltic republics and Byelorussian SSR. (In Russian). Gidrobiol. i ikhtiolog. vnutr. vod. Pribaltiki, Riga, p. 169-175.

Cukerzis, J.M. 1963. On restoration of Astacus astacus L. resources in the water bodies of Lithuania. (In Russian). Soveshch. po vospr. zapasov rechnykh rakov, Vilnius, p. 46-63.

Cukerzis, J.M. 1963. Restoration of crayfish resources in the water bodies of the Baltic republics and Byelorussian SSR. (In Russian). X nauch. conf. po vnutr. vod. Pribaltiki: Tez. dolk., Minsk, p. 24-25.

Cukerzis, J.M. 1964. On crayfish plague. (In Russian). Tr. AN LitSSR. Ser. V, vol. 1, p. 77-85.

Cukerzis, J.M. 1964. On interspecific relations between Astacus astacus L. and Astacus leptodactylus Esch. in the lakes of Eastern Lithuania. (In Russian, summ. in English). Zool. zhurn., vol. 43, issue 2, p. 172-177.

Cukerzis, J.M. 1964. Experiments on incubation of crayfish eggs in the ponds. (In Russian). Tr. AN LitSSR. Ser. V, vol. 1, p. 87-93.

Cukerzis, J.M. 1965. Breeding of Astacus astacus L. under artificial conditions. (In Russian). Tr. AN LitSSR. Ser. V, vol. 1, p. 155-164.

Cukerzis, J.M. 1965. Some eco-physiological characteristics of Astacus astacus L. and Astacus leptodactylus Esch. (In Russian). XII nauch. conf. po izuch. vnutr. vod. Pribaltiki: Tez. dolk., Vilnius, p. 88.

Cukerzis, J.M. 1965. Investigations into gas exchange in Astacus astacus L. and Astacus leptodactylus Esch. (In Russian). Physiol. osnovy ecol. vodn. zhivotnykh: Tez. dokl., Sevastopol, p. 103-104.

Cukerzis, J.M. 1966. Investigations into some eco-physiological characteristics of Astacus astacus L. and Astacus leptodactylus Esch. in connection with their interspecific relations. (In Russian). Tr. AN LitSSR. Ser. V, vol. 2, p. 279-286.

Cukerzis, J.M. 1966. An incubator as a cradle of crayfish juveniles. (In Russian). Nauka i tekhnika, Riga, No 7, p. 28-29.

Cukerzis, J.M. 1966. The temperature chosen by Astacus astacus L. and Astacus leptodactylus Esch. juveniles. (In Russian). Tez. dokl. XIII nauch. conf. po izuch. vnutr. vod. Pribaltiki, Tartu, p. 193.

Cukerzis, J.M. 1967. The dependence between gas exchanges and weight of the body in Astacus astacus L. and Astacus leptodactylus Esch. (In Russian, summ. in English). Tr AN LitSSR. Ser. V, vol. 3, p. 85-90.

Cukerzis, J.M. 1967. Adaptational reactions of Astacus astacus L. and Astacus leptodactylus Esch. in the light of their interspecific relations. (In Russian). Vid. i prirodno-klimat. adapt. org. zhivotnykh, Novosibirsk, p. 139-141.

Cukerzis, J. 1967. Eco-physiological investigations into interspecific relations of closely related species of crayfish. (In Polish). VII Zjazd Hydrobiol. polskich, Warszawa, p. 21-22.

Cukerzis, J.M. 1968. Interspecific relations between Astacus astacus L. and A. leptodactylus Esch. Ekologia Polska. Ser. A, vol. 16, No 31, p. 629-639.

Cukerzis, J.M. 1968. Distribution of Astacus astacus L. and Astacus leptodactylus Esch. juveniles in temperature gradient. (In Russian, summ. in English). Zhurn. obshch. biol., vol. 29, No 4, p. 459-462.

Cukerzis, J.M. 1968. On interspecific relations between closely related crayfish species. (In Russian). III zool. conf. BSSR: Tez. dokl., Minsk, p. 150-151.

Cukerzis, J.M. 1968. The effect of pH on gas exchange in Astacus astacus L. and Astacus leptodactylus Esch. (In Russian). Gidrobiol. i ikhtiolog. issled. vnutr. vod. Pribaltiki: Tr. XII nauch. conf., Vilnius, p. 189-190.

Cukerzis, J.M. 1969. Some eco-physiological characteristics of Astacus astacus L. and Astacus leptodactylus Esch. juveniles. (In Russion). Gidrobiol. i rybn. khoz. vnutr. vod. Pribaltiki, Tallin, vol. 5, p. 171-177.

Cukerzis, J.M. 1970. On eco-physiological investigations into adaptation to environmental conditions in closely related crayfish species. (In Russian, summ. in English). Adapt. org. cheloveka i zhivotnykh k experim. prirodnym phact. sredy: Mat. symp., Novosibirsk, p. 164-165.

Cukerzis, J.M. 1970. Present conditions and future prospects of reproduction of Astacus astacus L. resources in lakes of the Lithuanian SSR. (In Russian). Biol. ozior: Tr. Vses. symp., Vilnius, p. 212-220.

Cukerzis, J.M. 1970. Biological principles of artificial breeding of Astacus astacus L. (In Russian). Biol. protsessy v morskikh i continent. vod.: Tez. dokl. II syezda VGBO, Kishinev, p. 396-397.

Cukerzis, J.M. 1970. The biology of crayfish Astacus astacus L. (In Russian, summ. in English). Vilnius, 206 p.

Cukerzis, J.M. 1972. Biologische Grundlagen der Methode der Kunstlichen Aufzucht der Brut des Astacus astacus L. Freshwater Crayfish I, Lund, p. 187-201.

Cukerzis, J.M. 1972. On the prospects of acclimatization of Pacifastacus leniusculus Dana in the NW of the USSR. (In Russian). Aclimat. ryb i bespozv. v vod. SSSR, Frunze, p. 250-252.

Cukerzis, J. 1975. Die Zahl, Struktur und Produktivität der isolierten Population von Astacus astacus L. Freshwater Crayfish II, Baton Rouge, p. 513-527.

Cukerzis, J.M. 1975. On interspecific relations between closely related species at the juncture of their ranges. (In Russian, summ. in English). Issled. produktivnosti vida v areale, Vilnius, p. 56-58.

Cukerzis, J.M. 1976. On interspecific competition in closely related crayfish species (Astacus astacus L, Astacus leptodactylus Esch., Pacifastacus leniusculus Dana, Decapoda, Crustacea). (In Russian). DAN, vol. 229, No 1, p. 250-252.

Cukerzis, J.M. 1976. On interspecific relations between aboriginal and acclimatized species of crayfish. (In Russian). Mat. k II Vses. soveshch., Vilnius, p. 127-129.

Cukerzis, J.M. 1977. Investigations into acclimatization of Pacifastacus leniusculus Dana in some isolated lakes of Lithuania. (In Russian). Symp. po reaktsii vodn. eco-system na vseleniye novykh vidov, Tallin, p. 143-145.

Cukerzis, J. 1979. Jak zvysit stavy raka. Rybarstvi, Praha, No 3, p. 51.

Cukerzis, J.M. 1979. On acclimatization of Pacifastacus leniusculus Dana in an isolated lake. Freshwater Crayfish IV, Thonon-les-Bains, p. 445-450.

Cukerzis, J.M. 1979. Biological investigations into Astacus astacus L. (In Russian, summ. in English). Biologiya rechnykh rakov vod. Litvy, Vilnius, p. 20-27.

Cukerzis, J., J. Sestokas. 1964. Crayfish resources in the water bodies of the Lithuanian SSR and their restoration. (In Russian). XI nauch. conf. po izuch. vnutr. vod. Pribaltiki: Tez. dokl., Petrozavodsk, p. 67-68.

Cukerzis, J., J. Sestokas. 1968. Crayfish resources in northeastern Lithuania and means to increase them. (In Russian). Syryevye resursy vnutr. vod. Severo-Zapada: Mat. dokl. na XI nauch. conf., Petrozavodsk, p. 211-218.

Cukerzis, J., E. Tamkeviciene. 1972. Artificial breeding of Astacus astacus L. (In Russian). Voprosy razvedeniya ryb i rakoobraznykh v vod. Litvy, Vilnius, p. 219-228.

Cukerzis, J.M., E. Tamkeviciene. 1973. The role of Charophyta algae in the feeding of crayfish. (In Russian). Kharov. vodorosli i ikh ispolz. v issled. biolog. protsessov kletki, Vilnius, p. 164-166.

Cukerzis, J.M., J.V. Doroshenko. 1975. Behaviour of Astacus astacus L. during food intake. (In Russian). Povedeniye vodn. bespozv.: Mat. II Vses. symp., Borok, p. 96-98.

Cukerzis, J.M., M.J. Doroshenko. 1976. Domination and subordination in Astacus astacus L. (In Russian, summ. in English). Tr. AN LitSSR. Ser. V, vol. 1, p. 71-76.

Cukerzis, J.M., J. Sestokas. 1977. The ways of infection of aboriginal crayfish species with crayfish plague by Pacifastacus leniusculus Dana. (In Russian). Vses. nauch. conf. po ispolz. promysl. bespozv.: Tez. dokl., Moskva, p. 103-104.

Cukerzis, J.M., J. Sestokas. 1977. Embryonal diapause in Astacus astacus L. (In Russian, summ. in English). Zhurn. obshch. biol., vol. 38, No 6, p. 929-933.

Cukerzis, J.M., J. Sestokas. 1977. Shortening of diapause in Astacus astacus L. (In Russian). XIX nauch. conf. po izuch. i osv. vod. Pribaltiki i Belorussii: Tez. dokl., Minsk, p. 160-161.

Cukerzis, J., A. Terentjew. 1979. Acclimatation de Pacifastacus leniusculus Dana dans un lac isole. La pisciculture francaise, No 56, p. 13-16.

Cukerzis, J., J. Sestokas. 1980. Astacological investigations in Lithuania. (In Russian, summ. in English). Acta Hydrobiologica Lithuanica, Vilnius, vol. 1, p. 74-80.

Cukerzis, J.M., A. Terentyev, E. Tamkeviciene. 1968. The dependence between body weight and age in Astacus astacus L. (In Russian). Limnologiya: Mat. XIV conf. po izuch. vnutr. vod. Pribaltiki, Riga, vol. 3, part. 2, p. 124-129.

Cukerzis, J.M., J. Sestoka, A. Terentyev. 1971. An attempt to determine the size of an isolated population of Astacus astacus L. (In Russian). Mat. XVI conf. po izuch. vnutr. vod. Pribaltiki, Petrozavodsk, part 1, p. 89-90.

Cukerzis, J.M., J. Sestokas, A. Terentyev. 1972. The effects of temperature on growth and development of Astacus astacus L. juveniles in winter under artificial conditions. (In Russian). Energ. aspecty rosta i obmena vodn. zhivotnykh, Kiev, p. 237-238.

Cukerzis, J.M., J. Sestokas, A. Terentyev. 1972. Some peculiarities of territorial behaviour of Astacus astacus L. (In Russian). Povedeniye vodn. bespozv.: Mat. I Vses. symp., Borok, p. 139-147.

Cukerzis, J.M., J. Sestokas, A Terentyev. 1972. Size and structure of the isolated population of Astacus astacus L. in Lake Berziukas (Trakai district). (In Russian, summ. in English). Tr. AN LitSSR. Ser. V, vol. 3, p. 85-93.

Cukerzis, J.M., E. Tamkeviciene, A. Terentyev. 1973. The influence of temperature and lightening on feeding of crayfish. (In Russian). Biol. issled. na vnutr. vod. Pribaltiki: Tr. XV nauch. conf., Minsk, p. 240-241.

Cukerzis, J.M., J. Sestokas, A. Terentyev. 1973. The effect of water temperature on hatching and growth dynamics of Astacus astacus L. juveniles. (In Russian). Limnologiya Severo-Zapada SSSR, Tallin, vol. 3, p. 164-166.

Cukerzis, J.M., J. Sestokas, A. Terentyev. 1974. Typology of lakes rich in crayfish in Lithuanian SSR and a method for determination of Astacus astacus L. resources. (In Russian). Promysl. ikhtiologiya: Ref. inform., ser. 1, issue 9, p. 4.

Cukerzis, J.M., J.V. Doroshenko, L. Mickeniene. 1975. Some peculiarities of Astacus astacus L. behaviour. (In Russian). Osnovy bioproduktivnosti vnutr. vod. Pribaltiki: Mat. XVIII nauch. conf., Vilnius, p. 339-342.

Cukerzis, J.M., J. Sestokas, A. Terentyev. 1976. A new species of crayfish in water bodies of the Lithuanian SSR. (In Russian). Biol. osnovy osvoeniyz, reconstruktsii i okhrany zhivotnog mira Belorussii: Tez. IV zool. conf. BSSR, Minsk, p. 34-36.

Cukerzis, J.M., J. Sestokas, A. Terentyvey. 1977. Accelerated breeding of Astacus astacus L. juveniles. (In Russian). Vses. nauch. conf. po ispolz. promyslovykh bespozv.: Tez. dokl., Odessa, p. 102-103.

Cukerzis, J.M., J. Sestokas, A Burba. 1978. The role of food and shelter in crayfish behaviour under artificial conditions. (In Russian). Experiement. issled. povedeniya vodn. bespozv.: Tez. dokl. III Vses. symp., Borok, p. 52-54.

Cukerzis, J.M., J. Sheshtokas, A.L. Terentyev. 1979. Method for accelerated artificial breeding of crayfish juveniles. Freshwater Crayfish IV, Thonon-les-Bains, p. 451-458.

Cukerzis, J.M., J. Sestokas, E. Tamkeviciene. 1979. Artificial breeding of Astacus astacus L. juveniles in Lithuania. (In Russian, summ. in English). Biologiya rechnykh rakov vod. Litvy, Vilnius, p. 41-52.

Cukerzis, J.M., J. Sestokas, A. Mazylis, A.L. Terentyev. 1974. The role of extrinsic factors in ontogenesis of Astacus astacus l. (In Russian). Rol phactorov vneshney sredy v ontogeneze: Tez. dokl., Moskva, p. 48-49.

Cukerzis, J.M., J. Sestokas, A.L. Terentyev, J.V. Doroshenko. 1974. Productivity of the isolated population of Astacus astacus L. in Lake Berziukas. (In Russian, summ. in English). Tr. AN LitSSR. Ser. V, vol. 3, p. 141-150.

Cukerzis, J.M., J. Sestokas, E. Tamkeviciene, L. Mickeniene. 1977. Cannibalism among crayfish Astacus astacus L. (In Russian, summ. in English). Tr. AN LitSSR. Ser. V, vol. 3, p. 97-103.

Cukerzis, J.M., J. Sestokas, A. Mazylis, E. Tamkeviciene, G. Mackeviciene, L. Mickeniene, A.L. Terentyev. 1976. Some ecological, eco-physiological, biochemical and microbiological peculiarities of Astacus astacus L. (In Russian). III syezd Vses. gidrobiolog. obshchestva: Tez. dokl., Riga, vol. 2, p. 163-165.

Doroshenko, J.V. 1976. Multifunctional role of the chela in Astacus astacus L. behaviour. (In Russian). Tez. conf. molodykh uchion. instituta, Vilnius, p. 34-35.

Doroshenko, J.V. 1979. Formation of motive structures of behaviour of Astacus astacus L. juveniles. Freshwater Crayfish IV, Thonon-les-Bains, p. 459-464.

Doroshenko, J.V., A.L. Terentyev. 1975. Registering units of "daily" behaviour of Astacus astacus L. females. (In Russian). II Vses. Conf. molodykh uchion. po voprosam sravnit. morph. i ecol. zhivotnykh, Moskva, p. 144-145.

Doroshenko, J.V., J. Sestokas. 1975. Behaviour of Astacus astacus L. during reproduction. (In Russian). Povedeniye vodn. bespozv.: Mat. II Vses. symp., Borok, p. 19-20.

Doroshenko, J.V., J.M. Cukerzis. 1976. Behaviour of Astacus astacus L. juveniles during the early stages of post-embryonic development. (In Russian). Grupp. povedeniye zhivotnykh: Dolk. II Vses. conf., Moskva, p. 103-105.

Doroshenko, J.V., A. Mazylis, G.S. Kan, O.L. Belianin. 1979. On the creation of the model of the habitat for the investigation of behaviour of Astacus astacus L. (In Russian, summ. in English). Biologiya rechnykh rakov vod. Litvy, Vilnius, p. 128-135.

Eglit, P. 1913. Review of the present conditions of crayfish in Suvalkai province. (In Russian). Vestnik ryboprom., No 7/8, p. 192-215.

Girdwoyn M. 1883. Projekt gospodarstwa rybnego jeziorowego w dobrach Dukszty. Warszawa, 20 p.

Golubev, A.P., A. Mazylis. 1977. Growth and moult of Astacus astacus L. juveniles (In Byelorussian). Vestnik AN BSSR, Ser. biol. nauk, No 5, p. 100-103.

Hofmann, J. 1971. Die Flusskrebse: Biologie, Haltung und wirtschaftliche Bedeutung. Hamburg; Berlin, p. 38-41.

Kossakowski, J. 1966. Raki. -Warszawa, 292 p.

Mackeviciene, G. 1972. The content of mineral substances and proteins in Astacus astacus L. juveniles. (In Russian). Energet. aspekty rosta i obmena vodn. zhivotnykh: Mat. symp., Kiev, p. 136-137.

Mackeviciene, G. 1973. ATPase activity in the muscles of two closely related Astacus species. (In Russian). Limnologiya Severo-Zapada SSSR, Tallin, part 2, p. 139-142.

Mackeviciene, G. 1973. The comparative study of proteases in two closely related Astacus species. (In Russian). Khimiya proteolit. phermentov: Mat. Vses. symp., Vilnius, p. 137.

Mackeviciene, G. 1974. Metabolism of mineral substances and proteins in Astacus astacus L. juveniles on rearing them under artificial conditions. (In Russian). Biol. promysl. ryb i bespozv. na rannikh stadiyakh razvitiya: Vses. conf., Murmansk, p. 133-134.

Mackeviciene, G. 1975. Some pecularities of protein metabolism in Astacus astacus L. females during oogenesis. (In Russian). Osnovy bioproduktivnosti vnutr. vod. Pribaltiki: Mat. XVIII nauch. conf., Vilnius, p. 326-328.

Mackeviciene, G. 1975. Studies on the ionic, protein concentration and the protease activity in the juveniles of crayfish Astacus astacus L. Freshwater Crayfish II, Baton Rouge, p. 187-194.

Mackeviciene, G. 1977. Physiological and biochemical characteristics of Astacus astacus L. and Astacus leptodactylus Esch. in the Lithuanina SSR. (In Russian). Moskva, 21 p.

Mackeviciene, G. 1979. Some biochemical differences between two species of freshwater crayfish, Astacus astacus L. and Astacus leptodactylus Esch. in Lithuania. Freshwater Crayfish IV, Thonon-les-Bains, p. 465-470.

Mackeviciene, G. 1979. Dynamics of water, mineral substances and protein metabolism during the postembryonic development of Astacus astacus L. (In Russian, summ. in English). Biologiya rechnykh rakov vod. Litvy, Vilnius, p. 53-66.

Mackeviciene, G. 1979. some peculiarities of metabolism of Astacus astacus L. (In Russian, summ. in English). Biologiya rechnykh rakov vod. Litvy, Vilnius, p. 85-120.

Maldziunaite, S. 1960. Spread of otters in Lithuania and their food. (In Lithuanian). Tr. AN LitSSR. Ser. V, vol. 3, p. 181-189.

Maniukas, J., J. Virbickas, J. Sestokas, I. Gasiunas. 1973. Die Transformierung der Hydrofauna Litauens im letzten Jahrhundert. Internationale Vereinigung fur theoretische und angewandte Limnologie: Verhandlungen, Leningrad, vol. 18, part 3, p. 1610-1615.

Mazylis, A. 1973. Chemical composition and calorific value of the edible part of Astacus astacus L. (In Russian, summ. in English). Tr. AN LitSSR. Ser. V, vol. 4, p. 147-152.

Mazylis, A. 1973. On the infection of Astacus astacus L with branchiobdellae and control measures against them. (In Russian, summ. in English). Tr. AN LitSSR. Ser. V, vol. e, p. 107-113.

Mazylis, A. 1973. Toxic influence of chlorophos on crayfish organism. (In Russian). Biol. issled. na vnutr. vod. Pribaltiki: Tr. XV nauch. conf., Minsk, p. 121-123.

Mazylis, A. 1975. Diseases and parasites of Astacus astacus L. in some lakes of Lithuania. (In Russian). Osnovy bioproduktivnosti vnutr. vod. Pribaltiki: Mat. XVIII nauch. conf., Vilnius, p. 319-321.

Mazylis, A. 1979. On Astacus astacus L infected with Thelohania contejeani Henneguy. Freshwater Crayfish IV, Thono-les-Bains, p. 471-473.

Mazylis, A. 1979. Embryonic development of Astacus astacus L. (In Russian, summ. in English). Biologiya rechnykh rakov vod. Litvy, Vilnius, p. 28-40.

Mazylis, A., J. Sestokas. 1968. Crayfish plague in a natural water body. (In Russian). Limnologiya: Mat. XIV conf., Riga, vol. 3, part 2, p. 63-66.

Mazylis, A., J. Sestokas. 1971. Toxic influence of chemicals on crayfish. (In Russian). Mat. XVI conf. po izuch. vnutr. vod. Pribaltiki, Petrozavodsk, part 1, p. 323-325.

Mazylis, A., J.M. Cukerzis. 1975. The effect of electric current on growth and reproduction of crayfish. (In Russian). Energet. aspektyu rosta i razmnozh. vodn. bespozv.: Mat. symp., Minsk, p. 151-162.

Mazylis, A., G. Mackeviciene. 1975. Changes in gas exchange and activity of digestive enzymes in Astacus astacus L. depending on feeding. (In Russian). Osnovy bioproduktivnosti vnutr. vod. Pribaltiki: Mat. XVIII nauch. conf., Vilnius, p. 322-325.

Mazylis, A., A. Sruoga. 1977. Immunochemical studies of antigenic characteristics in Astacus astacus L. and its parasite. (In Russian). Genetika ir selekcija liaudies ukiui: II suv. tezes, Vilnius; Kaunas, p. 98-99.

Mazylis, A., A. Grigelis. 1979. On the diseases of Astacus astacus L. in some Lithuanian lakes. (In Russian, summ. in English). Biologiya rechnykh rakov vod. Litvy, Vilnius, p. 121-127.

Micha, O. 1901. Review and statistics of crayfish export from Russia. (In Russian). Vestnik ryboprom., No 11, p. 598-613.

Mickeniene, L. 1979. Influence of food and nourishment on the digestive tract microflora of Astacus astacus L. (In Russian, summ. in English). Biologiya rechnykh rakov vod. Litvy, Vilnius, p. 78-84.

Mickeniene, L., A. Lekeviciene. 1976. Microflora of the digestive tract of Astacus astacus L. (1. General amount of bacteria and amount of microorganisms of some physiological groups). (In Russian, summ. in English). Tr. AN LitSSR. Ser. V, vol. 1, p. 77-80.

Miseikyte, A. 1963. An attempt to determine crayfish resources in some lakes of Lithuania. (In Russian). Soveshch. po vosproizvodstvu zapasov rechnykh rakov, Vilnius, p. 42-45.

Petrauskiene, L. 1977. Investigations into electrotaxis of aquatic animals. (In Russian). Moskva, 18 p.

Petruszewski, J. 1951. To develop the production of crayfish. (In Russian). Rybnoye khoz., No 5, p. 64.

Silver, D., J.M. Cukerzis. 1964. The number of chromosomes in Astacus leptodactylus Esch. (In Russian). Tsitologiya, vol. 6, No 5, p. 631-633.

Smolian, K. 1920. Merkbuch der Binnenfischerei. Berlin, vol. 1, p. 226-241, 298.

Staniewicz, C. 1902. Lakes and rivers of North-Western or Lithuanian land. Vilna, p. 20, 33-35.

Sestokas, J. 1965. Some data of crayfish diseases in lakes of the Lithuanian SSR. (In Russian). XII nauch. conf. po izuch. vnutr. vod. Pribaltiki, Vilnius, p. 63-64.

Sestokas, J. 1967. Influence of some chemicals used in agriculture on crayfish. (In Lithuanian). Liet. TSR jaunuju moksl.-biologu ir biochemiku moksl. konf., Vilnius, p. 228-233.

Sestokas, J. 1968. Diseases of Astacus astacus L. and Astacus leptodactylus Esch. in water bodies of the Lithuanian SSR. (In Russian). Gidrobiol. i ikhtiolog. issled. vnutr. vod. Pribaltiki: Tr. XII nauch. conf., Vilnius, p. 219-221.

Sestokas, J. 1970. Distribution, present conditions of resources and reproduction of crayfish in the Lithuanian SSR. (In Russian). Vilnius, 25 p.

Sestokas, J. 1973. Distribution of crayfish in the Lithuanian SSR. (In Russian, summ. in English). Tr. AN LitSSR. Ser. V, vol. 3, p. 87-105.

Sestokas, J., J.M. Cukerzis. 1965. Quantitative registration of crayfish in water bodies of the Lithuanian SSR. (In Russian). Tr. AN LitSSR. Ser. V, vol. 1, p. 141-154.

Sestokas, J., J.M. Cukerzis. 1966. Present conditions of crayfish resources in artificially populated Lithuanian lakes. (In Russian). Tr. AN LitSSR. Ser. V, vol. 3, p. 145-153.

Sestokas, J., J.M. Cukerzis. 1966. Introduction of crayfish-sires into water bodies of the Lithuanian SSR. (In Russian). Tez. dokl. XIII nauch. conf. po izuch. vnutr. vod. Pribaltiki, Tartu, p. 200-201.

Sestokas, J., J.M. Cukerzis. 1968. The effect of some mineral fertilizers and pesticides on crayfish. (In Russian). Limnologiya: Mat. XIV conf. po izuch. vnutr. vod. Pribaltiki, Riga, vol. 3. part 3, p. 130-132.

Sestokas, J., J.M. Cukerzis. 1972. Influence of chemical preparations on Astacus astacus L. and Astacus leptodactylus Esch. (In Russian, summ. in English). Tr. AN LitSSR. Ser. V, vol. 4, p. 119-123.

Sestokas, J., J.M. Cukerzis. 1973. The effect of phenol on crayfish. (In Russian). Biol. issled. na vnutr. vod. Pribaltiki: Tr. XV nauch. conf., Minsk, p. 120-121.

Sestokas, J., J.M. Cukerzis. 1979. History of the investigations of crayfish in the Lithuanian inner waters. (In Russian, summ. in English). Biologiya rechnykh rakov vod. Litvy, Vilnius, p. 8-19.

Siurnaite-Grauziniene, P. 1940. Crayfish of Lithuania. (In Lithuanian). Kosmos, No 7-12, p. 337-352.

Sivickis, P. 1940. Crayfish and its stock. (In Lithuanian). Gyvoji gamta ir mes, Kaunas, p. 282-287.

Tamkeviciene, E. 1969. Growth indexes of Astacus astacus L. (0+) in basins. (In Lithuanian). Trumpi pranesimai: XV moksl. konf., Kaunas, p. 208-209.

Tamkeviciene, E. 1971. Growth and development of Astacus astacus L. juveniles. (In Russian). Vilnius, 17 p.

Tamkeviciene, E. 1972. The influence of temperature on food intake of Astacus astacus L. juveniles. (In Russian). Energ. aspekty rosta i obmena vodn. zhivotnykh: Mat. symp., Kiev, p. 212-213.

Tamkeviciene, E. 1972. Growth and development of Astacus astacus juveniles (In Russian). Voprosy razv. ryb i rakoobraznykh v vod. Litvy, Vilnius, p. 229-234.

Tamkeviciene, E. 1973. The influence of lightening on food intake of Astacus astacus L. juveniles. (In Russian). Limnologiya Severo-Zapada SSSR, Tallin, vol. 3, p. 123-124.

Tamkeviciene, E. 1979. Some quantative regularities of Astacus astacus L. feeding. (In Russian, summ. in English). Biologiya rechnykh rakov vod. Litvy, Vilnius, p. 67-77.

Tamkeviciene, E., J. Cukerzis. 1967. Rearing and development of Astacus astacus L. and Astacus leptodactylus Esch. juveniles. (In Lithuanian). Liet. TSR jaunuju moksl. biologu ir biochemiku moksl. konf., Vilnius, p. 233-238.

Tamkeviciene, E., J. Cukerzis. 1968. On rearing of Astacus astacus L. juveniles under artificial conditions. (In Russian). Mat. XIV conf. po izuch. vnutr. vod. Pribaltiki, Riga, vol. 3, part 2, p. 107-111.

Tamkeviciene, E., G. Mackeviciene. 1973. Output of Astacus astacus L. juveniles (0+) and some biochemical indexes in different stages of development of juveniles. (In Lithuanian). Respublikine jaun. moksl. konf., Vilnius, p. 120-122.

Tamkeviciene, E., J. Sestokas. 1975. The influence of density on growth of Astacus astacus L. juveniles. (In Russian). Osnovy bioproduktivnosti vnutr. vod. Pribaltiki: Mat. XVIII nauch. conf., Vilnius, p. 337-338.

Tamkeviciene, E., G. Mackeviciene, J. Cukerzis. 1971. Some data of feeding and chemical composition of Astacus astacus L. juveniles. (In Russian). Mat. XVI conf. po izuch. vnutr. vod. Pribaltiki, Petrozavodsk, p. 281-282.

Terentyev, A.L. 1979. Acclimatization of crayfish Pacifastacus leniusculus Dana in the Lithuanian SSR. (In Russian, summ. in English). Biologiya rechnykh rakov vod. Litvy, Vilnius, p. 136-143.

Tyshkevich, K. 1863. Collection of historical and statistical materials from Vilnius district. (In Russian). Vilna, p. 83-130.

Virbickas, J., A. Mazylis, D. Sineviciene. 1975. The effect of electric current on reproduction and development of crustacea. (In Russian). Osnovy bioproduktivnosti vnutr. vod. Pribaltiki: Mat. XVIII nauch. conf., Vilnius, p. 360-362.

Virbickas, J., A. Mazylis, D. Sinevicience. 1976. The results of the effect of electric fields on gametes and ontogenesis of aquatic animals. (In Russian). Ecologicheskaya physiologiya ryb: Tez. dokl. III Vses. conf., Kiev, part. 2, p. 179-180.

Zimnicki, W. 1935. On crayfish and its managment in the district of Vilnius. (In Polish). Vilnius, 76 p.

BIOCHEMICAL CHARACTERISTICS OF THE DNIESTER LONG-CLAWED CRAYFISH OF ASTACIDAE FAMILY

A.V. Suprunovich, R.P. Kadniuk, T.A. Petkevich, I.A. Stepaniuk
V.I. Lisovskaya, L.V. Antsupova
The Whole-Union Scientific Research Institute
Fisheries and Oceanography,
Moscow-140, V. Krasnoselskaya 17a
The Institute of South Marine Biology
UkrSSR AS, Odessa-11, Pushkinskaya 37

ABSTRACT

The content and distribution of the total organic substances, lipids, sterols, carotenoides, amino acids and microelements, was studied in organs and tissues of the Dniester long-clawed crayfish of different age groups, and also in its eggs and larvae.

Age and sex differences are shown in the content and accumulation of many biochemical components in the hepatopancreas and in some cases, the carapace.

The information obtained is of interest for elaboration of biotechniques for crayfish and the cultivation of other hybrobionts in our country.

INTRODUCTION

Among decopods (order Decapoda) the most valuable food objects in our country are long-clawed crayfishes of Astacidae family. In the USSR numerous commercial populations of long-clawed crayfish (Pontastacus leptodactylus) are distributed in the Ukrainian reservoirs including the Dniester estuary. The present annual catch of the crayfish in the estuary is 1200-1400 centners which is the highest yield of crayfish in the Ukrainian reservoirs.

For successful development of crayfish catches in the Dniester estuary and other crayfish reservoirs, and for their commercial cultivation, the biology of long-clawed crayfish must be studied. In particular, its biochemistry deserves careful consideration.

The study of biochemical composition of long-clawed crayfish enables us to make a correct estimation of its feeding within a reservoir. This makes it possible to assess in detail about distribution and stock. Furthermore, without knowledge of the main biochemical indices of crayfishes it is impossible to correctly determine their food components for commercial cultivation.

The number of works concerning the biochemistry of long-clawed crayfishes within commercial populations in the south of our country is quite limited (Balashova, 1974), though some biochemical characteristics of crayfish Astacus astacus from the northwest of the USSR have been studied (Tsukerzis, 1970; Matskiavichene et al., 1979).

The aim of the present work is the study of biochemical characteristics of the Dniester long-clawed crayfish which are necessary for
working out biotechnique of commercial crayfish cultivation in our
country.

METHODS AND MATERIALS

The work was carried out on the basis of biochemical study of the
white Dniester long-clawed crayfish* (<u>Pontastacus eichwaldi
bessarabicus</u>, Brodsky, 1967) caught in the Dniester estuary (northwest
of the Black Sea) in 1977-1979 (Figure 1).

Live crayfish were delivered to the biochemical laboratory of the
Odessa Branch of the Institute of South Marine Biology at the UkrSSR AC
(4-5 hours after they were caught) and were sorted there according to
sex and size.

The measuring of crayfish was made to within 1 mm with regard to
body length (between rostrum and telson). Weighting was done on a
technical balance to within 1 g. For estimation of biochemical indices
length was taken rather than mass, which changes depending on the
length of transportation period and storage in the laboratory.

Simulatenously, biochemical studies of the ovaries of reproducing
females in autumn, eggs taken from the females' pleopoda in spring
(May) and summer (June) and of larvae of the 1st development stage
taken from females' pleopoda in summer (June) were done.

In order to determine lipids, sterols, carotenoides, free and
combined amino acids, microelements, ashes (total mineral substances)
and total organic substances of crayfish individuals of different
sizes, 2,401 biochemical analyses were made of the ovaries, eggs,
larvae of the 1st development stage and 175 samples were studied.

The content of lipids was found according to Folch method (Folch
et al., 1951) in Bligh and Dyer (1959) modification.

Non-saponaceous residuum was determined according to the method of
F.M. Rzhavskaya and M.A. Alekseeva (1966), while its subdivision into
fractions was carried out with the help of thin-layer chromatography
(Lisboa, 1969).

Identification of sterols was made with the help of Liberman-
Burhard chemical reaction (Moore, Baumann, 1952) and ultra-violet
spectrometry on the registering spectrophotometre "Specord"
(Palamarchuk et al., 1974).

Carotenoid pigments were studied with the help of the colourimet-
ric (Godnev, Terentiev, 1956; Sapozhnikov et al., 1956; Davies, 1965)
and spectrophotometric (Sapozhnikov, 1964; Drujan et al., 1968; Davies,
1965; Jensen, A, Jensen S., 1959; Jensen A., 1960) methods.

---

* At the present time the systematic position of the white Dniester
long-clawed crayfish is <u>Astacus (Pontastacus) leptodactilus eichwaldi
natio bessarabicus Brodsky</u>. This crayfish is included into the Caspian
crayfish subspecies range (Brodsky, 1980).

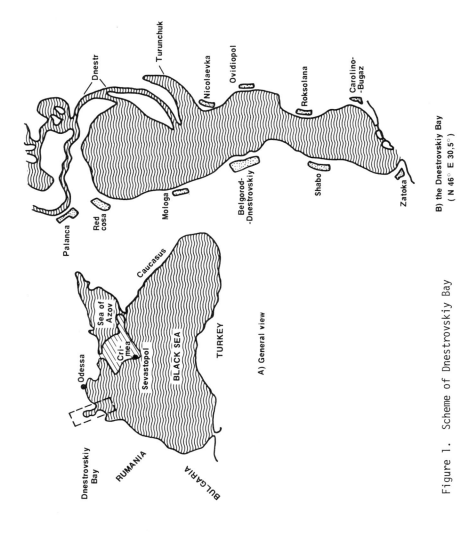

Figure 1.   Scheme of Dnestrovskiy Bay

A) General view

B) the Dnestrovskiy Bay
(N 46° E 30.5°)

Labels within figure:

Dnestr
Turunchuk
Nicolaevka
Ovidiopol
Roksolana
Carolino-
-Bugaz
Palanca
Red
cosa
Mologa
Belgorod-
-Dnestrovskiy
Shabo
Zatoka

Odessa
Sea of
Azov
Cri-
mea
Sevastopol
Caucasus
BLACK SEA
TURKEY
Dnestrovskiy
Bay
RUMANIA
BULGARIA

Amino acid composition was studied by distributions on chromatography paper with certain modifications (Filippovich, 1958). Chromatogrammes were deciphered with the help of standard amino acid mixtures placed next to the studied mixture. Every amino acid was determined separately while leucine and isoleucine were combined. Non-replaceable amino acids were found both as parts of free amino acids and those combined in proteins. Only non-replaceable amino acids were summarized and their total was calculated as percentage of the total amount of combined amino acids in order to study proportion of replaceable and non-replaceable amino acids in the proteins of crayfish.

Microelements were measured by emission and spectral analysis on the quartz spectrograph ISP-28 with quantitative results deciphered on microphotometre MF-2. Standard samples were prepared on artificial basis. Graduating graphs were built in ΔS, lg C co-ordinates (Mitchell, 1961; Gribovskaya, 1968).

The content of organic substance was determined by ashing.

Results of the study

Lipids. Energy balance of a living organism is partly determined by the content of lipids in different body parts and organs.

The maximum content of lipids for the Dniester crayfish females and males is registered in their hepatopancreas, where in some cases it reached 92.43% of dry substance. This is explained by the role of hepatopancreas as accumulator (depot) of many biochemical components. While studying the fraction composition of lipids the following classes were found: polar-phospholipids and neutral ones which include cholesterine, free fatty amino acids and triglycerides. Up to 15% of phospholipids were found in the total amount of lipids and about the same content of triglycerides. The most important share in the total amount of lipids belongs to cholesterine which makes up to 80%.

In lower amounts, (up to 38.62% of dry mass) lipids were found in ovaries, which are connected with reproductive function of this organ and the use of "energy material."

With growth of crayfish, the content of lipids in muscles of females and males increases insignificantly. For example males and females 10.1-11.0 cm long have the content of lipids 6.39 and 5.86 of dry mass, crayfish 15.1-17.1 cm long - 6.77 and 7.00% of dry mass respectively. Such redistribution of accumulated lipids in different organs and tissues of the Dniester crayfish is connected with its growth and development, though during moulting no great differences were found in the content of lipids between non-moulting crayfish (with hard chitin integument) and moulting ones (with soft chitin integument). It was 6.93 and 6.81% of dry mass, respectively. Similarly, no differences in fraction composition of non-moulting and moulting crayfishes were found. Phospholipids made 25% of the total amount (Table 1).

Sterols. Biologically active substances are very important for the vital life processes in an organism. They influence protein, carbohydrate, lipid and mineral metabolism.

493

Table 1. Content of lipids in body and tissues of the Dniester crayfish.

Sex	Body length cm	Body parts and organs	Content of lipids - % of dry mass	
			June	August
Males	9.1-10.0	muscles	5.89	7.14
		hepatopancreas	---	87.45
Females	9.1-10.0	muscles	6.04	7.40
		hepatopancreas	---	88.53
Males	10.1-11.0	muscles	5.86	5.50
		hepatopancreas	92.29	88.70
Females	10.1-11.0	muscles	6.39	8.63
		hepatopancreas	---	85.25
		ovary	---	38.62
Males (moulting)	10.1-11.0	muscles	6.93	---
		hepatopancreas	92.43	---
Females (moulting)	10.1-11.0	muscles	6.81	---
Males	15.1-17.0	muscles	6.77	---
		hepatopancreas	89.44	---
Females	15.1-17.0	muscles	7.00	5.57
		hepatopancreas	89.90	89.99
		ovary	---	28.57

Note: dash (-) means absence of analyses.

The total content of provitamins D in eggs and ovaries of females is nearly twice as high as in larvae of the first development stage (0.88% compared to 0.45% in non-saponaceous fraction, respectively) also, there is three times more cholesterine in ovaries than in eggs which is an important adaptive factor for reproduction. In eggs of one-sized females and in their larvae of the first development stage the amount of cholesterine is nearly on the same level.

The amount of provitamins D (total) in muscles of adult crayfish (males and females) is much lower than in muscles of immature ones (with body length less than 8.0 cm) since it has been spent for body growth and development. It was noticed that the content of provitamins D (total) and cholesterine is lower in females than in males. It particularly concerns cholesterine, the amount of which in females is nearly one order lower (1.58% of non-saponaceous fraction), than in males (10.63% of non-saponaceous fraction).

While determining sterols' content in non-saponaceous fraction (NF) extracted from the muscles of crayfish we have not found 7-dehydrocholecterol, provitamin of the vitamin D_3 and precursor of cholesterine (Table 2). Its role in the organism of crayfish is likely to be played by some other sterol, as yet not identified by us. Since cholesterine formation can develop in two different ways in animal organism: through 7-dehydrocholesterol or through desmosterole, it can be assumed that the precursor of cholesterine in the muscles of crayfish is desmosterole in our case.

Maximum content of provitamins D (total) and cholesterine in adult crayfish was in hepatopancreas with females' data exceeding those of males.

Sterols are known to play an important role in the formation of chitin integument of crustacea and regular moults provide for the growth of crayfishes. The analyses of moulting hydrobionts with soft carapace (males and females) showed that the amount of provitamins D and cholesterine is lower than in non-moulting crayfish with hard carapace, which is an indication of the role of provitamins D and cholesterine in the formation of the chitin intergument of crayfish (Table 3).

Carotenoides. Carotenoid pigments have different functions in animal organism than in plants: they take part in photosynthesis, reproduction, phototropic reaction of invertebrates and fishes. At the present time it is known that carotenoides together with hemoproteins and other respiratory enzymes form a special organelle inside animal cells which, according to N.V. Karnauhov (1973, 1979), is capable of depositing oxygen and producing the necessary energy under unfavourable conditions. Therefore, organisms with high carotenoid pigment content are the most resistant to pollution, which is of certain practical interest.

The content of carotenoides in different organs and tissues of long-clawed crayfish is not the same. Maximum content of carotenoid pigments was found in females' ovaries - 54.6-58.2 mkg/g.

During the growth process the content of carotenoid pigments changed insignificantly. The study of seasonal changes of carotenoides

Table 2. Sterols content in eggs, larvae of the first development stage and muscles of the Dniester crayfish.

Time of study	Object, organ	Mass of non-saponaceous fraction	Contained in non-saponaceous fraction in %			% of damp mass		
			total of pro-vitamin D	metho-stenol	chole-sterin	total of pro-vitamin D	metho-stenol	chole-sterin
8.06.1978	Eggs	0.0249	0.88	1.32	7.35	0.0049	0.0066	0.0366
	Larvae in the 1st stage of development	0.0272	0.45	0.87	6.17	0.0025	0.0048	0.0336
4.07.1978	Muscles of males with the length of (in cm):							
	9.1-10.0	0.0214	0.53	0.66	10.63	0.0023	0.0028	0.0455
	10.1-11.0	0.0235	0.77	1.42	12.45	0.0036	0.0067	0.0585
	11.1-12.0	0.0164	0.40	0.66	5.64	0.0013	0.0022	0.0185
4.07.1978	Muscles of females with the length of (in cm):							
	9.1-10.0	0.0042	0.35	0.48	1.58	0.0004	0.0005	0.0017
	10.1-11.0	0.0215	0.23	0.50	1.97	0.0010	0.0022	0.0085
	12.1-13.0	0.0155	0.29	0.52	5.08	0.0009	0.0016	0.0157
	14.1-15.0	0.0128	0.20	0.88	12.39	0.0005	0.0022	0.0317

Table 3. Content of sterols in hepatopancreas of the Dniester crayfish (4.07.1978).

Sex	Body length cm	SN mass g	Percent of NS	Contained in non-saponaceous fraction in %					% of damp mass				
				D	7-DHC	L	M	C	D	7-DHC	L	M	C
Crayfish with hard carapace (non-moulting)													
Males	11.1-12.0	0.0699	1.401	3.02	0.67	0.94	4.11	24.00	0.0423	0.0094	0.0134	0.0574	0.3355
	15.1-17.0	0.0834	1.671	2.74	1.09	0.13	4.04	24.90	0.0454	0.0182	0.0022	0.0674	0.4154
Females	10.1-12.0	0.0607	1.211	3.43	0.53	1.70	4.94	26.62	0.0416	0.0064	0.0206	0.0500	0.3232
	15.1-17.0	0.0801	1.881	3.12	0.86	0.39	4.49	26.07	0.0589	0.0162	0.0073	0.0846	0.4913
Crayfish with soft carapace (moulting)													
Males	10.1-12.0	0.0518	1.280	1.84	0.35	0.96	2.19	22.45	0.0217	0.0041	0.0114	0.0258	0.2643
Females	10.1-12.0	0.489	0.970	1.08	0.39	0.51	1.60	20.13	0.0105	0.0038	0.0050	0.0155	0.1957

Note: Symbols stand for: NS = non-saponaceous fraction
D = total of provitamins D
7-DHC = 7-dehydrochole-sterol
L = Lathosterol
M = Methosterol
C = Cholesterine

497

in different organs of one-sized males and females showed that in hepatopancreas, eyes and carapace of males and females the amount of pigments varied from 39.3 to 73.0 mkg/g of damp mass in summer (Table 4).

Muscles of males and females contained several times less carotenoides than other organs. The amount of carotenoides in them did not exceed 19.0 mkg/g. By the beginning of autumn a redistribution of carotenoides in organs was observed in crayfish females. The content of pigments in the muscles of females reduced drastically (2.7-4.5 mkg/g) while in ovaries it became quite substantial (125-152 mkg/g). In hepatopancreas, eyes, and carapace there was also observed the tendency toward reduction of carotenoid pigments. The content of carotenoides in males reduced at that time.

By the end of autumn the carotenoid pigments content in muscles was observed to increase. The content of carotenoides varied from 10.4 to 20.0 mkg/g of damp mass. In that period an increase of the amount of pigment in females' eyes and males' carapaces was observed. The content of carotenoides in females' ovaries remained high throughout autumn (123.0-168.6 mkg/g). The high content of carotenoides in females' ovaries indicates that they are important for reproduction.

Amino acids. Amino acids play an important role in protein metabolism of animals, behaviour and osmoregulation. They also determine the main life functions of an organism in ontogenesis.

In eggs (0.333% of damp mass) and larvae of the first development stage (0.468%) less amino acids than in males' muscles of individuals 15.1-17.0 cm long (0.671% of damp mass) and roed females 14.1-17.0 cm long (0.470% of damp mass) were found. This occurred not only in samples taken simultaneously, but in other crayfish samples too. Larvae are richer in free amino acids than are eggs, particularly in lysine, arginine, threonine, valine, leucines.

The content of combined amino acids in the larvae of the first development stage (15.681% of damp mass) muscles of crayfish of different sizes (9.1-17.0 cm), and both males and roed females (13.903-15.922% of damp mass). However, the percentage of non-replaceable amino acids in the total amount of combined acids is higher in eggs - 52% compared to 46% in larvae of the 1st development stage. Eggs contain much combined proline, glutamine acid and have non-replaceable amino acids that contain much lysine, valine and leucine. The content of the same amino acids is high in larvae bodies, while the amount of combined proline exceeds the content of glutamine acid.

The content of combined amino acids in males 9.1-10.0 cm long (15.922% of damp mass) and 15.1-17.0 cm (14.621% of damp mass) is higher than in females of the same sizes (14.676 and 13.303% of damp mass, respectively). The highest content of proline, lysine, leucines, glutamine acid and arginine among combined acids was registered.

The content of non-replaceable amino acids in muscles reached 52% in females 14.1-17.0 cm long (Table 5).

498

Table 4. Content of carotenoides in the Dniester crayfish.

Sex	Body length cm	Carotenoides, mkg/g damp mass				
		Muscles	Hepato-pancreas	eyes	carapace	ovaries
		Summer				
Females	9.1-10.0	11.8	57.7	62.4	53.1	56.2
	10.1-12.0	19.0	45.1	67.8	48.4	58.2
	15.1-17.0	15.9	39.3	65.9	49.2	54.6
Males	9.1-10.0	16.6	54.5	57.8	69.6	---
	10.1-12.0	18.3	72.2	39.8	66.1	---
	15.1-17.0	17.9	68.4	42.6	63.4	---
		Autumn (early)				
Females	7.1- 8.0	2.7	32.2	34.2	56.0	152.2
	9.1-10.0	2.8	40.4	41.3	57.3	135.8
	10.1-12.0	2.9	44.0	49.8	58.7	125.0
	15.1-17.0	4.5	43.6	25.6	68.0	136.0
Males	7.1- 8.0	1.9	49.9	33.8	38.2	---
	9.1-10.0	2.9	46.7	31.9	41.8	---
	10.1-12.0	2.3	34.1	21.3	43.6	---
	15.1-17.0	3.1	48.0	36.6	40.8	---
		Autumn (late)				
Females	9.1-10.0	18.4	24.1	61.3	32.6	128.8
	10.1-12.0	16.9	23.7	61.7	31.9	123.0
	15.1-17.0	12.0	23.1	63.0	29.3	168.6
Males	9.1-10.0	19.9	44.0	43.3	76.2	---
	10.1-12.0	10.4	74.5	29.5	72.6	---
	15.1-17.0	11.9	58.7	37.3	66.6	---

Table 5. Amino-acid composition of the Dniester crayfish in June-July 1978 (% of damp mass).

	eggs	larvae	muscles		eggs	larvae	muscles		
			free				combined		
Length, cm		0.8-1.2	14.1-17.0	15.1-17.0		0.8-1.2	9.1-10.0	14.1-17.0	15.1-17.0
Sex			♂	♀			♂+♀	♂	♀
Amino acids	2	3	4	5	6	7	8	9	10
1									
Lysine	0.017	0.031	0.046	0.053	1.030	1.430	1.208	1.420	1.380
Histidine	0.007	0.009	0.009	0.015	0.430	0.322	0.258	0.353	0.258
Arginine	0.020	0.044	0.060	0.115	0.720	0.970	1.190	1.080	1.280
Asparaginic acid	0.006	0.005	0.011	0.011	0.295	0.400	0.365	0.315	0.326
Serine	0.007	0.005	0.012	0.015	0.300	0.305	0.360	0.320	0.300
Glycine	0.023	0.038	0.055	0.042	0.580	0.740	0.670	0.680	0.860
Glutamine acid	0.039	0.041	0.018	0.031	1.200	1.370	1.545	1.350	1.070
Threonine	0.007	0.012	0.019	0.032	0.642	0.850	0.590	0.600	0.700
Alanine	0.064	0.092	0.067	0.090	0.890	1.040	0.690	1.120	1.060
Proline	0.043	0.049	0.096	0.090	1.370	2.800	2.540	2.800	2.800
Tyrosine	0.019	0.018	0.010	0.026	0.185	0.380	0.320	0.274	0.395
Tryptophan	0.001	0.012	traces	0.001	0.071	0.188	0.190	0.103	0.182
Methionine	traces	traces	traces	traces	0.212	0.346	0.310	0.560	0.340
Valine	0.025	0.039	0.022	0.048	0.890	1.020	1.000	0.910	0.925
Phenylalanine	0.014	0.015	traces	0.017	0.450	0.635	0.910	0.735	0.700
Leucines	0.040	0.057	0.016	0.044	1.140	1.460	1.900	1.320	1.320
Total	0.333	0.468	0.470	0.671	9.835	15.681	13.303	14.676	15.922
Percentage of non-replaceable amino acids	52.0	49.0	50.0	49.0	52.0	46.0	52.0	50.0	49.0

500

The muscles of long-clawed crayfish are very rich in free amino acids, particularly in proline and cystine. The content of free amino acids in muscles of the majority of different-sized crayfish was higher than that of females. Yet females 12.1-13.0 cm long contained more free amino acids (0.674% of damp mass) than males of the same size (0.475% of damp mass). Females 12.1-13.0 cm long are the most numerous reproducing group in the Dniester population of long-clawed crayfish.

The comparison of the amino-acid content of free amino acids in the muscles of different-sized crayfish (9.1-17.0 cm) showed that the amount of free amino acids in males gets higher (from 0.508% to 0.904% of damp mass) with growth except for males 12.1-13.0 cm long. They were found to have the lowest content of free amino acids in muscles. The change of content of free amino acids in females is uneven, i.e. their growth increases or diminishes drastically. The highest content of free amino acids was registered in females 15.1-17.0 cm long (0.790% of damp mass). The high content of free amino acids in muscles of the most numerous reproducing females in the Dniester population (12.1-13.0 and 15.1-17.0 cm) is likely to be connected with favourable energy metabolism of the animal which results in active growth, development and reproduction of the Dniester crayfish females (Table 6.). Among mature crayfish (body length over 8.0 cm), the minimum content of free amino acids was found in roed females 14.1-17.0 cm long and in their muscles (0.470% of damp mass) studied after eggs withdrawal. This is attributed to the loss of energy and plastic material by a roed female (including amino acids) (Table 5).

The content of free amino acids in muscles of moulting males with the length of 10.1-12.0 cm (0.661% of damp mass) and females of the same size (0.653% of damp mass) was the same. There were no considerable differences compared to hard crayfish either, though according to published date the content of free amino acids and protein in crustacea decreases after moult (Matskiavichene, 1979).

Dead crayfish 10.1-11.0 cm long (males and females) contained comparatively high amounts of free amino acids in their muscles (0.844% of damp mass), which may be connected with posthumous accumulation of free amino acids in tissues as a result of ended biosynthesis of proteins and the start of proteolysis. This corresponds to the published data concerning posthumous changes in tissues of crustacea (Kizavetter, 1973). There were found no sex differences in the content of free amino acids of dead males 10.1-12.0 cm long (0.661% of damp mass) and females of the same size (0.653%) (Table 6).

Hepatopancreas of different-sized crayfish (10.1-17.0 cm) contained 2-3 times more amino acids (1.231-2.021% of damp mass) than muscles of crayfish of the same size (0.475-0.904% of damp mass) both in males and females. Males 15.1-17.0 cm long (with hard chitin integument) and moulting males 10.1-12.0 cm long contained more free amino acids in hepatopancreas (2.021 and 1.609% of damp mass) than females of the same size 1.603 and 1.231% of damp mass, respectively) and females with hard chitin integument (non-moulting) 10.1-12.0 cm long contained somewhat more free amino acids (1.869% of damp mass) than moulting females of the same size (1.231% of damp mass).

In hepatopancreas of different-sized crayfish (10.1-17.0 cm) the same amino acids generally prevailed that did in muscles. The amount

Table 6. Content of free amino-acid in muscles of the Dniester crayfish (% of damp mass).

Sex	♂	♀	♂	♀	♂	♀	♂	♀	♂	♀	♂ + ♀
Length, cm	9.1-10.0		12.1-13.0		14.1-15.0		15.1-17.0		10.1-12.0		10.1-11.00
			non-moulting						moulting		dead
	2	3	4	5	6	7	8	9	10	11	12
Cystines	0.0581	0.028	0.047	0.043	0.041	0.040	0.060	0.073	0.064	0.064	0.076
Lysine	0.045	0.058	0.033	0.062	0.063	0.046	0.063	0.053	0.045	0.045	0.049
Histidine	0.016	0.014	0.013	0.019	0.013	0.026	0.016	0.021	0.019	0.024	0.019
Arginin	0.088	0.074	0.066	0.092	0.087	0.065	0.076	0.087	0.062	0.055	0.100
Aspoaraginic acid	0.010	0.007	0.005	0.009	0.010	0.004	0.002	0.024	0.024	0.021	0.027
Serine	0.008	0.008	0.005	0.011	0.011	0.006	0.054	0.033	0.027	0.019	0.037
Leucine	0.022	0.027	0.018	0.037	0.027	0.027	0.025	0.019	0.040	0.040	0.029
Glutamine acid	0.023	0.018	0.017	0.037	0.020	0.032	0.042	0.042	0.030	0.042	0.029
Threonine	0.022	0.021	0.034	0.036	0.036	0.034	0.055	0.049	0.031	0.031	0.047
Alanine	0.087	0.094	0.089	0.103	0.092	0.094	0.104	0.106	0.106	0.104	0.110
Proline	0.043	0.067	0.064	0.114	0.091	0.096	0.115	0.115	0.047	0.060	0.096
Tyrosine	0.013	0.009	0.010	0.009	0.012	0.009	0.023	0.027	0.023	0.022	0.026
Tryptophan	0.011	0.007	0.006	0.007	0.007	0.006	0.035	0.033	0.021	0.021	0.033
Methionine	0.006	0.009	0.010	0.009	0.012	0.009	0.023	0.027	0.023	0.022	0.026
Valine	0.029	0.031	0.026	0.037	0.044	0.030	0.062	0.025	0.034	0.031	0.051
Phenyla-lanine	0.001	0.002	0.001	0.011	0.001	0.001	0.023	0.017	0.019	0.017	0.020
Leucines	0.026	0.036	0.035	0.041	0.054	0.035	0.095	0.049	0.048	0.036	0.061
Total	0.508	0.510	0.475	0.674	0.618	0.564	0.904	0.790	0.661	0.653	0.844

of free amino acids in hepatopancreas of dead (1.819% of damp mass) and live (non-moulting) crayfish 10.1-12.0 cm long (1.760-1.869%) is similar. There were found no sex differences in amino-acid composition of hepatopancreas of dead crayfishes (Table 7).

As a whole, long-clawed crayfish of the Dniester estuary are characterized by high content of free and combined amino acids in muscles. In hepatopancreas the content of free amino acids is 2-3 times higher than in muscles. The amount of free amino acids in larvae and eggs is lower than in muscles. The content of combined amino acids in larvae is not lower than in muscles of adult individuals and in eggs there are 1.5 times less combined amino acids.

Long-clawed crayfish of the Dniester estuary (northwest part of the Black sea) are valuable food product owing to their non-replaceable amino acids.

Microelements. Microelements are found in organs and tissues of crayfish in different types and quanties proportional to changes in its growth and development and depending on ecologic inhabitance conditions, seasons and other biotic and abiotic factors.

The role of microelements in the life of crayfish and other animals is enormous and to a great extent determines the development of life process. Prevalence of certain microelements and absence of others in an animal organism may give an indication of reproduction, development and growth, lead to various diseases, ontogenesis disturbances and even to death.

Much manganese, aluminum, strontium, titanium, vanadium is concentrated in crayfish larvae compared to the content of the same microelements in mature animals. This shows the need for special food composition for larvae.

In July 1978 the content of microelements in organs and tissues of the Dniester crayfish differed depending on sex and age. Males 9.1-11.0 cm long contained more copper, iron, aluminum and nickel in their muscles than females of the same size. The content of other elements was close. In the livers of females 15.1-16.9 cm long, the concentration of manganese, iron and nickel turned out to be higher compared to smaller females (11.1-12.0 cm) and the amount of aluminum, titanium and lithium was lower. Thus, the content of a number of microelements both in muscles and livers of males 9.1-12.0 cm long was higher than in females of the same size. In the livers of "soft" crayfish the content of certain elements (aluminum, lithium, titanium) was lower and of some others (zinc, silve, cobalt) higher than in livers of "hard" crayfish. Copper, manganese, iron, aluminum, lithium, nickel, cobalt, silver concentrate more in livers than in other tissues. In crayfish carapace strontium, barium, zinc and vanadium concentrate while lead, lithium, silver and cobalt were not found. The content of lead and tin in different tissues and organs of crayfish is close and changes little. In claw muscles barium, strontium, zinc and vanadium are accumulated, i.e. the same elements as in body and claw carapace, while lead, cobalt and lithium were not found. In claw carapaces the amount of copper, manganese, iron, aluminum, zinc, vanadium and tin is lower than in body carapaces (Table 8).

503

Table 7. Content of free amino-acid in hepatopancreas of the Dniester crayfish (% of damp mass).

Sex	Males	Males	Females	Females	Females	Females	Males and Females
Length, cm	10.0-11.0	15.1-17.0	10.1-12.0	15.1-17.0	15.1-17.0	10.1-12.0	10.1-11.0
	non-moulting				moulting		dead
Cystines	0.067	0.100	0.074	0.081	0.056	0.034	0.135
Lysine	0.118	0.111	0.120	0.078	0.120	0.089	0.098
Histidine	0.096	0.058	0.070	0.039	0.062	0.048	0.084
Arginine	0.135	0.117	0.147	0.129	0.160	0.135	0.096
Asparaginic acid	0.074	0.116	0.107	0.110	0.113	0.073	0.110
Serine	0.105	0.120	0.100	0.096	0.097	0.048	0.093
Glycine	0.090	0.180	0.084	0.074	0.088	0.037	0.075
Glutamine acid	0.115	0.143	0.122	0.110	0.108	0.073	0.115
Threonine	0.087	0.102	0.095	0.078	0.086	0.059	0.098
Alanine	0.125	0.130	0.122	0.112	0.114	0.116	0.126
Proline	0.155	0.230	0.262	0.155	0.157	0.110	0.135
Tyrosine	0.054	0.065	0.049	0.046	0.059	0.036	0.087
Tryptophan	0.074	0.102	0.083	0.038	0.039	0.031	0.101
Methionine	0.042	0.040	0.044	0.056	0.035	0.028	0.051
Valine	0.130	0.142	0.120	0.128	0.123	0.095	0.113
Phenylalanine	0.103	0.105	0.090	0.090	0.085	0.087	0.102
Leucines	0.200	0.210	0.180	0.189	0.162	0.136	0.200
Total	1.770	2.021	1.869	1.609	1.664	1.231	1.819

Table 8. Content of microelements in organs and tissues of the long-clawed crayfish of the Dniester estuary in July 1978 (% of ash).

Length, cm	9.0 -12.0		10.1 - 12.0		15.1 - 16.0		15.0 - 16.0		10.0 - 12.0	
Organ, tissue	muscles		hepatopancreas		claw muscles		hepatopancreas		soft crayfish	
	♀	♂	♀	♂	♀	♂	♀	♂	♀ muscles	♂
Cu	0.1000	0.1300	>0.1	>0.1	0.12	0.065	>0.1	>0.1	0.060	0.069
Mn	0.023	0.030	0.28	0.045	0.048	0.24	0.24	0.31	0.012	0.022
Fe	0.061	0.116	0.45	0.75	0.045	0.036	1.4	1.5	0.080	0.093
Al	0.095	0.24	0.45	0.32	0.052	0.01	0.19	0.5	0.065	0.25
Zn	0.041	0.037	0.055	0.057	0.090	0.150	0.053	0.067	0.048	0.047
Ba	0.0022	0.0021	0.0027	0.0024	0.0130	0.0150	0.0024	0.0035	0.0023	0.0034
Sr	0.065	0.049	0.040	0.038	0.160	0.200	0.038	0.035	0.035	0.043
Pb	0.0029	0.0030	0.0029	0.0030	0	0	0.0030	0.0031	0.0033	0.0040
Ti	0.0060	0.012	0.030	0.016	0.0038	0.016	0.011	0.040	0.0072	0.021
V	0.0008	0.0007	0.0007	0.0007	0.0015	0.0016	0.0007	0.0007	0.0006	0.0008
Sn	0.0030	0.0029	0.0030	0.0030	0.0030	traces	0.0030	0.0030	0.0030	0.0029
Li	0.0033	0.0036	0.0076	0.0068	0	0	0.0066	0.0085	0	0
Ni	0.0033	0.0061	0.0074	0.0160	0.0028	0.0027	0.0110	0.0170	0.0029	0.0026
Ag	0.0007	0.0007	0.025	0.0066	0.0007	0.0007	0.0024	0.0025	0.0006	0.0008
Co	0	0	0.030	0.069	0	0	0.030	0.040	0	0
% of ashes per dry substance	6.16	6.27	2.90	1.27	8.08	8.88	2.27	3.78	5.12	3.54

(Continued)

Table 8 (continued).

Length, cm Organ, tissue	10.0 - 12.0 hepatopancreas of soft crayfish		10.0 - 12.0 dead crayfish		15.0 - 17.0 body carapace		15.0 - 17.0 claw carapace	
	♀	♂	muscles	hepatopancreas	♀	♂	♀	♂
Cu	>0.1	>0.1	0.13	>0.1	0.0030	0.0035	0.0019	0.0019
Mn	~0.1	0.270	0.052	0.45	0.059	0.058	0.049	0.037
Fe	0.47	0.42	0.070	0.27	0.042	0.056	0.036	0.033
Al	0.040	~0.1	0.10	0.45	0.045	0.12	0.025	~0.01
Zn	0.090	0.130	0.049	0.060	0.130	0.080	0.063	0.056
Ba	0.0024	0.0024	0.0021	0.0024	0.035	0.012	0.015	0.009
Sr	0.035	0.040	0.067	0.040	0.27	0.26	0.27	0.25
Pb	0.0030	0.0033	0.0029	0.0030	0	0	0	0
Ti	0.0034	0.0055	0.0060	0.017	0.0022	0.0048	0.0022	0.0022
V	0.0007	0.0007	0.0010	0.0008	0.0018	0.0025	0.0014	0.0015
Sn	0.0030	0.0030	0.0030	0.0030	0.0030	0.0030	traces	traces
Li	0	0	0	0.0069	0	0	0	0
Ni	0.011	0.016	0.0043	0.0075	0.0028	0.0028	0.0032	0.0032
Ag	0.0058	0.0070	0.0007	0.0072	0	0	0	0
Co	0.070	0.075	0	0.032	0	0	0	0
% of ashes per dry substance	2.94	1.38	6.96	2.68	-	41.05	52.07	53.62

In August sharp differences were also found in the quantity of microelements in muscles of different-sized crayfish of both sexes. There are sharp differences (2-3 times) in the content of manganese, iron, aluminum, strontium, lithium and nickel in females' muscles compared to those of males. However, in muscles of big females 15.1-17.0 cm long the content of manganese, iron, aluminum, strontium, titanium, lithium and nickel is considerably lower than in muscles of females 10.1-12.0 cm long. As the results of the study show, the distribution of microelements in muscles of mature crayfish (individuals over 8.0 cm) is quite pronounced and should be taken into consideration in biotechnical studies of the crayfish cultivation.

Similarly, hepatopancreas of female crayfish 10.1-12.0 cm long contain more (2-5 times) copper, manganese, iron, aluminum, zinc, lead, titanium, nickel and silver than larger females 15.1-117.0 cm long and more (1.5-4 times) copper, manganese, aluminum, zinc, strontium, lead, titanium, lithium and nickel than in hepatopancreas of males of the same size.

No considerable accumulation of microelements was noted in the ovaries of females 15.1-17.0 cm long compared with muscles of crayfish of the same size (except for aluminum - 0.33% per ashes). This may be explained by the less important role of microelements in the formation of ovicells in the ovaries of females taken after reproduction (August).

On the other hand, the reproductive ability of females 15.1-17.0 cm long is much lower compared to that of smaller females (10.1-12.0 cm). Therefore, the proportion of microelements in ovaries and muscles of large females may be identical (Table 9).

Tissues of soft (moulting or just moulted) crayfish show considerable differences in the microelement content compared to hard (moulted) crayfish, which demonstrates the need for correct combinations and proportions of microelements in crayfish diets.

Ash (summary mineral substances). Ash is one of the important indices of the biochemical analysis of an animal, as knowledge of the optimal mineral content is essential for biotechnical studies.

In July 1978 the content of ashes in the muscles of crayfish males and females 9.1-12.0 cm long was very similar. It also did not differ in muscles of dead crayfish (6.2-6.9% of dry substance). In muscles of moulted crayfish the content of ashes was lower compared to that of one-sized non-moulting crayfish. It was 5.1% for females and 3.5% for males, which may be connected with the loss of minerals during moult. In the hepatopancreas of dead crayfish the content of ashes did not differ from that of live ones and amounted to 1.3% for males 10.1-12.0 cm long and 2.9% for females of this size. For bigger individuals 15.1-16.0 cm, it was 3.8% and 2.3% of dry mass respectively. For dead ones 10.1-11.0 cm long it was 2.7% (Table 10). However, the hepatopancreas of soft crayfish contained less ash - 1.4%. The minerals of an organism are evidently mobilized for the formation of a new carapace.

The growth of crayfish is connected with regular moults, i.e. with regular formation of a new chitin integument (carapace). Therefore it is quite clear that maximum content of ashes is noted in the chitin

Table 9. Content of microelements in Dniester crayfish in August 1978 (% of ashes).

Length, cm	0.8 – 1.2	9.5 – 10.5		hepatopancreas	10.1 – 12.0	
Body parts, organs, larvae	larvae of the 1st stage	muscles			muscles	
		♂	♀	♂	♂	♀
Cu	0.015	0.016	0.015	0.1	0.019	0.019
Mn	0.052	0.013	0.015	0.1	0.016	0.042
Fe	0.33	0.18	0.17	0.75	0.11	0.30
Al	>0.30	0.10	0.1	0.22	~0.1	0.30
Zn	0.30	0.19	0.22	1.0	0.13	0.14
Sr	0.160	0.032	0.030	0.040	0.026	0.040
Pb	0.0018	0	0.0017	0.0019	0.0018	0.0017
Ti	0.0400	0.0073	0.0056	0.0150	0.0052	0.0200
V	0.0019	0.0010	0.0010	0.0010	<0.0010	<0.0010
Sn	0.0012	0.0010	0.0010	0.0012	0.0010	0.0012
Li	0	0.0063	0	0.0085	0	0.0090
Ni	0.0012	0.0012	0.0011	0.01	0.0012	0.0170
Ag	0	0.0004	0.0006	0.023	0.0004	0
% of ashes per dry substance	13.4	6.4	6.2	2.1	7.0	2.9

(Continued)

Table 9 (continued).

Body parts, organs, larvae	10.1 – 12.0		15.1 – 17.0			
	hepatopancreas		muscles		hepatopancreas	ovaries
	♂	♀	♂	♀	♀	♀
Cu	0.027	0.1	0.024	0.018	0.020	0.021
Mn	0.080	0.095	0.018	0.015	0.055	0.020
Fe	0.40	0.45	0.19	0.11	0.25	0.14
Al	0.18	0.30	0.25	0.09	0.21	0.33
Zn	0.35	1.00	0.28	0.18	0.25	0.09
Sr	0.026	0.042	0.034	0.026	0.030	0.045
Pb	0	0.0140	0	0	0	0
Ti	0.0048	0.0120	0.01	0.0052	0.01	0.0045
V	<0.0010	<0.0010	~0.001	0.001	0.001	0.001
Sn	0.0012	0.0012	0.0010	0.0012	0.0012	0.0011
Li	0	0.0079	0	0	0	0
Ni	0.0150	0.0400	0.0012	0.0012	0.0210	0.0013
Ag	0.0550	0.0100	0.0005	0.0004	0.0025	0.0007
% of ashes per dry substance	1.5	2.8	5.5	6.7	3.4	not found

Table 10. Content of ashes in organs and tissues of the Dniester crayfish.

| Sex | Length, cm | % if ashes per mass | |
		dry	damp
		Chitin integument (carapace) of claws	
Males	15.1-16.0	53.62	not determined
Females	15.1-16.0	52.07	not determined
		Chitin integument (carapace of body	
Males	15.1-16.0	41.05	not determined
		Body Muscles	
Females	9.1-10.0	6.01	1.40
	11.1-12.0	6.38	1.50
		Dead crayfish	
		6.96	1.86
		Soft crayfish	
Males	10.1-12.0	3.54	not determined
Females	10.1-12.0	5.12	not determined
		Claw muscles	
Males		8.88	not determined
Females		8.08	not determined
		Hepatopancreas	
Males	10.1-11.0	1.50	0.87
	11.1-12.0	1.08	0.68
Females	11.1-12.0	2.90	0.89
Males	15.1-16.0	3.78	1.30
Females	15.1-16.0	2.72	0.97
		Dead crayfish	
Males and females	10.1-11.0	2.68	1.28
		Soft crayfish	
Males	10.1-11.0	1.38	not determined
Females	10.1-11.0	2.94	not determined

July 4, 1978.

(Continued)

510

Table 10 (Continued).

		August 28	
Sex	Length, cm	% of ashes per dry substance	
		Body muscles	
Males	9.5-10.5	6.4	
Females	9.5-10.5	6.2	
Males	10.1-12.0	7.0	
Females	10.1-12.0	2.9	
Feamles	15.1-17.0	6.7	
		Hepatopancreas	
Males	9.5-10.5	2.1	
Males	10.1-12.0	1.5	
Females	10.1-12.0	2.8	

		November 20	
		% of ashes per mass	
Sex	Length, cm	dry	damp
		Body muscles	
Males	9.5-10.5	7.1	1.4
Males	10.1-12.0	7.0	1.4
Females	10.1-12.0	7.5	1.7
Females	15.1-17.0	8.2	1.6
Females	15.1-17.0	6.7	1.6
		Hepatopancreas	
Males	9.5-10.5	4.8	1.5
Males	10.1-12.0	5.0	1.3
Females	10.1-12.0	6.0	1.7
Males	15.1-17.0	5.8	1.4
Females	15.1-17.0	8.5	2.0
		Ovaries	
Females	10.1-12.0	3.7	1.7

integument of claws (52-53% per dry mass) and body (41% per dry mass) compared to muscles of claws (8% per dry mass) and body muscles (6.01-6.38% per dry mass) in females 9.1-12.0 cm long.

This distribution of minerals in different parts of chitin integument and in adjoining muscles is quite clear and explainable since the most powerful part of the animal's body is its claws during its feeding, movement and defense. Because of the high content of ashes, they are robust and strong.

There are much less ashes in hepatopancreas of males and females, which is quite natural biologically for many animals (Table 10).

Organic substance (summary). There were no considerable differences in the content of organic substance in males and females of different age groups. In females' muscles the content of organic substance varied from 18.6 to 22.7% per damp mass and from 95.1 to 97.4% per dry mass. The amount of organic substance in males' muscles was somewhat lower (16.9-17.3% per damp mass and 92.8-99.7% per dry mass).

The content of organic substance in ovaries of females was much higher than in muscles (43.6-48.9% against 18.6-22.7% per damp mass) and approximately the same taken per dry mass (Table 11).

The content of organic substance in different age groups of males and females of long-clawed crayfish was high enough, which is of practical use.

CONCLUSIONS

1. Maximum content of lipids (up to 92.43% of dry mass) in mature males and females of the Dniester crayfish (body length over 8.0 cm) was registered in hepatopancreas, which is explained by the role of the latter as accumulator of many biochemical components.

2. Reduction of the amount of lipids in females' ovaries (up to 38.62% of dry mass) is observed during their reproduction, which emphasized the role of lipids as an energy source for reproduction.

3. In the study of lipid content in muscles of non-moulting crayfish (with hard chitin integument), both males and females 9.1-17.0 cm long (5.89-7.00% of dry mass) and moulting (soft) ones 9.1-10.0 cm long (6.81-6.93% of dry mass) showed insignificant changes in the muscles of crayfishes while they grow and develop.

4. Lipids were found to contain phospholipids, cholesterine, free fatty acids and triglycerides. The proportion of phospholipids in muscles reaches 25% of the total amount of lipids and 15% in liver. The biggest fraction in liver is cholesterine, which reaches 80% of the total amount.

5. The content of cholesterine in ovaries of one-sized females is 3 times higher than in their eggs and larvae of the first development stage (7.35% of non-saponaceous fraction in eggs and 6.17% in

Table 11. Content of organic substance in long-clawed crayfish.

| Size, cm | Females | | Males |
	Ovaries	Muscles	Muscles
	Percentage per damp mass		
7.1- 8.0	22.1±0.8	44.4±1.6	16.9±0.4
9.1-10.0	21.4±0.7	43.6±1.4	17.0±0.4
10.1-12.0	22.7±0.8	45.0±1.6	17.1±0.4
15.0-17.0	18.6±0.5	48.9±1.8	17.3±0.5
	Percentage of dry mass		
7.1- 8.0	97.2±4.3	92.4±3.9	99.7±5.1
9.1-10.0	96.9±4.1	90.5±3.1	92.6±4.1
10.1-12.0	97.4±4.5	88.2±3.2	92.8±4.3
15.1-17.0	95.1±4.2	95.4±4.3	92.8±4.2

larvae) which emphasizes the role of cholesterine in the formation of ovaries and maturation of sexual products (ovicells).

6. The amount of cholesterine in muscles of females 9.1-10.0 cm long (1.58% of non-saponaceous fraction) is much lower than in males of the same size (10.63% of non-saponaceous fraction) which may be connected with lower mobility of reproducting females and concentration of cholesterine in ovaries and hepatopancreas.

7. Sterols in non-saponaceous fraction extracted from the muscles of crayfish contained non 7-dehydrocholesterol, provitamin of the vitamin D_3 and precursor of cholesterol. Cholesterine in crayfish's muscles could be therefore synthesized through desmosterol.

8. The amount of provitamins D (summary) 3.02-3.43% of non-saponaceous fraction and cholesterine - 24.00-26.62% of non-saponaceous fraction is higher in non-moulting crayfish 11.1-17.0 cm long than in moulting crayfish of the same size (1.08-1.84 and 20.13-22.45% of non-saponaceous fraction, respectively). This demonstrates the role of provitamins D and cholesterine in the formation of chitin integument of crayfish.

9. In autumn changes in carotenoid content were observed. Minimum content (1.9-4.5 mkg/g of damp mass) is observed in muscles, maximum content 123.0-168.6 mkg/g) is observed in ovaries indicating their important role in the development of oocytes.

10. In hepatopancreas of non-moulting crayfish (with hard chitin integument) 10.1-17.0 cm long the content of free amino acids (1.609-2.021% of damp mass) is much higher than in muscles of crayfish (males and females) of the same size (0.475-0.904%) due to the accumulation of different organic and mineral substances in hepatopancrease responsible for metabolic processes.

11. The amount of free amino acids in eggs (0.333%) of Dniester crayfish and their larvae on stage I (0.468%) is lower than in females 15.1-17.0 cm long (0.671%) due to the morbility of mature females and their energy consumption.

12. The amount of amino acids (combined) in the muscles of the Dniester crayfish (13.303-15.922% of damp mass) is the same as in larvae of the 1st development stage (15.681% of damp mass) and in eggs taken from pleopoda of different-sized females 9.1-10.0 cm and 14.1-17.0 cm long. This shows their redistribution during reproduction and development of juveniles.

13. Larvae of the Dniester crayfish were found to contain much more manganese and aluminum than in mature crayfish with body length over 8.0 cm, which shows the need of specific selection microelements for larvae food composition.

14. During growth and development of the Dniester crayfish, redistribution of microelements in muscles of males and females is quite distinct, with lower content of manganese, iron, aluminum, strontium, titanium, lithium, nickel in muscles of females 15.1-17.0 cm long compared to muscles of females 10.1-12.0 cm long. However,

the content of certain elements was 3-4 times higher in larger males of the same size group, which should be taken into account while selecting diets for different sized crayfish.

15. The hepatopancreases of females 10.1-12.0 cm long were found to contain higher levels of copper, manganese, iron, aluminum, zinc, titanium, nickel and silver than in hepatopancreas of females 15.1-17.0 cm long. This reveals the greater requirement by younger females for microelements. This is connected with more intensive digestion and metabolism by females.

16. The higher content of copper, manganese, aluminum, zinc, strontium, tin, titanium, lithium and nickel in hepatopancreas of females 15.1-17.0 cm long compared to that of males of the same size is likely to be connected to reproduction and the influence of these elements over the regularity and intensiveness of moults.

17. Similar accumulations of microelements in ovaries of females 15.1-17.0 cm long and in their muscles may be explained by the less important role of microelements in the formation of ovicells in the females' ovaries after their reproduction (August). It is quite possible that the reproductive potential of big females 15.1-17.0 cm long is much lower.

18. The maximum values for ashes (summary mineral substance) are observed in chitin integument of crayfish (52-53% per dry mass in claws and 41% per dry mass in body) compared to claw muscles (8% per dry mass) and body (6.01-6.38% per dry mass) as well as hepatopancreas (1.50-2.90% per dry mass) which is quite natural for many crustacea.

19. No great differences were observed in the content of organic substances in different sized males and females. The amount of organic substance varied from 16.9 to 22.7% per damp mass and 22.8-99.7% per dry mass. Maximum values of organic substance were noted for females' ovaries (43.6-48.9% of damp mass).

The biochemical analysis of the Dniester long-clawed crayfish showed a number of biological peculiarities of crayfish during its growth, development and reproduction as well as values for the content of lipids, sterols, carotenoides, amino acids, micro-elements and ashes. The obtained material may be utilized in the study of food composition for commercial crayfish cultivation.

LITERATURE CITED

Armitage, K.B., A.L. Buikema, N.S. Willems, Jr. 1972. Organic constitutents in the animal cycle of the crayfish Orconectes nais (Faxan) Comp. Biochem. and Physiol., v. 41A, No. 4, pp 825-842.

Balashova, M.N. 1971. Biochemical study of river crayfishes. Synopsis on commercial ichthyology. Ser. 1, issue 9, pp 11-12.

Bligh, E., W. Dyer. 1959. A rapid method of total lipid extraction and purification. Can. J. Biochem. Physiol. v. 37, No. 8, pp 911-917.

Brodsky, S.J. 1980. Mixohaline decapoda as cultivation object. "Marine biology," No. 4, pp 61-67.

Cowey, C.B. 1961. The non-protein nitrogenous constitutents of the tissues of the freshwater crayfish Astacus pallipes Lereb. Compar. Biochem. a Physiol., No. 3, pp 173-180.

Davies, B.H. 1965. Analysis of carotenoid pigments. In "Chemistry and Biochemistry of plant pigments." T. Goodwin (Ed.) N.Y. London. Acad. Press; 489-533.

Drujan, B., R. Castillon, E. Guerrero. 1968. Application fluorimetry in determination in vitamin A. Anal. Biochem., v. 23, No. 1, pp 44-52.

Filippovich, J.B. 1958. Quantitative assessment of amino acids by chromatography on paper. Scien. papers, Moscow Pedagog. Inst. of V.I. Lenin, v. 61, issue 9, pp 147-212.

Folch, J. et al. 1951. Preparation of lipid extracts from brain tissue. Biol. Chem., v. 191, No. 3, pp 833-841.

Godnev, T.N. and V.M. Terentiev. 1956. Of quantitative assessment of chlorophyll and some carotenoides. Works of the Institute of Plant Physiology of K.A. Timiriazev, v-3, No. 5, pp 124-129.

Gribovskays, I.F. 1968. Preparation of soil and plant samples for spectral assessment of the microelements' content. Biolog. Sciences, No 1, pp 136-144.

Jensen, A. 1960. Chromatographic separation of carotenoides and other chloroplast pigments on aluminum oxide containing paper. Acta chem. scand., v. 16, No. 9, pp 2051.

Jensen A., S. Jensen. 1959. Quantitative paper chromatography of carotenoides. Acta chem. scand., v. 13, No. 9, p. 1863.

Karnaukkov, V.M. 1973. Carotenoides' functions in animal cells. M., Science, p 103.

Karnaukkov, V.M. 1978. Role of molluscs with high carotenoides content in protection of aquatic environment from pollution. Puschino, 1978, preprint, 74.

Kizavetter, I.F. 1973. Biochemistry of aquatic raw materials. "Food Industry," m., 1973, 423.

Lisboa, B. 1969. Thin-layer chromatography of steroids, sterols and related compounds. In "methods in enzymology." New York, London, Acad. Press, v. 15, pp 3-158.

Matskiavichene, G.I. 1979. Modification of the level of water-salt and protein metabolism during postembryonic development of broad-clawed crayfish. In "Lithuanian river crayfish biology," "Mosklas," Vilnius, 1979, pp 53-66.

Mitchel, R.L. 1961. Determining of element's traces in plants and other biologic objects. In "Analysis of elements' traces," Il., M., 1961.

Moore, P.R., C.A. Baumann. 1952. Skin sterols I Colorimetric determination of cholesterol and other sterols in skin. J. Biochem., v. 195, No. 2, pp 615-621.

Palamarchuk, V.G., V.G. Koval, V.P. Vendt. 1974. Himichni ta biologichni vlastivosti rechovini steroidnoi prirodi z maximuman poglinannia zadovzhini hvili 245 nm. Ukr. Biochemical journal, v. 46, No. 6, pp 736-740.

Rzhavskaya, F.M., M.A. Alekseeva. 1966. Method of determining non-saponaceous substances in fish and marine mammals' fats. "Fisheries," No. 6, pp 66-67.

Sapozhnikov, D.N. (ed.). 1964. Pigments of plastida of green plants and method of their study. "Science" publishers, M.-L.

Sapozhnmikov, D.I., D.A. Krasnovska, A.N. Maevskaya. 1956. Quantitative assessment of the main carotenoides of green leaf with the help of paper chromatography. Works of the Institute of plant phasiology of K.A. Timiriazev, v. 3, No. 5, pp 46-52.

Tsukerzis, J.M. 1970. Biology of broad-clawed crayfish. "Mintis," Vilnius, p 208.

VIII
DESCRIPTIVE PAPERS

RESEARCH METHODOLOGY AT THE UNIVERSITY OF
SOUTHWESTERN LOUISIANA CRAWFISH RESEARCH CENTER

Donald Gooch
Department of Animal Science
University of Southwestern Louisiana
Lafayette, Louisiana

ABSTRACT

Long standing culture problems such as water quality, over-population, inconsistent annual crops, harvesting, forage utilization, accurate data collection, insecure and low quality bait sources and by-product uses are discussed. Related research and possible solutions are given along with recent innovations in data collection.

INTRODUCTION

Crawfish farming in Louisiana is considered a profitable enterprise (LaCaze 1975, Roberts 1980), at least in well managed, open ponds. However, accurate records are rare, and many people have entered crawfish farming over-optimistic about profits and under-educated in good pond management, and have terminated their ventures in disappointment.

Pond failures and inconsistent annual crops are probably more common than most would believe. Researchers have pointed out the major problems for years (Jaspers and Avault 1969, LaCaze 1970, Avault et al. 1974, Huner 1980, Huner and Barr 1980), but economical solutions have been very slow in becoming a reality. Harvesting innovations have solved a major problem (Gooch 1975, Gooch 1977) for the present by lowering harvest costs from 50% of the gross to under 15% when used properly. A peeling machine now for lease (Huner 1981) will undoubtedly revolutionize the processing portion of the industry within a few years. Excellent ongoing research at universities and government experiment stations assures forthcoming answers to many of the problems that persist.

The U.S.L. Crawfish Center is concentrating its efforts on pond management and crawfish population structure. It now takes 12 months to realize a crawfish crop using present methods, but U.S.L. continues to research a hatchery concept that would allow a harvest of 1,000 lbs/acre in four months. This has been discussed elsewhere (Gooch 1980). Waste utilization possibilities and alternate sources of bait are also being considered and researched as funding becomes available.

POND MANAGEMENT

There are several different types of crawfish ponds in Louisiana (Gooch 1977, Huner and Barr 1980), and they present specific problems and/or advantages to farmers. Over half of the present acreage in Louisiana consists of rice field ponds, so for the sake of simplicity the statements made herein will apply expressly to rice field culture. The major problems of water quality, overpopulation and forage utilization are actually uniquely interconnected and must be approached with this in mind.

Crawfish ponds in Louisiana are usually flooded around September 15 to October 15, although the range is September 1 to November 15. At this time the water covers the straw left by the rice harvest. Water temperatures are very warm in South Louisiana at this time, and decomposition of the straw via microbial activity begins quickly. When the C:N ratio drops to below 17:1 the straw becomes crawfish food, but because of the decomposition process oxygen commonly drops to near lethal levels at this time, especially for newly hatched crawfish (Avault et al. 1974). In many ponds of 10-40 hectares in size no amount of (economical) pumping can alleviate the problem in all areas of the pond. This is usually because of poor circulation within these larger ponds. But two other problems develop simultaneously. One is that the decomposition process prevents colloidal formation and crystal-clear water is the result. Since crawfish are extremely nocturnal in clear water, activity (foraging and growth) is severely limited to nighttime hours. But because of the clear water, oxygen levels drop drastically during the night, further restricting activity. The second problem encountered involves forage utilization. Although crawfish farmers may generate 8800 pounds (dry weight) per acre of vegetation (Huner and Barr 1980), rice field ponds are generally completely out of food by late season. Because most of the straw begins decomposition at the same time, much of it "goes to pieces" before the crawfish can benefit from it directly. This is confirmed by processors who watch tail meat yields from rice field crawfish drop from 17-19% in February-March to 11-13% by late April and May. A pond that has too little food invariably leads to overcrowded and stunted crawfish. Crawfish in this condition are generally inferior and extremely hard to market. If the food can be made to last longer, then stunting problems may be alleviated (Romaire et al. 1978) and solve the economic problems of transporting supplemental food into a pond (Rivas et al. 1978).

Flooding at later (cooler) dates would partially solve the problem, but farmers want to take advantage of the warm weather for crawfish growth to produce earlier, higher priced crops. The U.S.L. Crawfish Research Center has adopted the following possible solutions for the aforementioned interrelated problems.

Once the rice is cut we recommend baling the straw before flooding. Preliminary results have shown that this effectively and drastically decreases O_2 problems at the most critical time by not allowing all of the straw to begin decomposition at the same time. Breaking up the bales as needed throughout the season prevents the pond from running out of food before the end of the season. Rice field crawfish pond waters normally remain clear from flood date until November-December (2-3 months), but by using the baling method we have achieved turbidity within several days following flooding. Once the water is fairly turbid, oxygen problems, especially at night and early in the morning, are greatly reduced. Also, crawfish now become active during the daylight hours under the mask of the turbid water.

The other problem mentioned above is water circulation. The U.S.L. Crawfish Center believes that this problem can be greatly overcome by utilizing interior baffle levees to ensure good water circulation in every area of the pond. A recirculating pond design is even better (Fig. 1) because a much smaller water well can be used. Many potential crawfish farmers in Louisiana never enter the business

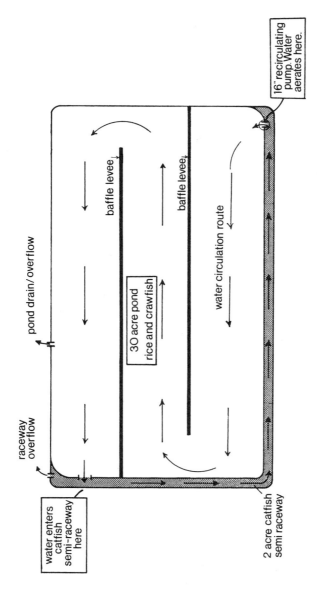

Figure 1. USL Crawfish Research Center, Experimental Pond, Cade, LA.

because of the high cost of drilling a deep water well large enough to handle pond requirements (Craft 1980). Water cost in many places in the United States is a limiting factor in aquaculture ventures. Overuse of underground water has also become a national concern. But with a recirculating system a well can be much smaller since it is only used to fill the pond and to replace water lost to seepage and evaporation. Once the pond is filled, a much larger pump costing only 5% of what a 10-12" deep water well would cost is used for aeration. The baffle levees insure good water circulation in every area of a pond. In a recent U.S.L. study 12 area crawfish ponds of commercial size were monitored for a nine-week period beginning in February and ending in April. Oxygen levels in the U.S.L. Cade Pond shown in Figure 1 were significantly higher than any of the other ponds which did not utilize interior baffle levees.

CRAWFISH POPULATION STRUCTURE IN PONDS

Crawfish population structure has been a difficult subject because accurate samples in heavily vegetated ponds are extremely hard to obtain. Population densities of adult crawfish have been mentioned in ponds and natural habitats by several writers (Huner and Romaire 1978, Konikoff 1977, Lowery and Mendes 1977, etc.). Size distributions likewise have been noted (Huner and Romaire 1978, Konikoff 1977). Brood stock counts (burrow counts) in drained pond floors and on levees throughout the year have not been published. Indeed, de la Bretonne and Avault (1977) conclude that burrow counting is unacceptable due to erosion of the mud caps over a period of time. However, burrow counts can be made and adult burrows distinguished by hole diameter if the counts are made during the first few weeks after draining, when vegetation is not too dense, and before rain activity erodes the chimneys and/or mud plugs. In fact, the red crawfish characteristically "reworks" the burrow following rains in the summertime. The fresh clay brought up during reworking periods enhances the capability of spotting the burrows. U.S.L. has utilized metal rods welded together into a square meter to count burrows. By randomly tossing this apparatus in different areas of a pond, brood stock estimates can be made. Counts over the past few years have shown brood stock present in the floor of Louisiana crawfish ponds ranging from 50-400 lbs/acre. Levee reproduction during the months when the pond is flooded has been mentioned often by Louisian researchers. Two to three waves of young-of-the-year recruitment originating from levee burrows have apparently been noted (de la bretonne and Avault 1977, Huner and Barr 1980). However, no burrow counts on levees have been published. In other words, no crawfish farmer in Louisiana actually knows how much (or little) his brood stock is for any given year. The U.S.L. Crawfish Center hopes to clarify and verify past observations by counting burrows in several commercial ponds and then relating these numbers to actual production the following season.

Size distributions in flooded ponds are also being investigated in spite of difficulties in sampling. Seines, where they can be used, and dip-net samples are taken each week in several area ponds. The crawfish samples are poured on top of a series of four boxes with decreasing wire mesh sizes. The top box has a bottom screen of 19 mm size mesh which retains only market-size crawfish; smaller crasfish drop through this screen onto the next size mesh (12 mm). Even smaller crawfish fall through to the third box with 6 mm mesh wire, and finally

524

there is a solid bottom box in which everything smaller remains.

By making burrow counts in the pond floor in the summer and on the levees throughout the season, combined with seine samples of the entire population during the flooded portion of the year and daily- and per-acre-catch data from farmers' records, we will eventually see a clear picture of crawfish populations in culture ponds.

OTHER RESEARCH

Finally, though not fully funded, the Crawfish Center is conducting indicator-type studies on the bait problems facing the industry. The bait industry in Louisiana is a separate business, with bait sources and species insecure and unpredictable. In years when the "wild" Atchafalaya swamp crop materializes, bait for ponds becomes a great concern to pond owners (Bean and Huner 1978). In unpublished studies (U.S.L.) wet crawfish waste and crawfish waste combined with a binder caught 70% as many crawfish as fresh gizzard shad (Dorosomo cepedianum) using equal amounts (weight) in each trap. Crawfish waste is known to be valuable (Desselle 1980, Barry 1980, Meyers and Rutledge 1971), but dehydration costs and other factors have not made it feasible to use (Huner and Barr 1980). Processing plants must find a way to utilize this waste which makes up over 80% of the total weight before processing. Since most processors are also bait dealers, the possibility of using waste as an ingredient in an artificial bait seems promising. This would also mean a source of bait wherever this species is cultured.

In the bait category U.S.L. is also conducting research on the pond culture of carp (Cyprinus carpio), which under the right circumstances is considered a top bait fish by pond operators and commercial crawfishermen. It is also probably the easiest of fish species to culture (Bardach et al. 1972).

At a production rate of 5000 lbs/acre, 1 acre of carp would yield enough bait (10 traps/acre 0.5 lb bait/trap/day, 100 fishing days/season) for 15 acres of crawfish.

ACKNOWLEDGMENTS

Research reported herein has been supported by the Louisiana Department of Wildlife and Fisheries, the Louisiana Board of Regents, and the University of Southwestern Louisiana.

LITERATURE CITED

Avault, J. W. Jr., L. W. de la Bretonne and J. V. Huner. 1974. Two major problems in culturing crayfish in ponds: oxygen depletion and overcrowding, pp. 139-144. In J. W. Avault (ed.) Freshwater Crayfish II, Louisiana, U.S.L. (1975). 676 pp.

Bardack, J. E., J. H. Ryther and W. O. McLarney. 1972. Aquaculture -- The Farming and Husbandry of Freshwater and Marine Organisms. Wiley-Interscience, New York. 868 pp.

Barry, J. R. 1980. Utilization of crawfish peeling plant waste as a soil amendment for vegetable crop production. Proceedings of the First National Crawfish Culture Workshop, University of Southwestern Louisiana, Lafayette, 3-4 March 1980.

Bean, R. A. and J. V. Huner. 1979. An evaluation of selected crawfish traps and trapping methods, pp. 141-152. In P. J. Laurent (ed.) Freshwater Crayfish IV, Thonon-les-Bains, France (1978). 473 pp.

Craft, B. R. 1980. Some basic considerations in crawfish pond construction. Proceedings of the First National Crawfish Culture Workshop, University of Southwestern Louisiana, Lafayette, 3-4 March 1980.

de la Bretonne, L. W. and J. W. Avault, Jr. 1977. Egg development and management of Procambarus clarkii (Girard) in a south Louisiana crayfish pond, pp. 133-140. In O. V. Lindqvist (ed.) Freshwater Crayfish III, Kuopio, Finland (1976). 504 pp.

Deselle, L.J. 1980. The Value of crawfish waste as a lime source and soil amendment. Proceedings of the First National Crawfish Culture Workshop, University of Southwestern Louisiana, Lafayette, 3-4 March 1980.

Gooch, D. M. 1975. Crawfish boat method harvesting in Louisiana marshes. Aquatic Studies Program, University of Southwestern Louisiana, Lafayett. Crawfish research Center Publication USL-ASP-75-04.

Gooch, D. M. 1977. Boat method harvesting of crawfish in open ponds. Aquatic Studies Program, University of Southwestern Louisiana, Lafayette, Crawfish Research Center Publication USL-ASP-77-06.

Gooch, D. M. 1980. Crawfish farming in Louisiana. Proceedings of the First National Crawfish Culture Workshop, University of Southwestern Louisiana, Lafayette, 3-4 March 1980.

Huner, J. V. 1980. Red crawfish population dynamics in culture ponds: a summary. Proceedings of the First National Crawfish Culture Workshop, University of Southwestern Louisiana, Lafayette, 3-4 March 1980.

Huner, J. V. 1981. A report on a crawfish peeling machine. Aquaculture Digest 6:7:33.

Huner, J. V. and J. E. Barr. 1980. Red swamp crawfish: biology and exploitation. Louisiana State University Center for Wetland Resources, Baton Rouge. Sea Grant Publication No. LSU-T-80-001.

Huner, J. V. and R. P. Romaire. 1979. Size at maturity as a means of comparing populations of Procambarus clarkii (Gerard) (crustacea, decapoda) from different habitats, pp. 53-64. In P. J. Laurent (ed.) Freshwater Crayfish IV, Thonon-les-Baines, France (1978). 473 pp.

Jaspers, E. and J. W. Avault, Jr. 1969. Environmental conditions in burrows and ponds of the red swamp crawfish, Procambarus clarkii. Proceedings, 23rd Annual Conference, Southeastern Association Game and Fish Commissioners 23:592-605.

Konikoff, M. 1977. Study of the life history and ecology of the red swamp crayfish, Procambarus clarkii, in the lower Atchafalaya Basin floodway. Final report for the U.S. Fish and Wildlife Service, Department of Biology, University of Southwestern Louisiana, Lafayette.

LaCaze, C. G. 1970. Crawfish farming. Louisiana Wildlife and Fisheries Commission Fisheries Bulletin No. 7. 27 pp.

LaCaze, C. G. 1976. Crawfish farming (revised). Louisiana Wildlife and Fisheries Commission Fisheries Bulletin No. 7. 27 pp.

Lowery, R. S. and A. J. Mendes. 1977. Procambarus clarkii in Lake Naivasha, Kenya, and its effects on established and potential fisheries. Aquaculture 11:111-121.

Meyers, S. P. and J. E. Rutledge. 1971. Economic utilization of crustacean meals. Feedstuffs 43(43):16.

Rivas, R., R. P. Romaire, J. W. Avault, Jr., and M. Giamalva. 1979. Agricultural forages and by-products as feed for crawfish, Procambarus clarkii, pp. 337-342. In P. J. Laurent (ed.) Freshwater Crayfish IV, Thonon-les-Bains, France (1978). 473 pp.

Roberts, K. J. 1980. Louisiana crawfish farming: an economic view. Proceedings of the First National Crawfish Culture Workshop, University of Southwestern Louisiana, Lafayette, 3-4 March 1980.

Romaire, R. P., J. S. Forester and J. W. Avault, Jr. 1979. Growth and survival of stunted red swamp crawfish (Procambarus clarkii) in a feeding-stocking density experiment in pools, pp. 331-336. In P. J. Laurent (ed.) Freshwater Crayfish IV, Thonon-les-Bains, France (1978). 473 pp.

CRAYFISH SPECIES PLAN FOR THE UNITED STATES: AQUACULTURE[1]

James W. Avault, Jr.
Fisheries Section
School of Forestry and Wildlife Management
Louisiana State University
Baton Rouge, Louisiana 70803 USA

ABSTRACT

The National Aquaculture Plan prepared by the U.S. Joint Subcommittee on Aquaculture is comprised of a number of species plans including one on crayfish. The purpose of the crayfish species plan is to identify major requirements for enhancing aquaculture in the United States. Two groups of crayfish are cultured in the United States - Procambarus (P. acutus and P. clarkii) and Orconectes (O. immunis, O. virilis, and O. rusticus). Pacifastacus ssp. (principally P. leniusculus) show good culture potential. In general, major research needed includes that for marketing, harvesting, economics, food/forages, water quality, and multiple cropping with land crops. The crayfish species plan also covers information on status, potential, and constraints of crayfish aquaculture; technology transfer; and research facilities required.

INTRODUCTION

On a world basis, fishery products from aquaculture came to 6 million metric tons in 1975. By comparison, fishery products harvested from oceans and inland waters amounted to 71 million MT. Of this total, roughly one-third was reduced to fish meal and oil. Commercial harvest from our oceans on a world basis has leveled off at about 70 million MT for over the past decade, yet production from aquaculture continues to rise. In 1976, the Food and Agricultural Organization of the United Nations held a global meeting in Kyoto, Japan to discuss the status, potential, and constraints of aquaculture. The overall potential is excellent, and many countries such as the Philippines have embraced aquaculture as a national priority for providing low-cost animal protein.

In the U.S.A., aquaculture has excellent potential, but motivations for expansion vary. With only 6% of the world's population, we are the world's leading market for fishery products. In 1978 the U.S. imported 60% of the fishery products consumed domestically. That year the $2.62 billion trade deficit for fishery products amounted to almost 10% of the national trade deficit and 28% of the deficit for non-petroleum products. World demand is expanding, and this will limit the amount of fishery products exported to the U.S. and make them more expensive. For this and other reasons the U.S. has in recent years paid very close attention to aquaculture and has fostered measures for

[1] Permission to summarize highlights of the crayfish species plan was granted by the Joint Subcommitte on Aquaculture.

its expansion. The Eastland Fisheries Survey (a report to congress) and NOAA Aquaculture Plan assessed aquaculture in the U.S. Former President Carter by signing the Omnibus Farm Bill on 29 September 1977 gave the U.S. Department of Agriculture a mandate to include aquaculture among its responsibilities, and the USDA responded by drafting a USDA Aquaculture Plan in 1980. The USDA Plan encouraged each state to develop a state plan for aquaculture. President Carter signed into law the National Aquaculture Act in 1980, which called for a National Aquaculture Plan.

NATIONAL AQUACULTURE PLAN

The National Aquaculture Plan is a document prepared by members of the Joint Subcommittee on Aquaculture (JSA) and by experts from all parts of the U.S., with assistance and advice of persons from industry, universities, and state and federal governments. The USA consists of 13 members, the major members being the USDA, the U.S. Department of Commerce, and the U.S. Department of Interior. The goal of the Plan is "to provide an adequate supply of aquatic foods, to increase recreational benefits and to encourage new industry by the conservation, development, and utilization of land and water through aquaculture." The objective of the Plan is "to improve the economic, educational, financial, technical, scientific, and institutional base needed to promote the orderly development of aquaculture, and to facilitate early application of research results." The Plan is not a proposal for programs to be embarked upon by the federal government. It does suggest, however, appropriate roles for government, university, private and industrial components. The Plan, then, is designed as a guide to the formation of national policy, and to identify the actions that should be taken by federal and state governments, universities, and industry to make aquaculture an economically viable industry which can contribute substantial benefits to the nation.

The heart of the Plan consists of species plans for 12 groups of aquatic species which show the most promise for aquaculture. The 12 groups include: baitfish, catfish, clams, crayfish, freshwater prawns, largemouth bass, marine shrimp, mussels, oysters, salmon, smallmouth bass and trout. Each species plan covers research needs, requirements for facilities, economic data, markets, financial assistance, advocacy for land and water use, multiple-use conflicts, technical assistance and transfer of information, legal constraints, and jurisdictional overlaps.

CRAYFISH SPECIES PLAN

The following discussion is intended to give the essence of the Plan. A draft of the crayfish species plan was promulgated in May 1980 (Avault 1980), so production figures and the like are not necessarily current.

Current Status

Two groups of crayfish are cultured in the United States -- Procambarus (P. a. acutus and P. clarkii) and Orconectes (O. immunis, O. virilis, and O. rusticus). Pacifastacus spp. (principally P. leniusculus) show good culture potential. Captive fisheries exist for each group. In Louisiana, approximately 60% of the total crayfish crop is harvested from natural waters, principally the Atchafalaya Basin. The Basin is unpredictable, however, and may produce only two good crayfish crops every 5 years. Total annual crayfish production in Louisiana (Procambarus spp.) from ponds and natural waters combined has ranged up to 45 million pounds (20,250 metric tons). Data are lacking on capture fisheries of Orconectes spp. The estimated annual fishery catch for Pacifastacus spp. is approximately 500,000 pounds (227 metric tons).

Procambarus spp. are cultured for food on approximately 50,000 acres (20,000 ha) in Louisiana, 3,000 acres (1,200 ha) in Texas, and 40 acres (16 ha) in South Carolina. Tremendous numbers of P. clarkii live in California rice fields, but the crayfish neither are intentionally cultured nor utilized commercially. Mississippi has expressed keen interest in culturing P. clarkii for food. P. clarkii and P. a. acutus probably could be cultured in states where they occur naturally. P. clarkii occurs naturally in 16 states, and has been introduced into others such as South Carolina. P. a. acutus is found naturally in 30 states. Europeans culture P. clarkii, and possible P. a. acutus, for food in warm climates such as Southern Spain. They also are harvested for fish bait. In most cases, the crayfish are a by-product from minnow and catfish fingerling ponds. Enterprises in Arkansas, Mississippi, and California sell Procambarus spp. as bait, and some businesses in Arkansas and Mississippi sell the crayfish as food. In Louisiana, California, and perhaps other states, P. clarkii are sold to biological supply houses.

Orconectes culture occurs on a low scale, primarily for fish bait. These crayfish live throughout the mid-upper Mississippi River Valley, but the industry is too diffuse to quantify. Wisconsin harvests approximately 50,000 pounds (23 metric tons) annually. Missouri produces approximately 72,000 pounds (33 metric tons) annually.

P. leniusculus, found on the West Coast, is not cultured currently, although a captive fishery exists with food markets in Europe. There is a great deal of interest in culture of this species for food in France, Sweden, Finland, Austria, West Germany and elsewhere in Europe. Based on successful culture in Europe, this crayfish may be grown in the U.S. to satisfy rapidly expanding domestic and foreign markets.

Potential for Development

The potential for expansion and development is excellent. In Louisiana, aquaculture acreage increased from 6,000 acres (2,400 ha) in 1966, to 18,000 acres (7,200 ha) in 1970, to 45,000 acres (18,000 ha) in 1977, to approximately 50,000 acres (20,000 ha) in 1979. These figures include wooded, open, and rice-field ponds. Texas, a relative newcomer to crayfish culture, already has 3,000 acres (1,214 ha) in production, and an estimated 300,000 acres (121,408 ha) of marginal lands (generally not suitable for agriculture) available for aquaculture. Approximately the same acreage is available for expansion in Louisiana. Marginal lands are available for crayfish culture in other states, and surveys of such areas need to be conducted.

A major trend in crayfish culture involves double-cropping of rice and crayfish. Soybeans also may be rotated with crayfish. Louisiana grows 500,000 to 600,000 acres (220,000-240,000 ha) of rice. If 10% of this land were used for double-cropping, acreage devoted to crayfish raising would increase dramatically. Arkansas, with approximately 850,000 acres (340,000 ha) of rice, has a large potential for double-cropping rice and crayfish. Approximately 150,000 acres (60,000 ha) in Mississippi, and 500,000 acres (200,000 ha) in Texas could be used for this purpose. Rice acreage in California is significant, as is the potential for double-cropping. The U.S. grows approximately 2.5 to 2.8 million acres (1.01-1.13 million ha) of rice on farms that possess the major ingredients for growing crayfish--water, levees, and a source of food (decaying rice hay).

P. clarkii are cultured in closed systems by two commercial ventures -- Monterey Bay Hydroculture Lab and the Newberry Crayfish Hatchery. Successful culture of this species in tanks gives crayfish culture an extra dimension. Newly-hatched individuals cultured on rice stubble grow to harvestable size about as well as crayfish fed a special crustacean diet. These preliminary results underscore the fact that crayfish do not have larval stages that require special foods such as Artemia. The results also demonstrate that crayfish (detrital feeders) can be successfully grown on agricultural wastes and aquatic vegetation.

Crayfish reproduce in tanks or ponds, and obtaining young is no problem. P. clarkii thrives in brackish water up to 8 o/oo salinity; therefore, its aquacultural range can be extended into coastal regions. The market potential for crayfish as food and bait seems excellent, but marketing studies and promotional efforts are required. This is discussed elsewhere in the plan.

In 1978-79 Louisiana crayfish-farmers received an average of 52 cents a pound ($1.14/kg) live-weight, and Texas farmers received 92 cents a pound ($2.20/kg) live-weight. Pacifastacus brought more than 85 cents a pound ($1.87/kg) to commercial trappers. Prices obtained by farmers have provided incentive for expansion of acreage. Prices paid for crayfish as fish bait vary, but they may exceed $1 per dozen, retail, in some areas.

In 1978, more than 25 states and 14 countries requested information about crayfish culture from Louisiana State University. Knowledge generated by research should have wide application throughout the U.S. A total of 300 species and subspecies occur in North America, and virtually every state has some potential to culture native crayfish for food, bait, or other puposes.

At the 2nd International Crayfish Symposium, held in 1974 in Louisiana, over 20 foreign countries were represented, and at least six countries reported on crayfish-culture efforts. Most dealt with P. leniusculus. California will host the 5th International Symposium in 1981.

Crayfish constitute the most important freshwater fishery in Sweden and Finland. In August, 1978, crayfish averaged $3 each in European restaurants. However, the market is unstable, and Europeans are highly exacting about the type product that they require.

Research Requirements

The Plan includes a review of research priorities. Attention was focused on established production of P. clarkii as food. The potential for culture of Pacifastacus spp. and Orconectes spp. was clearly identified but is in the earlier stages. Marketing studies and harvesting research were listed as first priority. Although markets are established in south Louisiana, little information exists on how crayfish as food will be accepted elsewhere. Prerequisite for expansion of the industry are marketing studies. Currently crayfish farmers spend 40 to 60% of their gross income on harvesting crayfish. Typically, traps baited with fish are emptied of crayfish daily. Bait, traps, and labor are expensive. Many farmers lease trapping rights to professionals for a percentage of the gross. Ideally, crayfish need to be harvested mechanically without the use of traps or bait.

Next in order of priority are studies needed on economics, water quality, alternating crayfish and land crops, and forages and feed. Only two published reports exist on crayfish production costs and returns and both were based on interviews. Reliable figures are required to aid crayfish farmers in making management decisions and to provide entrepreneurs with investment data. Water quality management is paramount to production of crayfish crops. Maintaining adequate ($\geq$ 3 ppm) levels of dissolved oxygen, particularly at fall flooding, is an absolute necessity. Perhaps the greatest potential for expansion of crayfish farming is with multiple cropping of rice and crayfish. Fundamental techniques have been developed by Louisiana State University. Major research is still required on the proper use of pesticides in growing rice without harm to crayfish. The need for good feed (forage) delivery system is of major importance. Planted rice (grain not harvested) is currently the best forage for crayfish culture, but this plant decomposes and is eaten before the crayfish growing season is over. Other forage plants with differing decomposition rates must be found that are acceptable to crayfish.

Listed in the third category of priority are studies on stocking and population dynamics, genetics, engineering (pond construction, mechanical aeration, etc.), diseases, seed production, general biology, growth and behavior, effluent control, and processing. Last in priority are studies in predation and mortality, product quality control, introduction of nonindigenous species, and use of new crayfish species for aquaculture.

Facilities

Louisiana State University has an in-place crayfish research facility consisting of 150 ponds and comprising 38 ha. As interest in other states grows, new facilities should be developed.

Technology Transfer

Technology transfer should be accomplished by utilizing the in-place Cooperative Extension Service at Land Grant Universities. Short courses for extension agents and crayfish farmers would enhance technology transfer. The Plan also calls for workshops to be held for researchers at LSU in order to exchange ideas, research findings, and to evaluate priorities. As crayfish farming spreads, workshops should be hosted by other states. Finally, Information Center(s) need to be developed for a two-fold purpose. First, data pertaining to crayfish as a farm commodity should be coordinated by the USDA's in-place program for other crops. Second, data on published information of a scientific nature should be catalogued and programmed at regional centers. LSU has already embarked on development of such a center.

LITERATURE CITED

Avault, J.W., Jr. Compiler. 1980. Crawfish species plan. In the National Aquaculture Plan. Joint Subcommittee on Aquaculture U.S. Department of Commerce, NOAA.

CRAYFISH RESEARCH AND INDUSTRY ACTIVITIES IN AUSTRALIA, NEW GUINEA AND NEW ZEALAND

N. M. Morrissy
Western Australial Marine Research Laboratories
(State Department of Fisheries and Wildlife)
P.O. Box 20 North Beach 6020, Western Australia

ABSTRACT

The upsurge in interest in parastacology in Australia, New Guinea and New Zealand over the past decade or so is documented for kindred astacologists. The species receiving most attention are the eastern Australian yabbie, Cherax destructor, and the south-western Australian marron, C. tenuimanus, because of their aquaculture potential.

INTRODUCTION

Australia is the bastion of the parastacid crayfish fauna of the Southern Hemisphere since it has the greatest diversity of genera (10) and species (approaching 100) (Riek 1969, 1972). Madagascar has a single genus and species, South America has 8 species in 2 genera, New Zealand a single genus (Paranephrops) with two species (Hopkins 1970), and New Guinea and associated islands about 12 species in the single genus Cherax (Holthuis 1949, 1950). This classic, southern distribution is common to many groups of Australian plants and animals, supporting the now well-established theory of a former, single landmass (Gondwanaland) fragmenting under continental drift (Francois 1967). To the north, Australia has been protected from invasion by the oriental fauna moving down from the Malay Archipelago by deep oceanic trenches (Holthuis 1949).

Academic and entrepreneurial interest in Australian crayfish remained at a low level until a decade or so ago. Anon (1975) lists the scattered older references to, mainly, taxonomic and natural history observations. The latter-day upsurge in attention has accompanied the general worldwide acknowledgement of the promise of aquaculture, the lure of the well-publicized Scandanavian market for imported crayfish, and local demand from the sophisticated restaurant trade in the few major coastal cities, which house the majority of the Australian population. Maclean (1975) reviewed the emerging, and Morrissy (1980b) the more recent, aquaculture scene including crayfish. Morrissy (1978a) described the growth of commercialization of crayfish in Australia and the long ago comments of the early European crayfish taxonomists (Huxley, McCoy, and Smith) as to the gourmet qualities and possibility for aquaculture of the prominent species.

New Guinea

Holthuis (loc. cit., 1958) described a major capture fishery for crayfish in the Wissel Lakes District of West Irian where crayfish are an important source of animal protein for the native population (Fig. 1). More recently, expatriate fisheries officers at Daru in Papua New Guinea have encouraged the collection of crayfish from the Fly River area for export to Europe (Anon 1978, Dr. Ray Moore, pers. comm.).

New Zealand

There is little exploitation of the native crayfishes (Koura) since fishing rights are confined to the Maori population, for example in the Lake Taupo area. Research has been limited (Hopkins 1966, 1967a, b, Wong and Freeman 1976, Jones 1981a, b) and government research on aquaculture feasibility recently ceased (Dr. Brian Jones, pers. comm.).

Tasmania

Astacopsis gouldi, the largest crayfish in the world, occurs in streams of north-west Tasmania. Exploitation is confined to regulated sport fishing, periodically mentioned in Annual Reports of the Inland Fisheries Commission, Hobart, and the species cannot be considered to have any potential for aquaculture (Lynch 1967, 1970), there being a ban on export to conserve this oddity.

Considerable research has come from the Zoology Department of the University of Tasmania, Hobart, on Parastacoides and Engaeus stimulated by conservation issues stemming from damming of wilderness catchments for hydro-electric power. Dr. P. S. (Sam) Lake was an initiator of this work which as been continued by Dr. A. A. Richardson (e.g., Lake and Newcombe 1975, Mills and Lake 1975, Newcombe 1975, Suter 1977, Ritchie 1978, Sumner 1978, Richardson and Swain 1980).

Mainland Australia

Crayfish species occupy most of the eastern half of Australia, ranging from the vast central arid region to the northern tropics and down the eastern seaboard, and the Great Dividing Range to subalpine zones in the south-east. Except for the temperate extreme south-west, crayfish are absent from the western half, most of which is extremely arid; the large euryhaline Macrobrachium rosenbergii occurs in the tropical north-west (Kimberleys). The most diverse genera in Australia are Cherax, Euastacus and Engaeus. Most interest is centered upon the eastern yabbie, C. destructor, and the south-western marron, C. tenuimanus, with perhaps the large Euastacus armatus, because of their aquaculture potential. C. destructor is a thinly-shelled apomorphic type with a highly developed ability to burrow and the capability of repeated breeding, given favourable temperatures, the year around. Over most of its distribution in the inland, there are long periods of drought followed by sudden large-scale flooding. In the parlance of ecosystem strategies, C. destructor is an "r-selection" species characteristic of relatively impermanent, and often highly eutrophied, still waters, although with the present species description its range extends into subalpine permanent, and often torrential, waters. C. tenuimanus and E. armatus, on the other hand, are heavily shelled plesiomorphic types characteristic of larger, permanent rivers. C. tenuimanus, at least, is a "K-selection" species with large, stable populations in low nutrient waters, little burrowing ability, a long annual period of ovary development, with a single springtime breeding season, and a propensity for breeding failure in eutrophied waters.

However, C. tenuimanus, in contrast to the other, still larger, plesiomorphic species, E. armatus and A. gouldi, and C. destructor, has, in commercial processing terms, a high "tail recovery rate."

535

Edgar Riek, who combined epicurean predilictions with his Australia-wide sampling of crayfish for his monumental taxonomic works, described C. tenuimanus as the finest flavored crustacean he had ever tasted.

Euastacus armatus (Murray River crayfish)

Australia's major river, the Murray, bordering N.S.W. and Victoria in its upper reaches, has long been fished commercially for the Murray River crayfish. Paul O'Connor at the N.S.W. State Fisheries Inland Research Station at Narranderra on the Murrumbidgee, a larger tributary, has commenced study of this popular capture fishery.

Cherax destructor (yabbie)

C. destructor occurs over five States of Australia (Fig. 1) and it, or a closely related type (C. albidus), has, unfortunately, been introduced into the south-west of W.A. There is obviously a question of species definition, at least in terms of physiological tolerances, arising from a consideration of this enormous range of distance, climate and habitat.

A small trap fishery for C. destructor on the lower Murray River and its lakes boomed in the late sixties under the spur of the sudden demand from Sweden for foreign crayfish. The annual catch rose from 30 tonnes in 1968 to 273 tonnes in 1974, but then declined rapidly to less than 1 tonne in 1979, for reasons not yet known to the South Australian State Fisheries Department (Olsen 1981). Over the same period, a number of hardy entrepreneurs, seeking to supply the same, supposedly lucrative, overseas market, explored the vast inland areas of Victoria, and more especially N.S.W. and Queensland, seeking the boom populations of yabbies following floods. Neither yabbies, nor the other species of foreign crayfish (from Turkey, Iran, etc.) have realized the high price per kilogram paid by Swedes for their native "Kraftor". And there have been problems with the blotchy appearance of cooked yabbies, the inferior dill variety available in Australia for flavouring and the difficulty of guaranteeing a large supply for the Swedish crayfish festival.

While other entrepreneurs have turned to aquaculture, the potentially enormous capture fishery for inland stocks of yabbies remains largely unexplored commercially and a fascinating, though arduous, research topic.

Possibly the first commercial crayfish farm in Australia was operated in Victoria near Kyneton during the early seventies by Ian L. Carstairs, and Dr. Jack Frost was experimenting with yabbie culture techniques near Leeton on the Murrumbidgee (Frost 1975). Notable yabbie farms were also established by Bob Andrews near Deniliquin, N.S.W. on the Murray and by Ed Davis near Bordertown in the upper south-east of South Australia. However, to this date, while such endeavours are extremely common now from Queensland through to South Australia, there has been no report of such a venture being successful in a real sense in biological or economic terms. Successful commercial ventures, such as "Yabbie City" - a restaruant on the lower Murray River near Goolwa in South Australia, are based upon fishing of wild yabbies.

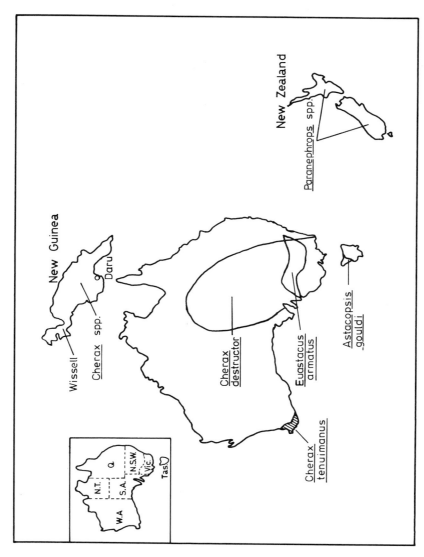

Figure 1. Distribution of prominent Australian parastacids.

537

The Inland Fish Farmers Association of Australia (Registered Office: RMB 626 Sturt Highway, Wagga Wagga N.S.W. 2650) has a policy of lobbying government for research and industry protection. The President, Pat P. McLaren, operates Murray Cod Hatcheries of Australia at Wagga Wagga, and consults on yabbie culture. The Association has produced a high quality biannual journal, "Australian Aquaculture," since 1977 when the cover of the first issue featured a yabbie.

Peter N. Carroll, a lecturer at the Animal Production Centre of the Hawkesbury Agricultural College, Richmond, N.S.W. 2753, has established an annual (January) freshwater crayfish school -- the third was held in 1981 -- where talks are given by fish farmers, professional fishermen, fish retailers and scientists. The annual general meeting of the recently formed Australian Freshwater Crayfish Association is held at this school. H. T. (Tom) Johnston, who formerly researched yabbies at the Narranderra Inland Fisheries Research Station, and others have provided information review papers. In addition to this extension work, there are pond research facilities at Hawkesbury. A summary is given by Carroll (1981).

A private advisory and news letter service also operates in Queensland for various species (Freshwater Australian Crayfish Traders, 42 Tasman Street, Stafford, Brisbane Qld 4053).

The first notable population study on yabbies was carried out in farm dams by Dr. D. J. Woodland at the University of New England, Armidale N.S.W. (Woodland 1967).

Dr. Stan J. Edmonds of the Zoology Department, University of Adelaide, supervised some of the early physiological and behavioural research on yabbies (Fradd 1974, Coombe 1976, Lewis 1976). More recently under Professor W. (Bill) D. Williams (formerly at Monash University, Victoria, with Dr. Ian A. E. Bayley, the fathers of Australian limnology), Dr. Mike C. Geddes and Dr. Keith F. Walker have continued this work (Kaines 1979, Aerfeldt 1980, Mills and Geddes 1980).

Dr. Brian J. Mills, of the South Australian State Fisheries, has been conducting intensive research on yabbies with an aquacultural intent and publication has commenced (loc. cit.).

Dr. Peter Greenaway has supervised a recent study of "boom and bust" yabbie populations in farm dams at the University of N.S.W., Fowlers Gap Research Station near Broken Hill (Reynolds 1980).

Dr. Sam Lake, now at Monash University, Victoria, has transferred his crayfish interest to yabbies (Kretser 1979).

Australian crayfish are known to be highly susceptible to Aphanomyces astaci (Unestam 1975) which has not been advertently introduced so far on any, now illegal, imports of foreign crayfish. However, there is some concern over microsporidians (Carstairs 1979, Jones 1980). Dr. John Copland, of the Regional Veterinary Laboratory of the Victorian Department of Agriculture at Benalla, has taken up this problem.

Yabbies have also been increasingly utilized as research animals for pure research of a biochemical or neurophysiological nature (Leonard and MacDonald 1963, McKay and Jenkin 1970, McKay, Jenkin and Tyson 1973, Tyson and Jenkin 1973, Fellows and Hird 1979a,b, Jeffrey 1979, Ache and Sanderman 1980).

Cherax tenuimanus (Marron)

Besides the tropical north-west, the temperate extreme south-west (roughly delineated by a line from Perth to Albany) is the only other part of Western Australia receiving a consistent annual rainfall and having permanently flowing rivers, with highly unusual topography (Morrissy 1974a, 1979a) of any significance. The diversity of fish fauna is very low and it is not surprising that, apart from a nocturnal eel-tailed catfish and sparsely distributed introduced trout, the large and abundant marron is the basis for a highly esteemed sport fishery (Morrissy 1978a). The State Department of Fisheries and Wildlife licenses amateur fishermen and has monitored catch and effort by logbooks since 1971. The original natural distribution of marron has been considerably modified by man (Morrissy 1978b).

Wild stocks of marron in public fishing waters have been protected from commercial use since 1955. But sale of "domesticated" marron from privately owned dams and ponds has been permitted under license since 1976. The history and rationale of this development is described by Morrissy (1978a). Currently there are sixteen licensed marron farms of which seven actively sell young stock (locally, interstate and overseas) and larger sizes (minimum legal size 120 g) to Perth restaurants. Marron farmers may be contacted through the Marron Growers Association of W.A. via the Department of Fisheries and Wildlife, P.O. Box 20, North Beach 6020, W.A.

Research on marron has been carried out since 1967 by the Fisheries Research Branch of the State Department based at the Waterman (North Beach) Laboratories near Perth and with pond and laboratory facilities at the Pemberton Fish Hatchery, 300 km south of Perth.

Apart from the amateur fishery, three levels of aquaculture have been considered: 1) Use of existing farm dams, providing a low yield (300-400 kg/ha) but with little management and no feeding, has been satisfactorily explored (Morrissy 1970a, 1974b, 1980a). 2) A high yield (3000 kg/ha) can be expected from specially constructed ponds in areas with warmer winters and by employing intensive management and heavy feeding (Morrissy 1976a, 1976b, 1979b). In practice this high yield based upon data from small experimental ponds has been rarely approached in large ponds because of the difficulty of water quality management with the present food types employed. Economic feasibility is highly dependent upon water supply for pond flushing during the dry summers (Morrissy 1979c). 3) Culture of individually housed marron (superintensive battery culture) is being explored in cooperation with the Commonwealth Department of Science and Technology, which has acquired the development rights to a number of battery principles patented by Western Australian, Peter Brinkworth. The requirements of this project have initiated a major study for a suitable food for marron culture.

Satisfying results have been obtained on the general problem of estimating the absolute numbers of crayfish in dams and ponds by sampling (Morrissy 1973, 1975, 1981). Spawning of marron has received attention (Morrissy 1970b, 1974c, 1976b). And the inverse relationship between the wide ranges of growth rate and density has been resolved (Morrissy 1974c, 1980a).

Dr. Brenton Knott of the Zoology Department, University of W.A., has supervised an important study clarifying the taxonomic status of Cherax spp. in the south-west by employing the electrophoretic technique (Austin 1979). Chris Austin is continuing this approach on the other Australian Cherax, and on variation within C. tenuimanus, for a Ph.D.

LITERATURE CITED

Ache, B. W. and D. C. Sanderman. 1980. Olfactory-introduced central neutral activity in the Murray crayfish, Eucastacus armatus. J. Comp. Physiol. 140:295-301.

Aerfeldt, P. J. 1980. Growth of the Yabby, Cherax destructor Clark, under high density conditions. B.Sc. Honours Thesis, University of Adelaide.

Anon. 1975. Bibliography of Australian Fisheries. Aust. Fish. Pap. No. 12. Aust. Gov. Publ. Serv., Canberra.

Anon. 1978. Fisheries Research Annual Report for 1976. Fisheries Division, Department of Primary Industry, Port Moresby.

Austin, C. M. 1979. Biochemical systematics of the genus Cherax (Decapoda: Parastacidae). B.Sc. Honours Thesis, University of Western Australia.

Carroll. P. N. 1981. Aquaculturists' enthusiasm for yabbies highlights potential beyond the problems. Aust. Fisheries 40(6):23-31.

Carstairs, I. L. 1979. Report of microsporidial infestation of the freshwater crayfish, Cherax destructor. Freshwater Crayfish 4:343-347.

Coombe, D. R. 1976. Studies of aggression in the yabbie Cherax destructor. B.Sc. Honours Thesis, University of Adelaide.

Devcich, A. A. 1979. An ecological study of Paranephrops planifrons White. (Decapoda Parastacidae) in Lake Rotoiti, North Island. Ph.D. Thesis, University of Waikato, N.S.

Fellows, F. C. I. and F. J. R. Hird. 1979a. Nitrogen metabolism and excretion in the freshwater crayfish Cherax destructor. Comp. Biochem. Physiol. 64B:235-238.

Fellows, F. C. I. and F. J. R. Hird. 1979b. From whence comes ammonia in crustaceans? P. Aust. Biol. 12:37 (abstract).

Fradd, P. J. 1974. The effect of temperature on the respiration of the yabbie, Cherax destructor. B.Sc. Honours Thesis, University of Adelaide, South Australia.

Francois, D. D. 1967. Freshwater crayfishes. In: D. F. McMichael (ed.) A Treasury of Australian Wildlife. Ure Smith, Sydney, London.

Frost, J. V. 1975. Australia crayfish. Freshwater Crayfish 2:87-95.

Holthius, L. B. 1949. Decapoda Macrura, with a revision of the New Guinea Parastacidae. Zoological results of the Dutch New Guinea expedition 1939. No. 3. Nova Guinea (n.s.) 5:289-328.

Holthius, L. B. 1950. The Crustacea Decapoda Macrura collected by the Archbold New Guinea Expeditions. Am. Mus. Novit. 1961:1-17.

Holthius, L. B. 1958. Freshwater crayfish in Netherlands New Guinea Mountains. S.P.C. Quarterly Bulletin April 1958.

Hopkins, C. L. 1966. Growth in the freshwater crayfish, Paranephrops planifrons White. N.Z. J. Sci. 9:50-56.

Hopkins, C. L. 1967a. Breeding in the freshwater crayfish Paranephrops planifrons White. N.Z. J. Mar. Freshwater Res. 1:51-58.

Hopkins, C. L. 1967b. Growth rate in a population of the freshwater crayfish, Paranephrops planifrons White. N.Z. J. Mar Freshwater Res. 1:464-474.

Hopkins, C. L. 1970. Systematics of the New Zealand Freshwater crayfish (Paranephrops) (Crustacea: Decapoda: Parastacidae). N.Z. J. Mar. Freshwater Res. 4:278-291.

Jeffrey, P. D. 1979. ; Hemocyanin from the Australian freshwater crayfish Cherax destructor -- electron microscopy of native and reassembled molecules. Biochem 18:2508-2513.

Jones, J. B. 1980. Freshwater crayfish Paranephrops planifrons infected with the microsporidian Thelohania. N.Z. J. Mar. Freshwater Res. 14:45-46.

Jones, J. B. 1981a. Growth of two species of freshwater crayfish (Paranephrops spp.) in New Zealand. N.Z. J. Mar. Freshwater Res. 15:15-20.

Jones, J. B. 1981b. The aquaculture potential of New Zealand freshwater crayfish. N.Z. Agricultural Science 15:21-23.

Kaines, R. J. 1979. Osmoregulation and respiration in the freshwater crayfish Euastacus armatus (Von Martens) and Cherax destructor Clark. B.Sc. Honours Thesis, University of Adelaide.

Kretser, E. Dane de. 1979. Aspects of the population ecology of the yabbie (Cherax destructor) in two Victorian farm dams. B.Sc. Honours Thesis, Monash University, Victoria.

Lake, P. S. and K. J. Newcombe. 1975. Observations on the ecology of the crayfish Parastacoides tasmanicus (Decapoda: Parastacidae) from south-western Tasmania. Aust. J. Zool. 18:197-214.

Leonard, G. J. and K. MacDonald. 1963. Homarine (N-methyl picolinic acid) in muscles of some Australian Crustacea. Nature 200(4901):78.

Lewis, R. B. 1976. Aspects of the life history, growth, respiration and feeding efficiencies of the yabbie Cherax destructor Clark with regard to potential for aquaculture. B.Sc. Honours Thesis, University of Adelaide, South Australia.

Lynch, D. D. 1967. Synopsis of biological data on Astacopsis gouldi. Australian/N.Z. Meeting on Decapod Crustacea, Sydney, October 24-28, 1967.

Lynch, D. D. 1970. The giant freshwater Crayfish of Tasmania, Astacopsis gouldi. Bull. Aust. Soc. Limnol. 2:20-21.

Maclean, J. L. 1975. The potential of aquaculture in Australia. Aust. Fish. Pap. No. 21. Aust. Gov. Publ. Serv., Canberra.

McKay, D. and C. R. Jenkin. 1970. Immunity in the invertebrates. Correlation of phagocytic activity of haemocytes with resistance to infection in the crayfish (Parachaeraps bicarinatus)* to infection. J. Infect. Dis. 128: S165-169.

Mills, B. J. and M. C. Geddes. 1980. Salinity tolerance and osmoregulation of the Australian freshwater crayfish Cherax destructor Clark (Decapoda: Parastacidae). Aust. J. Mar. Freshwater Res. 31:667-676.

Mills, B. J. and P. S. Lake. 1975. Setal development and moult staging in the crayfish Parastacoides tasmanicus. Erichson. Aust. J. Mar. Freshwater Res. 26:103-107.

Morrissy, N. M. 1970a. Report on marron in farm dams. Fish. Rept. West. Aust. 5:1-34.

Morrissy, N. M. 1970b. Spawning of marron, Cherax tenuimanus (Smith) (Decapoda: Parastacidae) in Western Australia. Fish. Bull. West. Aust. 10:1-23.

Morrissy, N. M. 1973. Normal (Gaussian) response of juvenile marron Cherax tenuimanus (Smith) (Decapoda: Parastacidae), to capture by baited sampling units. Aust. J. Mar. Freshwater Res. 24:183-195.

Morrissy, N. M. 1974a. Reversed longitudinal salinity profile of a major river in the south-west of Western Australia. Aust. J. Mar. Freshwater Res. 25:327-335.

———

*Parachaeraps bicarinatus = Cherax destructor

Morrissy, N. M. 1974b. The ecology of marron Cherax tenuimanus (Smith) introduced into some farm dams near Boscabel in the Great Southern area of the Wheatbelt Region of Western Australia. Fish. Bull. West. Aust. 12:1-55.

Morrissy, N. M. 1974c. Spawning variation and its relationship to growth rate and density in the marron Cherax tenuimanus (Smith). Fish. Res. Bull. West. Aust. 16:1-32.

Morrissy, N. M. 1975. The influence of sampling intensity on the catchability of marron, Cherax tenuimanus (Smith) (Decapoda: Parastacidae). Aust. J. Mar. Freshater Res. 26:47-73.

Morrissy, N. M. 1976a. Aquaculture of marron, Part 1. Site selection and the potential of marron for aquaculture. Fish. Res. Bull. West. Aust. 17:Pt. 1, 1-27.

Morrissy, N. M. 1976b. Aquaculture of marron. Part 2. Breeding and early rearing. Fish. Res. Bull. West. Aust. 17:Pt. 2, 1-32.

Morrissy, N. M. 1978a. The amateur marron fishery in south-western Australia. Fish. Res. Bull. West. Aust. 21:1-44.

Morrissy, N. M. 1978b. The past and present distribution of marron in Western Australia. Fish. Res. Bull. West. Aust. 22:1-38.

Morrissy, N. M. 1979a. Inland (non-estuarine) halocline formation in a Western Australian River. Aust. J. Mar. Freshwater Res. 19:343-353.

Morrissy, N. M. 1979b. Experimental pond production of marron, Cherax tenuimanus (Smith) (Decapoda: Parastacidae). Aquaculture 16:319-244.

Morrissy, N. M. and R. R. House. 1979c. Economic feasibility of intensive outdoor pond culture of freshwater crayfish in Australia. 38 pp. + figures. (Held in Dept. of Fisheries and Wildlife Library).

Morrissy, N. M. 1980a. Production of marron in W.A. farm dams. Fish. Res. Bull. West. Aust. 24:1-80.

Morrissy, N. M. 1980b. Aquaculture, Chapt. 21. In: W. D. Williams (ed.) An Ecological Basis for Water Resource Management. ANU Press, Canberra.

Morrissy, N. M. and N. Caputi. 1981. Use of catchability equations for population estimation of marron, Cherax tenuimanus (Smith) (Parastacidae). Aust. J. Mar. Freshwater Res. 32:213-225.

Newcombe, K. J. 1975. The pH tolerance of the crayfish Parastacoides tasmani- cus (Erickson) (Decapoda, Parastacidae). Crustaceana 29:231-234.

Olsen, A. M. 1981. The decline in wild populations of yabbies. South Australian Department of Fisheries 5:25-27.

Reynolds, K. M. 1980. Aspects of the biology of the freshwater crayfish, Cherax destructor in farm dams in far-western N.S.W. M.Sc. Thesis, University of N.S.W.

Riek, E. F. 1969. The Australian freshwater crayfish (Crustacea: Decapoda: Parastacidae), with descriptions of new species. Aust. J. Zool. 17: 855-918.

Riek, E. F. 1972. The phylogeny of the Parastacidae (Crustacea: Astacoidea), and description of a new genus of Australian freshwater crayfishes. Aust. J. Zool. 20:369-389.

Ritchie, M. E. 1978. Circadian rhythms, and their measurement by microcomputer, in Geocherax gracilis. B.Sc. Honours Thesis, University of Tasmania.

Sumner, C. E. 1978. A revision of the genus Parastacoides Clark (Crustacea: Decapoda: Parastacidae). Aust. J. Zool. 26:809-821.

Suter, P. J. 1977. The biology of two species of Engaeus (Decapoda: Parastacidae) in Tasmania. III. Habitat, food, associated fauna and distribution. Aust. J. Mar. Freshwater Res. 28:95-103.

Tyson, C. J. and C. R. Jenkin. 1973. The importance of opsonic factors in the removal of bacteria from the circulation of the crayfish (Parachaeraps bicarinatus). Aust. J. Exp. Biol. Med. Sci. 51:609-615.

Unestam, T. 1975. Defense reactions and susceptibility of Australian and New Guinea freshwater crayfish to European-crayfish-plague fungus. Aust. J. Exp. Biol. Med. Sci. 53:349-359.

Wong, T. M. and R. F. H. Freeman. 1976. Osmotic and ionic regulation in different populations of the New Zealand freshwater crayfish Paranephrops zealandicus. J. Exp. Biol. 64:645-663.

Woodland, D. J. 1967. Population study of a freshwater crayfish, Cherax albidus Clark. Ph.D. Thesis, University of New England, N.S.W. (C. destructor).

EPIZOOTIOLOGY OF THE CRAYFISH PLAGUE (APHANOMYCOSIS) IN SPAIN

L. Cuellar and M. Coll
Ministerio de Agriculltura y Pesca
Icona Gran Via de San Francisco, 35
Madrid 5, Espana

In the past few years, the crayfish plague has shown a disturbing increase in incidence, causing moderate to severe mortality in different Spanish regions. The plague has decimated and even caused extinction of the crayfish populations in many rivers and water masses of the Iberian Peninsula.

The first observation of Aphanomycosis was reported in Duero River (Valladolid) in 1958. The origin and method of transmission was unknown. Seven years later, in 1965, another area of high mortality was detected in Ucero River (Soria), probably caused by infection from crayfish of the species Astacus astacus (Linn.) coming from Germany: Nevertheless, it must be noted that the German crayfish survived and bred for two years in the Pisciculture centre "Ucero" without any sign of abnormality.

In the last cases, the diagnosis of the "pest" couldn't be confirmed through laboratory analysis; the ethiologic agent couldn't be observed nor cultivated.

In the following years, several sporadic mortalities located in some water courses took place and always coincided with "stress" caused by deficient water conditions, water pollution, storm etc. One of these die-offs, located in Iregua River (Logrono), was checked in situ in August 1975 and September 1976, but the laboratory tests did not show the patogenic fungus.

During 1977 the crayfish of the species Astacus leptodactylus (Esch.) reared in Cabezon de Pisuerga Centre (Valladolid) suffered a total mortality and the culture was abandoned.

In April 1978 a great crayfish die-off appeared in Riaza River passing through Segovia and Burgos provinces. Research carried out by Icona's Veterinary Services and Ecopathology Laboratory allowed the first definate diagnosis of aphanomycosis based on the identification of the characteristic reticulated micellium of the fungus Aphanomyces astaci in the cuticle injuries of some specimens.

The origin of the fungi rests unknown, but the plague has very probably progressed from an unauthorized nursery (previously reported by the National Forest Guard) where Turkish crayfish of the species Astacus leptodactylus (Eschs.) were placed. These crayfish were probably introduced through France, since the importation of live crayfish of any origin or species was forbidden in Spain eight years ago.

During June 1978, with well known observation methods, the presence of the plague was confirmed in specimens from Guadiana River in Ciudad Real Province; the epizoa, spread later to other rivers in the same region.

In the last days of August 1978, the Spanish Delegation at the International Conference on Astacology (Thonon-les-Bains-France) reported the appearance of the Aphanomycosis in Spain. The spanish diagnosis was confirmed by Unestam through cuticle samples sent to Sweden.

From the end of spring to the beginning of autumn 1979, the agent causing the aphanomycosis extended to several regions of Spain. It was identified in provinces and rivers during the months pointed out in Table 1.

During the same year 1979, we were informed without specifications of the appearance of massive crayfish deaths in different rivers, probably due to mycotic pest.

In 1980, the Veterinarian Services of Icona detected and identified the different focus of Aphanomycosis shown in Table 2. We are sure that very few of the Provinces involved will escape from this pest, and we worry about the evolution of the epizootic wave during summer and autumn 1981. In almost all the affected Provinces, the permanence of healthy populations that took refuge in the small lagoon that remain in brooks and streams during summer low water and in the cold high water courses, was observed.

Given the data available, very poor prospects for our spanish Austrapotamobius pallipes populations can be deduced. The initial focus of the epizootic wave and its evolution are shown on the map of Spain (Figure 1).

The Ministery of Agriculture and Fisheries, worried by the problem and foreseeing the consequences that could result from the fast spreading of Aphanomycosis, has been taking different measures for four years. Among them the building of two Crayfish Repopulation Icona centers in Ciudad Real and Burgos Provinces, and another project in Guadalajara Province. These centers are dedicated to native crayfish culture (Austropotamobius pallipes), under controlled conditions (reproduction, feeding and health). On the other hand, different highly controlled attempts with species Pacifastacus leniusculus Dana, in several rivers of Soria, Quenca and Burgos Provinces are being undertaken.

In 1981, a series of official measures against pest have been put into practice; they consist of:

- Emergency quarantine between two and five years, once the Aphanomycosis diagnosis is confirmed.

- Crayfish and other ill or dead crustaceans must be destroyed by incineration or quicklime burial.

- Disinfection of boots, nets, lamps, tools, baskets, containers, capture and transportation equipments etc. ... in all fishing operations.

- It is recommended: a) Live baits not be used. b) Movement from one place to another during the same day should be avoided. c) Immediate cooking of captured crayfish.

Table 1. Places, rivers and months of pest appearances in 1979.

Province	River	Month
Vizcaya	Cadagua	June
Guipuzcoa	Leizaran y Araxes	June
Alava	Omecillo, Bairax y Ayuga	June
Navarra	Ega, Cidacos, Araquil, Leizaran, Elorza, Salazar, Erro y larrain	July and August
Valladolid	Pisuerga	July
Teruel	Guadalaviar, Jiloca y Alfambra	July
Zaragoza	Ebro and affluents	August
Logrono	Iregua	August and September

Table 2. Places and months of pest appearance in 1980.

Province	River	Month
Albacete	Jucar	May
Cuidad Real	Guadiana	May
Toledo	Guadarrama	May
Zamora	Duero	June
Palencia	Carrion	June
Leon	Bernesga	June
Valladolid	Eresma and Esgueva	June-July
Alava	Omecillo and Ayuga	July-August
Navarra	Ega, Cidacos and Erro	August-September
Logrono	Iregua	August-September
Segovia	Riaza and Duraton	August-September-October
Burgos	Riaza and Esgueva	September-October
Zaragoza	Duero and Jalon	September
Soria	Duero and Ucero	September

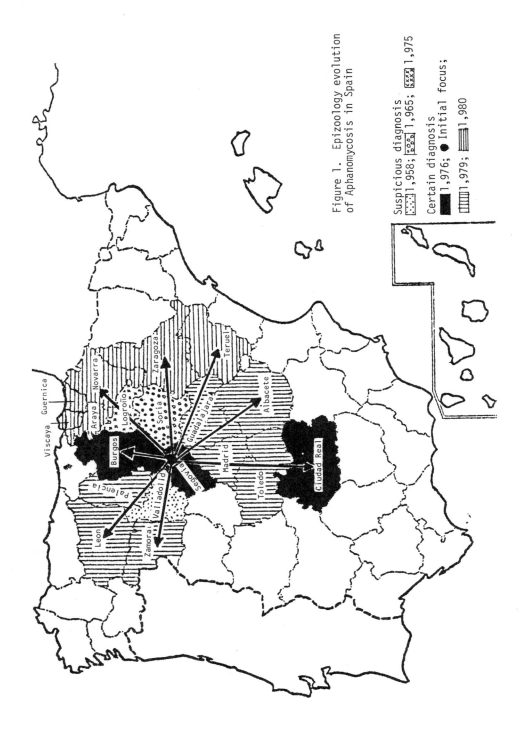

Figure 1. Epizoology evolution of Aphanomycosis in Spain

Suspicious diagnosis
1,958; 1,965; 1,975

Certain diagnosis
1,976; ● Initial focus;
1,979; 1,980

SOME OBSERVATIONS ON CRAWFISH FARMING IN SPAIN

Andres Salvador Habsburgo Lorena
Fuentemilanes 2
Madrid 35 Spain

Ground water table and burrows:

Water for flooding of rice fields is pumped up to approximately eight meters, with the topsoil several meters above the water table. From observations, it was noted that crawfish burrowing and the presence of chimneys depend on the consistency of the soil and depth of the water table. (See Figure 1). Where the water table is constant and not too deep, crawfish burrows with chimneys are found. In rice fields, crawfish may dig down to the hardpan soil. Here water is trapped, making conditions suitable for crawfish survival. If the water table is close to the surface, the chimney will be taller. When the water is drained quickly out of the rice field and the water table is deep, the crawfish will dig to reach the ground water and will not construct a chimney.

Burrowing activity:

If we calculate the cubic centimeters of the tube made by the crawfish through the soil leading from the soil surface to the refuge, we will find, depending on the depth, that a large amount of soil material was displaced by the crawfish. As all this soil is never needed for the construction of the chimney nor is this soil material found nearby, the conclusion is that the crawfish is not emptying all the soil but displacing it by pushing it aside. Whether this activity is carried out by using the chelas or using the uropods has not been observed. It was noted that the soil along the duct has a harder consistancy than the soil elsewhere. Also it was observed that the inner side of the duct was full of "footprints" of crawfish which may be a way of compacting the soil which has been removed.

A method of protection against damages by crawfish burrows:

The burrowing habits of crawfish may cause damage in certain situtation. One situation where it may occur with certain frequency is along tubes, such as along pipes passing through a levee or under a road. The holes may weaken the dam, and a heavily loaded truck will cause the levee to collapse. The protection developed to prevent such damages is as follows. Before the installation of such a tube or pipe, two ditches are dug with a back hoe vertically and at a right angle to the projected pipe or tube. These ditches have to be as deep as the deeper water level. Then each ditch is filled with sand, and the tube or pipe is installed and also surrounded with sand. The ditches are then covered with whatever material is desired.

These two sand cores protect the tube against the burrowing activity of crawfish. When a crawfish tries to dig through the levee he reaches the first sand core and further digging is of no avail, so the crawfish cannot advance more. As sand will not avoid seepage there must be an inner core of impermeable materials to give protection aginst water seepage. The second sand core serves as protection

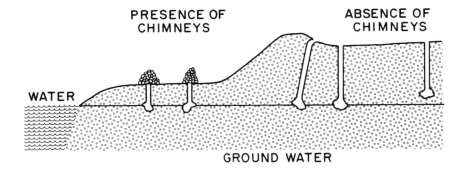

Figure 1.

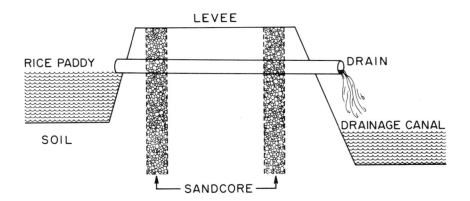

Figure 2.

550

against crawfish trying to dig holes from the other side of the levee. (See Figure 2.)

Social Behaviour

When the weather is suitable in the spring, rice farmers prepare the soil for planting. During plowing it was observed that in some parts of the rice field the concentration of crawfish burrows is much higher than in other parts of the same rice field. Two burrows per square meter versus twelve burrows per square meter were noted. It was not determined if crawfish were present in the burrows. The soil in the rice field is homogeneous, so the factor causing the different burrow concentration can only be a longer retention of water after draining the paddy or a higher oxygenation of the water due to less or more rice stalks in this part of the rice field.

Herd movement:

During the first autumn rains when it is still warm and humid, a massive herdlike emigration of crawfish has been observed. The emigration is so pronounced that roads often become covered with the crushed bodies of crawfish. This phenomenon is of short duration. It was not noted if crawfish involved in this were of a specific age or sex. This phenomenon causes some people a great deal of amusement and others a great deal of worry.

Eel and crawfish:

In the province of Cadiz there is a lagoon which normally floods annually. During flooding elvers (young eels) come up the lagoon. Due to increased pollution and drought, elvers could not reach the lagoon in recent years. so fishermen, who have leased the waterbody, stocked elvers and crawfish. The lagoon has a very hard substrate, and crawfish could not burrow. In the fall when fishermen began trapping, they noticed that both, eels and crawfish had fared very well. The eel preyed on the excess crawfish young. By thinning down the number of small crawfishs, the remaining crawfish had an ample food supply and grew well and stunting of crawfish did no occur.

Crawfish quality:

In an industrial testing trial, one kilo of Procambarus clarkii gave the same intensity of taste as four kilos of Astacus leptodactylus. This characteristic makes it possible for the red crawfish to be highly accepted and consequently the same high prices are paid for all crayfish species in Europe, with exception of the Pacifastacus leniusculus because of its larger size and handsome appearance.

SOCIOECONOMIC ASPECTS OF THE CRAWFISH INDUSTRY IN SPAIN

Andres Salvador Habsburgo Lorena
Fuentemilanos 2
Madrid 35 Spain

Two events took place in 1972 which are of importance for our meeting today and for this paper. One was that in Hinterthal, Austria some of the persons present here met for the first time. There they spoke about the need of relating and interchanging ideas about crawfish as an important part of aquaculture. The other important event of this year for us, was the publication of the book titled "Aquaculture" written by John E. Bardach (1972). This book was nominated "as the best Science Book" in 1973. In this book Bardach writes, "In fact, the nearly worldwide neglect of crayfish by aquaculturists is amazing. Nevertheless, there is money to be made, with hard work, from crayfish farming, and it would seen that crayfish are overdue to assume an important role in aquaculture."

Well, I am pleased to inform you that both events of that year have had a positive influence on our activity and that the efforts have produced rewarding results.

In recent years, Spain has become more and more aware of the potential of aquaculture. The deteriorating value of money through inflation and the reduction of the commercial fisheries due to political factors have all been reasons for enhancement of aquaculture. Moreover, the scarcity of suitable land, the cost involved in construction of ponds, the need to obtain an adequate supply of water are all factors to consider.

The introduction of Procambarus clarkii to southern Spain is of economic importance. In late 1980 officials of the Spanish Central Government, the fishermen's union, the local regional government, rice farmers, lawyers, food health personnel, and businessmen met in Sevilla. Fourteen points were established and approved regarding the incipient crawfish industry. The first point recognized the economic importance of crawfish. Meanwhile interest in crawfish spread to the provinces of Huelva and Cadiz, since both rice fields and natural waters had potential to grow crawfish.

The increasing pollution of the Guadalquivir River near Sevilla, has resulted in a decline of natural fisheries. Sturgeon no longer entered the river, and a station for extracting eggs had to close. Then during the 1940s eel fishermen had prosperous times. Four or five eel-buying stations were established, and the eel fishermen formed a union. Until the 1960s there was booming harvesting and exporting business, since virtually no eels are consumed in Spain. Meanwhile rice farming, which began in the 1940s, burgeoned, and run off of pesticides greatly diminished the elver population. Eel production dropped dramatically, and eel fishermen lost interest. In 1974 crawfish were stocked in protected rice fields in southern Spain. During two years the introduction was proclaimed a failure by some. However, after crawfish were sold in the market for a good price, interest suddenly grew. Fishermen and other interested parties began stocking crawfish into other areas including natural water bodies.

They did not want that one rice farmer or the rice field owner to hold a monopoly on crawfish and they wished to restock impoverished natural water bodies. Officials did not prevent the indiscriminate spreading of crawfish. During 1977 and 1978 some fishermen earned a great deal of money by this new crop. One fisherman, for example, earned around $40,000 in one month. This sudden wealth revitalized the freshwater fishing industry. At first only those who were considered fishermen were allowed to harvest crawfish. Some confusion existed because previous trapping regulations which were based on capture of Astacus species, the native species decimated by the plague. These regulations allowed 21 days for harvesting in August. In order to avoid these regulations, many fishermen claimed that they were eel fishing and if crawfish entered their eel traps, they could not help it. The crawfish captures resulted in great wealth for some. Money allowed purchase of new traps and other harvesting gear, not to mention cars and houses. Eventually crawfish trapping became crowded and friction developed. Sports fishermen made matters worse. As public streams became crowded with trappers, people began harvesting crawfish from privately owned rice fields, and the situation worsened. During 1979 and 1980, some fishermen began building their own installations for purchasing and processing crawfish, but suddenly the price dropped as crawfish flooded the market. Large numbers of crawfish were lost due to glutted markets. As overproduction began, fishermen had to be satisfield with only a quarter of the price formerly received. Some municipalities protected their fishing waters, allowing only local people to fish. With overproduction, buying centers were established and helped to stabilize the disarray. Out of seven buying centers, three processing plants developed.

Now, fishermen harvest and sell the crop to a buying center. He may harvest daily or several days a week, the best capture coming during rain. Establishment of processing plants allowed surplus crawfish to be frozen, thus avoiding a market glut.

At present, 1981, two fishermens unions exist with restricted membership. Only two new members are accepted each year. Three buying and depuration centers converted their installations into processing plants. Some rice farmers found that crawfish is more profitable than rice. The crawfish industry revitalized related business. Women in villages knit crawfish trapnets while chatting. Shop owners may earn more money selling line and other crawfish materials than from selling sports and hunting gear. Even crawfish souvenirs can be bought.

Social order has now been established in the newly formed crawfish industry. Factory buildings and processing plants can be seen as an integral part of the industry; the Louisiana red swamp crawfish can be found painted on walls of these industries. Fishery unions have been reestablished. The name may be changed from eel to crawfish union, and bylaws modified. Members of a union run their own buying center and jobs have been created other than simply trapping. Some people prepare sacks of crawfish for market, others are occupied with other related work.

At present it can be stated that 408 fulltime fishermen make their living year round from this new field of aquaculture specialized in crawfish. During main trapping seasons, March, April, May, September and October, some thousand men who are living from can obtain their daily salary with crawfish fishing. The improved harvesting methods make it easy for one man to trap ten, or more kilos per day. The overall money turnover made by the crawfish industry in the south of Spain during 1980, was around $7,600,000--U.S.

To finish my paper I would like to cite another author who wrote in 1906 an article about "The future of the crayfish industry." E.A. Andrew wrote; "No doubt, in time, the demand for crayfish will exceed the natural supply and this industry will tend to run the same retrograde course as that of the lobster, oyster, clam and many more important fisheries till real, or assumed, value of the crayfish as food, warrants legislative control and scientific aid such as alone makes possible the continuance of more and more of our once "inexhaustible" food supplies." How right he was.

LITERATURE CITED

Bardach, John E., John H. Ryther, and William O. McLarney. 1972. Aquaculture; the farming and husbandry of freshwater and marine organisms, New York, Wiley-Interscience. 868 pp.

AN ATTEMPT TO RAISE JUVENILE CRAYFISH PACIFASTACUS LENIUSCULUS DANA

Jozef Kossakowski
Marian Mnich
Inland Fisheries Institute
Olsztyn - Kortowo, Poland

Gustaw Kossakowski
State Fish Farm Ostroda

ABSTRACT

During investigations on crayfish Pacifastacus leniusculus Dana introduction into Polish waters, an attempt was undertaken to raise juvenile crayfish, immediately after their arrival from Simontorps Akvatiska Avelslaboratorium in Blentarp, Sweden.

On 15 August 1979 about 5800 individuals of small crayfish were placed in two metal troughs, previously used for raising the fry of rainbow trout. Total volume of both troughs was 3.3 m^2. The troughs were placed in a station producing fish stocking material. Constant water flow was assured of "natural" temperature, i.e., with no regulation. Hiding places for crayfish were prepared in the troughs. Crayfish were fed with water plants (Elodea, Charales) and fish meat. The lowest water temperatures of 0.5°C were noted in winter. Those noted in summer, did not surpass 21°C. The troughs were tightly covered with plastic net. The crayfish were left in the troughs for the period of two years. Sporadic observations revealed significant numbers of crayfish, in condition and of differentiated size. Intensive feeding and moulting periods were also observed.

On 1 June 1981 crayfish were fished out of the troughts. A total of 316 individuals was obtained, or 5.4% of the initial stock. Of this 154 were males (48.7%) and 162 were females (51.3%). The ratio between males and females was as 1:1.05.

Body length of all crayfish was measured, as also the length of the right propus in 120 individuals. Length of crayfish was highly differentiated (Figure 1). For both sexes calculations were made of the range of body length, average body length, and variation coefficient (Table 1).

Crayfish raised in the troughs were characterised by a slightly slower growth than crayfish in a small fish pond. In this pond, at significant crayfish mortality, singular individuals reached 9.0 cm during the period July 1974-May 1976 (Freshwater Crayfish April, 1978:195). Lower growth rate in the troughs might have been connected with significant population density, especially in the initial period, as well as with relatively lower water temperatures than in the pond.

Crayfish obtained from the troughs will now be bred in two concrete tanks of 12 m^2 capacity, at the same Center. We plan to carry out this experiment for at least another year.

555

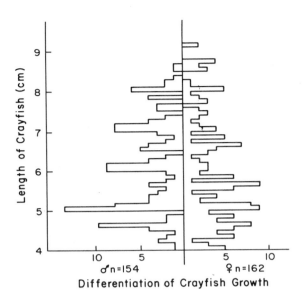

Differentiation of Crayfish Growth

Table 1 -- Characteristic of Crayfish Size

Feature		Males	Females
Number (n)		154	162
Range of body length (cm)		4.0 - 8.6	4.0 - 9.2
Average body length (cm)		6.0	5.9
Variation coefficient σ		1.17	1.04
	V%	19.5	17.6

THE INTRODUCTION OF THE SIGNAL CRAYFISH INTO THE UNITED KINGDOM AND ITS DEVELOPMENT AS A FARM CROP

Kenneth J. Richards
Riversdale Farm
Stour Provost
Gillingham, Dorset, U.K.

When we purchased our first 1,000 Signal Crayfish juveniles from Simontorp in 1976 with the simple idea of farming these little creatures we certainly did not know what we were in for - but I think I can tell you now! Quite simply, you start to farm something that, for most of the time doesn't seem to be there. You are thoroughly embarrassed by the disbelief and pathos on the faces of people who turn up almost daily to see your invisible livestock and then, if all goes well, after some two or three years you have the satisfaction and excitement of catching a few splendid specimens that you promptly return to the water. Two or three years pass and then, if all continues to go well, your catch increases and you have something to sell. It all starts as an act of faith and you have to live through what we now call the "credibility gap," which is the time between stocking juveniles and seeing your first adult crays. Without the constant support and help of our President Elect, fellow members, let me reveal now that we would not have survived these years - and probably that goes for our bank manager as well.

Firstly, I would like to tell you something of the husbandry that has developed from our experience over the years. Ours is a small mixed farm and from some twenty tanks and stew ponds we produce mirror carp, trout and crayfish. The quantities are small, for we are limited in both area and water supply, and so we have striven to become a demonstration farm, where visitors are welcomed and advised on the development of their own water resources.

In June 1976 the arrival of our first juveniles coincided with one of the worst draughts that the U.K. has ever experienced - in our particular area the rainfall over 18 months was in the region of 30 cm, less than a quarter of normal. Our springs were low, the ponds deoxygenated and many of our trout had succumbed; and so we decided to keep our juveniles initially in trout egg trays in the hatchery, having lined them with ample hides made from pipes of a variety of sizes, until the rains came. This was our first mistake. The animals were healthy enough, we never found a dead one, but it was soon clear that the number was going down daily. The water was kept at 10°C and growth rate was slow but like ambitious politicians, the crayfish were consuming each other with relish. The draught had not broken but there was no alternative but to put the survivors out into our weedy, almost stagnant ponds. From that moment, although we did not realize it, the crayfish began to thrive. Nothing was seen, of course, until a year later when, accidentally and quite forlornly we pulled out some weeds and were amazed to find sexually mature adults, a good 10 cm long, browsing away in the watercress.

This has been our experience in many ponds since. Crayfish will survive with oxygen levels as low as 3 ppm, and in the U.K. the most important factor is to achieve high water temperatures during the summer. In many lakes and ponds this may mean actually reducing the water flow. Initially, this often encourages excessive weed growth but over the years an increasing crayfish population will deal with this problem.

Our first pond, some 40 metres by 5 and 1/2 metres deep, was stocked with 200 juveniles in 1976 and 1977. The estimated population is now 3,000, giving a useful commercial return. The summers since 1976-1977 have not been good and growth rate has not since equalled those halcyon years. However, we have found that crayfish in the main reach sexual maturity after two summers even if growth rate has been rather poor - and females of only 5 cm body length have carried fertile eggs through to the spring. The only food introduced into this pond has been a mixture of 25% trout pellets and 75% chicken-layer pellets, intended for the carp. Some six months ago I was convinced that food shortage was an additional factor in the retarded growth rate. However, other ponds where ample food was clearly present produced crayfish of the same age/size - so confounding, in part, my first conclusion. However, I was not entirely convinced and as trout guts are a by-product of the farm and no live trout are present in this and similar carp/crayfish ponds I have been throwing guts into the pond and the crayfish certainly feed on this material. In the middle of the day, if we set traps with guts as bait, we can find up to 10 crays per trap in a matter of an hour or so.

This pond has clay sides and so the crayfish make their own cover. Smaller, one-summer crayfish can be pulled out of the emergent grass along the shoreline at almost any time of the year. The larger animals, we find, tend to live nearer the bottom, presumably the territory where the most prized food is found. This is quite a convenient arrangement for our trapping - for traps set on the bottom tend to catch the larger specimens. Some 300 to 400 crayfish have already been marketed from this pond this year and during the autumn virtually all of the marketable size, i.e. 3 summers, will be sold off to give as much room as possible for the following generations.

To store adults ready for marketing, sometimes for as long as 10 days, we use old spring-fed raceways filled with short lengths of plastic pipe and find casualty rates here are vitually nil.

There has been much controversy in the U.K. on the advisability of introducing juveniles into water stocked with trout. Clearly there could be both ecologiccal and economic advantages if a poly-culture was feasible and so into this trout stew in June 1980 we introduced 100 juveniles. They shared this pond, 20 x 5 metres, with two thousand rainbow trout and had to survive seine-netting carried out at almost fortnightly intervals. Pulling out emergent grass in the autumn of that year we came across 18 one-summer crayfish and this spring, without trapping, we have pulled out a further 20. There has certainly, therefore, been a survival rate of 38% and I would lay a bet that this figure is soon found to be over 50% - quite remarkable in a pond so densely stocked with trout. Clearly, if we had poured the juveniles into the top of the pond, almost all would have been taken as they tried to swim for cover. This, of course, we did not do. We introduced them in an implantation tube in the emergent weed and let them crawl out themselves, trusting to their good sense that they would not go swimming around in the middle of the pond in the daytime. Clearly

558

they did not and there is a good colony present today.

There is one other point of husbandry that I would mention and that is control of predators. We keep a constant watch for both mink and eels, setting traps for them as frequently as we can. Being alongside a river they can, and do, turn up at any time and both these animals will decimate a crayfish population if they are given half a chance.

Turning now from husbandry to the introduction of Signal juveniles in other parts of the U.K., I must return again to that wonderful summer of 1976.

As pioneers of what is, in our country, a new farm crop we soon found that we had innocently aroused the interest of the media, long before we had anything to show and even before we had very much to say. Almost every national newspaper, TV and radio station, as well as a variety of magazines, ran features on our crayfish. As I mentioned we are quite a small farm and we were soon swamped with uninvited visitors, phone calls and letters. Exaggerated claims were made of the profit potential and the possible yields of crayfish production; and misprints too, confounded the issue: advising people, for instance, to throw broken tyres instead of tiles into their lakes to provide crayfish hides.

The major problem from our point of view was that we had nothing to sell these enthusiasts and the obvious remedy to this situation was to turn to Simontorp. Stellan Karlsson came over and together we visited lakes and ponds and rivers around the country and initially, with great caution and for one year only, Stellan and Maja Abrahamsson agreed to give us a franchise to sell their juveniles in the U.K. That was some six years ago and, I am happy to say, we are still at it and growing in experience and confidence as each day passes. We have had our setbacks, our detractors and our disappointments, but now we have thriving colonies of crayfish from one end of the U.K. to the other. Since 1977 we have stocked some 245 lakes and ponds with juveniles as well as supplying 10 universities and research stations with crayfish of all ages.

It is, of course, a great responsibility introducing a foreign species into a new environment and we were determined to follow the best rules of the game. Here we were greatly helped by my son, Christopher, a partner in the farm who holds a doctorate from Oxford specializing in Ecology.

Christopher was well aware of the criteria for the responsible release of animals into new environments put forward by the World Wildlife Fund and the International Union for Conservation of Nature and Natural Resources. Now, Britain with its partial isolation from Continental Europe has a relatively impoverished fauna and this fauna has the benefit of an influential conservationist lobby. However, the economic reasons for introducing Signal crayfish are well-known and compelling. Our only native crayfish is Austropotamobius pallipes and this now co-exists with Astacus astacus, which was probably introduced about 400 years ago for its superior eating qualities. British crayfish populations tend to crash every 12-15 years and this is probably due to the crayfish plague (Aphanomyces astaci) since fish stocks are not affected (Duffey, 1933). The pattern of infection is thus different from that in continental Europe, probably because British wild crayfish have not been fished commercially. Only disease free stock have been brought into Britain and implantations have been planned carefully to minimize possible disruption to established ecosystems. Together with the extensive crayfish research carried out in California and in Sweden, this has enabled us to follow the WWF recommendations to a large extent.

And to be doubly sure, we make a point of not implanting juveniles where native crayfish are likely to be found and, in almost every case, into closed water systems. This also, of course, makes economic sense to the farmer who buys juveniles, for very few would wish to stock an open water system and find their crop being harvested by their neighbours!

The waters that we have stocked have varied from small ponds to lakes covering up to 100 acres. Not surprisingly, perhaps, the initial success was greatest in the small clay ponds and one happy owner of a half acre pond told us that his crayfish owed him nothing after only 3 years. However, as time progresses the larger lakes are beginning to prove themselves and the water boards, too, are now taking an interest. Here I should particularly mention the Thames Water Authority where Mr. John Hogger, a scientific officer in the Thames Conservancy division, has been monitoring the waters that we have stocked in his area.

All 7 of the sites that he has trapped have yielded varying numbers of crayfish - the maximum being 36 in 25 traps in one night at Kingsmead, a disused gravel pit near London Airport. Breeding has definitely occurred at three of these sites and I would presume that it either has or will occur at the other four.

In a lake at Stratfield Saye, the home of the Duke of Wellington, growth rate has also been monitored (Figure 1) - and it has been shown to be considerably ahead of Canadian, Swedish and even Oregon growth rates. This lake was first stocked in 1977 and breeding has occurred in each year since 1979, when 78% of females caught were berried.

Farmers, of course, are not quite as bucolic by nature as some would have you believe and many have come up with promising ideas of their own.

To help us evaluate the potential of different waters for crayfish production, in 1979 we produced a test-data sheet that gives us, I believe, basic information from which we can make useful assessments. I hope, before the end of this conference, that some additional factors will emerge and that we may update this sheet - without changing its practical bias.

You will see that after details of the area, depth and shoreline of the lake, we note the nature of the lake bottom, the weed cover present, existing fauna and likely temperatures. Every stretch of water has its own individual features - as varied as the human thumbprint, - and with experience we have learned to ask many questions that give us further clues in venturing our final assessment.

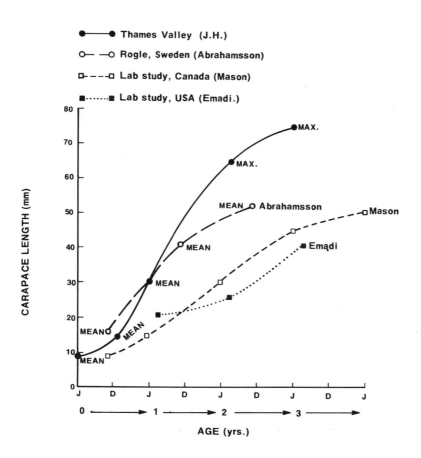

Figure 1. Maximum growth of P. leniusculus in the U.K.
Abrahamsson, S.A.A. : Freshwater Crayfish I. Lund 1972. p 27-40.
Emadi H.: Ph.D Thesis; Oregon State University. 1974. (Unpublished).
Mason, J.C.: Fish. Res. Bd. Canada. Tech. Report 440. 1974.

Riversdale Farm

STOUR PROVOST Nr. GILLINGHAM
DORSET

Tel.: East Stour (0747 85) 495

TO: _____   Date: _____

_____   Time: _____

_____

Area of Water: .... sq.m. approx   Perimeter .... m. approx

Average Depth: ....

Tests and observations made to determine the suitability of the above water for Crayfish indicate as follows:

1. Nature of Lake Bottom: ....

2. Nature of Weed Cover: ....

3. Existing Fauna: ....

4. Estimated weeks of year $8 - 12°c$ ....
   with water temp of $0 - 7°$ c ....

   (Growth rate is directly related to temperature)

5. Calcium Content: (Parts per million)

   0 10 20 30 40 50 60 70 80 90 100 110 120 +

   Poor   Fair   Good

   (Ca, normally in the form of lime is necessary for moulting Crayfish to make new hard carapaces (shells). A high Ca content gives a 'Buffering' capacity to ensure pH remains neutral or alkaline.)

6. pH:

   4.5   5   5.5   6   6.5   7   7.5   +

   Poor   Fair   Good

   (A high pH is necessary for good Crayfish production)

7. Nitrite and Nitrate:

   Inlet   Nitrite   Outlet   Nitrite   (These tests can give warning
            Nitrate   Nitrate   of organic pollution.)

   5   4   3   2   1   0

   Poor   Fair   Good

8. Oxygen Content: (3ft depth)

   1   2   3   4   5   6   7   8   9   (Oxygen content varies with time of day,
                                                   temperature, sunlight, etc. Crayfish tolerate
   Poor   Fair   Good   low levels for short periods.)

Recommendations:

Figure 2

As most waters in the South of England produce sexually mature crayfish at the end of the second summer, we generally advise implantation of juveniles for two successive years. In Scotland and the North of England, a three-year cycle is normally followed.

Sensibly enough, the qustion that came up from farmers and landowners most frequently was: "OK, now let's suppose the little devils thrive, then how do we sell 'em?" It was a good point! Crayfish have not featured in the food culture of the U.K. for perhaps a hundred years. Obviously a little promotion and education was needed. To this end we at Riversdale and seven other farmers decided to sponsor a feasibility study and with a 60% grant from the CCAHC, a government body, we commissioned a young agricultural economist to undertake this work. We were fortunate in our choice. Nick Stamp had completed similar studies on prawns on this side of the Atlantic and the excellent report that he produced encouraged us to form the British Crayfish Marketing Association. We hadn't yet got a crop to market mind you, and once again we found that attendant publicity soon created a mountain out of a mole hill and stimulated a market that we could not satisfy. We took a lot of stick about that one too but I still believe that sometimes there are advantages in putting the cart before the horse - and certainly, I'm told that General Motors do a lot of consumer research before they produce a new cart. And that, in a minor way, is what we have done. The initial aim of the Marketing cooperative is to identify markets, suggest guide prices and standards for the product. It's income, we trust, will be from a membership fee and no commission will be charged on sales. Eventually we hope to produce marketing boxes with our own logo and perhaps to have a small computer listing producers and consumers in each area. On the whole, agricultural marketing certainly of minor crops, is fairly chaotic in the U.K., our farmers are individualists and more inclined, however innocently, to cut their neighbours throats, than to co-operate - and I think our scheme will be seen by many to be a triumph of hope over reason. But it's worth a go; here we have a new crop, so let's see if we can create a new and ordered way of marketing it. I have a copy of our feasibility study here with me if any should wish to see it.

At the end of July some of you may have heard of the celebrations that took place in the U.K. when our Prince and Princess of Wales were married. The first days of their honeymoon were spent at Broadlands, the home of the late Earl Mountbatten. Now, this had us at Riversdale rather worried, for we had only recently stocked the lake at Broadlands with juveniles. How would it be, we wondered, if the young couple went for a swim and were nipped by an American crayfish, presumably with Republican tendencies? We had visions of spending our days in the Tower rather than enjoying ourselves here at this conference. But fortunately all was well and here we are.

563

ON THE NON-EXISTENCE OF AN INDIGENOUS SPECIES OF CRAYFISH ON THE CONTINENT OF AFRICA

Duro Adegboye
Department of Biological Sciences
Ahmadu Bello University
Zaria, Kaduna State, Nigeria

ABSTRACT

Apart from recent moves to introduce the true crayfish into East African waters (Huner, 1977), there are no decapod crustaceans with the description of the crayfish (Family: Astacidae) on the continent of Africa. The crustaceans that have been wrongly tagged with the name "crayfish" in Nigeria are none other than the prawns and shrimps which belong to the families Palaemonidae and Atyidae of the sub-order Natantia. Possible explanations for the non-existence of crayfish on continental Africa are given.

INTRODUCTION

The crayfish, which is also referred to as crawfish, crawdad, stonecrab, creekcrab and some other traditional names, looks like the lobster, the shrimp and the prawn. The crayfish is a versatile animal with about 500 known species; 50 percent of these species are found in North America alone while there are no known species of crayfish on the continent of Africa (Fawcett, 1970; Avault, 1975; Huxley, 1906).

With the exception of continental Africa, each continent has its own indigenous species of crayfish. The fact that there are no true species of crayfish (family Astacidae) in Nigerian waters is a bitter pill for students, fisheries officers as well as market women to swallow. Since the arrival of the white-man in Nigeria, the local shrimps, prawns and lobsters have been incorrectly referred to as "kayrayfish" or Crayfish (Bradbury, 1957; Olaniyan, 1968; Nte, 1978; Omatete and Nte, 1979 among others).

Numerous reports on the indigenous crayfishes from America, Europe and Australia have been given at various international crayfish conferences (Abrahamsson, 1973; Avault, 1975; Lindqvist, 1977; Laurent, 1979). However, no report of an indigenous African species of crayfish can be given since 1906 (Huxley, 1906). The non-existence of crayfish on the continent cannot be said to be due to inadequate environmental conditions that prevail in Africa, since foreign species of crayfish have been known to survive and even thrive in the waters of East Africa (Huner, 1977; Mikkola, 1978). The red Louisiana crayfish <u>Procambarus clarkii</u> have been successfully cultured in Kenya, Sudan and Uganda. This success story about the survival of crayfish in African waters coupled with the existence of an "indigenous" species of crayfish on the island of Madagascar (<u>Astacus madagascariensis</u>) off the coast of Africa would suggest that the lack of sufficient calcium ions in African waters could not be the sole reason for the lack of indigenous species of crayfish in Africa. Indeed, over 21 species of decapod crustaceans with calcareous exoskeleton have been described in Nigeria alone (Powell and Wilcox, 1980; Marioghae, 1980).

In this paper, we are presenting four different hypotheses, each of which deserves our due consideration for two main reasons. In the first place, such consideration could lead to the completion of the crayfish history of the world. Secondly, a clear understanding of the probable reasons for the non-existence of the economically important crayfish in Africa, might encourage the policymakers of the nations of Africa to consider the possibilities, problems and probable solutions to the idea of introducing revenue-yielding crayfish into the fish-farming programes on the continent.

THEORIES

1. "African Crayfish Holocaust" Theory

Simply stated, the "African Crayfish Holocaust" Theory holds that once upon a time, there were indigenous crayfish species on all continents including Africa. However, due to some disaster of one kind or the other, all the species of crayfish on the continent got wiped out.

The holocaust theory is supported by the fact that in Europe, the crayfish population was almost completely wiped out by the crayfish plague Aphanomyces astaci (N.A.S., 1976; Huner, 1977; Brink, 1977; Furst, 1977; see Lindqvist, 1977, 1978). It is hardly likely that the species of crayfish on the African continent got wiped out due to over exploitation, and even, there is no evidence from fossil records to indicate either this notion, or the theory itself.

This continental holocaust would readily explain why there is an indigenous species of crayfish on the Island of Madagascar and none on the African continent. The sea separating the Malagasy Republic from the African coast might have been responsible for putting a stop to the eastward spread of the hypothetical "African-crayfish-plague" into Madagascar. Hence, if the theory holds, all the species of crayfish on the continent got wiped out while Madagascar retained her plague-free crayfish.

Incidentally, in the recent past, crayfish was not found in all the 50 states of the United States, and dry land is an effective barrier to the migration of crayfish (Pennak, 1953). Thus, perhaps, if an indigenous crayfish existed at all in Africa, it must have been restricted to very small areas, such as the East African coast, where it was supposedly wiped out by the "plague." (cf. Lake Hjalmaren in Sweden where crayfish was completely wiped out -Abrahamsson, 1973).

2. The "No-Never" Theory of African Crayfish

As the name implies, this theory states that never in the history of Africa was there any indigenous species of crayfish on that continent. If this theory holds, then one of two other theories must be proposed to show why crayfish exists on Madagascar - and not in Africa.

Meanwhile, by looking at the history of the family: Astacidae, one might get an insight into the problem of non-occurrence of Africa. Crayfish are believed to have originated in Europe (Avault, 1975), and the five American species of Astacus are believed to have migrated from "old world" (Pennak, 1953).

Lindroth (1957) was quoted by Per Brinck (1977) as having analysed the faunal connections between Europe and America. Lindroth suggested that most of the species occur on both sub-continents were carried there by human agency. The Berring Strait between Asia and North America was once the Isthmus of Berring, across which rivers bearing crayfish could have flowed into North America. This explanation could serve for the occurrence of crayfish in North America. Similarly, the land bridges connecting New Guinea and Australia to mainland Asia and a possible one linking Madagascar to Asia were responsible for the spread of the crayfish to the islands. (Read Villee et al., 1973, pp. 250-252).

Given the Mediterranean Sea, the Sahara Desert and the Arabian Desert, one could easily see why the European or the Turkish crayfish A. leptodactylus (Erecin and Koksal, 1977) had no chance of migrating into the continent of Africa. The fact of non-migration due to geographical barrier is well illustrated by the observation that the crayfish Astacus madagascariensis had never crossed the sea onto the east coast of Africa. Since the conditions on the continent of Africa has been shown to be conducive to crayfish growth and reproduction, (Huner, 1977) the only simple explanation for the non-radiation of an indigenous crayfish on the continent of Africa is that the crayfish never got to Africa from Europe (where the crayfish had its beginnings).

3. "Transplantation Theory" of Madagascar's Crayfish

If the two theories, namely of "African Crayfish Holocaust" and the "No-Never" theories are doubtful, then there has to be an explanation for the existence of Astacus madagascariensis on Madagascar and the non-existence of crayfish on continental Africa. The "Faunal Connection" theory of Lindroth (1957) might also hold for Europe and Madagascar on the one hand, and Asia and Madagascar on the other. It is plausible that the Madagascar's crayfish was introduced into the island from elsewhere through human agency. Such human influence on animal migration is widespread and it is responsible for fauna connections between continents. There is no evidence to support this theory, except that it could have happened as was the case when humans introduced crayfish into Sweden from Germany in the 16th century (Abrahamsson, 1973).

4. "Land-Bridge Theory" for Madagascar's Crayfish

If human agency was not in any way responsible for the existence of crayfish on Madagascar, then either Madagascar was "Center of Origin" for the crayfish or Madagascar obtained her crayfish via a land-bridge that connected the island to mainland Asia where crayfish exists. If a set of crayfish did originate on Madagascar, then the isolating conditions of the island prevented the spread of the new species to continental Africa. In North America, the crayfish had certain centers of origin; Procambarus in Mexico, Orconectes at the Ohio-Mississippi-Missouri junction, Cambarus in southern Appalachians, Cambarellus in Mexico, and Astacus in Euro-Asian region (Pennak, 1953). From these centers, there was a general spread to other areas. Therefore, if Madagascar was a center of origin for crayfish, the surrounding waters of the Pacific ocean prevented the spread of the crayfish into Africa (see Villee et al., 1973). In a similar vein,

Westman (1973) believed that Astacus could not have invaded Finland across the Litorina sea because of the high salinity of the water. If Astacus did spread into Finland from the then freshwater Baltic Sea, it must have done so via the Isthmus of Karelia. The high salinity of the Sea of Madagascar must have been responsible for the isolation of Astacus madagascariensis on the island. The case of the crayfish is not the only faunal difference between Africa and Madagascar. There are several examples of plants and animals that exist in Malagasy Republic without any representatives or relatives on continental Africa. The reverse is also true. A classical example is the Dodo, an extinct land bird, which was restricted to the island of Madagascar. The species differences between Madagascar and Africa, are so great that many people hold the view that Madagascar was never "African" by nature. This is reflected in the tilting of conventions such as "Organization of African and Malagasy states." Thus, the geographical isolation of Madagascar must have been responsible for the non-existence of crayfish on the continent of Africa.

5. Which Theory

Several theories could be proposed for the non-existence of true crayfish on the continent of Africa. Whichever theory one believes in, is a matter of conjecture and personal opinion. And such theories do not alter the fact that no taxonomic crayfish exists on continental Africa. The aquatic productivity of a continent is not judged on the basis of the presence or absence of just one species of water organism. What Africa lacks in terms of crayfish has been perhaps made up for by the huge populations of animals that sometimes form part of the export to zoos and zoological gardens in other parts of the world.

The interest in crayfish has been generated by the fact that a carefully planned culture of these animals in Africa could lead to a diversification of fishery practices followed by the unavoidable boost in the economic sector. (NSA, 1976; Adegboye, 1980).

In conclusion, we would like to quote Jay V. Huner (1977) who said:

"The morality of Procambarus clarkii (crayfish) introductions can be disputed, but the species is now a natural resource beyond its natural range and can be of great value especially in food-poor countries."

He suggests that professional astacologists should be called in to advise and assist the government on the control and utilization of the species. What Huner (1977) said about Procambarus clarkii should hold true for other commercially important species of crayfish.

Should we then introduce the crayfish into the waters of West Africa, and Nigeria in particular? This is the subject worthy of further discussion. (NSA, 1976).

ACKNOWLEDGMENTS

The author is grateful to the members of the International Association of Astacology for their constant encouragement. My gratitude also goes to the National Academy of Sciences Washington D.C., for their interest in my job. Mrs. M.O. Obilana was kind enough to type the manuscript.

LITERATURE CITED

Abrahamsson, S. 1973. (Ed.) Freshwater Crayfish Vol. 1, 252 pp. Papers from the First International Symposium on Freshwater Crayfish, Austria, 1972. Student literature, Lund, Sweden.

Adegboye, D. 1980. "Kayrayfish": The Nigerian Decapod Crustaceans; Their Potentialities and Importance in the Fisheries Programmes of the Nineteen Eighties. Paper presented at the National Conference on Proposals and Strategies for increasing food production in Nigeria Port Harcourt, River State, Nigeria.

Avault, J.W. Jr. 1975. Crayfish Farming in the United States. Freshwater Crayfish Vol. 1:239-250.

Avault, J.W. Jr. 1975. (Ed.) Freshwater Crayfish Vol. 2. Papers from the 2nd International Symposium on Freshwater Crayfish, Baton Rouge, Louisiana, U.S.A. 1974.

Bradbury, R.E. 1957. The Benin Kingdom and the Edo-speaking peoples of southwest Nigeria. In West Africa XIII. London International African Institute. pp. 134, 176.

Brinck, P. 1977. Developing crayfish populations. Freshwater Crayfish Vol. 3:211-228.

Erecin, Z., and G. Koksal. 1977. On the crayfish A. leptodactylus in Anatolia. Freshwater Crayfish Vol. 3:187-192.

Fawcett, G. 1970. 'Crawfishin': A Report from the Capital. Acadiana Profile. Vol. 2(1):23-25.

Furst, M. 1977. Introduction of Pacifastacus leniusculus (Dana) into Sweden: Methods, results and management. Freshwater Crayfish Vol. 3:229-248.

Hobbs, H. 1972. Crayfishes of North and Middle America. U.S. Government Printing Office, Washington, D.C.

Huner, J.V. 1977. Introductions of the Louisiana red swamp crayfish, Procambarus clarkii (Girard): An update. Freshwater Crayfish. Vol. 3:193-201.

Huxley, T.H. 1906. The Crayfish. 2nd ed. London.

Laurent, P.J. 1979. (Ed.) Freshwater Crayfish Vol. 4. Papers from the 4th International Symposium on Freshwater Crayfish, Thonon, France. 1978.

Lindrothe, C.H. 1957. Quoted by Brinck. The faunal connections between Europe and North America. Stockholm 334 pp.

Lindqvist, OV. 1977. (Ed.) Freshwater Crayfish Vol. 3. Papers from the 3rd International Symposium on Freshwater Crayfish, Kuopio, Finland. 1976. University of Kuopio.

Lindqvist, O.V. & H. Mikkola. 1978. On the etiology of the muscle wasting disease in Procambarus clarkii in Kenya. Freshwater Crayfish 4:363-372.

Marioghae, I.E. 1980. Review of Research on Penaeid Shrimps in Nigeria. Abstracts: Workshop on the Niger Delta Mangrove Ecosystem. University of Port Harcourt. No. 31.

Mikkola, H. 1978. Ecological and social problems in the use of the crayfish, Procambarus clarkii in Kenya. Freshwater Crayfish 4:197-206.

National Academy of Sciences. 1976. Making Aquatic Weeds Useful: Some perspectives for Developing Countries. National Academy of Sciences, Washinton, D.C. Part I (4) The Crayfish pp. 41-48.

Nte, V.I. 1978. Determination of drying rate curve and drier design for crayfish. B.Sc. Thesis. Chemical Engineering Department, University of Lagos, 1978.

Olaniyan, C.I.O. 1968. An Introduction to West African Animal Ecology. Ondon: Heinemann.

Omatete, O.O., and V.I. Nte. 1979. Drying Rate Curve and Cottage Dryer Design for Hard Shelled Crayfish (Palaemon paucidens) Chemical Technology for Developing Countries, (Ed.) K.A.C.M. Beanackers. Chemical Engineering, A.B.U., Zaria, Nigeria.

Pennak, R.W. 1953. Freshwater Invertebrates of the United States. The Ronald Press Company New York. pp. 435-469.

Powell, C.B. and B. Wilcox. 1980. (Eds.) Abstracts: Workshop on the Niger Delta Mangrove Ecosystem. University of Port Harcourt.